INTRODUCTION TO THE NUMERICAL SOLUTION OF MARKOV CHAINS

INTRODUCTION TO THE NUMERICAL SOLUTION OF MARKOV CHAINS

William J. Stewart

Princeton University Press
Princeton, New Jersey

Copyright © 1994 by Princeton University Press
Published by Princeton University Press, 41 William Street,
Princeton, New Jersey 08540
In the United Kingdom: Princeton University Press,
Chichester, West Sussex

Library of Congress Cataloging-in-Publication Data

Stewart, William J., 1946–
Introduction to the numerical solution of Markov chains / William J.
Stewart.
p.   cm.
Includes bibliographical references and index.
ISBN 0-691-03699-3
1. Markov processes—Numerical Solutions.   I. Title.
QA274.7.S74   1994
519.2'33—dc20   94-17465

This book has been composed in Computer Modern

The publisher would like to acknowledge the author of this volume for
providing the camera-ready copy from which this book was printed

Princeton University Press books are printed on acid-free paper and meet
the guidelines for permanence and durability of the Committee on
Production Guidelines for Book Longevity of the Council on
Library Resources

Printed in the United States of America

10 9 8 7 6 5 4 3 2 1

To
Kathie,
Nicola, Stephanie, Kathryn, and William

Contents

Preface and Acknowledgments

The purpose of this book is to provide an introductory, yet systematic and detailed, treatment of the numerical solution of Markov chains. Markov chain modelling is used in many diverse areas, not only in computer science and engineering but also in other disciplines such as mathematics, probability and statistics, operations research, industrial engineering, electrical engineering, biology, genetics and agriculture, economics and demographics, education, and so on. Markov chains may be used to locate bottlenecks in communication networks; to assess the benefit of increasing the number of CPUs in multiprocessor systems; and to quantify the effect of scheduling algorithms on throughput. They may be used in reliability modelling to estimate the mean time to failure of components in systems as diverse as software systems and aerospace systems and to model the performance of fault-tolerant computers. Markov chains have also shown themselves to be a valuable analysis tool in a variety of economic models, from population forecasting to financial planning. They have been and continue to be the method of choice for modelling very many other systems. In view of the fact that computer technology has reached the point that numerical computation is now readily within the reach of everyone, and the fact that analytic solutions are simply not available for models that incorporate the characteristics modellers seek, we find ourselves in a phase of rapidly rising interest in solving Markov chains numerically. Since there is no other book currently available that treats this subject, this book seeks to fill that important gap.

It is often possible to represent the behavior of a physical system by describing all the different states that it can occupy and by indicating how it moves from one state to another in time. If the future evolution of the system depends only on its current state, the system may be represented by a Markov process. The term *Markov chain* is employed when the state space is discrete. The information that is most often sought from such a model is the probability of being in a given state or subset of states at a certain time after the system becomes operational. Often this time is taken to be sufficiently long that all influence of the initial starting state has been erased. The probabilities thus obtained are referred to as the *long-run* or *stationary probabilities*. Probabilities at a particular time t are called *transient probabilities*. When the number of states is small, it is relatively easy to obtain transient and stationary solutions quickly and accurately and from these to predict the behavior of the system. However, as models become more complex — and this is increasingly the trend — the process of obtaining these solutions becomes much more difficult. It is to this aspect that the current text is geared. It is the purpose of *Introduction to the Numerical Solution of Markov Chains* to explore and explain all aspects of numerically computing solutions of Markov chains, especially when the state space is huge.

Organization

The book is organized into 10 chapters, as follows. In Chapter 1 the basic definitions, properties, and theorems of discrete-time and continuous-time Markov chains are presented. These are used throughout the remainder of the book and are collected together in one place for ease of reference. Additionally, this chapter provides sufficient background on queueing networks to permit readers whose interests lie mainly in numerical solution procedures to understand how and why queueing network models give rise to large-scale Markov chains.

Chapters 2 through 4 deal respectively with the three major classes of solution procedures, namely, direct methods, iterative methods, and projection methods. Direct methods are seen to be variants of the ubiquous Gaussian elimination algorithm. When these are applied to Markov chains, provision must be made to handle the zero pivot that results from the singularity of the system, and different approaches are examined. Implementation considerations, with special emphasis on compact storage schemes and the GTH variant, are discussed. Chapter 2 closes with a discussion of stability, conditioning, and error analysis as they pertain to the application of direct methods to Markov chain problems.

Chapter 3 examines the major single-vector iterative methods: the power method, Jacobi, Gauss–Seidel, and successive overrelaxation (SOR). Block variants as well as preconditioning procedures are given. Like the previous chapter, Chapter 3 also provides implementation details. It terminates with a discussion of known convergence results.

The application of projection methods to the solution of Markov chains is still relatively new. However, these methods have been used with much success in other domains and have the potential of being very well suited to solving Markov chain problems. They are the subject of Chapter 4. Given that this material is likely to be new to many readers, the basic projection processes are examined in some detail before consideration is given to specific algorithms. These algorithms include simultaneous iteration and the methods of Arnoldi and GMRES. Due to the large measure of success that this latter algorithm has achieved, a detailed description of its implementation is given. Chapter 5 closes with a discussion of the methods of Lanczos and Conjugate Gradients and includes implementation considerations.

The three chapters that follow deal with Markov chains whose underlying transition matrices possess a special structure. When this matrix is block upper Hessenberg (or even better, block tridiagonal), it may be possible to use block recursive methods. This provides the substance of Chapter 5. Great care must be taken with recursive methods, because they have been known to give very unusual answers when used inappropriately. The stability of recursive algorithms is considered in detail. We also look at the matrix geometric approach made popular by Neuts, provide the most recent procedures for computing his famous R matrix, and consider instances in which this matrix may be computed explicitly.

Stochastic matrices that are nearly completely decomposable appear relatively frequently in Markov chain modelling and in general create enormous difficulties in computing their stationary probability distributions. They are considered in Chapter 6, where the problem is posed, the theoretical concepts of stochastic complementation introduced, and the only type of method found to bring any satisfaction — iterative aggregation/disaggregation — is analyzed in detail. The major methods and several variants are considered, and some results relating to convergence properties and behavior are provided.

In Chapter 7, we consider the case of Markov chains whose transition matrices possess a periodic or cyclic structure. We begin by considering the saving that this affords to the standard direct and iterative methods; then we proceed to consider methods specifically designed for matrices having this periodic property. In particular, we describe reduced schemes that work on only one portion of the matrix, but that may be used to compute the complete solution very efficiently. Also, stochastic matrices that are cyclic possess certain properties that allow us to develop block SOR–type methods for which close to optimal relaxation factors can be computed rather easily.

The methods in Chapters 2 through 7 all concern the computation of stationary probabilities. The computation of transient solutions is considered in Chapter 8. The method most often used in this context is that of uniformization, and it is with this method that the chapter begins. This is followed by an examination of methods that are applicable when the state space is small. A substantial part of Chapter 8 is devoted to the numerical solution of systems of linear ordinary differential equations (ODEs), for it is well known that the probability distribution of a Markov chain at any time t may be defined by such a system. This approach is treated in considerable detail, because the numerical solution of ODEs is perhaps not as well exploited in the Markov chain community as are numerical linear algebraic techniques for stationary distributions. Finally, we conclude this chapter with a novel approach for obtaining transient solutions, one that shows enormous promise, a method based on Krylov subspaces that exploits the methods previously given for small state spaces.

Stochastic automata networks are the subject of Chapter 9. Among all the modelling paradigms that give rise to Markov chains, this is perhaps the most general, and it has the potential of becoming the basis for future developments in software geared to the automatic generation and solution of Markov chain models. After discussing the underlying concepts of stochastic automata networks, the effects of both synchronizing events and functional transitions on the complexity of computing solutions are examined. It is shown how the stationary probability vector can be obtained from a compact tensor description of the network. However, the constraints imposed by this very compact description are severe and restrict the choice of possible solution method.

The final chapter of this book provides information on some of the software that is currently available for developing, manipulating, and solving Markov chain models, as well as software that may be useful in developing specific methods, such as that found in libraries of numerical linear algebra routines.

It has not been possible to include everything that I would have wished to include into a single, reasonably sized book. For example, I have omitted material on bounding methods and techniques for state space reduction. These are not the only topics omitted. Although I would be the first to admit that the content is biased by my personal experiences and preferences, these and other topics have been omitted, not because I think they are unimportant, but rather because of limitations of space and a desire to bring the book to market in a timely fashion. If I were to continue to extend the book by incorporating all the new and important results that become available almost on a daily basis, this book would never appear.

Acknowledgments

I am indebted to the administration of North Carolina State University and to the faculty and staff of its Computer Science Department for their support of my efforts in writing this book and for providing all the necessary facilities that we all too often take for granted. Also, I am indebted to the administration and staff of IMAG, *Informatique et Mathématiques Appliquées de Grenoble*, France. During the 1993–94 academic year, they provided me with the yearlong refuge from teaching and administrative duties that allowed me the time to formulate my classroom notes into a coherent whole. I am most grateful to the National Science Foundation for their generous financial support of my research under a variety of grants dating back to 1979.

Many of my friends and colleagues have contributed to this book through collaborative research efforts, through technical or philosophical discussions of different parts of it, or simply through their encouragement. They will see their influence throughout this book. I am deeply indebted to them and offer them my sincere gratitude.

First, let me begin by expressing my thanks to the many teachers who have influenced me so profoundly throughout the years of my education. I was fortunate in that for a precious few years at St. Thomas's Secondary School, Belfast, I had the privilege of having the renowned Irish novelist, Michael McLaverty, as my mathematics teacher. My great regret is that it was only later in life that I came to understand and appreciate his influence on me. It was he, perhaps more than anyone else, who sent me skipping down the magical yellow brick road that is mathematics.

The influence of my M.Sc. and Ph.D. advisors on the work that is described in this book is second to none. Alan Jennings first introduced me to the matrix eigenvalue problem and to simultaneous iteration methods, while Tony Hoare introduced me to Markov chains and their application to the modelling of computer operating systems. These were the origins for the research in which I have been engaged for some twenty years. It is a pleasure to acknowledge their guidance and to thank them for it.

After completing my Ph.D. degree, I moved to France, where I fell under the influence of Erol Gelenbe, Jacques Lenfant, Jean-Pierre Verjus, and, somewhat later, Raymond Marie. I would like to thank all four for their support and encouragement during those early days. Since then I have had the good fortune to be associated with a large number of friends and colleagues, all of whom have had a profound effect on me and on my research. It would be impossible and perhaps inappropriate to try to list them all. Rather let me just single out a few from among the many. They include Don Bitzer, Pierre-Jacques Courtois, Yves Dallery, David McAllister, Carl Meyer, Arne Nilsson, Harry Perros, Bob Plemmons, Bernard Philippe, Brigitte Plateau, Guy Pujolle, Yousef Saad, Ken Sevcik, Pete Stewart, Sandy Stidham, Kishor Trivedi, and Mladen Vouk. To these, and to all my colleagues who have helped and encouraged me in my research, I offer my sincere thanks.

This book grew out of a graduate-level course on aspects of performance modelling that I taught at irregular intervals at North Carolina State University. It therefore follows, as surely as day follows night, that the students who enrolled in this course contributed significantly to its form and content. I am thankful to all of them for their patience and their input in this endeavor. Some of these students continued on to do M.Sc. or Ph.D. research projects with me and I thank them in particular. The intellectual and philosophical discussions that I have had in working with students everywhere has had a profound effect on the contents and organization of the book. These students include Hassan Allouba, Karim Atif, Francois Bonhoure, Wei-Lu Cao, Munkee Choi, Tugrul Dayar, Steve Dodd, Larry Hodges, Halim Kafeety, Rick Klevans, Kimon Kontovasilis, Bob Koury, Butch Lin, David Lin, David Michael, Blaise Sidje, Patricia Snyder, Wayne Stohs, Abderezak Touzene, and Wei Wu.

Among my students, Tugrul Dayar, Halim Kafeety, and Kimon Kontovasilis were exposed to preliminary versions of this text. They provided many helpful comments on one or more draft chapters. For this, I offer them a special word of gratitude. I also wish to single out Wei-Lu Cao for particular thanks for the wonderful job he did in meticulously reading through a final draft of the entire manuscript, searching for all kinds of errors, and making numerous corrections and suggestions for improvement.

I wish to express my profound thanks to my parents, William and Mary Stewart, for their constant encouragement and the selfless sacrifices that they made in allowing me, the eldest of their eight children, to pursue my education at a time when I would normally have been expected to contribute financially to the family budget. Without their farsightedness, this book would never have seen the light of day.

Last and most of all, I offer a special word of thanks to my wife, Kathie, and to our four children, Nicola, Stephanie, Kathryn, and William. Despite the numerous evenings and weekends devoted to writing this book and excursions near and far devoted to developing the material, they provided the loving family environment that afforded me the tranquility and peace of mind that made writing it possible. This book is dedicated to them.

INTRODUCTION TO THE NUMERICAL SOLUTION OF MARKOV CHAINS

Chapter 1

Markov Chains

1.1 Introduction

It is often possible to represent the behavior of a physical system by describing all the different states the system may occupy and by indicating how the system moves from one state to another in time. If the time spent in any state is exponentially distributed, the system may be represented by a *Markov process*. Even when the system does not possess this exponential property explicitly, it is usually possible to construct a corresponding implicit representation. Examples of the use of Markov processes may be found extensively throughout the biological, physical, and social sciences as well as in business and engineering.

Associated with every Markov process is a (possibly infinite) set of states. The system being modelled by the process is assumed to occupy one and only one of these states at any moment in time. The evolution of the system is represented by transitions of the Markov process from one state to another. These transitions are assumed to occur instantaneously; in other words, the actual business of moving from one state to another consumes zero time. The fundamental property of a Markovian system, referred to as the *Markov property*, is that the future evolution of the system depends only on the current state of the system and not on its past history.

As an example, consider the behavior of a frog who leaps from lily pad to lily pad on a pond or lake. The lily pads constitute the states of the system. Where the frog jumps to depends only upon what information he can deduce from his current lily pad — he has no memory and thus recalls nothing about the states he visited prior to this current state, nor even the length of time he has been on the present lily pad. The assumption of instantaneous transitions is justified by the fact that the time the frog spends in the air between two lily pads is negligible compared to the time he spends sitting on a lily pad.

The information we would like to obtain from the system is a knowledge of the probabilities of being in a given state or set of states at a certain time after

3

the system becomes operational. Often this time is taken to be sufficiently long that all influence of the initial starting state has been erased. Other measures of interest include the time taken until a certain state is reached for the first time. For example, in the case of the frog and the lily pond, if one of the lily pads were not really a lily pad but the nose of an alligator waiting patiently for breakfast, we would be interested in knowing how long the frog would survive before being devoured by the alligator. In Markov chain terminology, such a state is referred to as an *absorbing state*, and in such cases we may wish to determine the mean time to absorption.

1.2   Markov Chains

A *stochastic process* is defined as a family of random variables $\{X(t), t \in T\}$ defined on a given probability space and indexed by the parameter t, where t varies over some *index set (parameter space)* T. T is a subset of $(-\infty, +\infty)$ and is usually thought of as the time parameter set. As such, T is sometimes called the time range; $X(t)$ denotes the observation at time t. If the index set is discrete, e.g., $T = \{0, 1, 2, \ldots\}$, then we have a *discrete-(time) parameter* stochastic process; otherwise, if T is continuous, e.g., $T = \{t : 0 \leq t \leq +\infty\}$, we call the process a *continuous-(time) parameter* stochastic process. In many models, the statistical characteristics in which we are interested are often independent of the time t at which we begin to observe the system. The stochastic process is said to be *stationary* when it is invariant under an arbitrary shift of the time origin. The values assumed by the random variable $X(t)$ are called *states*, and the set of all possible states forms the *state space* of the process. The state space may be discrete or continuous (real-valued). Examples of a continuous state space might be the level of water in a dam or the temperature inside a nuclear reactor. An example of a discrete state space is the number of customers at a service facility.

A *Markov process* is a stochastic process whose conditional probability distribution function (PDF) satisfies the so-called "Markov property." Specifically, let $X(t)$ be a continuous-time stochastic process that denotes the state of a system at time t. Then $X(t)$ is a continuous-time Markov process if, for all integers n and for any sequence $t_0, t_1, \ldots, t_n$ such that $t_0 < t_1 < \cdots < t_n < t$, we have

$$\text{Prob}\{X(t) \leq x | X(t_0) = x_0, X(t_1) = x_1, \ldots, X(t_n) = x_n\}$$

$$= \text{Prob}\{X(t) \leq x | X(t_n) = x_n\}.$$

Thus, the fact that the system was in state x_0 at time t_0, in state x_1 at time t_1 and so on, up to the fact that it was in state x_{n-1} at time t_{n-1} is completely irrelevant. The state in which the system finds itself at time t depends only on where it was at time t_n. The state $X(t_n)$ contains all the relevant information concerning the history of the process. This does not imply that the transitions are not allowed to depend on the actual time at which they occur. When the transitions out of state $X(t)$ depend on the time t, the Markov process is said to be *nonhomogeneous*. However, throughout most of our discussions in later chapters, we shall assume that the transitions are independent of time. In this case, the Markov process is said to be *homogeneous*.

If the state space of a Markov process is *discrete*, the Markov process is referred to as a *Markov chain*. The state space of a Markov chain is usually taken to be the set of natural integers $\{0, 1, 2, \ldots\}$ or a subset of it. We shall consider only discrete-time Markov chains (DTMC) and continuous-time Markov chains (CTMC).

To satisfy the "Markov property," the time spent in a state of a Markov chain (generally referred to as the *sojourn* time) must exhibit the *memoryless* property: at any time t, the remaining time that the chain will spend in its current state must be independent of the time already spent in that state. If the time parameter is continuous, this means that the sojourn time must be exponentially distributed; for a discrete-time Markov chain, the sojourn time must be geometrically distributed. These are the only distributions that possess the memoryless property.

1.3   Discrete-Time Markov Chains

1.3.1   Definition

For a discrete-time Markov chain (DTMC), we observe the state of a system at a discrete set of times. We may therefore represent the discrete parameter space T by the set of natural numbers $\{0, 1, 2, \ldots\}$ without loss of generality. The successive observations define the random variables, $X_0, X_1, \ldots, X_n, \ldots$, at time steps $0, 1, \ldots, n, \ldots$, respectively. A discrete-time Markov chain satisfies the following relationship for all natural numbers n and all states x_n.

$$\text{Prob}\{X_{n+1} = x_{n+1} | X_0 = x_0, X_1 = x_1, \ldots, X_n = x_n\}$$

$$= \text{Prob}\{X_{n+1} = x_{n+1} | X_n = x_n\}.$$

The conditional probabilities $\text{Prob}\{X_{n+1} = x_{n+1} | X_n = x_n\}$ are called the *single-step transition probabilities*, or just the transition probabilities, of the Markov chain. They give the conditional probability of making a transition from state x_n to state x_{n+1} when the time parameter increases from n to $n + 1$. They are denoted by

$$p_{ij}(n) = \text{Prob}\{X_{n+1} = j | X_n = i\}. \tag{1.1}$$

In a homogeneous DTMC these probabilities are independent of n and are consequently written as

$$p_{ij} = \text{Prob}\{X_{n+1} = j | X_n = i\}, \quad \text{for all } n = 0, 1, \ldots.$$

homogeneous Markov chains are time independent

The matrix P, formed by placing p_{ij} in row i and column j, for all i and j, is called the *transition probability matrix* or *chain matrix*. Note that the elements of the matrix P satisfy the following two properties:

$$0 \le p_{ij} \le 1$$

and, for all i,

$$\sum_{\text{all } j} p_{ij} = 1.$$

When the Markov chain is nonhomogeneous, the elements p_{ij} must be replaced with $p_{ij}(n)$ and hence, the matrix P with $P(n)$.

Example: Consider a homogeneous, discrete-time Markov chain that describes the daily weather changes in Belfast, Northern Ireland (well known for its prolonged periods of rainy days). We simplify the situation by considering only three types of weather pattern: rainy, cloudy, and sunny. These three weather conditions describe the three states of our Markov chain: state 1 represents a rainy day; state 2, a cloudy day; and state 3, a sunny day. The weather is observed daily. On any given rainy day, the probability that it will rain the next day is estimated at 0.8; the probability that the next day will be cloudy is 0.15, while the probability that the morrow will be sunny is only 0.05. Similar probabilities may be assigned when a particular day is cloudy or sunny. Let us assume that the transition probability matrix P for this Markov chain is as follows:

$$P = \begin{pmatrix} 0.8 & 0.15 & 0.05 \\ 0.7 & 0.2 & 0.1 \\ 0.5 & 0.3 & 0.2 \end{pmatrix}. \tag{1.2}$$

Note that the elements in P represent *conditional probabilities*. For example, the element p_{32} tells us that the probability that tomorrow is cloudy, *given* that today is sunny, is 0.3.

1.3.2   The Chapman–Kolmogorov Equations

We may generalize a single-step transition probability matrix to an n-step transition probability matrix whose elements are $p_{ij}^{(n)} = \text{Prob}\{X_{m+n} = j | X_m = i\}$. These elements may be obtained from the single-step transition probabilities. From equation (1.1), we have

$$P^{(n)}(m, m+1, \ldots, m+n) = P(m)P(m+1)\cdots P(m+n).$$

For a homogeneous DTMC we may write

$$p_{ij}^{(n)} = \text{Prob}\{X_{m+n} = j | X_m = i\}, \quad \text{for all } m = 0, 1, 2, \ldots,$$

from which it may be seen that $p_{ij} = p_{ij}^{(1)}$. From the Markov property, we may establish the following recursive formula for calculating the $p_{ij}^{(n)}$:

$$p_{ij}^{(n)} = \sum_{\text{all } k} p_{ik}^{(l)} p_{kj}^{(n-l)}, \quad \text{for } 0 < l < n.$$

[handwritten: $P^{(n)} = P^{(l)} P^{(n-l)}$]

This is called the *Chapman–Kolmogorov* equation for the Markov chain.

Proof For a homogeneous DTMC, we have

$$
\begin{aligned}
p_{ij}^{(n)} &= \text{Prob}\{X_n = j | X_0 = i\} \\
&= \sum_{\text{all } k} \text{Prob}\{X_n = j, X_l = k | X_0 = i\}, \quad \text{for } 0 < l < n. \\
&= \sum_{\text{all } k} \text{Prob}\{X_n = j | X_l = k, X_0 = i\}\text{Prob}\{X_l = k | X_0 = i\}
\end{aligned}
$$

Applying the Markov property, we have

$$
\begin{aligned}
p_{ij}^{(n)} &= \sum_{\text{all } k} \text{Prob}\{X_n = j | X_l = k\}\text{Prob}\{X_l = k | X_0 = i\} \\
&= \sum_{\text{all } k} p_{kj}^{(n-l)} p_{ik}^{(l)}, \quad \text{for } 0 < l < n.
\end{aligned}
$$

$\square$

In matrix notation, the Chapman–Kolmogorov equations are written as

$$P^{(n)} = P^{(l)} P^{(n-l)}.$$

This relation states that it is possible to write any n-step homogeneous transition probability as the sum of products of l-step and $(n-l)$-step transition probabilities. To go from i to j in n steps, it is necessary to go from i to an intermediate state k in l steps, and then from k to j in the remaining $n-l$ steps. By summing over all possible intermediate states k, we consider all possible distinct paths leading from i to j in n steps. Note in particular that

$$P^{(n)} = P P^{(n-1)} = P^{(n-1)} P.$$

Hence, the matrix of n-step transition probabilities is obtained by multiplying the matrix of one-step transition probabilities by itself $(n-1)$ times. In other words, $P^{(n)} = P^n$.

Example: If we return to the model of the weather in Belfast, the elements of the matrix P^2 tell us what the weather will be like the day after tomorrow, given today's weather. We have

$$P^2 = \begin{pmatrix} .770 & .165 & .065 \\ .750 & .175 & .075 \\ .710 & .195 & .095 \end{pmatrix}.$$

Thus, given that today the weather is sunny, the probability that it will be sunny the day after tomorrow is $(P^2)_{33} = 0.095$. For this example, it may be shown that successive powers of the matrix P converge to

$$P^\infty = \begin{pmatrix} .76250 & .16875 & .06875 \\ .76250 & .16875 & .06875 \\ .76250 & .16875 & .06875 \end{pmatrix}. \tag{1.3}$$

1.3.3 Classification of States

We now proceed to present a number of important definitions concerning the states of discrete-time Markov chains. In particular, we wish to distinguish between states that are *recurrent*, meaning that the DTMC is guaranteed to return to these states infinitely often, and states that are *transient*, meaning that there is a nonzero probability that the DTMC will never return to such a state. This does not mean that transient states cannot be visited many times, only that the probability of never returning to the state is nonzero. Let us define

$$\begin{aligned} f_{jj}^{(n)} &= \text{Prob \{first return to state } j \text{ occurs exactly } n \text{ steps after leaving it\}} \\ &= \text{Prob } \{X_n = j, X_{n-1} \neq j, \ldots, X_1 \neq j | X_0 = j\}, \quad \text{for } n = 1, 2, \ldots \end{aligned}$$

This should not be confused with $p_{jj}^{(n)}$, which is the probability of returning to state j in n steps, without excluding the possibility that state j was visited at one or more intermediate steps. We have $f_{jj}^{(1)} = p_{jj}$, and by applying the theorem of total probability and using $p_{jj}^{(0)} = 1$, it may be shown that $f_{jj}^{(n)}$, $n \geq 1$ may be calculated recursively from

$$p_{jj}^{(n)} = \sum_{l=1}^{n} f_{jj}^{(l)} p_{jj}^{(n-l)}, \qquad n \geq 1.$$

The probability of ever returning to state j is denoted by f_{jj} and is given by

$$f_{jj} = \sum_{n=1}^{\infty} f_{jj}^{(n)}.$$

If $f_{jj} = 1$, then state j is said to be *recurrent*; the Markov chain is guaranteed to return to this state sometime in the future. In this case, we have $p_{jj}^{(n)} > 0$ for some $n > 0$. If $f_{jj} < 1$, then state j is said to be *transient*. There is a nonzero probability that the chain will never return to this state.

When state j is recurrent, i.e., when $f_{jj} = 1$, we define the *mean recurrence time*, M_{jj}, of state j as

$$M_{jj} = \sum_{n=1}^{\infty} n f_{jj}^{(n)}.$$

This is the average number of steps taken to return to state j for the first time after leaving it. A recurrent state j for which M_{jj} is finite is called a *positive-recurrent* state or a *recurrent nonnull* state. If $M_{jj} = \infty$, we say that state j is a *null-recurrent* state. In a *finite* Markov chain, each state is either positive-recurrent or transient, and furthermore, at least one state must be positive-recurrent. In a finite Markov chain, none of the states can be null-recurrent.

Corresponding to $f_{jj}^{(n)}$, let us define $f_{ij}^{(n)}$ for $i \neq j$ as the probability that, starting from state i, the first passage to state j occurs in exactly n steps. We then have $f_{ij}^{(1)} = p_{ij}$, and

$$p_{ij}^{(n)} = \sum_{l=1}^{n} f_{ij}^{(l)} p_{jj}^{(n-l)}, \qquad n \geq 1.$$

This equation may be rearranged to obtain

$$f_{ij}^{(n)} = p_{ij}^{(n)} - \sum_{l=1}^{n-1} f_{ij}^{(l)} p_{jj}^{(n-l)}$$

which is somewhat more convenient for finding $f_{ij}^{(n)}$. The probability f_{ij} that state j is ever reached from state i is given by

$$f_{ij} = \sum_{n=1}^{\infty} f_{ij}^{(n)}.$$

If $f_{ij} < 1$, the process starting from state i may never reach state j. When $f_{ij} = 1$, the expected value of the sequence $f_{ij}^{(n)}$, $n = 1, 2, \ldots$ of first passage probabilities for a fixed pair i and j ($i \neq j$) is called the *mean first passage time* and is denoted by M_{ij}. We have

$$M_{ij} = \sum_{n=1}^{\infty} n f_{ij}^{(n)}, \quad \text{for } i \neq j.$$

The M_{ij} uniquely satisfy the equation

$$M_{ij} = p_{ij} + \sum_{k \neq j} p_{ik}(1 + M_{kj}), \tag{1.4}$$

since the system in state i either goes to state j in one step (with probability p_{ij}), or else (with probability p_{ik}) goes first to some intermediate state k in one step and then eventually on to j. If $i = j$, then M_{ij} is the mean recurrence time of state i. Since equation (1.4) also holds when $i = j$, we may write it in matrix form as

$$M = E + P(M - \text{diag}\{M\}), \qquad (1.5)$$

where $\text{diag}\{M\}$ is a diagonal matrix whose i^{th} diagonal element is M_{ii}, $E = ee^T$ with $e = (1, 1, \ldots, 1)^T$, and $(\cdot)^T$ denotes transposition (thus E is a matrix whose elements are all equal to 1). The diagonal elements of M are the mean recurrence times, whereas the off-diagonal elements are the mean first passage times. The matrix M may be obtained iteratively from the equation

$$M^{(k+1)} = E + P(M^{(k)} - \text{diag}\{M^{(k)}\}), \quad \text{with } M^{(0)} = E. \qquad (1.6)$$

As a later example will show, not all of the elements of $M^{(k+1)}$ need converge to a finite limit. The mean recurrence times of certain states as well as certain mean first passage times may be infinite, and the elements of $M^{(k+1)}$ that correspond to these situations will diverge as $k \to \infty$. Divergence of some elements in this context is an "acceptable" behavior for this iterative approach. Later we shall see a more efficient way to find mean recurrence times, M_{jj}, but not mean first passage times.

A state j is said to be *periodic with period p*, or *cyclic of index p*, if on leaving state j a return is possible only in a number of transitions that is a multiple of the integer $p > 1$. In other words, the period of a state j is defined as the greatest common divisor of the set of integers n for which $p_{jj}^{(n)} > 0$. A state whose period is $p = 1$ is said to be *aperiodic*.

A state that is positive-recurrent and aperiodic is said to be *ergodic*. If all the states of a Markov chain are ergodic, then the Markov chain itself is said to be ergodic.

Example: Consider a homogeneous DTMC whose transition probability matrix is

$$P = \begin{pmatrix} a & b & c \\ 0 & 0 & 1 \\ 0 & 1 & 0 \end{pmatrix}. \qquad (1.7)$$

Note that, in this example,

$$\begin{aligned} p_{11}^{(n)} &= a^n, & \text{for } n = 1, 2, \ldots; \\ p_{22}^{(n)} &= p_{33}^{(n)} = 0, & \text{for } n = 1, 3, 5, \ldots; \\ p_{22}^{(n)} &= p_{33}^{(n)} = 1, & \text{for } n = 0, 2, 4, \ldots; \end{aligned}$$

and that

$$p_{12}^{(n)} = a p_{12}^{(n-1)} + b \times 1_{\{n \text{ is odd}\}} + c \times 1_{\{n \text{ is even}\}}$$

where $1_{\{.\}}$ is an *indicator function*, which has the value 1 when the condition inside the braces is true and the value 0 otherwise. Then

$$
\begin{aligned}
f_{11}^{(1)} &= p_{11}^{(1)} = a \\
f_{11}^{(2)} &= p_{11}^{(2)} - f_{11}^{(1)}p_{11}^{(1)} = a^2 - a \times a = 0 \\
f_{11}^{(3)} &= p_{11}^{(3)} - f_{11}^{(1)}p_{11}^{(2)} - f_{11}^{(2)}p_{11}^{(1)} = p_{11}^{(3)} - f_{11}^{(1)}p_{11}^{(2)} = 0.
\end{aligned}
$$

It immediately follows that $f_{11}^{(n)} = 0$ for all $n \geq 2$, and thus the probability of ever returning to state 1 is given by $f_{11} = a < 1$. State 1 is therefore a transient state. Also,

$$
\begin{aligned}
f_{22}^{(1)} &= p_{22}^{(1)} = 0 \\
f_{22}^{(2)} &= p_{22}^{(2)} - f_{22}^{(1)}p_{22}^{(1)} = p_{22}^{(2)} = 1 \\
f_{22}^{(3)} &= p_{22}^{(3)} - f_{22}^{(1)}p_{22}^{(2)} - f_{22}^{(2)}p_{22}^{(1)} = p_{22}^{(3)} = 0,
\end{aligned}
$$

and again it immediately follows that $f_{22}^{(n)} = 0$ for all $n \geq 3$. We then have $f_{22} = \sum_{n=1}^{\infty} f_{22}^{(n)} = f_{22}^{(2)} = 1$, which means that state 2 is recurrent. Furthermore, it is positive-recurrent, since $M_{22} = \sum_{n=1}^{\infty} n f_{22}^{(n)} = 2 < \infty$. In a similar fashion, it may be shown that state 3 is also positive-recurrent.

Now consider $f_{12}^{(n)}$:

$$
\begin{aligned}
f_{12}^{(1)} &= b \\
f_{12}^{(2)} &= p_{12}^{(2)} - f_{12}^{(1)}p_{22}^{(1)} = p_{12}^{(2)} = ap_{12}^{(1)} + c = ab + c \\
f_{12}^{(3)} &= p_{12}^{(3)} - f_{12}^{(1)}p_{22}^{(2)} - f_{12}^{(2)}p_{22}^{(1)} = p_{12}^{(3)} - f_{12}^{(1)} \\
&= (a^2b + ac + b) - b = a^2b + ac.
\end{aligned}
$$

Continuing in this fashion, we find

$$
\begin{aligned}
f_{12}^{(4)} &= a^3b + a^2c \\
f_{12}^{(5)} &= a^4b + a^3c
\end{aligned}
$$
etc.,

and it is easy to show that in general we have

$$
f_{12}^{(n)} = a^{n-1}b + a^{n-2}c.
$$

It follows that the probability that state 2 is ever reached from state 1 is

$$
f_{12} = \sum_{n=1}^{\infty} f_{12}^{(n)} = \frac{b}{1-a} + \frac{c}{1-a} = 1.
$$

Similarly, we may show that $f_{13} = 1$. Also, it is evident that

$$f_{23}^{(1)} = f_{32}^{(1)} = 1$$

and

$$f_{23}^{(n)} = f_{32}^{(n)} = 0, \quad \text{for } n \geq 2$$

so that $f_{23} = f_{32} = 1$. However, note that

$$f_{21}^{(n)} = f_{31}^{(n)} = 0, \quad \text{for } n \geq 1,$$

and so state 1 can never be reached from state 2 or from state 3.

To examine the matrix M of mean first passage times (with diagonal elements equal to the mean recurrence times), we shall give specific values to the variables a, b and c. Let

$$P = \begin{pmatrix} 0.7 & 0.2 & 0.1 \\ 0.0 & 0.0 & 1.0 \\ 0.0 & 1.0 & 0.0 \end{pmatrix} \quad \text{and take} \quad M^{(1)} = \begin{pmatrix} 1.0 & 1.0 & 1.0 \\ 1.0 & 1.0 & 1.0 \\ 1.0 & 1.0 & 1.0 \end{pmatrix}.$$

Then, using the iterative formula, (1.6), we find

$$M^{(2)} = \begin{pmatrix} 1.3 & 1.8 & 1.9 \\ 2.0 & 2.0 & 1.0 \\ 2.0 & 1.0 & 2.0 \end{pmatrix}; \quad M^{(3)} = \begin{pmatrix} 1.6 & 2.36 & 2.53 \\ 3.0 & 2.0 & 1.0 \\ 3.0 & 1.0 & 2.0 \end{pmatrix}; \quad \text{etc.}$$

The iterative process tends to the matrix

$$M^{\infty} = \begin{pmatrix} \infty & 11/3 & 12/3 \\ \infty & 2 & 1 \\ \infty & 1 & 2 \end{pmatrix},$$

and it may be readily verified that this matrix satisfies equation (1.5). Thus, the mean recurrence time of state 1 is infinite, as are the mean first passage times from states 2 and 3 to state 1. The mean first passage time from state 2 to state 3 or vice versa is given as 1, which must obviously be true since on leaving either of these states, the process immediately enters the other. The mean recurrence time of both state 2 and state 3 is 2. One final comment that should be made concerning this example is that states 2 and 3 are periodic. On leaving either of these states, a return is possible only in a number of steps that is an integer multiple of 2.

1.3.4 Irreducibility

We say that a DTMC is *irreducible* if every state can be reached from every other state, i.e., if there exists an integer m for which $p_{ij}^{(m)} > 0$ for every pair of states i and j. Let S be the set of all states in a Markov chain, and let S_1 and S_2 be two subsets of states that partition S. The subset of states S_1 is said to be *closed* if no one-step transition is possible from any state in S_1 to any state in S_2. More generally, any nonempty subset S_1 of S is said to be closed if *no state* in S_1 leads to any state outside S_1 (in any number of steps), i.e.,

$$p_{ij}^{(n)} = 0, \quad \text{for } i \in S_1, \ j \notin S_1, \ n \geq 1.$$

If S_1 consists of a single state, then that state is called an *absorbing* state. A necessary and sufficient condition for state i to be an absorbing state is that $p_{ii} = 1$. If the set of all states S is closed and does not contain any proper subset that is closed, then the Markov chain is irreducible. On the other hand, if S contains proper subsets that are closed, the chain is said to be *reducible*. Notice that the matrix given as an example of the weather in Belfast (1.2) is irreducible, but the matrix given by equation (1.7) contains a proper subset of states (states 2 and 3) that is closed. Often closed subsets may be studied as independent entities.

Some important theorems concerning irreducible DTMCs are now presented without proof.

Theorem 1.1 *The states of a finite, aperiodic, irreducible Markov chain are ergodic.*

Note that the conditions given are *sufficient* conditions only. The theorem must not be taken as a definition of ergodicity.

Theorem 1.2 *Let a Markov chain C be irreducible. Then C is positive-recurrent or null-recurrent or transient; i.e.,*

- *all the states are positive-recurrent, or*

  finite Mean visitation time

  infinite mean visitation times.

- *all the states are null-recurrent, or*

- *all the states are transient.*

Furthermore, all states are periodic with the same period, p, or else all states are aperiodic.

Theorem 1.3 *Let C be an irreducible Markov chain, and let k be a given state of C. Then C is recurrent iff for every state j, $j \neq k$, $f_{jk} = 1$.*

Theorem 1.4 *Let k be a given state of an irreducible, recurrent Markov chain. Then the set of mean first passage times M_{jk}, $j \neq k$, uniquely satisfies the system of equations*

$$M_{jk} = 1 + \sum_{i \neq k} p_{ji} M_{ik}, \qquad j \neq k,$$

and the mean recurrence times satisfy

$$M_{kk} = 1 + \sum_{i \neq k} p_{ki} M_{ik}.$$

Note that these equations are obtained by a rearrangement of the terms of equation (1.4).

1.3.5   Probability Distributions

We now turn our attention to distributions defined on the states of a discrete-time Markov chain. In studying discrete-time Markov chains we are often interested in determining the probability that the chain is in a given state at a particular time step. We shall denote by $\pi_i(n)$ the probability that a Markov chain is in state i at step n, i.e.,

$$\pi_i(n) = \text{Prob}\{X_n = i\}.$$

In vector notation, we let $\pi(n) = (\pi_1(n), \pi_2(n), \ldots, \pi_i(n), \ldots)$. Note that the vector π is a row vector.[1] The state probabilities at any time step n may be obtained from a knowledge of the initial state distribution (at time step 0) and the matrix of transition probabilities. We have, from the theorem of total probability,

$$\pi_i(n) = \sum_{\text{all } k} \text{Prob}\{X_n = i | X_0 = k\} \pi_k(0).$$

The probability that the Markov chain is in state i at step n is therefore given by

$$\pi_i(n) = \sum_{\text{all } k} p_{ki}^{(n)} \pi_k(0), \tag{1.8}$$

which in matrix notation becomes

$$\pi(n) = \pi(0) P^{(n)} = \pi(0) P^n,$$

where $\pi(0)$ denotes the initial state distribution.

---

[1] We shall adopt the convention that all probability vectors are row vectors. All other vectors will be considered to be column vectors unless specifically stated otherwise.

Definition 1.1 (Stationary distribution) *Let P be the transition probability matrix of a DTMC, and let the vector z whose elements z_j denote the probability of being in state j be a probability distribution; i.e.,*

$$z_j \in \Re, \quad 0 \le z_j \le 1, \quad \text{and} \quad \sum_{\text{all } j} z_j = 1.$$

Then z is said to be a stationary distribution if and only if $zP = z$.

In other words, if z is chosen as the initial state distribution, i.e., $\pi_j(0) = z_j$ for all j, then for all n, we have $\pi_j(n) = z_j$.

Definition 1.2 (Limiting distribution) *Given an initial probability distribution $\pi(0)$, if the limit*

$$\lim_{n \to \infty} \pi(n),$$

exists, then this limit is called the limiting distribution, and we write

$$\pi = \lim_{n \to \infty} \pi(n).$$

Example: Consider the four-state Markov chain whose transition probability matrix is

$$P = \begin{pmatrix} 0 & 1 & 0 & 0 \\ 0 & 0 & 1 & 0 \\ 0 & 0 & 0 & 1 \\ 1 & 0 & 0 & 0 \end{pmatrix}.$$

It may readily be verified that the vector $(.25, .25, .25, .25)$ is a stationary distribution, but that no matter which state the system starts in, there is no limiting distribution.

Thus, the stationary distribution is *not necessarily* the limiting probability distribution of a Markov chain. An *irreducible, positive-recurrent* Markov chain has a unique stationary probability distribution π, whose elements are given by

$$\pi_j = 1/M_{jj}, \tag{1.9}$$

but does not necessarily have a limiting probability distribution. Equation (1.9) is readily verified by multiplying both sides of equation (1.5) by the vector π. We get

$$\pi M = \pi E + \pi P(M - \text{diag}\{M\}) = e + \pi(M - \text{diag}\{M\}) = e + \pi M - \pi \text{diag}\{M\}$$

and thus $\pi \text{diag}\{M\} = e$.

In an *irreducible and aperiodic* Markov chain it may be shown that the *limiting distribution* always exists and is independent of the initial probability distribution. Moreover, exactly one of the following conditions must hold:

1. All states are transient or all states are null-recurrent, in which case $\pi_j = 0$ for all j, and there exists *no stationary distribution* (even though the limiting distribution exists). The state space in this case must be infinite.

2. All states are positive-recurrent (which, together with the irreducibility and aperiodicity properties, makes them ergodic), in which case $\pi_j > 0$ for all j, and the set π_j is a stationary distribution. The π_j are *uniquely* determined by means of

$$\pi_j = \sum_{\text{all } i} \pi_i p_{ij} \quad \text{and} \quad \sum_j \pi_j = 1.$$

1.3.6 Steady-State Distributions of Ergodic Markov Chains

Recall that in an ergodic DTMC all the states are positive-recurrent and aperiodic. In such a Markov chain the probability distribution $\pi(n)$, as a function of n, always converges to a limiting stationary distribution π, which is independent of the initial state distribution. The limiting probabilities of an ergodic Markov chain are often referred to as the *equilibrium* or *steady-state* probabilities, in the sense that the effect of the initial state distribution $\pi(0)$ has disappeared. It follows from equation (1.8) that

$$\pi_j(n+1) = \sum_{\text{all } i} p_{ij} \pi_i(n),$$

and taking the limit as $n \to \infty$ of both sides gives

$$\pi_j = \sum_{\text{all } i} p_{ij} \pi_i.$$

Thus, the equilibrium probabilities may be uniquely obtained by solving the matrix equation

$$\pi = \pi P, \quad \text{with} \quad \pi > 0 \quad \text{and} \quad \|\pi\|_1 = 1. \tag{1.10}$$

Other useful and interesting properties of an ergodic Markov chain are as follows:

- As $n \to \infty$, the rows of the n-step transition matrix $P^{(n)} = P^n$ all become identical to the vector of steady-state probabilities. Letting $p_{ij}^{(n)}$ denote the ij^{th} element of $P^{(n)}$, we have

$$\pi_j = \lim_{n \to \infty} p_{ij}^{(n)}, \quad \text{for all } i.$$

This property may be observed in the example of the Markov chain that describes the evolution of the weather in Belfast. The matrix given in equation (1.3) consists of rows that are all identical and equal to the steady-state probability vector. We have $\pi = (.76250, .16875, .06875)$.

- The average time spent by the chain in state j in a fixed period of time τ at steady-state, $\nu_j(\tau)$, may be obtained as the product of the steady-state probability of state j and the duration of the observation period:

$$\nu_j(\tau) = \pi_j \tau.$$

The steady-state probability π_i itself may be interpreted as the proportion of time that the process spends in state i, averaged over the long run. Returning to the weather example, the mean number of sunny days per week is only 0.48125, while the average number of rainy days is 5.33750.

- We say that the process visits state i at time n if $X_n = i$. The average time spent by the process in state i at steady-state between two successive visits to state j, denoted by ν_{ij}, may be shown to be equal to the ratio of the steady-state probabilities of states i and j:

$$\nu_{ij} = \pi_i / \pi_j.$$

The quantity ν_{ij} is called the *visit ratio*, since it indicates the average number of visits to state i between two successive visits to state j. In our example, the mean number of rainy days between two sunny days is 11.091.[2]

1.4 Continuous-Time Markov Chains

1.4.1 Definitions

If the states of a Markov process are discrete and the process may change state at any point in time, we say that the process is a continuous-time Markov chain (CTMC). Mathematically, we say that the stochastic process $\{X(t),\ t \geq 0\}$ forms a continuous-time Markov chain if for all integers n, and for any sequence $t_0, t_1, \ldots, t_n, t_{n+1}$ such that $t_0 < t_1 < \ldots < t_n < t_{n+1}$, we have

$$\text{Prob}\{X(t_{n+1}) = x_{n+1} | X(t_0) = x_0, X(t_1) = x_1, \ldots, X(t_n) = x_n\}$$

$$= \text{Prob}\{X(t_{n+1}) = x_{n+1} | X(t_n) = x_n\}.$$

We may define transition probabilities analogously to those in the discrete-time case. If a continuous-time Markov chain is nonhomogeneous, we write

$$p_{ij}(s, t) = \text{Prob}\{X(t) = j | X(s) = i\},$$

---

[2]We would like to point out to our readers that the weather in Northern Ireland is not as bad as the numbers in this example, set up only for illustrative purposes, would have us believe. In fact, I have happy memories of spending long periods of glorious sunny summer days there.

where $X(t)$ denotes the state of the Markov chain at time $t \geq s$. When the CTMC is homogeneous, these transition probabilities depend on the difference $\tau = t - s$. In this case, we simplify the notation by writing

$$p_{ij}(\tau) = \text{Prob}\{X(s + \tau) = j | X(s) = i\}.$$

This denotes the probability of being in state j after an interval of length τ, given that the current state is state i. It depends on τ but not on s or t. It follows that

$$\sum_{\text{all } j} p_{ij}(\tau) = 1 \quad \text{for all values of } \tau.$$

1.4.2   Transition Probabilities and Transition Rates

Whereas a discrete-time Markov chain is represented by its matrix of *transition probabilities*, P, a continuous-time Markov chain is represented by a matrix of *transition rates*. It therefore behooves us to examine the relationship between these two quantities in the context of Markov chains.

The *probability* that a transition occurs from a given source state depends not only on the source state itself but also on the length of the interval of observation. In what follows, we shall consider a period of observation $\tau = \Delta t$. Let $p_{ij}(t, t + \Delta t)$ be the probability that a transition occurs from state i to state j in the interval $(t, t + \Delta t)$. As the duration of this interval becomes very small, the probability that we observe a transition also becomes very small. In other words, as $\Delta t \rightarrow 0$, $p_{ij}(t, t + \Delta t) \rightarrow 0$ for $i \neq j$. It then follows from conservation of probability that $p_{ii}(t, t + \Delta t) \rightarrow 1$ as $\Delta t \rightarrow 0$. On the other hand, as Δt becomes large, the probability that we observe a transition increases. As Δt becomes even larger, the probability that we observe multiple transitions becomes nonnegligible. In our models, we would like to ensure that the durations of our observation intervals are sufficiently small that the probability of observing two or more transitions within this period is negligible. By negligible, we mean that the probability of observing multiple transitions is $o(\Delta t)$, where $o(\Delta t)$ is the "little oh" notation and is defined as a quantity for which

$$\lim_{\Delta t \rightarrow 0} \frac{o(\Delta t)}{\Delta t} = 0.$$

In other words, $o(\Delta t)$ tends to zero faster than Δt.

A *rate* of transition does not depend on the length of an observation period; it is an *instantaneously* defined quantity that denotes the number of transitions that occur per unit time. Let $q_{ij}(t)$ be the rate at which transitions occur from state i to state j at time t. Note that for a nonhomogeneous Markov chain the rate of transition, like the probability of transition, may depend on the time t. However, unlike the transition probability, it does not depend on a time *interval* Δt. We have

$$q_{ij}(t) = \lim_{\Delta t \rightarrow 0} \left\{ \frac{p_{ij}(t, t + \Delta t)}{\Delta t} \right\}, \quad \text{for } i \neq j. \tag{1.11}$$

It then follows that

$$p_{ij}(t, t + \Delta t) = q_{ij}(t)\Delta t + o(\Delta t), \quad \text{for } i \neq j, \tag{1.12}$$

which in words simply states that, correct to terms of order $o(\Delta t)$, the probability that a transition occurs from state i at time t to state j in the next Δt time units is equal to the rate of transition at time t multiplied by the length of the time period, Δt.

From conservation of probability and equation (1.11),

$$1 - p_{ii}(t, t + \Delta t) = \sum_{j \neq i} p_{ij}(t, t + \Delta t) \tag{1.13}$$

$$= \sum_{j \neq i} q_{ij}(t)\Delta t + o(\Delta t). \tag{1.14}$$

Dividing by Δt and taking the limit as $\Delta t \to 0$, we get

$$q_{ii}(t) \equiv \lim_{\Delta t \to 0} \left\{ \frac{p_{ii}(t, t + \Delta t) - 1}{\Delta t} \right\} = \lim_{\Delta t \to 0} \left\{ \frac{-\sum_{j \neq i} q_{ij}(t)\Delta t + o(\Delta t)}{\Delta t} \right\}$$

i.e.,

$$q_{ii}(t) = -\sum_{j \neq i} q_{ij}(t). \tag{1.15}$$

The fact that the $q_{ii}(t)$ are negative should not be surprising. This quantity denotes a transition *rate* and as such is defined as a derivative. Given that the system is in state i at time t, the probability that it will transfer to a different state j increases with time, whereas the probability that it remains in state i must decrease with time. It is appropriate in the first case that the derivative at time t be positive, and in the second that it be negative.

Substituting equation (1.15) into equation (1.14) provides the analog of (1.12). For convenience we write them both together,

$$p_{ij}(t, t + \Delta t) = q_{ij}(t)\Delta t + o(\Delta t), \quad \text{for } i \neq j,$$
$$p_{ii}(t, t + \Delta t) = 1 + q_{ii}(t)\Delta t + o(\Delta t).$$

The matrix $Q(t)$ whose ij^{th} element is $q_{ij}(t)$ is called the *infinitesimal generator matrix* or *transition rate matrix* for the continuous-time Markov chain. In matrix form, we have

$$Q(t) = \lim_{\Delta t \to 0} \left\{ \frac{P(t, t + \Delta t) - I}{\Delta t} \right\},$$

where $P(t, t + \Delta t)$ is the transition probability matrix, its ij^{th} element is $p_{ij}(t, t + \Delta t)$, and I is the identity matrix. When the CTMC is homogeneous, the transition rates q_{ij} are independent of time, and the matrix of transition rates is written simply as Q.

1.4.3   The Embedded Markov Chain

In a homogeneous, continuous-time Markov chain, $X(t)$, the time spent by the system in any state is exponentially distributed. In state i, the parameter of the exponential distribution of the sojourn time is given by $-q_{ii} = \sum_{j \neq i} q_{ij}$. If we ignore the time actually spent in any state and consider only the transitions made by the system, we may define a new, discrete-time, Markov chain called the *embedded Markov chain (EMC)*. Let the nth state visited by the continuous-time Markov chain be denoted by Y_n. Then $\{Y_n, \ n \geq 0\}$ is the embedded Markov chain of the CTMC $X(t)$.

Many of the properties of a CTMC can be deduced from those of its corresponding EMC. In a CTMC, we say that state i can reach state j, for any $i \neq j$, if there is a $t \geq 0$ such that $p_{ij}(t) > 0$. This allows us to define the concept of irreducibility in a CTMC. It is then easy to show that a CTMC is irreducible if and only if its EMC is irreducible. Other results follow similarly. For example, a state j is recurrent in a CTMC if and only if it is recurrent in its EMC. However, a certain amount of care must be exercised, for it is possible for a state j to be positive-recurrent in a CTMC and null-recurrent in its corresponding EMC, and vice versa! Also, a state may be aperiodic in a CTMC and periodic in its EMC.

For an ergodic CTMC the one-step transition probabilities of its EMC, denoted by s_{ij}, (i.e., $s_{ij} \equiv \text{Prob} \{Y_{n+1} = j | Y_n = i\}$), are given by

$$
\begin{aligned}
s_{ij} &= \frac{q_{ij}}{\sum_{j \neq i} q_{ij}}, \quad j \neq i \\
&= 0, \qquad\quad i = j.
\end{aligned}
$$

If we let S denote the transition probability matrix of the EMC, then in matrix terms we have

$$ S = I - [\text{diag}\{Q\}]^{-1} Q. $$

Note that all the elements of S satisfy $0 \leq s_{ij} \leq 1$ and that $\sum_{j, j \neq i} s_{ij} = 1$ for all i. Thus S possesses the characteristics of a transition probability matrix for a discrete-time Markov chain. Note also that for $i \neq j$, $s_{ij} = 0$ if and only if $q_{ij} = 0$, and thus, since Q is irreducible, S is irreducible.

Given a splitting of the infinitesimal generator Q,

$$ Q = D_Q - L_Q - U_Q $$

where D_Q is the diagonal of Q, $-L_Q$ is the strictly lower triangular part of Q, and $-U_Q$ is the strictly upper triangular part of Q, it is possible to characterize the matrix S as

$$ S = (I - D_Q^{-1} Q) = D_Q^{-1}(D_Q - Q) = D_Q^{-1}(L_Q + U_Q). $$

We shall see later that this defines S to be the iteration matrix associated with the method of Jacobi applied to the system of equations $\pi Q = 0$.

The problem of finding the stationary probability vector of the original continuous-time Markov chain is therefore that of finding the solution ϕ of the homogeneous system of linear equations

$$\phi(I - S) = 0 \quad \text{with} \quad \phi > 0 \quad \text{and} \quad \|\phi\|_1 = 1$$

and then computing the stationary solution of the CTMC from

$$\pi = \frac{-\phi D_Q^{-1}}{\|\phi D_Q^{-1}\|_1}.$$

Example: Consider a homogeneous CTMC whose transition rate matrix is given as

$$Q = \begin{pmatrix} -4 & 4 & 0 & 0 \\ 3 & -6 & 3 & 0 \\ 0 & 2 & -4 & 2 \\ 0 & 0 & 1 & -1 \end{pmatrix}.$$

The transition probability matrix corresponding to the embedded Markov chain is

$$S = \begin{pmatrix} 0 & 1 & 0 & 0 \\ .5 & 0 & .5 & 0 \\ 0 & .5 & 0 & .5 \\ 0 & 0 & 1 & 0 \end{pmatrix}.$$

It is easy to check that the stationary probability vector of the embedded Markov chain is given by

$$\phi = (1/6, 1/3, 1/3, 1/6)$$

Since

$$D_Q = \begin{pmatrix} -4 & 0 & 0 & 0 \\ 0 & -6 & 0 & 0 \\ 0 & 0 & -4 & 0 \\ 0 & 0 & 0 & -1 \end{pmatrix},$$

we have

$$-\phi D_Q^{-1} = (1/6, 1/3, 1/3, 1/6) \begin{pmatrix} 1/4 & 0 & 0 & 0 \\ 0 & 1/6 & 0 & 0 \\ 0 & 0 & 1/4 & 0 \\ 0 & 0 & 0 & 1 \end{pmatrix} = (1/24, 1/18, 1/12, 1/6)$$

which when normalized yields the stationary probability vector of the CTMC as

$$\pi = (.12, .16, .24, .48).$$

Note that $\pi Q = 0$. In this example the CTMC is nonperiodic, whereas the EMC is periodic with period $p = 2$.

1.4.4   The Chapman–Kolmogorov Equations

The Chapman–Kolmogorov equations for a nonhomogeneous CTMC may be obtained directly from the Markov property. They are specified by

$$p_{ij}(s,t) = \sum_{\text{all } k} p_{ik}(s,u)p_{kj}(u,t) \quad \text{for } i,j = 0,1,\ldots, \text{ and } s \leq u \leq t.$$

In passing from state i at time s to state j at time t $(s < t)$, we must pass through some intermediate state k at some intermediate time u. When the continuous-time Markov chain is homogeneous, the Chapman–Kolmogorov equation becomes

$$p_{ij}(\tau) = \sum_{\text{all } k} p_{ik}(\tau - \alpha)p_{kj}(\alpha), \quad \text{for } 0 \leq \alpha \leq \tau. \qquad (1.16)$$

From equation (1.16), we have

$$
\begin{aligned}
p_{ij}(t + \Delta t) - p_{ij}(t) &= \sum_{\text{all } k} p_{ik}(t + \Delta t - \alpha)p_{kj}(\alpha) - \sum_{\text{all } k} p_{ik}(t - \alpha)p_{kj}(\alpha) \\
&= \sum_{\text{all } k} [p_{ik}(t + \Delta t - \alpha) - p_{ik}(t - \alpha)]p_{kj}(\alpha).
\end{aligned}
$$

As $\alpha \to t$, the term $p_{ik}(t - \alpha) \to 0$ if $i \neq k$ and $p_{ii}(t - \alpha) \to 1$. It then follows that dividing both sides by Δt and taking the limits as $\Delta t \to 0$ and $\alpha \to t$, we get

$$\frac{dp_{ij}(t)}{dt} = \sum_{\text{all } k} q_{ik}(t)p_{kj}(t), \quad \text{for } i,j = 0,1,\ldots$$

These are called the *Kolmogorov backward equations*. In matrix form they are written as

$$\frac{dP(t)}{dt} = Q(t)P(t).$$

In a similar manner, we may derive the *Kolmogorov forward equations*, which are

$$\frac{dp_{ij}(t)}{dt} = \sum_{\text{all } k} q_{kj}(t)p_{ik}(t), \quad \text{for } i,j = 0,1,\ldots$$

or, in matrix form,

$$\frac{dP(t)}{dt} = P(t)Q(t).$$

1.4.5   Probability Distributions

Consider a system that is modelled by a continuous-time Markov chain. Let $\pi_i(t)$ be the probability that the system is in state i at time t, i.e.,

$$\pi_i(t) = \text{Prob}\{X(t) = i\}.$$

Then

$$\pi_i(t + \Delta t) = \pi_i(t) \left(1 - \sum_{\text{all } j \neq i} q_{ij}(t)\Delta t\right) + \left(\sum_{\text{all } k \neq i} q_{ki}(t)\pi_k(t)\right)\Delta t + o(\Delta t).$$

Since $q_{ii}(t) = -\sum_{\text{all } j \neq i} q_{ij}(t)$, we have

$$\pi_i(t + \Delta t) = \pi_i(t) + \left(\sum_{\text{all } k} q_{ki}(t)\pi_k(t)\right)\Delta t + o(\Delta t),$$

and

$$\lim_{\Delta t \to 0} \left(\frac{\pi_i(t + \Delta t) - \pi_i(t)}{\Delta t}\right) = \lim_{\Delta t \to 0} \left(\sum_{\text{all } k} q_{ki}(t)\pi_k(t) + o(\Delta t)/\Delta t\right),$$

i.e.,

$$\frac{d\pi_i(t)}{dt} = \sum_{\text{all } k} q_{ki}(t)\pi_k(t).$$

In matrix notation, this gives

$$\frac{d\pi(t)}{dt} = \pi(t)Q(t).$$

When the Markov chain is homogeneous, we may drop the dependence on time and simply write

$$\frac{d\pi(t)}{dt} = \pi(t)Q.$$

It follows that the solution $\pi(t)$ is given by

$$\pi(t) = e^{Qt} = I + \sum_{n=1}^{\infty} \frac{Q^n t^n}{n!}$$

which unfortunately may be rather difficult and unstable to compute [101]. We shall take up this topic again in Chapter 8. In the meantime, we turn our attention to steady-state distributions.

We shall not list, as we did for the discrete-time case, conditions that are necessary for the existence of a steady-state probability distribution. These may often be inferred from the corresponding properties of the embedded Markov chain. As in the discrete case, the existence of a limiting distribution depends on the structure of the matrix Q, i.e., on the row and column position of its nonzero elements, as well as on the individual properties of the states themselves. Readers requiring further information should consult one of the standard texts. That of Ronald W. Wolff [179] is particularly recommended.

If a steady-state distribution exists, then it necessarily follows that the rate of change of $\pi(t)$ at steady-state is zero, i.e., $d\pi(t)/dt = 0$, and therefore, for a homogeneous CTMC,

$$\pi Q = 0. \tag{1.17}$$

In the queueing theory literature these equations are called the *global balance equations*. The vector π (now written as independent of t) is the equilibrium or long-run probability vector (π_i is the probability of being in state i at statistical equilibrium) and may be obtained by applying equation-solving techniques to the homogeneous system of equations (1.17). In a manner similar to that for a discrete-time Markov chain, we define a stationary probability vector as any vector z for which $zQ = 0$. For an ergodic CTMC, a stationary distribution is also the steady-state distribution.

Equation (1.17) may be written as

$$\pi P = \pi, \tag{1.18}$$

where $P = Q\Delta t + I$, and now matrix eigenvalue/eigenvector techniques may be applied to equation (1.18) to determine the stationary probability vector π as a left-hand eigenvector corresponding to a unit eigenvalue of P. When we set $P = Q\Delta t + I$, we essentially discretize or *uniformize* the continuous-time Markov chain to get a discrete-time Markov chain whose transitions take place at intervals Δt, Δt being chosen sufficiently small that the probability of two transitions taking place in time Δt is negligible, i.e., of order $o(\Delta t)$. In this case, the matrix P is stochastic. The stationary probability vector π of the continuous-time Markov chain (obtained from $\pi Q = 0$) is obviously identical to that of the discretized chain (obtained from $\pi P = \pi$).

We caution the reader whose interest is in transient solutions (i.e., probability distributions at an arbitrary time t) that those of the discrete-time chain (represented by the transition probability matrix, P) are not the same as those of the continuous-time chain (represented by the infinitesimal generator, Q). This subject is raised in the chapter on transient solutions.

1.5   Nonnegative Matrices

1.5.1   Definition

Let A be an $\Re^{n \times m}$ matrix whose elements a_{ij} satisfy

$$a_{ij} \geq 0.$$

Then A is said to be a *nonnegative* matrix of real-valued elements. When the elements of A satisfy

$$a_{ij} > 0,$$

then A is said to be *positive*. Continuing in this sense, we write

$$A \geq B$$

when $a_{ij} \geq b_{ij}$ for all i and j, and

$$A > B$$

when $A \geq B$ and $A \neq B$. Thus, the inequalities hold elementwise. Given that 0 represents the matrix whose elements are all equal to 0, then A is a nonnegative matrix if $A \geq 0$ and A is positive if $A > 0$.

If A is a matrix of complex-valued elements, we shall denote by A^* the matrix obtained by replacing each element of A by its modulus. Thus, the ij^{th} element of A^* is $|a_{ij}|$.

1.5.2 Nonnegative Decomposable Matrices

A square nonnegative matrix A is said to be *decomposable* if it can be brought by a symmetric permutation of its rows and columns to the form

$$A = \begin{pmatrix} U & 0 \\ W & V \end{pmatrix}, \tag{1.19}$$

where U and V are square, nonzero matrices and W is, in general, rectangular and nonzero. We may relate this concept of decomposability with that of reducibility in the context of Markov chains; in fact, the two terms are often used interchangeably. If a Markov chain is reducible, then there is an ordering of the state space such that the transition probability matrix has the form given in (1.19). If the matrices U and V are of order n_1 and n_2 respectively ($n_1 + n_2 = n$, where n is the order of A), then the states s_i, $i = 1, 2, \ldots, n$ of the Markov chain may be decomposed into two nonintersecting sets

$$B_1 = \{s_1, s_2, \ldots, s_{n_1}\}$$

and

$$B_2 = \{s_{n_1+1}, \ldots, s_n\}.$$

The nonzero structure of the matrix A in equation (1.19) shows that transitions are possible from states of B_2 to B_1 but not the converse, so that if the system is at any time in one of the states $s_1, \ldots, s_{n_1}$, it will never leave this set. Let us assume that $W \neq 0$ and the states of B_2 communicate with each other. Then, if the system is initially in one of the states of B_2, it is only a question of time until it eventually finishes in B_1. The set B_1 is called *isolated* or *essential*, and B_2 is called *transient* or *nonessential*. Since the system eventually reduces to B_1, it is called *reducible* and the matrix of transition probabilities is said to be *decomposable* [129]. If $W = 0$, then A is completely decomposable, and in this case both sets, B_1 and B_2, are essential.

It is possible for the matrix U itself to be decomposable, in which case the set of states in B_1 may be reduced to a new set whose order must be less than n_1. This process of decomposition may be continued until A is reduced to the form, called the *normal* form of a decomposable nonnegative matrix, given by

$$
A = \begin{pmatrix}
A_{11} & 0 & 0 & \cdots & 0 & 0 & \cdots & 0 \\
0 & A_{22} & 0 & \cdots & 0 & 0 & \cdots & 0 \\
\vdots & \vdots & \vdots & \ddots & \vdots & \vdots & \ddots & \vdots \\
0 & 0 & 0 & \cdots & A_{kk} & 0 & \cdots & 0 \\
A_{k+1,1} & A_{k+1,2} & \cdots & \cdots & A_{k+1,k} & A_{k+1,k+1} & \cdots & 0 \\
\vdots & \vdots & \vdots & \ddots & \vdots & \vdots & \ddots & \vdots \\
A_{m,1} & A_{m,2} & \cdots & \cdots & A_{m,k} & A_{m,k+1} & \cdots & A_{m,m}
\end{pmatrix}. \quad (1.20)
$$

The submatrices A_{ii}, $i = 1, \ldots, m$ are square, nonzero, and nondecomposable. All submatrices to the right of the diagonal blocks are zero, as are those to the left of A_{ii} for $i = 1, 2, \ldots, k$. As for the remaining blocks, for each value of $i \in [k+1, m]$ there is at least one value of $j \in [1, i-1]$ for which $A_{ij} \neq 0$. The diagonal submatrices are as follows:

- A_{ii}, $i = 1, \ldots, k$, isolated and nondecomposable,

- A_{ii}, $i = k+1, \ldots, m$, transient, and again, nondecomposable.

If B_i is the set of states represented by A_{ii}, then states of B_i for $i = 1, \ldots, k$, have the property that once the system is in any one of them, it must remain in it. For sets B_i, $i = k+1, \ldots, m$, any transition out of these sets is to one possessing a lower index only, so that the system eventually finishes in one of the isolated sets $B_1, \ldots, B_k$. If all of the off-diagonal blocks are zero, $(k = m)$, then A is said to be a *completely decomposable* nonnegative matrix. Sometimes the term *separable* is used to describe a completely decomposable matrix.

1.5.3   The Theorem of Perron–Frobenius

Some of the major results concerning nonnegative matrices are contained in the theorem of Perron–Frobenius (e.g., Cox and Miller [33], p. 120; Varga [170], p. 30; Berman and Plemmons [10]). Before stating this theorem we introduce the *spectral radius* of a matrix A as the number $\rho(A) = \max_s |\lambda_s(A)|$ where $\lambda_s(A)$

denotes the sth eigenvalue of A. The *spectrum* of a matrix is the set of its eigenvalues. It is known that for any natural matrix norm $\| \cdot \|$, $\rho(A) \leq \|A\|$. Furthermore, the matrix infinity norm, defined as

$$\|A\|_\infty = \max_j \left(\sum_{\text{all } k} |a_{jk}| \right),$$

and the matrix one norm,

$$\|A\|_1 = \max_k \left(\sum_{\text{all } j} |a_{jk}| \right),$$

are natural matrix norms. These norms are equal to the maximum absolute row sum and maximum absolute column sum, respectively.

Theorem 1.5 (Perron–Frobenius) *Let $A \geq 0$ be a nondecomposable square matrix of order n. Then,*

1.
    - *A has a positive real eigenvalue, λ_1, equal to its spectral radius.*
    - *To $\rho(A)$ there corresponds an eigenvector $x > 0$, i.e.,*

    $$Ax = \lambda_1 x \quad \text{and} \quad x > 0.$$

    - *$\rho(A)$ increases when any entry of A increases.*
    - *$\rho(A)$ is a simple eigenvalue of A, i.e., λ_1 is a simple root of*

    $$|\lambda I - A| = 0.$$

2. *Let S be a matrix of complex-valued elements and S^* obtained from S by replacing each element by its modulus. If $S^* \leq A$, then any eigenvalue μ of S satisfies*

    $$|\mu| \leq \lambda_1.$$

    Furthermore, if for some μ, $|\mu| = \lambda_1$, then $S^* = A$. More precisely, if $\mu = \lambda_1 e^{i\theta}$, then

    $$S = e^{i\theta} DAD^{-1}$$

    where $D^* = I$.

3. *If A has exactly p eigenvalues equal in modulus to λ_1, then these numbers are all different and are the roots of the equation*

    $$\lambda^p - \lambda_1^p = 0.$$

When plotted as points in the complex plane, this set of eigenvalues is invariant under a rotation of the plane through the angle $2\pi/p$ but not through *smaller angles. When $p > 1$, then A can be symmetrically permuted to the following cyclic form*

$$A = \begin{pmatrix} 0 & A_{12} & 0 & \cdots & 0 \\ 0 & 0 & A_{23} & \cdots & 0 \\ \vdots & \vdots & \vdots & \ddots & \vdots \\ 0 & 0 & 0 & \cdots & A_{p-1,p} \\ A_{p1} & 0 & 0 & \cdots & 0 \end{pmatrix},$$

in which the diagonal submatrices A_{ii} are square and the only nonzero submatrices are A_{12}, A_{23}, $\ldots$, A_{p1}.

A primary reason for the introduction of this theorem is its application to stochastic matrices, a topic to which we now turn.

1.6   Stochastic Matrices

1.6.1   Definition

A matrix $P \in \Re^{n \times n}$ is said to be a *stochastic* matrix if it satisfies the following three conditions:

1. $p_{ij} \geq 0$ for all i and j.

2. $\sum_{\text{all } j} p_{ij} = 1$ for all i.   fwd t propogation

3. At least one element in each column differs from zero.

We have just seen that matrices that obey condition (1) are called nonnegative matrices, and stochastic matrices form a proper subset of them. Condition (2) implies that a transition is guaranteed to occur from state i to at least one state in the next time period (that may be state i again). Condition (3) specifies that since each column has at least one nonzero element, there are no *ephemeral* states, i.e., states that could not possibly exist after the first time transition. In much of the literature on stochastic matrices this third condition is omitted (being considered trivial). In the remainder of this text we also shall ignore this condition.

1.6.2   Some Properties of Stochastic Matrices

Property 1.1 *Every stochastic matrix has an eigenvalue equal to unity.*

Proof   Since the sum of the elements of each row of P is 1, we must have

$$Pe = e$$

where $e = (1, 1, \ldots, 1)^T$. It immediately follows that P has a unit eigenvalue. $\square$

Corollary 1.1 *Every infinitesimal generator matrix Q has at least one zero eigenvalue.*

Proof This follows immediately from the fact that $Qe = 0$. □

Property 1.2 *The eigenvalues of a stochastic matrix must have modulus less than or equal to 1.*

Proof To prove this result we shall use the fact that for any matrix A,

$$\rho(A) \leq \|A\|_\infty = \max_j \left(\sum_{\text{all } k} |a_{jk}| \right).$$

For a stochastic matrix P, $\|P\|_\infty = 1$, and therefore we may conclude that

$$\rho(P) \leq 1.$$

Hence, no eigenvalue of a stochastic matrix P can exceed 1 in modulus. □

This result may also be proven using Gerschgorin's theorem, which states that the eigenvalues of an $(n \times n)$ matrix A must lie within the union of the n circular disks with center a_{ii} and radius $\sum_{\text{all } j \neq i} |a_{ij}|$, $i = 1, 2, \ldots, n$. Notice that this property, together with Property 1.1, implies that

$$\rho(P) = 1. \tag{1.21}$$

Property 1.3 *The stochastic matrix of an irreducible Markov chain possesses a simple unit eigenvalue.*

Proof This follows directly from Properties 1.1 and 1.2 and Part 1 of the Perron–Frobenius theorem. □

Property 1.4 *The right-hand eigenvector corresponding to a unit eigenvalue $\lambda_1 = 1$ of a stochastic matrix P is given by e where $e = (1, 1, \ldots)^T$.*

Proof Since the sum of each row of P is 1, we have $Pe = e = \lambda_1 e$. □

Property 1.5 *The vector π is a stationary probability vector of a stochastic matrix P iff it is a left-hand eigenvector corresponding to a unit eigenvalue.*

Proof By definition a stationary probability vector does not change with time. Since the effect of a single-step transition on a distribution x is given by xP, we must have

$$\pi = \pi P,$$

and thus π satisfies the eigenvalue equation

$$\pi P = \lambda_1 \pi \quad \text{for } \lambda_1 = 1.$$

The converse is equally obvious. □

Property 1.6 *For any irreducible Markov chain with stochastic transition probability matrix P, let*

$$P(\alpha) = I - \alpha(I - P) \tag{1.22}$$

where $\alpha \in \Re' \equiv (-\infty, \infty) \setminus \{0\}$ (i.e., the real line with zero deleted). Then 1 is a simple eigenvalue of every $P(\alpha)$, and associated with this unit eigenvalue is a uniquely defined positive left-hand eigenvector of unit 1-norm, which is precisely the stationary probability vector π of P.

Proof Notice that the spectrum of $P(\alpha)$ is given by $\lambda(\alpha) = 1 - \alpha(1 - \lambda)$ for λ in the spectrum of P. Furthermore, as can be verified by substituting from equation (1.22), the left-hand eigenvectors of P and $P(\alpha)$ agree in the sense that

$$x^T P = \lambda x^T \quad \text{if and only if} \quad x^T P(\alpha) = \lambda(\alpha) x^T \quad \text{for all } \alpha.$$

In particular, this means that regardless of whether or not $P(\alpha)$ is a stochastic matrix, $\lambda(\alpha) = 1$ is a simple eigenvalue of $P(\alpha)$ for all α, because $\lambda = 1$ is a simple eigenvalue of P. Consequently, the entire family $P(\alpha)$ has a unique positive left-hand eigenvector of unit 1-norm associated with $\lambda(\alpha) = 1$, and this eigenvector is precisely the stationary distribution π of P.   □

Example: As an example of this last property, consider the 3×3 stochastic matrix given by

$$P = \begin{pmatrix} .99911 & .00079 & .00010 \\ .00061 & .99929 & .00010 \\ .00006 & .00004 & .99990 \end{pmatrix}. \tag{1.23}$$

Its eigenvalues, as computed by MATLAB[3], are 1.0; .9998 and .9985. The maximum value of α that we can choose so that $P(\alpha)$ is stochastic is $\alpha_1 = 1/.00089$. The resulting stochastic matrix that we obtain is given by

$$P(\alpha_1) = \begin{pmatrix} .0 & .88764 & .11236 \\ .68539 & .20225 & .11236 \\ .06742 & .04494 & .88764 \end{pmatrix},$$

and its eigenvalues are 1.0; .77528 and $-.68539$. Yet another choice, $\alpha_2 = 10,000$, yields a $P(\alpha_2)$ that is not stochastic. We find

$$P(\alpha_2) = \begin{pmatrix} -7.9 & 7.9 & 1.0 \\ 6.1 & -6.1 & 1.0 \\ 0.6 & 0.4 & 0.0 \end{pmatrix}.$$

Its eigenvalues are 1.0; -1.0 and -14.0. The left-hand eigenvector corresponding to the unit eigenvalue for all three of the above matrices is given by

$$\pi = (.22333, .27667, .50000),$$

the stationary probability vector of the original stochastic matrix.

---

[3]MATLAB is a registered trademark of The MathWorks, Inc.

1.6.3 Effect of the Discretization Parameter, Δt

Let us now ask for what values of $\Delta t > 0$ the matrix $P = Q\Delta t + I$ is stochastic. Consider a two-state Markov chain whose infinitesimal generator is

$$Q = \begin{pmatrix} -q_1 & q_1 \\ q_2 & -q_2 \end{pmatrix},$$

with $q_1, q_2 \geq 0$. The transition probability matrix is then

$$P = \begin{pmatrix} 1 - q_1\Delta t & q_1\Delta t \\ q_2\Delta t & 1 - q_2\Delta t \end{pmatrix}.$$

Obviously the row sums are equal to unity. To ensure that $0 \leq q_1\Delta t \leq 1$ and $0 \leq q_2\Delta t \leq 1$, we require that $0 \leq \Delta t \leq q_1^{-1}$ and $0 \leq \Delta t \leq q_2^{-1}$. Let us assume, without loss of generality, that $q_1 \geq q_2$. Then $0 \leq \Delta t \leq q_1^{-1}$ satisfies both these conditions. To ensure that $0 \leq 1 - q_1\Delta t \leq 1$ and $0 \leq 1 - q_2\Delta t \leq 1$ again requires that $\Delta t \leq q_1^{-1}$. Consequently the maximum value that we can assign to Δt subject to the condition that P is stochastic is $\Delta t = q_1^{-1}$.

Similar results hold for a general stochastic matrix $P = Q\Delta t + I$ to be stochastic, given that Q is an infinitesimal generator. As before, for any value of Δt, the row sums of P are unity, since by definition the row sums of Q are zero. Therefore we must concern ourselves with the values of Δt that guarantee that the elements of P lie in the interval $[0,1]$. Let q be the size of the largest off-diagonal element:

$$q = \max_{i,j\ i\neq j}(q_{ij}) \quad \text{and} \quad q_{ij} \geq 0, \quad \text{for all } i,j.$$

Then $0 \leq p_{ij} \leq 1$ holds if $0 \leq q_{ij}\Delta t \leq 1$, which is true if $\Delta t \leq q^{-1}$. Now consider a diagonal element $p_{ii} = q_{ii}\Delta t + 1$. We have

$$0 \leq q_{ii}\Delta t + 1 \leq 1$$

or

$$-1 \leq q_{ii}\Delta t \leq 0.$$

The right-hand inequality holds for all $\Delta t \geq 0$, since q_{ii} is negative. The left-hand inequality $q_{ii}\Delta t \geq -1$ is true if $\Delta t \leq -q_{ii}^{-1}$, i.e., $\Delta t \leq (|q_{ii}|)^{-1}$.

It follows then that if $0 \leq \Delta t \leq (\max_i |q_{ii}|)^{-1}$, the matrix P is stochastic. (Since the diagonal elements of Q equal the negated sum of the off-diagonal elements in a row, we have $\max_i |q_{ii}| \geq \max_{i\neq j}(q_{ij})$.) Thus, Δt must be less than or equal to the reciprocal of the absolute value of the largest diagonal element of Q.

The choice of a suitable value for Δt plays a crucial role in some iterative methods for determining the stationary probability vector from equation (1.18).

The rate of convergence is intimately related to the magnitude of the eigenvalues of P. As a general rule, the closer the magnitudes of the subdominant eigenvalues are to 1, the slower the convergence rate. We would therefore like to maximize the distance between the largest eigenvalue, $\lambda_1 = 1$, and the subdominant eigenvalue (the eigenvalue that in modulus is closest to 1). Notice that, as $\Delta t \to 0$, the eigenvalues of P *all* tend to unity. This would suggest that we choose Δt to be as large as possible, subject only to the constraint that P be a stochastic matrix. However, note that choosing $\Delta t = (\max_i |q_{ii}|)^{-1}$ does not necessarily guarantee the maximum separation of dominant and subdominant eigenvalues. It is simply a good heuristic.

Consider the (2×2) case as an example. The eigenvalues of P are the roots of the characteristic equation $|P - \lambda I| = 0$, i.e.,

$$\begin{vmatrix} 1 - q_1 \Delta t - \lambda & q_1 \Delta t \\ q_2 \Delta t & 1 - q_2 \Delta t - \lambda \end{vmatrix} = 0$$

These roots are $\lambda_1 = 1$ and $\lambda_2 = 1 - \Delta t(q_1 + q_2)$. As $\Delta t \to 0$, $\lambda_2 \to \lambda_1 = 1$.

Note that the left eigenvector corresponding to the unit eigenvalue, λ_1, is independent of the choice of Δt. We have

$$\left(q_2/(q_1 + q_2) \quad q_1/(q_1 + q_2) \right) \begin{pmatrix} 1 - q_1 \Delta t & q_1 \Delta t \\ q_2 \Delta t & 1 - q_2 \Delta t \end{pmatrix} = \left(q_2/(q_1 + q_2) \quad q_1/(q_1 + q_2) \right).$$

This eigenvector is, of course, the stationary probability vector of the Markov chain and as such must be independent of Δt. The parameter Δt can only affect the speed at which matrix iterative methods will converge to this vector. As we mentioned above, it is often advantageous to choose Δt to be as large as possible, subject only to the constraint that the matrix P be a stochastic matrix. Intuitively, we may think that by choosing a large value of Δt we are marching more quickly toward the stationary distribution. With small values we essentially take only small steps, and therefore it takes longer to arrive at our destination, the stationary distribution. It was for this reason that Wallace and Rosenberg in their "Recursive Queue Analyzer, RQA-1" [174] chose the value $\Delta t = 0.99 \times (\max_i |q_{ii}|)^{-1}$. The factor 0.99 was used rather than 1.0 to ensure that P would not be periodic. This choice forces the diagonal elements of P to be nonzero.

1.6.4   Eigenvalues of Decomposable Stochastic Matrices

In Section 1.5.3 we enunciated the theorem of Perron–Frobenius with respect to nonnegative matrices. Now we wish to examine the consequences of parts (1) and (2) of this theorem as they apply to decomposable *stochastic* matrices. Let P be a decomposable stochastic matrix of order n written in normal decomposable form as

$$P = \begin{pmatrix} P_{11} & 0 & 0 & \cdots & 0 & 0 & \cdots & 0 \\ 0 & P_{22} & 0 & \cdots & 0 & 0 & \cdots & 0 \\ \vdots & \vdots & \vdots & \ddots & \vdots & \vdots & \ddots & \vdots \\ 0 & 0 & 0 & \cdots & P_{kk} & 0 & \cdots & 0 \\ P_{k+1,1} & P_{k+1,2} & \cdots & \cdots & P_{k+1,k} & P_{k+1,k+1} & \cdots & 0 \\ \vdots & \vdots & \vdots & \ddots & \vdots & \vdots & \ddots & \vdots \\ P_{m,1} & P_{m,2} & \cdots & \cdots & P_{m,k} & P_{m,k+1} & \cdots & P_{m,m} \end{pmatrix}. \quad (1.24)$$

The submatrices P_{ii}, $i = 1, \ldots, k$ are all nondecomposable stochastic matrices. From Property 1.3, nondecomposable stochastic matrices possess a unique unit eigenvalue. Thus each of the submatrices P_{ii}, $i = 1, 2, \ldots, k$ possesses one and only one unit eigenvalue.

Part (2) of the Perron–Frobenius theorem may be used to show that if one or more elements, p_{ij}, of an irreducible stochastic matrix is reduced by a positive real number ϵ such that $p_{ij} - \epsilon \geq 0$, then the maximum eigenvalue of the new matrix is strictly less than 1 in modulus. Thus, the largest eigenvalue of each of the matrices P_{ii}, $i = k + 1, \ldots, m$ of (1.24) is strictly less than unity in modulus, since each of these matrices contains at least one row that does not sum to 1. These P_{ii} are said to be *substochastic*.

Finally, it is known that the set of eigenvalues of a block (upper or lower) triangular matrix (such as P in Equation 1.24) is given by the union of the sets of eigenvalues of the individual blocks [54]. This, in conjunction with the above remarks, means that a decomposable stochastic matrix of the form (1.24) has exactly as many unit eigenvalues as it has isolated subsets; i.e., k.

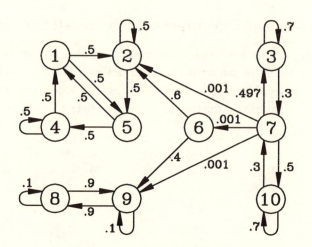

Figure 1.1: A Markov chain.

Example: Consider the Markov chain defined diagrammatically by Figure 1.1. The corresponding stochastic matrix is given by

$$
P = \begin{pmatrix}
 & 0.5 & & & 0.5 & & & & & \\
 & 0.5 & & & 0.5 & & & & & \\
 & & 0.7 & & & & 0.3 & & & \\
0.5 & & & 0.5 & & & & & & \\
0.5 & & & 0.5 & & & & & & \\
 & 0.6 & & & & & & & 0.4 & \\
0.001 & 0.497 & & & & 0.001 & & & 0.001 & 0.5 \\
 & & & & & & & 0.1 & 0.9 & \\
 & & & & & & & 0.9 & 0.1 & \\
 & & & & & & 0.3 & & & 0.7
\end{pmatrix},
$$

which has no clearly defined form. However, if the rows and columns are permuted according to

New State #	1	2	3	4	5	6	7	8	9	10
Old State #	1	2	4	5	8	9	6	3	7	10

the matrix becomes

$$
\begin{pmatrix}
& 0.5 & & 0.5 & \vdots & & & \vdots & & \vdots & & & \\
& 0.5 & & 0.5 & \vdots & & & \vdots & & \vdots & & & \\
0.5 & & 0.5 & & \vdots & & & \vdots & & \vdots & & & \\
0.5 & & 0.5 & & \vdots & & & \vdots & & \vdots & & & \\
\cdots & \cdots & \cdots & \cdots & \vdots & \cdots & \cdots & \vdots & \cdots & \vdots & \cdots & \cdots & \cdots \\
& & & & \vdots & 0.1 & 0.9 & \vdots & & \vdots & & & \\
& & & & \vdots & 0.9 & 0.1 & \vdots & & \vdots & & & \\
\cdots & \cdots & \cdots & \cdots & \vdots & \cdots & \cdots & \vdots & \cdots & \vdots & \cdots & \cdots & \cdots \\
& 0.6 & & & \vdots & & 0.4 & \vdots & 0.0 & \vdots & & & \\
\cdots & \cdots & \cdots & \cdots & \vdots & \cdots & \cdots & \vdots & \cdots & \vdots & \cdots & \cdots & \cdots \\
& & & & \vdots & & & \vdots & & \vdots & 0.7 & 0.3 & \\
& 0.001 & & & \vdots & & 0.001 & \vdots & 0.001 & \vdots & 0.497 & & 0.5 \\
& & & & \vdots & & & \vdots & & \vdots & & 0.3 & 0.7
\end{pmatrix},
$$

which is in the form (1.24). Clearly it may be seen that once the system is in the set B_1 consisting of reordered states $\{1, 2, 3, 4\}$, or in B_2 with reordered states $\{5, 6\}$, it will never leave it. If it is initially in any other set, e.g., $B_3 = \{7\}$ or $B_4 = \{8, 9, 10\}$, it will in the long run leave it and move into B_1 or B_2. The set of eigenvalues of this matrix is

$$\{1.0, 1.0, 0.9993, -0.8000, 0.7000, -0.2993, 0.0000, 0.0000, 0.0000, 0.0\}$$

containing two unit eigenvalues and with the other eigenvalues having modulus strictly less than unity. It possesses one eigenvalue that is identically zero and three that are less than 10^{-5} but nonzero.

1.6.5 Eigenvectors of Decomposable Stochastic Matrices

Let the stochastic matrix P be decomposable into the form (1.24), and let y_l, $l = 1, \ldots, k$ be a set of right-hand eigenvectors corresponding to the k unit eigenvalues. Then by definition

$$Py_l = y_l, \qquad l = 1, 2, \ldots, k.$$

Let the y_l be partitioned into m disjoint parts corresponding to the m different sets of essential and transient states, i.e.,

$$y_l = \left(y_l^1, y_l^2, \ldots, y_l^i, \ldots, y_l^m \right)^T.$$

If n_i is the number of states in the ith set, then y_l^i consists of elements $1 + \sum_{j=1}^{i-1} n_j$ through $\sum_{j=1}^{i} n_j$ of the eigenvector y_l. We have

$$
\begin{pmatrix}
P_{11} & 0 & 0 & \cdots & 0 & 0 & \cdots & 0 \\
0 & P_{22} & 0 & \cdots & 0 & 0 & \cdots & 0 \\
\vdots & \vdots & \vdots & \ddots & \vdots & \vdots & \ddots & \vdots \\
0 & 0 & 0 & \cdots & P_{kk} & 0 & \cdots & 0 \\
P_{k+1,1} & P_{k+1,2} & 0 & \cdots & P_{k+1,k} & P_{k+1,k+1} & \cdots & 0 \\
\vdots & \vdots & \vdots & \ddots & \vdots & \vdots & \ddots & \vdots \\
P_{m,1} & P_{m,2} & 0 & \cdots & P_{m,k} & P_{m,k+1} & \cdots & P_{m,m}
\end{pmatrix}
\begin{pmatrix}
y_l^1 \\
y_l^2 \\
\vdots \\
y_l^k \\
y_l^{k+1} \\
\vdots \\
y_l^m
\end{pmatrix}
=
\begin{pmatrix}
y_l^1 \\
y_l^2 \\
\vdots \\
y_l^k \\
y_l^{k+1} \\
\vdots \\
y_l^m
\end{pmatrix}
$$

and therefore $P_{ii} y_l^i = y_l^i$ for $i = 1, 2, \ldots, k$, which implies that y_l^i is a right eigenvector corresponding to a unit eigenvalue of P_{ii}. But for values of $i = 1, 2, \ldots, k$, P_{ii} is a nondecomposable stochastic matrix and has a unique unit eigenvalue. The subvector y_l^i is therefore the right eigenvector corresponding to the unique unit eigenvalue of P_{ii}, and as such all of its elements must be equal. But this is true for all $i = 1, 2, \ldots, k$. This allows us to state the following theorem:

Theorem 1.6 *In any right-hand eigenvector corresponding to a unit eigenvalue of the matrix P, the elements corresponding to states of the same essential class have identical values.*

The above reasoning applies only to the states belonging to essential classes. When the states of the Markov chain have not been ordered, it might be thought that the values of components belonging to transient states might make the analysis more difficult. However, it is possible to determine which components correspond to transient states. These are the only states that have zero components in *all* of the *left-hand* eigenvectors corresponding to the k unit eigenvalues. These states must have a steady-state probability of zero. Consequently these states may be detected and eliminated from the analysis of the right-hand eigenvectors that is given above.

Notice finally that it is possible to choose linear combinations of the right-hand eigenvectors to construct a new set of eigenvectors y_i', $i = 1, 2, \ldots, k$, in which the components of eigenvector i corresponding to states of essential set i are all equal to 1, while those corresponding to states belonging to other essential subsets are zero.

Example: Referring back to the previous example, the eigenvectors corresponding to the three dominant eigenvalues are printed below.

Eigenvalues:

$$\lambda_1 = 1.0 \qquad \lambda_2 = 1.0 \qquad \lambda_3 = 0.9993$$

Right-hand eigenvectors:

$$
y_1 = \begin{pmatrix} -.4380 \\ -.4380 \\ -.4380 \\ -.4380 \\ 0.0000 \\ 0.0000 \\ -.2628 \\ -.2336 \\ -.2336 \\ -.2336 \end{pmatrix}, \quad
y_2 = \begin{pmatrix} .4521 \\ .4521 \\ .4521 \\ .4521 \\ -.1156 \\ -.1156 \\ .2250 \\ .1872 \\ .1872 \\ .1872 \end{pmatrix}, \quad
y_3 = \begin{pmatrix} 0.0 \\ 0.0 \\ 0.0 \\ 0.0 \\ 0.0 \\ 0.0 \\ 0.0 \\ 0.5778 \\ 0.5765 \\ 0.5778 \end{pmatrix}
$$

Left-hand eigenvectors:

$$
x_1 = \begin{pmatrix} -0.5 \\ -0.5 \\ -0.5 \\ -0.5 \\ 0.0 \\ 0.0 \\ 0.0 \\ 0.0 \\ 0.0 \\ 0.0 \end{pmatrix}, \quad
x_2 = \begin{pmatrix} 0.0 \\ 0.0 \\ 0.0 \\ 0.0 \\ 0.7071 \\ 0.7071 \\ 0.0 \\ 0.0 \\ 0.0 \\ 0.0 \end{pmatrix}, \quad
x_3 = \begin{pmatrix} 0.1836 \\ 0.1828 \\ 0.1836 \\ 0.1833 \\ 0.3210 \\ 0.3207 \\ -0.0003 \\ -0.5271 \\ -0.3174 \\ -0.5302 \end{pmatrix}
$$

Since 0.9993 is not exactly equal to unity, we must consider only the right-hand eigenvectors y_1 and y_2. In y_1, the elements belonging to the essential set B_2 are zero. We may readily find a linear combination of y_1 and y_2 in which the elements of the essential set B_1 are zero. Depending upon the particular software used to generate the eigenvectors, different linear combinations of the vectors may be produced. However, it is obvious from the vectors above that regardless of the combination chosen, the result will have the property that the components related to any essential set are all identical. One particular linear combination of interest is $\alpha y_1 + \beta y_2$, with

$$\alpha = \frac{(1 - \frac{.4521}{-.1156})}{-.4380} \quad \text{and} \quad \beta = \frac{1}{-.1156},$$

which gives $\alpha y_1 + \beta y_2 = e$.

Notice that in this example, the elements of the right-hand eigenvectors corresponding to the transient class B_4 are also identical. With respect to the left-hand eigenvectors, notice that regardless of the linear combination of x_1 and x_2 used, the components belonging to the two transient classes will always be zero.

1.6.6 Nearly Decomposable Stochastic Matrices

If the matrix P given by (1.24) is now altered so that the off-diagonal submatrices that are zero become nonzero, then the matrix is no longer decomposable, and P will have only one unit eigenvalue. However, if these off-diagonal elements are small compared to the nonzero elements of the diagonal submatrices, then the matrix is said to be *nearly decomposable* in the sense that there are only weak interactions among the diagonal blocks. In this case there will be eigenvalues close to unity.

In the simplest of all possible cases of completely decomposable matrices, we have

$$P = \begin{pmatrix} 1.0 & 0.0 \\ 0.0 & 1.0 \end{pmatrix}.$$

The eigenvalues are, of course, both equal to 1.0. When off-diagonal elements ϵ_1 and ϵ_2 are introduced, we have

$$P = \begin{pmatrix} 1 - \epsilon_1 & \epsilon_1 \\ \epsilon_2 & 1 - \epsilon_2 \end{pmatrix},$$

and the eigenvalues are now given by $\lambda_1 = 1.0$ and $\lambda_2 = 1.0 - \epsilon_1 - \epsilon_2$. As $\epsilon_1, \epsilon_2 \to 0$, the system becomes completely decomposable, and $\lambda_2 \to 1.0$.

In a strictly nondecomposable system a subdominant eigenvalue close to 1.0 is often indicative of a nearly decomposable matrix. However, the existence of eigenvalues close to 1.0 is *not* a sufficient condition for the matrix to be nearly decomposable. When P has the form given by (1.24), the characteristic equation $|\lambda I - P| = 0$ is given by

$$P(\lambda) = P_{11}(\lambda) P_{22}(\lambda) \cdots P_{mm}(\lambda) = 0, \tag{1.25}$$

where $P_{ii}(\lambda)$ is the characteristic equation for the block P_{ii}. Equation (1.25) has a root $\lambda = 1.0$ of multiplicity k. Suppose a small element ϵ is subtracted from P_{ii}, one of the first k blocks, i.e., $i = 1, \ldots, k$, and added into one of the off-diagonal blocks in such a way that the matrix remains stochastic. The block P_{ii} is then strictly substochastic, and from part two of the Perron–Frobenius theorem, its largest eigenvalue is strictly less than 1. The modified matrix now has a root $\lambda = 1.0$ of multiplicity $(k - 1)$, a root $\lambda = 1 - O(\epsilon)$, and $(m - k)$ other roots of modulus < 1.0; i.e., the system is decomposable into $(k - 1)$ isolated sets. Since the eigenvalues of a matrix are continuous functions of the elements of the matrix, it follows that as $\epsilon \to 0$, the subdominant eigenvalue tends to unity and the system reverts to its original k isolated sets.

If we consider the system represented by the previous figure, it may be observed that the transient set $B_4 = \{8, 9, 10\}$ is almost isolated, and as expected, the eigenvalues include one very close to unity.

Nearly completely decomposable (NCD) Markov chains arise frequently in modelling applications [28]. In these applications the state space may be partitioned into disjoint subsets with strong interactions among the states of a subset but with weak interactions among the subsets themselves. Efficient computational procedures exist to compute the stationary distributions of these systems. They are considered in detail in a later chapter.

1.7 Cyclic Stochastic Matrices

1.7.1 Definition

In an irreducible DTMC, when the number of single-step transitions required on leaving any state to return to that same state (by any path) is a multiple of some integer $p > 1$, the Markov chain is said to be *periodic of period p*, or *cyclic of index p*. One of the fundamental properties of a cyclic stochastic matrix of index p is that it is possible by a permutation of its rows and columns to transform it to the form, called the *normal cyclic form*

$$P = \begin{pmatrix} 0 & P_{12} & 0 & \cdots & 0 \\ 0 & 0 & P_{23} & \cdots & 0 \\ \vdots & \vdots & \vdots & \ddots & \vdots \\ 0 & 0 & 0 & \cdots & P_{p-1,p} \\ P_{p1} & 0 & 0 & \cdots & 0 \end{pmatrix}, \tag{1.26}$$

in which the diagonal submatrices P_{ii} are square, zero, and of order n_i and the only nonzero submatrices are $P_{12}, P_{23}, \ldots, P_{p1}$. This corresponds to a partitioning of the states of the system into p distinct subsets and an ordering imposed on the subsets. The ordering is such that once the system is in a state of subset i, it must exit this subset in the next time transition and enter a state of subset

$(i \bmod p) + 1$. The matrix P is said to be *cyclic of index p*, or *p-cyclic*. When the states of the Markov chain are not periodic, (i.e., $p = 1$), then P is said to be *primitive*.

1.7.2   Eigenvalues of Cyclic Stochastic Matrices

The matrix given in equation (1.26) possesses p eigenvalues of unit modulus. Indeed, it follows as a direct consequence of part (3) of the Perron–Frobenius theorem that a cyclic stochastic matrix of index p possesses, among its eigenvalues, all the roots of $\lambda^p - 1 = 0$. Thus a cyclic stochastic matrix of

- index 2 possesses eigenvalues $1, -1$;
- index 3 possesses eigenvalues $1, -(1 \pm \sqrt{3}i)/2$;
- index 4 possesses eigenvalues $1, -1, i, -i$;
- etc.

Notice that a cyclic stochastic matrix of index p has p eigenvalues equally spaced around the unit circle. Consider what happens as p becomes large. The number of eigenvalues spaced equally on the unit circle increases. It follows that a cyclic stochastic matrix will have eigenvalues, other than the unit eigenvalue, that will tend to unity as $p \to \infty$. Such a matrix is *not* decomposable or nearly decomposable. As pointed out previously, eigenvalue(s) close to unity do not necessarily imply that a Markov chain is nearly decomposable. On the other hand, a nearly completely decomposable Markov chain *must* have subdominant eigenvalues close to 1.

1.7.3   Example: A Decomposable System with Cyclic Subgroups

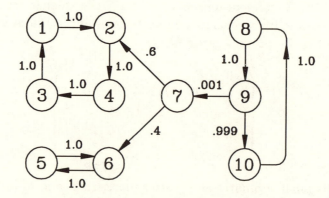

Figure 1.2: Modified Markov chain.

Consider the model shown in Figure 1.2, where the states are already ordered in the manner in which the subsystems are expected to arise. It may be observed that there are two essential subsets, $B_1 = \{1, 2, 3, 4\}$ and $B_2 = \{5, 6\}$, and two transient subsets, $B_3 = \{7\}$ and $B_4 = \{8, 9, 10\}$, as in the previous example. For this Markov chain the stochastic matrix P is

$$P = \left(\begin{array}{cccc:cc:c:ccc}
1.0 & & & & & & & & & \\
& 1.0 & & & & & & & & \\
& & 1.0 & & & & & & & \\
1.0 & & & & & & & & & \\
\hdashline
& & & & & 1.0 & & & & \\
& & & & 1.0 & & & & & \\
\hdashline
0.6 & & & & & 0.4 & & & & \\
\hdashline
& & & & & & & & 1.0 & \\
& & & & & & 0.001 & & & 0.999 \\
& & & & & & & 1.0 & &
\end{array}\right).$$

The submatrices comprising sets B_1 and B_2 are both of the form (1.26) and therefore cyclic; the first is of index 4 and the second of index 2. We should expect to find among the eigenvalues of P the values

$$1.0, -1.0, i, -i; \text{ and } 1.0 \text{ and } -1.0,$$

which is in fact true (the roots being ± 1.0, ± 1.0, $\pm i$, 0.99967, $0.4998 \pm 0.8657i$, and 0.0). It may be observed that subset B_4 is almost an essential subset, and further that it is almost cyclic of index 3. This is emphasized by the closeness of the roots 0.99967, $-0.4998 \pm 0.8657i$ to the cube roots of unity. In fact, analogous to the concept of nearly decomposable stochastic matrices, we may define nearly cyclic stochastic matrices. These matrices possess among their eigenvalues all the roots of the equation $\lambda^p - 1.0 + \epsilon = 0.0$, where ϵ is small and p is an integer indicating the (near-) cyclicity of the system.

In some cases, it may be possible to form an estimate of the near decomposability and/or near cyclicity of a system by an examination of the subdominant eigenvalues. However, great care must be taken, for the closeness of a subdominant eigenvalue to 1.0 may also depend on other factors. We have already mentioned the case of cyclic matrices of index p as $p \rightarrow \infty$. There are other factors that should be considered. For example, a matrix of order 100,000 will have 100,000 eigenvalues whose modulus lies between 0 and 1, so that it might be supposed that some of these may be close to 1 even though the matrix might be far from being decomposable or cyclic. Or again, consider the $(n \times n)$ stochastic matrix

$$P = \begin{pmatrix} 0.5 & 0.5 & & & & & \\ 0.25 & 0.5 & 0.25 & & & & \\ & 0.25 & 0.5 & 0.25 & & & \\ & & \ddots & \ddots & \ddots & & \\ & & & & 0.25 & 0.5 & 0.25 \\ & & & & & 0.5 & 0.5 \end{pmatrix}.$$

Its roots are given by

$$\lambda_k = cos^2 \frac{k\pi}{2(n-1)}, \quad k = 0, 1, \ldots, n-1$$

and are thus more tightly clustered toward the unit eigenvalue than toward zero. This matrix is definitely *not* nearly completely decomposable.

1.8   Indicators of State Clustering

1.8.1   Significance of Subdominant, Right-Hand Eigenvectors

It is known that a *left-hand* eigenvector corresponding to a unit eigenvalue is a stationary probability vector. As yet, no physical significance has been ascribed to the left-hand eigenvectors corresponding to eigenvalues different from unity.

The situation is otherwise for the set of right-hand eigenvectors. When the Markov chain under consideration is irreducible and noncyclic, its stochastic matrix has a single eigenvalue of modulus 1; i.e., the unit eigenvalue. In this case some information concerning the tendencies of the states to form groups may be obtained from an examination of the right-hand eigenvectors corresponding to the *subdominant* eigenvalues, i.e., the eigenvalues with modulus closest to but strictly less than 1.0. The reason is as follows:

The equilibrium position of the system is defined by the stationary probability vector, i.e., the left-hand eigenvector corresponding to the unit eigenvalue. With each state of the system can be associated a real number, which determines its "distance" from this equilibrium position. This distance may be regarded as the number of iterations (or the length of time) required to reach the equilibrium position if the system starts in the state for which the distance is being measured. Such measurements are, of course, only relative, but they serve as a means of comparison among the states.

Let the row vector $w_i^{(1)} = (0, 0, \ldots, 1, \ldots, 0)$ with i^{th} component equal to 1, denote that initially the system is in state i. We shall assume that P possesses a full set of n linearly independent eigenvectors. Similar results may be obtained when eigenvectors and principal vectors are used instead. Let $x_1, x_2, \ldots, x_n$ be the left-hand eigenvectors of P (i.e., $x_j^T P = \lambda_j x_j^T$ for all $j = 1, 2, \ldots, n$), arranged into descending order according to the magnitude of their corresponding eigenvalues. Writing $w_i^{(1)}$ as a linear combination of these eigenvectors, we have

$$w_i^{(1)} = c_{i1} x_1^T + c_{i2} x_2^T + \ldots + c_{in} x_n^T$$

where $c_{i1}, c_{i2}, \ldots, c_{in}$ are the constants that define the linear combination. Repeated postmultiplication of $w_i^{(1)}$ by P yields the steady-state probability vector. We have

$$w_i^{(1)} P = c_{i1} x_1^T P + c_{i2} x_2^T P + \ldots + c_{in} x_n^T P \tag{1.27}$$

$$= c_{i1} x_1^T + c_{i2} \lambda_2 x_2^T + \ldots + c_{in} \lambda_n x_n^T = w_i^{(2)}, \tag{1.28}$$

and in general

$$w_i^{(k+1)} = c_{i1} x_1^T + c_{i2} \lambda_2^k x_2^T + \ldots + c_{in} \lambda_n^k x_n^T.$$

If the system initially starts in some other state $j \neq i$, we have

$$w_j^{(k+1)} = c_{j1} x_1^T + c_{j2} \lambda_2^k x_2^T + \ldots + c_{jn} \lambda_n^k x_n^T.$$

Since only the constant coefficients differ, the difference in the length of time taken to reach the steady state from any two states i and j depends only on these constant terms. Further, if λ_2 is of strictly larger modulus than $\lambda_3, \lambda_4, \ldots$, then for large k, $\lambda_2^k \gg \lambda_l^k$ for $l \geq 3$, and it is the terms c_{i2} and c_{j2} in particular that contribute to the difference. Considering all possible starting states, we obtain

$$\begin{pmatrix} w_1^{(k+1)} \\ w_2^{(k+1)} \\ \vdots \\ w_n^{(k+1)} \end{pmatrix} = \begin{pmatrix} c_{11} x_1^T + c_{12} \lambda_2^k x_2^T + \ldots + c_{1n} \lambda_n^k x_n^T \\ c_{21} x_1^T + c_{22} \lambda_2^k x_2^T + \ldots + c_{2n} \lambda_n^k x_n^T \\ \vdots \\ c_{n1} x_1^T + c_{n2} \lambda_2^k x_2^T + \ldots + c_{nn} \lambda_n^k x_n^T \end{pmatrix},$$

i.e.,

$$W^{(k+1)} = \begin{pmatrix} c_{11} & c_{12} & \ldots & c_{1n} \\ c_{21} & c_{22} & \ldots & c_{2n} \\ \vdots & \vdots & \ddots & \vdots \\ c_{n1} & c_{n2} & \ldots & c_{nn} \end{pmatrix} \begin{pmatrix} 1 & & & \\ & \lambda_2^k & & \\ & & \ddots & \\ & & & \lambda_n^k \end{pmatrix} \begin{pmatrix} x_1^T \\ x_2^T \\ \vdots \\ x_n^T \end{pmatrix} \equiv C \Lambda^k X^T.$$

To obtain the matrix C, consider the following: the matrix $W^{(1)} = (w_1^{(1)}, w_2^{(1)},$ $\ldots, w_n^{(1)})^T$ was originally written in terms of the set of left-hand eigenvectors as

$$W^{(1)} = C X^T,$$

but since $W^{(1)} = I$, we obtain

$$I = C X^T,$$

i.e., $C = (X^T)^{-1} = Y$, the set of right-hand eigenvectors of P. Therefore, it is from the second column of the matrix C, i.e., the subdominant right-hand eigenvector of the matrix P, that an appropriate measure of the relative distance of each state from the stationary probability vector may be obtained. The third and subsequent columns may be employed to obtain subsidiary effects.

States whose corresponding component value in this vector is large in magnitude are, in a relative sense, far from the equilibrium position. Also, the states corresponding to component values that are relatively close together form a cluster, or a subset, of states. If, for example, the components of this vector are close either to $+1$ or to -1, then it may be said that those states corresponding to values close to $+1$ form a subset of states that is far from the remaining states, and vice versa. In this manner, it may be possible to determine which states constitute near essential and near cyclic subsets of states.

1.8.2   A Resource Pool Example

Consider as an example a process that may take (or return) units of resource, one at a time, from a pool (called the *free resource pool*). Let n be the total number of units of resource in the system, and let $(k, n - k)$ represent the state in which the process currently possesses k of these units. Further, suppose this process has either:

- Centrifugal tendencies, or

- Centripetal tendencies.

If it has the first, then its tendency is to possess all the resources or none at all; i.e., if $k > n/2$, the process has greater probability of taking some more units of resource than it has of returning some. Conversely, if $k < n/2$, the process tends to give its k units back to the pool. Hence, the process drifts to the extremities of the system, so that it is most likely to be either in state $(n, 0)$ or $(0, n)$. Centrifugal tendencies thus imply *degenerate tendencies*, since we would expect those states in the neighborhood of $(n, 0)$, i.e., $(n - 1, 1)$, $(n - 2, 2)$, etc. to form a subset very distinct from the states neighboring $(0, n)$. This is a phenomenon that we should expect to detect from an examination of the right subdominant eigenvector.

If, on the other hand, the process has centripetal tendencies, it tends to distribute the units of resource equally between itself and the pool; i.e., if $k > n/2$, it wants to return units of resource; if $k < n/2$, it wants to take some more. Thus, if n is even, the process tends to stabilize in the state $(n/2, n/2)$, but if n is odd, the process sets up a two-cycle, swapping the extra unit of resource in and out of the resource pool. In other words, for odd n, centripetal tendencies imply *cyclic tendencies*. We should expect this two-cycle tendency to be detectable from the subdominant right eigenvector.

For the purposes of this example, let us assume that there are nine units of resource. The tendency to be either centrifugal or centripetal is represented by the variable T. If T is greater than 1.0, then the tendency will be centrifugal; if less than 1.0, then the process will have centripetal tendencies. The propensity to be one or the other is measured by the magnitude of T. This parameter T may be thought of as a rate of transition, in which case the states and transitions are given diagrammatically by Figure 1.3.

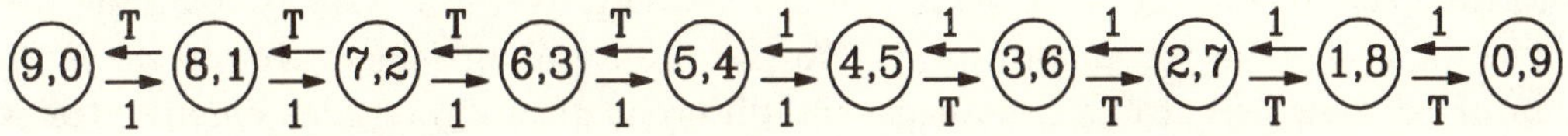

Figure 1.3: States and transitions in resource pool.

Several different values for T are considered, ranging from 20.0 to 0.1, including $T=1.0$. From the point of view of this example the most interesting results are the values of the subdominant eigenvalues and the corresponding right-hand eigenvectors, since we expect the eigenvalues to be indicative of the near-decomposability/near-cyclicity of the system, and the components of the eigenvector to determine which states constitute which subsets. Table 1.1 gives the values of the positive and negative eigenvalues of greatest modulus that are not equal to unity (denoted λ_+ and λ_- respectively) for different values of T.

An examination of the values of λ_+ shows that for $T > 1.0$ this eigenvalue is close to unity, and that as T increases, $\lambda_+ \to 1.0$. But subdominant eigenvalues that are close to unity may be indicative of nearly decomposable systems. It is precisely for such T that the process has centrifugal tendencies, and therefore decomposable tendencies.

For values of $T < 1.0$, λ_+ ceases to be important. However, for these values the eigenvalue λ_- assumes importance. In this range $(0 < T < 1)$ this root is the subdominant eigenvalue, and as $T \to 0, \lambda_- \to -1.0$. Such eigenvalues are to be found in nearly cyclic stochastic matrices. Again, values of $T < 1.0$ imply centripetal tendencies, and hence cyclic tendencies.

When $T = 1.0$, both λ_+ and λ_- are equal in modulus. If T varies even slightly on either side, then one of $\lambda_+, |\lambda_-|$ increases and the other decreases.

Table 1.1: Subdominant Eigenvalues of P for Different T Values

T	λ_+	λ_-
20.0	0.9999995	-0.356737
10.0	0.9999866	-0.488002
5.0	0.999714	-0.644595
3.0	0.997864	-0.763593
1.0	0.951057	-0.951057
0.33333	0.763593	-0.997868
0.2	0.644595	-0.999714
0.1	0.488002	-0.999986

The previous discussion relates the magnitude of the eigenvalues to the degree of near decomposability or cyclicity of the system. We now wish to examine the right-hand eigenvector corresponding to the subdominant eigenvalue to see if we can discover which states constitute the different groupings. This eigenvector is given in Table 1.2 for all the different values of T considered.

Table 1.2: Right-Hand Eigenvector Corresponding to Subdominant Eigenvalue

State	$T = 20$	$T = 10$	$T = 5$	$T = 3$	$T = 1$	$T = 0.33$	$T = 0.2$	$T = 0.1$
(9, 0)	1.0000	1.0000	1.0000	1.0000	1.0000	0.5092	0.6680	0.8182
(8, 1)	0.9999	0.9998	0.9983	0.9915	0.9021	-0.8472	-0.9349	-0.9819
(7, 2)	0.9998	0.9982	0.9880	0.9574	0.7159	0.9574	0.9880	0.9982
(6, 3)	0.9952	0.9819	0.9349	0.8472	0.4596	-0.9915	-0.9983	-0.9998
(5, 4)	0.9048	0.8182	0.6680	0.5092	0.1584	1.0000	1.0000	1.0000
(4, 5)	-0.9048	-0.8182	-0.6680	-0.5092	-0.1584	-1.0000	-1.0000	-1.0000
(3, 6)	-0.9952	-0.9819	-0.9349	-0.8472	-0.4596	0.9915	0.9983	.9998
(2, 7)	-0.9998	-0.9982	-0.9880	-0.9574	-0.7159	-0.9574	-0.9880	-0.9982
(1, 8)	-0.9999	-0.9998	-0.9983	-0.9915	-0.9021	0.8472	0.9349	0.9819
(0, 9)	-1.0000	-1.0000	-1.0000	-1.0000	-1.0000	-0.5092	-0.6680	-0.8182

It may be observed that for values of $T > 1.0$ these vector components are skew-symmetric about the center. Components 1 to 5 have the same sign, which is opposite to that of components 6 to 10. Furthermore, components 1 to 5 are, for the most part, clustered near 1.0, while those of 6 to 10 are clustered near -1.0. Consequently, from Section 1.8.1, it may be assumed that states (9, 0), ..., (5, 4) are near each other, that states (4, 5), ..., (0, 9) are near each other, but that both sets of states are far apart, i.e., the system is almost decomposable into two subsets, the first containing states 1 to 5 and the second, states 6 to 10.

For such values of T (i.e., $T > 1.0$), this decomposition may be demonstrated by considering the stochastic matrix that has rows and columns arranged such that if component i of the subdominant eigenvector is greater than component j, then the row/column of state i precedes the row/column of state j. The matrix is given as follows:

States	$(9,0)$	$(8,1)$	$(7,2)$	$(6,3)$	$(5,4)$	:	$(4,5)$	$(3,6)$	$(2,7)$	$(1,8)$	$(0,9)$
$(9,0)$	$\frac{T}{T+1}$	$\frac{1}{T+1}$				:					
$(8,1)$	$\frac{T}{T+1}$		$\frac{1}{T+1}$			:					
$(7,2)$		$\frac{T}{T+1}$		$\frac{1}{T+1}$		:					
$(6,3)$			$\frac{T}{T+1}$		$\frac{1}{T+1}$	:					
$(5,4)$				$\frac{T}{T+1}$		:	$\frac{1}{T+1}$				
...	...	...	...	...	...	...	...	...	...	...	...
$(4,5)$					$\frac{1}{T+1}$	:		$\frac{T}{T+1}$			
$(3,6)$						:	$\frac{1}{T+1}$		$\frac{T}{T+1}$		
$(2,7)$						:		$\frac{1}{T+1}$		$\frac{T}{T+1}$	
$(1,8)$						:			$\frac{1}{T+1}$		$\frac{T}{T+1}$
$(0,9)$						:				$\frac{1}{T+1}$	$\frac{T}{T+1}$

This is almost in the form of equation (1.20). The only elements in the off-diagonal matrices that are nonzero are those that result from the interaction between states (5, 4) and (4, 5) ($= 1/(T+1)$), and this is small for large values of T.

Thus, we see that the right-hand eigenvector corresponding to the subdominant eigenvalue may indeed be employed in determining which state belongs to which subset in a nearly decomposable subsystem. To put it another way, if the rows/columns of the stochastic matrix are rearranged into the order specified by this eigenvector, then the matrix will be in the form (1.20), with only small additional off-diagonal elements.

For values of $T < 1.0$, the subdominant eigenvalues indicate a nearly cyclic system of index 2, and therefore we should expect an ordering of the states according to the components of the subdominant eigenvector to yield a matrix of the form (1.26). An examination of these eigenvectors shows that, due to the reciprocity of the system, the elements that constitute the vector for which $T = \alpha$ are all present in the vector for which $T = 1/\alpha$. For example, the elements of the vector for which $T = 5.0$ are the same as the elements for which $T = 0.2$.

However, the order of the elements is not the same. For values of $T < 1.0$, states (5, 4) and (4, 5) have the components of largest modulus (it is the converse for $T > 1.0$). Also, although the elements of the vector are still skew-symmetric about the center, the elements on one side no longer have the same sign. Instead, the signs alternate. Thus, state (5, 4) is closely associated with states (7, 2) and (3, 6), i.e., its neighbors' neighbors, rather than with states (6, 3) and (4, 5). This would imply that states 1, 3, 5, 7 and 9 form one of the cyclic subsets and states 2, 4, 6, 8 and 10, the other, so that each transition moves the system from a state of the first set to one of the second. Rearranging the stochastic matrix according to the components of the subdominant right-hand eigenvector, we obtain

States	(5,4)	(3,6)	(7,2)	(1,8)	(9,0)	⋮	(0,9)	(8,1)	(2,7)	(6,3)	(4,5)
(5,4)						⋮				$\frac{T}{T+1}$	$\frac{1}{T+1}$
(3,6)						⋮			$\frac{T}{T+1}$		$\frac{1}{T+1}$
(7,2)						⋮		$\frac{T}{T+1}$		$\frac{1}{T+1}$	
(1,8)						⋮	$\frac{T}{T+1}$		$\frac{1}{T+1}$		
(9,0)					$\frac{T}{T+1}$	⋮		$\frac{1}{T+1}$			
...	...	...	...	...	...	...	...	...	...	...	...
(0,9)				$\frac{1}{T+1}$		⋮	$\frac{T}{T+1}$				
(8,1)			$\frac{1}{T+1}$		$\frac{T}{T+1}$	⋮					
(2,7)		$\frac{1}{T+1}$		$\frac{T}{T+1}$		⋮					
(6,3)	$\frac{1}{T+1}$		$\frac{T}{T+1}$			⋮					
(4,5)	$\frac{1}{T+1}$	$\frac{T}{T+1}$				⋮					

In this matrix all the diagonal elements, which denote the probabilities of the system remaining in the same state, are zero except those corresponding to (9, 0) and (0, 9). However, these are both given by $T/(T+1)$, which is small for small T, so that the matrix is in fact almost of the form of equation (1.26), as anticipated.

It may be observed that for values of $T > 1.0$, $T/(T + 1)$ is not small, and hence the diagonal submatrices can no longer be considered close to zero. The matrix is not nearly cyclic for such T, even though each actual transition of the system is from one of {(5, 4) (3, 6) (7, 2) (1, 8) (9, 0)} to {(0, 9) (8, 1) (2, 7) (6, 3) (4, 5)} or vice versa. This is because there is a high probability of the system remaining either in state (9, 0) or in state (0, 9).

1.9 Queueing Models

Often it happens that the Markov chains we analyze arise from queueing network models. It is therefore appropriate to devote some time to an examination of queueing network models and the characteristics of the Markov chains that stem from these models. Queueing models impose on the infinitesimal generator a structure that plays an important role in the success or failure of numerical methods destined to solve Markov chains.

Queueing network models have been widely used and are currently the focus of much research. They have been applied with a large measure of success to the evaluation of computer systems, to flexible manufacturing systems, and to communication networks, to name just a few of the areas of current interest. In computer system performance evaluation, for example, they provide the basis upon which new designs may be evaluated and a means to estimate the change in performance due to an increase in capacity of one or several of the system components. The effect of changes in both workload characteristics and system software can be determined from appropriate queueing network models.

If the queueing network satisfies *local balance* [20], then there exist efficient computational algorithms, called *product-form algorithms*, to compute the parameters of interest. When this is not the case, it is sometimes possible to modify the model so that it satisfies the conditions of local balance. It is a curious fact that the results obtained are often remarkably accurate, despite the unrealistic assumptions that are imposed on the model. However, the main difficulty with the product-form algorithms is their limited range of application.

In some cases, methods that compute approximate solutions to queueing networks have been used. These include the iterative techniques of Chandy, Herzog, and Woo [21] and of Marie [90] and the diffusion methods of Gelenbe [51] and Kobayashi [83]. Often, there is no rigorous mathematical justification for these methods, nor are there error bounds to indicate the magnitude of the error made by the approximation. Nevertheless, when applicable, the value of these approximate methods is that the results are obtained efficiently and appear to be sufficiently accurate to make them useful.

Given the limitations of the above approaches, it may be necessary to turn to strictly numerical procedures to obtain the solution of general queueing networks. It is always possible to obtain a Markov chain from a general queueing network. Phase-type representations of probability distributions [110] allow us to approximate, arbitrarily closely, any general probability distribution. In addition, features such as blocking, priority service disciplines, and state-dependent routing, which are simply not possible in the above analytical approaches, may be incorporated into a Markov chain representation — although the effect of doing so will increase the size of the state space. The solution of the Markov chain representation may then be computed and the desired performance characteristics, such as queue length distributions, utilizations, and throughputs, obtained directly from the stationary probability distribution vector.

1.9.1   Specification

1.9.1.1   Customers and Queues

Queueing models are generally described in terms of *customers*, who move among a set of *service stations*. The customers (also referred to as clients, jobs, etc.) are distinguishable by class. However, within a class they are considered to be identical. The stations (service centers) contain both servers and a place for customers to wait (the queue) until they are served. This waiting area is sometimes assumed to be sufficiently large to hold any number of customers that arrive, in which case it is said to have *infinite capacity*. Otherwise it is finite. Note that the number of customers at a particular station includes those who are waiting as well as those currently receiving service. We shall use the graphical representations shown in Figure 1.4.

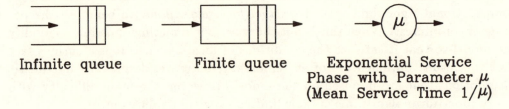

Infinite queue   Finite queue   Exponential Service
                                                Phase with Parameter μ
                                                (Mean Service Time $1/\mu$)

Figure 1.4: Symbols used in queueing model representations.

An *infinite-server* station is one in which the number of servers is sufficiently large that an arriving customer is immediately taken into service. There is always a server available to serve an arrival. The servers are all identical, and it is often assumed that the service they provide is exponentially distributed. Sometimes the term *ample server* or *pure delay* is used to describe such a station. A station may consist of any number of servers and any number of queues. Figures 1.5 through 1.7 illustrate some commonly employed stations.

1.9.1.2   Arrival and Service Distributions

The arrival pattern, usually specified in terms of the interarrival time distribution $A(t)$, characterizes the statistical properties of arrivals to a station. Commonly a Poisson distribution with parameter (rate) λ (alternatively, mean interarrival time $= 1/\lambda$) is used. The arrival of customers in bulk is sometimes important, in which case it is necessary to specify the distribution of the number of customers

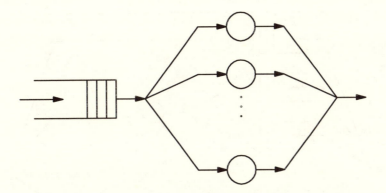

Figure 1.5: Single-queue, multiple-server station.

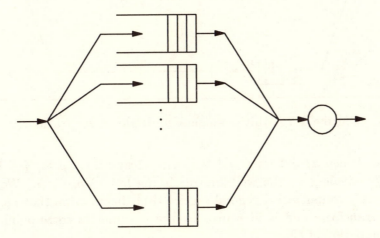

Figure 1.6: Multiple-queue, single-server station.

at arrival instants.

The service time distribution is denoted by $B(x)$ and the mean service time by $1/\mu$. Again, the most commonly used distribution is the exponential distribution. When this is insufficient, a convenient representation for more general distributions is the Coxian formulation. This formulation, by means of fictitious stages, allows the duration of service at each server to be described by a linear combination of exponential stochastic variables. Thus, a service is a continuous succession of phases, each having an exponential service time distribution of rate μ_j, $j = 1, 2, \ldots, k$. After phase j, a customer leaves the server with probability $(1 - a_j)$. Such a server is shown graphically in Figure 1.8. In this figure, do not confuse the dashed box with a station. Rather, it represents a single server (of which there may be many in a station), which can serve only one customer at a time. Therefore, there can be at most one customer within the dashed region

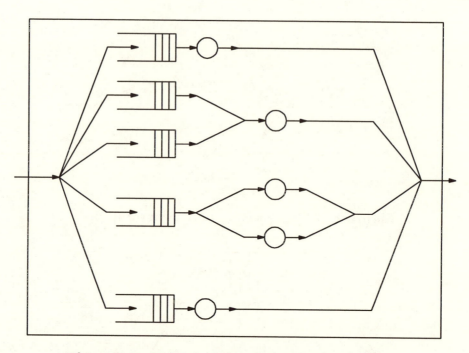

Figure 1.7: Multiple-queue, multiple-server station.

at any time. When $a_j = 1$ for $j = 1, 2, \ldots, k-1$ and $\mu_j = \mu$ for $j = 1, 2, \ldots, k$, this law is equivalent to the well-known Erlang law of order k. We shall use the notation C_k to indicate a general service time distribution that has a rational Laplace transform and is identified by the parameters corresponding to the general formulation of Cox.

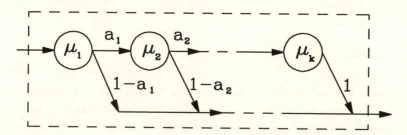

Figure 1.8: Coxian server.

1.9.1.3   Customer Classes and Scheduling Disciplines

Associated with each station is a *scheduling discipline*, which determines the order in which customers are served. Furthermore, different classes of customers may have different priorities, and the service discipline enforced may cause a low-priority customer to be preempted on the arrival of a higher-priority customer. Depending on the system being modelled, the preempted customer might be able to continue from the point at which it was stopped (preempt-resume), or it may have to redo at least a portion of the work previously accomplished (preempt-restart). Common scheduling rules are

- First-come, first-served (FCFS, or FIFO, First-In, First-Out);
- Last-come, first-served preempt-resume (LIFO-PR);
- Service in random order (SIRO);
- Round robin (RR);
- Processor sharing (PS).

We shall discuss these further as and when they are needed.

1.9.1.4 Routing Probability Matrices

It remains to specify the different paths the customers may take among the stations. This information is provided by the probability routing matrix, of which there is one for each class. This matrix specifies the probabilistic routing taken by the customers of a class among the various stations of the network. The ij element of this matrix denotes the probability that a customer on leaving station i will next proceed to station j. This routing may be dependent (or independent) on the state of the queueing network. The transition itself is considered to be instantaneous; if this is unrealistic in a certain system, a delay mechanism, consisting of an *infinite* server, can be interposed between the two stations.

Care must be taken not to confuse this transition *routing matrix*, whose order is equal to the number of stations in the network, with the transition *probability matrix*, whose order is equal to the number of states generated by the queueing model.

A final characteristic of queueing networks is that they may be *open, closed,* or *mixed*. A queueing network is open if customers arrive to the network from a point external to the network and eventually exit from the network; it is closed if there are no external arrivals or departures, in which case a fixed number of customers cycle perpetually among the stations of the network. The network is said to be mixed if some classes of customers arrive from outside and eventually depart, while customers belonging to the remaining classes stay forever in the network.

1.9.2 Markov Chain Analysis of Queueing Networks

It is apparent from the proceding sections that a number of steps must be completed in order to analyze a queueing network as a Markov chain. These involve

1. Choosing a state space representation;

2. Enumerating all the transitions that can possibly occur among the states;

3. Generating the transition rate (infinitesimal generator) matrix; and finally,

4. Computing appropriate probability vectors of the Markov chain, from which *measures of effectiveness* of the queueing network are derived.

The *state descriptor* is a vector whose components completely describe the state of the system at any point in time. This means that the state of each of the queueing elements that constitute the queueing network must be completely described. For example, for a single-server queue with exponential service time, a single indicator — the number of customers present — suffices to describe the state of the station completely. Since the service time is exponential and endowed with the memoryless property, no specific parameter relating to the state of the server is needed. However, for a single-server station with a Coxian service time distribution, an additional parameter, which indicates the phase of service, must also be specified. Further, if the server may become blocked, yet another parameter must be used, to specify whether the server is free or blocked. It may also be necessary to indicate the particular finite queue that caused the server to become blocked. Thus each queueing element may require several parameters to specify its state completely. Concatenating these yields a state descriptor for the queueing network.

Items 2 and 3 in the foregoing list are often completed simultaneously. As the state space is enumerated, the states the system may move to in a single step from each state are listed, and the transition rate matrix is generated on a row-by-row basis. Since the number of states is often large, this matrix is usually stored in a compact form; that is, only the nonzero elements and the positions of these elements are recorded. The last item involves the computation of the stationary probability vector and/or transient probabilities and is the major subject of discussion of this text.

1.9.3 Example 1: Two Stations in Tandem

To conclude this section, we consider two simple queueing systems and show the steps involved in generating the underlying Markov chain. The first example system is shown in Figure 1.9.

1.9.3.1 Model Description

- The arrival process is Poisson with rate λ.

- Each station consists of a single server.

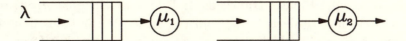

Figure 1.9: Simple queueing system.

- Service time distribution is exponential with mean service time $1/\mu_i$ at server i.

- Scheduling discipline is FCFS.

- There is only one class of customer.

- Each of the two queues has infinite capacity.

1.9.3.2 State Space Representation and Transitions

In this example we may denote the state of the system by the pair (n_1, n_2), where n_1 represents the number of customers in station 1 and n_2 denotes the number in station 2 (including the customers in service). For any nonnegative integer values of n_1 and n_2, (n_1, n_2) represents a feasible state of this queueing system. In generating the Markov chain we need to consider the single-step transitions that can occur between any two states. The ij^{th} element of the infinitesimal generator matrix Q denotes the rate of transition from the ith state to the jth state. Some of these states and the possible transitions among them are shown in Figure 1.10.

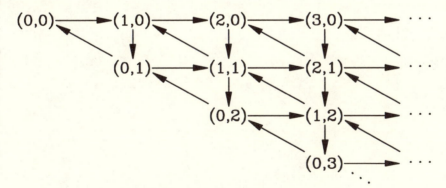

Figure 1.10: Transition diagram.

The arrows indicate the only transitions that are possible for this model. Transitions from left to right represent a change of state due to an arrival to the system. This occurs at rate λ. Transitions from top to bottom represent a change of state due to a customer completing service at station 1 and proceeding to station 2.

This occurs at rate μ_1. Finally, transitions indicated by diagonal arrows represent a change of state due to departures from the system. The transitions that are possible, and the rates at which these transitions occur, are therefore:

$$
\begin{aligned}
(n_1, n_2) &\to (n_1 + 1, n_2); &&\text{rate } \lambda \\
(n_1, n_2) &\to (n_1 - 1, n_2 + 1); &&\text{rate } \mu_1 \\
(n_1, n_2) &\to (n_1, n_2 - 1); &&\text{rate } \mu_2
\end{aligned}
$$

1.9.3.3   The Transition Rate Matrix

The structure of the transition rate matrix of this Markov chain depends on the order imposed on these states. One convenient ordering is shown in Table 1.3. This results in the following infinitesimal generator.

$$
\begin{pmatrix}
* & \vdots & \lambda & \vdots & & & \vdots & & & & \vdots \\
\cdots & \cdots & \cdots & \cdots & \cdots & \cdots & \cdots & \cdots & \cdots & \cdots & \cdots \\
 & \vdots & * & \mu_1 & \vdots & \lambda & & \vdots & & & \vdots \\
\mu_2 & \vdots & & * & \vdots & & \lambda & \vdots & & & \vdots \\
\cdots & \cdots & \cdots & \cdots & \cdots & \cdots & \cdots & \cdots & \cdots & \cdots & \cdots \\
 & \vdots & & & \vdots & * & \mu_1 & \vdots & \lambda & & \vdots \\
 & \vdots & \mu_2 & & \vdots & & * & \mu_1 & \vdots & \lambda & \vdots \\
 & \vdots & & \mu_2 & \vdots & & & * & \vdots & \lambda & \vdots \\
\cdots & \cdots & \cdots & \cdots & \cdots & \cdots & \cdots & \cdots & \cdots & \cdots & \cdots \\
 & \vdots & & & \vdots & & & \vdots & * & \mu_1 & \vdots \\
 & \vdots & & & \vdots & \mu_2 & & \vdots & & * & \mu_1 & \vdots \\
 & \vdots & & & \vdots & & \mu_2 & \vdots & & & * & \mu_1 & \vdots \\
 & \vdots & & & \vdots & & & \mu_2 & \vdots & & & * & \vdots \\
\cdots & \cdots & \cdots & \cdots & \cdots & \cdots & \cdots & \cdots & \cdots & \cdots & \cdots \\
 & \vdots & & & \vdots & & & \vdots & & & & \vdots
\end{pmatrix}
$$

The diagonal element in any row (denoted by an asterisk) is equal to the negated sum of the off-diagonal elements in that row. Note that the matrix may be partitioned into block tridiagonal form (the blocks are {1}, {2,3}, {4,5,6}, {7,8,9,10}, ...), in which superdiagonal blocks contain only arrival rates, subdiagonal blocks contain only departure rates, and diagonal blocks (with the exception of the diagonal elements themselves) contain only internal transition rates. Note that diagonal block i contains all the states in which the number of customers in the network is equal to $i - 1$. This type of structure arises frequently in Markov chains derived from queueing models and may be exploited by numerical methods to determine the stationary probability vector.

Table 1.3: State Ordering for Queueing Model

State number	State description
1	$(0, 0)$
2	$(1, 0)$
3	$(0, 1)$
4	$(2, 0)$
5	$(1, 1)$
6	$(0, 2)$
7	$(3, 0)$
8	$(2, 1)$
9	$(1, 2)$
10	$(0, 3)$
$\vdots$	$\vdots$

1.9.4 Example 2: One Coxian and Two Exponential Servers

This system is shown in Figure 1.11.

1.9.4.1 Model Description

- Each station consists of a single server and queue.

- Two stations provide service that is exponentially distributed.

- Station 1 contains a two-phase Coxian server.

- The number of customers in the network is fixed and is equal to N.

- Scheduling discipline is FCFS.

- There is only a single customer class.

- All queues have infinite capacity.

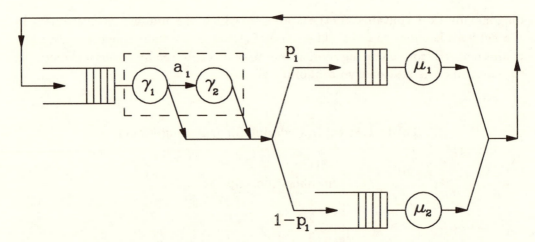

Figure 1.11: Queueing model containing one Coxian and two exponential servers.

1.9.4.2 State Space Representation and Transitions

In specifying a state of this system we must specify the number of customers at each station and, in addition, the phase of service (1 or 2) of the customer being served in station 1. Thus, a valid state representation is (n_1, k, n_2, n_3), where $n_i \in \{1, 2, \ldots N\}$ and $k \in \{1, 2\}$. (If $n_1 = 0$, then the value of k is irrelevant.)

The possible states of this system, of which there is a finite number, viz. $(N + 1)^2$, are shown in Table 1.4. The following transitions are possible:

$$
\begin{array}{llllll}
n_1 \neq 0 & (n_1, 1, n_2, n_3) & \rightarrow & (n_1, 2, n_2, n_3) & \text{at rate} & a_1\gamma_1 \\
 & & \rightarrow & (n_1 - 1, 1, n_2 + 1, n_3) & " & p_1(1-a_1)\gamma_1 \\
 & & \rightarrow & (n_1 - 1, 1, n_2, n_3 + 1) & " & (1-p_1)(1-a_1)\gamma_1 \\
 \\
n_1 \neq 0 & (n_1, 2, n_2, n_3) & \rightarrow & (n_1 - 1, 1, n_2 + 1, n_3) & " & p_1\gamma_2 \\
 & & \rightarrow & (n_1 - 1, 1, n_2, n_3 + 1) & " & (1-p_1)\gamma_2
\end{array}
$$

and, for $k = 1, 2$,

$$
\begin{array}{llllll}
n_2 \neq 0 & (n_1, k, n_2, n_3) & \rightarrow & (n_1 + 1, k, n_2 - 1, n_3) & \text{at rate} & \mu_1 \\
n_3 \neq 0 & (n_1, k, n_2, n_3) & \rightarrow & (n_1 + 1, k, n_2, n_3 - 1) & " & \mu_2
\end{array}
$$

Thus, for any given state, the maximum number of transitions possible is 5.

1.9.4.3 The Transition Rate Matrix

Again, the structure of the infinitesimal generator depends upon the ordering assigned to the different states. Consider as an example the case when $N = 2$

Table 1.4: Possible States in Queueing Model

$(N,$	$1,$	$0,$	$0)$
$(N,$	$2,$	$0,$	$0)$
$(N-1,$	$1,$	$1,$	$0)$
$(N-1,$	$1,$	$0,$	$1)$
$(N-1,$	$2,$	$1,$	$0)$
$(N-1,$	$2,$	$0,$	$1)$
$(N-2,$	$1,$	$2,$	$0)$
$(N-2,$	$1,$	$1,$	$1)$
$(N-2,$	$1,$	$0,$	$2)$
$(N-2,$	$2,$	$2,$	$0)$
$\vdots$	$\vdots$	$\vdots$	$\vdots$
$(1,$	$2,$	$0,$	$N-1)$
$(0,$	$-,$	$N,$	$0)$
$(0,$	$-,$	$N-1,$	$1)$
$(0,$	$-,$	$1,$	$N-1)$
$(0,$	$-,$	$0,$	$N)$

and the states are ordered as indicated in Table 1.4. The states are listed in Table 1.5, and the infinitesimal generator is

$$
\begin{pmatrix}
* & a_1\gamma_1 & \vdots & p_1a_2\gamma_1 & p_2a_2\gamma_1 & & & \vdots & & & \\
 & * & \vdots & p_1\gamma_2 & p_2\gamma_2 & & & \vdots & & & \\
\cdots & \cdots & \cdots & \cdots & \cdots & \cdots & \cdots & \cdots & \cdots & \cdots & \cdots \\
\mu_1 & & \vdots & * & & a_1\gamma_1 & & \vdots & p_1a_2\gamma_1 & p_2a_2\gamma_1 & \\
\mu_2 & & \vdots & & * & & a_1\gamma_1 & \vdots & & p_1a_2\gamma_1 & p_2a_2\gamma_1 \\
 & \mu_1 & \vdots & & & * & & \vdots & p_1\gamma_2 & p_2\gamma_2 & \\
 & \mu_2 & \vdots & & & & * & \vdots & & p_1\gamma_2 & p_2\gamma_2 \\
\cdots & \cdots & \cdots & \cdots & \cdots & \cdots & \cdots & \cdots & \cdots & \cdots & \cdots \\
 & & \vdots & \mu_1 & & & & \vdots & * & & \\
 & & \vdots & \mu_2 & \mu_1 & & & \vdots & & * & \\
 & & \vdots & & \mu_2 & & & \vdots & & & *
\end{pmatrix}
$$

Here $p_2 = (1 - p_1)$ and $a_2 = (1 - a_1)$. Note, once again, the block tridiagonal form, which is achieved by grouping states according to the number of customers in station 1, and that subdiagonal blocks contain only μ_1 and μ_2, while super-diagonal blocks contain the γ_1 and γ_2 terms.

Table 1.5: State Ordering in Queueing Model, $N = 2$

State number	State description
1	$(2, 1, 0, 0)$
2	$(2, 2, 0, 0)$
3	$(1, 1, 1, 0)$
4	$(1, 1, 0, 1)$
5	$(1, 2, 1, 0)$
6	$(1, 2, 0, 1)$
7	$(0, -, 2, 0)$
8	$(0, -, 1, 1)$
9	$(0, -, 0, 2)$

Chapter 2

Direct Methods

2.1 Introduction

Numerical methods that compute solutions of mathematical problems in a fixed number of operations are generally referred to as *direct* methods. Thus, for example, Gaussian elimination on a full $(n \times n)$ nonhomogeneous system of linear equations with nonsingular coefficient matrix computes a result in exactly $(n^3/3 + n^2/2 - 5n/6)$ multiplications and additions and $(n^2/2 + n/2)$ divisions. In contrast, iterative methods begin from some initial approximation and produce a sequence of intermediate results, which are expected to eventually converge to the solution of the problem. Most often it is not known how many iterations are required before a certain accuracy is achieved, and hence it is impossible to know in advance the number of numerical operations that will be needed. Nevertheless, iterative methods of one type or another are the most commonly used methods for obtaining the stationary probability vector of large-scale Markov chains. There are several important reasons for this choice.

- First, an examination of the iterative methods usually employed shows that the only operation in which the matrices are involved is a multiplication with one or more vectors. This operation does not alter the form of the matrix, and thus, compact storage schemes, which reduce the amount of memory required to store the matrix and are also well suited to matrix multiplication, may be conveniently implemented. Since the matrices involved are usually large and very sparse, the savings made by such schemes can be very considerable. With direct equation-solving methods, the elimination of one nonzero element of the matrix during the reduction phase often results in the creation of several nonzero elements in positions that previously contained zero. This is called *fill-in*, and not only does it make the organization of a compact storage scheme more difficult (since provision must be made for the deletion and the insertion of elements), but in addition, the amount of fill-in can often be so extensive that available memory is quickly exhausted. A successful direct method must incorporate a means of overcoming these difficulties.

- Iterative methods may make use of good initial approximations to the solution vector, and this is especially beneficial when a series of related experiments is being conducted. In such circumstances the parameters of one experiment often differ only slightly from those of the previous; many will remain unchanged. Consequently, it is to be expected that the solution to the new experiment will be close to that of the previous, and it is advantageous to use the previous result as the new initial approximation. If indeed there is little change, we should expect to compute the new result in relatively few iterations.

- An iterative process may be halted once a prespecified tolerance criterion has been satisfied, and this may be relatively lax. For example, it may be wasteful to compute the solution of a mathematical model correct to full machine precision when the model itself contains errors of the order of 5 to 10%. A direct method is obliged to continue until the final specified operation has been carried out.

- Finally, with iterative methods the matrix is never altered; hence, the buildup of rounding error is, to all intents and purposes, nonexistent.

For these reasons, iterative methods have traditionally been preferred to direct methods. However, iterative methods have a major disadvantage in that they often require a very long time to converge to the desired solution. More advanced iterative techniques, such as GMRES (the Generalized Minimum Residual method) and the method of Arnoldi, have helped to alleviate this problem, but much research still remains to be done, particularly in estimating *a priori* the number of iterations, and hence the time, required for convergence. Direct methods have the advantage that an upper bound on the time required to obtain the solution may be determined before the calculation is initiated. More important, for certain classes of problem, direct methods often result in a more accurate answer being obtained in less time [156]. Since iterative methods will in general require less memory than direct methods, direct methods can be recommended only if they obtain the solution in less time.

In this chapter we introduce the LU and LDU decompositions and show their relation to the ubiquitous Gaussian elimination method. We consider the method of inverse iteration, which, because of our prior knowledge of the dominant eigenvalue of the matrix, requires only one iteration to converge to full machine accuracy and should therefore be considered as a direct method rather than an iterative one as its name implies. This method essentially reduces to an LU decomposition. After discussing the direct techniques that are available, we consider some implementation problems, specifically the choice of an appropriate data structure in which to store the matrix and in which to perform the requisite operations. In Section 2.6 we describe how direct methods may be used to advantage to obtain the solution of two different queueing models — the first by a straightforward application of the techniques described in this chapter, and the second by considering the specific structure of the transition rate matrix in great detail. Finally, we provide some necessary background for a stability analysis and show that Gaussian elimination is a stable algorithm for obtaining the stationary probability vector of a Markov chain. In this context we develop error bounds on the computed solution and examine the issue of conditioning.

2.2 Direct Methods

We begin by considering direct methods in the general context of a nonhomogeneous system of n linear equations in n unknowns,

$$Ax = b.$$

We shall assume that the coefficient matrix A is nonsingular and that n is finite. Later, in Section 2.3, we shall specialize these results to the particular case of direct methods for the solution of Markov chains.

2.2.1 Gaussian Elimination

Gaussian elimination may be viewed as a transformation from a system of equations $Ax = b$ to an equivalent system $Ux = d$ in which the matrix U is upper triangular. The solution x is then obtained by back substitution, viz.,

$$x_n = d_n/u_{nn},$$

$$x_i = \left(d_i - \sum_{k=i+1}^{n} u_{ik}x_k\right)/u_{ii}, \qquad i = n-1, \ldots, 2, 1.$$

The procedure of obtaining U from A is called the *reduction phase* and is accomplished in $n-1$ steps. The ith step eliminates all nonzero elements below the ith diagonal element by adding a suitable multiple of the ith equation into each equation below the ith. Adding a multiple of one equation into another is an *elementary row operation* and does not alter the solution of the system of equations. The matrix at the end of the ith step is shown diagrammatically in Figure 2.1.

Let us use the superscripts $^{(i)}$; $i = 1, 2, \ldots, n-1$ to denote the elements obtained *after* i steps of this procedure. Then the elements $a_{kl}^{(i)}$ are given by

$$a_{kl}^{(i)} = a_{kl}^{(i-1)} \quad \text{for } k \le i \quad \text{and} \quad l = 1, 2, \ldots, n,$$

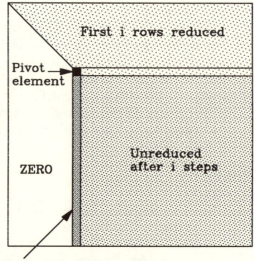

Figure 2.1: Gaussian elimination.

i.e., elements in the first i rows are unchanged, and

$$a_{kl}^{(i)} = a_{kl}^{(i-1)} - m_{ki}a_{il}^{(i-1)} \quad \text{for } k > i \quad \text{and} \quad l = 1, 2, \ldots, n$$

where the multipliers m_{ki} are given by $m_{ki} = a_{ki}^{(i-1)}/a_{ii}^{(i-1)}$. Obviously, for $k > i$, $a_{ki}^{(i)} = a_{ki}^{(i-1)} - m_{ki}a_{ii}^{(i-1)} = 0$, which means that elements below the diagonal element are set to zero. Elements to the left of column i were zero before step i and remain so after it. The elements $a_{ii}^{(i)}$ are called the *pivots* and must be nonzero if the algorithm is to terminate satisfactorily. For purposes of stability, it is generally necessary to interchange the rows of the matrix so that the pivotal element is the largest in modulus in its column in the unreduced portion of the matrix (called *partial pivoting*). This ensures that the absolute values of the multipliers do not exceed 1. For certain problems it is necessary to interchange both rows and columns so that the pivotal element is the largest among all elements in the unreduced part of the matrix (*full pivoting*).

Assuming that the algorithm can be completed without pivoting (which is, as we shall see in Section 2.7.4, the case for irreducible Markov chains), we may present the elimination procedure in matrix form as follows. Set $A^{(0)} \equiv A$ and let $A^{(i)}$ be the reduced matrix obtained after i steps of Gaussian elimination.

Then $A^{(n-1)} = U$. Define

$$M_i = \begin{pmatrix} 1 & & & & & \\ & \ddots & & & & \\ & & 1 & & & \\ & & -m_{i+1,i} & 1 & & \\ & & \vdots & & \ddots & \\ & & -m_{n,i} & & & 1 \end{pmatrix} \tag{2.1}$$

to be a unit lower triangular matrix whose only nonzero off-diagonal elements are to be found below the ith diagonal element. Since $m_{ki} = a_{ki}^{(i-1)}/a_{ii}^{(i-1)}$, it immediately follows that

$$M_i A^{(i-1)} = A^{(i)}.$$

In other words, premultiplying $A^{(i-1)}$ with M_i forces the elements below the ith diagonal position to zero, which is precisely the operation carried out during the ith step of Gaussian elimination. For this reason the matrix M_i is called a *Gauss transformation* or an *elementary lower triangular matrix* of index i. The vector m_i defined as

$$m_i \equiv (0, \ldots, 0, m_{i+1\,i}, \ldots, m_{ni})^T,$$

in which the first i elements are zero, is called a *Gauss vector*, and we have $M_i = I - m_i e_i^T$. Its inverse is given by

$$M_i^{-1} = I + m_i e_i^T = \begin{pmatrix} 1 & & & & & \\ & \ddots & & & & \\ & & 1 & & & \\ & & m_{i+1,i} & 1 & & \\ & & \vdots & & \ddots & \\ & & m_{n,i} & & & 1 \end{pmatrix}.$$

The complete reduction phase of the Gaussian elimination procedure may be represented as

$$M_{n-1} \ldots M_2 M_1 A = A^{(n-1)} \equiv U$$

Notice that for any two Gauss transformations M_i and M_j we have $M_i \times M_j = M_i + M_j - I$, and in general,

$$M_{n-1} \ldots M_2 M_1 = M_{n-1} + \ldots + M_2 + M_1 - (n-2)I.$$

Setting $L^{-1} = M_{n-1} \ldots M_2 M_1$, or equivalently $L = M_1^{-1} M_2^{-1} \ldots M_{n-1}^{-1}$, it follows that

$$A = LU,$$

where L is a unit lower triangular matrix whose elements below the diagonal are the multipliers, m_{ij}; $i > j$; and U is the upper triangular matrix to be used in the back-substitution phase.

2.2.2   The LU Decomposition

If a coefficient matrix A can be written as the product of a nonsingular lower triangular matrix L and a nonsingular upper triangular matrix U, i.e.,

$$A = LU \tag{2.2}$$

then the solution x of the system of linear equations $Ax = b$ is

$$x = U^{-1}L^{-1}b$$

and is obtained by calculating z from $z = L^{-1}b$ (by solving $Lz = b$ by forward substitution) and then computing x from $x = U^{-1}z$, this time by solving $Ux = z$ by backward substitution. Later in this chapter we shall turn our attention to conditions under which an LU decomposition exists for Markov chain problems. For the present, it suffices to say that it always exists if the Markov chain is irreducible. We now consider the problem of determining this decomposition.

From equation (2.2) it is apparent that finding an LU decomposition of a matrix A is equivalent to solving the following set of n^2 equations,

$$\sum_{k=1}^{n} l_{ik}u_{kj} = a_{ij}; \qquad i, j = 1, 2, \ldots, n \tag{2.3}$$

for the $n^2 + n$ unknowns

$$l_{ik}, \qquad i = 1, 2, \ldots n; \qquad k = 1, 2, \ldots, i,$$

and

$$u_{kj}, \qquad j = 1, 2, \ldots, n; \qquad k = 1, 2, \ldots, j.$$

Given that we have n^2 equations in $n^2 + n$ unknowns, we may assign specific values to n of the unknowns. Normally these values are chosen to make the determination of the remaining n^2 unknowns as simple as possible. It is customary to set

$$l_{ii} = 1, \qquad 1 \le i \le n \quad \text{(Doolittle Decomposition)}$$

or

$$u_{ii} = 1, \qquad 1 \le i \le n \quad \text{(Crout Decomposition)}.$$

For general systems of equations, some form of pivoting is usually required to ensure the stability of the decomposition procedure. Although the LU decomposition does not yield itself readily to complete pivoting, partial pivoting may be performed without too much difficulty. However, we shall see later that for irreducible Markov chains no pivoting at all is needed. To obtain an LU decomposition in the absence of pivoting, we return to equation (2.3) and consider first the case in which $i > j$. This case is shown in Figure 2.2.

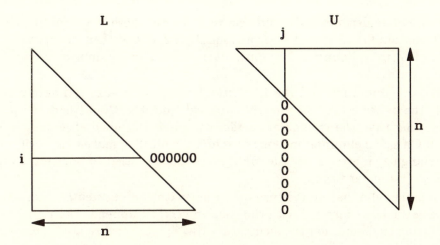

Figure 2.2: *LU* decomposition.

It may be observed that to form the inner product of the ith row of L with the jth column of U $\left(\sum_{k=1}^{n} l_{ik} u_{kj}\right)$, it is unnecessary to continue beyond u_{jj}. Thus, the summation $\sum_{k=1}^{n}$ may be replaced by $\sum_{k=1}^{j}$. Similarly, if $i \leq j$, the summation in equation (2.3) may be replaced by $\sum_{k=1}^{i}$. Then we have

$$i \leq j: \qquad \sum_{k=1}^{i} l_{ik} u_{kj} = a_{ij}$$

$$i > j: \qquad \sum_{k=1}^{j} l_{ik} u_{kj} = a_{ij}.$$

If we assume that the Doolittle decomposition is being performed, then

$$\sum_{k=1}^{i-1} l_{ik} u_{kj} + u_{ij} = a_{ij} \quad \text{for } i \leq j \quad \text{since } l_{ii} = 1,$$

and

$$\sum_{k=1}^{j-1} l_{ik} u_{kj} + l_{ij} u_{jj} = a_{ij} \quad \text{for } i > j.$$

From the first of these equations we obtain

$$u_{ij} = a_{ij} - \sum_{k=1}^{i-1} l_{ik} u_{kj} \quad \text{for } i \leq j, \tag{2.4}$$

while from the second

$$l_{ij} = \left[a_{ij} - \sum_{k=1}^{j-1} l_{ik} u_{kj} \right] / u_{jj} \quad \text{for } i > j. \tag{2.5}$$

The Doolittle decomposition procedure is to use these two equations successively to obtain first a row of U (from Equation 2.4) and then the corresponding column of L (from Equation 2.5) until all the rows and columns of U and L have been computed.

Gaussian elimination and the Doolittle decomposition are intimately related. In fact, the upper triangular matrix obtained from the Doolittle decomposition is none other than the triangular coefficient matrix obtained after the reduction phase of Gaussian elimination. Furthermore, the off-diagonal elements of the unit lower triangular matrix L are the multipliers that are used to reduce the original matrix to triangular form.

In Gaussian elimination it is usual to work with the *augmented matrix* $[A : b]$, i.e., the coefficient matrix A with the column vector b appended to its right-hand side. During the reduction the elementary row operations are also applied to the elements of b. These operations on b result in the vector z, which is the vector obtained as the solution of the triangular system $Lz = b$. The advantage of the LU decomposition approach is that the matrix L is available, and this is convenient for solving systems with different right-hand sides. Another more subtle and more important advantage of the LU decomposition is that its basic operations, the summations in equations (2.4) and (2.5), are inner product calculations. In certain computers, inner products may be accumulated in double-precision, which is a highly accurate means of forming this product.

2.2.3   The LDU Decomposition

In an LDU decomposition the coefficient matrix A is written as

$$A = LDU.$$

Here, L and U are chosen to be *unit* lower and *unit* upper triangular matrices respectively; i.e., both L and U have diagonal elements all equal to 1, and D is a diagonal matrix. The solution of $Ax = b$ is computed as $x = U^{-1}D^{-1}L^{-1}b$.

If an LU decomposition is available, then the LDU decomposition may easily be obtained. Suppose the Doolittle decomposition is already available (i.e., the diagonal elements of L are unity). Then setting $D = \text{diag}\{U\}$ and dividing each element on the ith row of U with u_{ii}, for $i = 1, 2, \ldots, n$ and $u_{ii} \neq 0$, yields the corresponding LDU decomposition. If the LDU decomposition is given, it is easy to write the Doolittle decomposition as $L(DU)$ and the Crout decomposition as $(LD)U$.

2.2.4   Inverse Iteration

Although this may sound rather like a contradiction in terms, we shall see that inverse iteration, when applied to an infinitesimal generator matrix Q to obtain the stationary probability vector π of an irreducible Markov chain, requires only a single iteration to compute π to machine accuracy. In fact, this method simply reduces to the standard LU decomposition method with special treatment of the zero pivot and the right-hand side vector.

In contrast with the direct methods just discussed, which are designed to obtain solutions for systems of linear equations, the method of inverse iteration is oriented toward finding eigenvectors corresponding to eigenvalues for which approximations are known. Suppose we wish to find the right-hand eigenvector corresponding to an approximate eigenvalue μ of an $(n \times n)$ matrix A. We begin by considering an iterative scheme based on the relation

$$x^{(k)} = (A - \mu I)^{-1} x^{(k-1)}. \tag{2.6}$$

Let the initial vector $x^{(0)}$ be an arbitrary column vector that can be written as a linear combination of the right-hand eigenvectors of A; i.e.,

$$x^{(0)} = \sum_{i=1}^{n} \beta_i v_i,$$

where the vectors v_i are the right-hand eigenvectors corresponding to the eigenvalues λ_i of A; i.e.,

$$A v_i = \lambda_i v_i; \qquad i = 1, 2, \ldots, n.$$

Using the well-known fact that if λ_i is an eigenvalue of A, then $\lambda_i - \mu$ is an eigenvalue of $A - \mu I$, we find

$$
\begin{aligned}
x^{(k)} &= (A - \mu I)^{-k} x^{(0)} = \sum_{i=1}^{n} \beta_i (\lambda_i - \mu)^{-k} v_i \\
&= (\lambda_r - \mu)^{-k} \left[\beta_r v_r + \sum_{i \neq r}^{n} \beta_i (\lambda_r - \mu)^k (\lambda_i - \mu)^{-k} v_i \right].
\end{aligned} \tag{2.7}
$$

Consequently, if for all $i \neq r$, $|\lambda_r - \mu| \ll |\lambda_i - \mu|$, then convergence to the eigenvector v_r is rapid, since $[(\lambda_r - \mu)/(\lambda_i - \mu)]^k$ will rapidly tend to zero. The fact that $(A - \mu I)$ is nearly singular when $\mu \approx \lambda_r$ will *not* affect the accuracy of $x^{(k)}$ [178]. No numerical difficulties arise when μ is exactly equal to λ_r. In the limit, convergence is achieved in a single iteration. If the initially chosen vector $x^{(0)}$ does not lie in the subspace spanned by the right-hand eigenvectors of A, the above analysis no longer holds from a rigorous mathematical point of view. Indeed, if the initial vector is deficient in the rth vector, i.e., if $\beta_r = 0$, the above analysis does not hold. However, from a practical point of view this is extremely unlikely to happen [178]. Practitioners will sometimes choose the elements of their initial vector to be randomly generated numbers to reduce the likelihood of this event even further.

In implementing inverse iteration, the preferred approach is not to form the inverse of the shifted matrix $(A - \mu I)$ and then postmultiply it with the current approximation, as indicated by the recurrence formulation (2.6), but rather to apply the following iteration formula:

$$(A - \mu I)x^{(k)} = x^{(k-1)}.$$

This involves finding the solution to a set of linear equations at each iteration, and it is obviously identical to the original formulation in equation (2.6). If μ is not an eigenvalue of A, then $(A - \mu I)$ is nonsingular, and an LU decomposition of $(A - \mu I)$ may be computed. Note that the actual decomposition of $(A - \mu I)$ need only be performed once, no matter how many iterations are required. Given the decomposition $(A - \mu I) = LU$, the method reduces to the following algorithm:

Algorithm: Inverse Iteration

1. Choose arbitrary initial approximation $x \neq 0$.

2. Do until convergence

   - Solve $Lz = x$ for z.
   - Solve $Uy = z$ for y.
   - Normalize y and test for convergence.
   - Set $x = y$.

If μ is an eigenvalue of A, then $(A - \mu I)$ is singular. In this case the zero pivot, which theoretically arises during the LU decomposition, should be replaced by *machine epsilon*. This is defined as the smallest representable number ϵ such that $1 + \epsilon > 1$ on the particular computer being used. A rigorous error analysis [178] shows that normalizing the vector obtained after a single iteration will yield a very accurate eigenvector.

2.3   Direct Methods and Markov Chains

We are concerned with obtaining the stationary probability vector π from the equations

$$\pi Q = 0, \qquad \pi \geq 0, \qquad \pi e = 1,$$

where Q is the $(n \times n)$ infinitesimal generator matrix corresponding to an irreducible Markov chain, and $e = (1, 1, \ldots, 1)^T$. Notice that if we try to apply direct methods to the alternate formulation

$$\pi P = \pi,$$

we need to rewrite this as

$$\pi(I - P) = 0,$$

and in both cases we need to solve a homogeneous system of n linear equations in n unknowns. A homogeneous system of n linear equations in n unknowns has a solution other than the trivial solution ($\pi_i = 0$ for all i) if and only if the determinant of the coefficient matrix is zero, i.e., if and only if the coefficient matrix is singular. Since the determinant of a matrix is equal to the product of its eigenvalues, and since Q (and $I - P$) possesses a zero eigenvalue, the singularity of Q, and hence the existence of a nontrivial solution, follows.

2.3.1 Handling the Singularity

In Section 2.2 we considered direct methods for the solution of systems of linear equations in the standard form $Ax = b$. To pose our Markov chain problem in this same format, we set $x = \pi^T$ and consider the application of direct methods to $Q^T x = 0$. Later, in Section 2.3.2, we shall examine the application of the methods directly to $\pi Q = 0$, i.e., without first transposing the infinitesimal generator.

2.3.1.1 The "Zero Pivot" Approach

In this approach, we proceed as for the nonhomogeneous case and make special arrangements for the zero pivot. We require the LU decomposition of $A = Q^T$. It is known that when the matrix Q is irreducible, there exist lower and upper triangular matrices L and U such that $Q^T = LU$. If Q is reducible, the problem may be simplified to a number of smaller irreducible problems.

Consider as an example, the infinitesimal generator matrix

$$Q = \begin{pmatrix} -2.0 & 1.0 & 1.0 \\ 1.0 & -3.0 & 2.0 \\ 2.0 & 0.0 & -2.0 \end{pmatrix}.$$

Following the guidelines of Section 2.2.2, we obtain the Doolittle LU decomposition of Q^T as follows:

From (2.4): $u_{1j} = a_{1j};\ \ j = 1, 2, 3,$ $\implies u_{11} = -2.0;\ \ u_{12} = 1.0;\ \ u_{13} = 2.0.$
From (2.5): $l_{i1} = a_{i1}/u_{11};\ \ i = 2, 3,$ $\implies l_{21} = -0.5;\ \ l_{31} = -0.5.$
From (2.4): $u_{2j} = a_{2j} - l_{21}u_{1j};\ \ j = 2, 3$ $\implies u_{22} = -2.5;\ \ u_{23} = 1.0.$
From (2.5): $l_{i2} = (a_{i2} - l_{i1}u_{12})u_{22};\ \ i = 3,$ $\implies l_{32} = -1.0.$
From (2.4): $u_{j3} = a_{j3} - l_{31}u_{1j} - l_{32}u_{2j};\ \ j = 3,$ $\implies u_{33} = 0.0.$

This gives the decomposition

$$\begin{pmatrix} 1.0 & 0.0 & 0.0 \\ -0.5 & 1.0 & 0.0 \\ -0.5 & -1.0 & 1.0 \end{pmatrix} \begin{pmatrix} -2.0 & 1.0 & 2.0 \\ 0.0 & -2.5 & 1.0 \\ 0.0 & 0.0 & 0.0 \end{pmatrix} = \begin{pmatrix} -2.0 & 1.0 & 2.0 \\ 1.0 & -3.0 & 0.0 \\ 1.0 & 2.0 & -2.0 \end{pmatrix}$$

in which the last row of U is equal to zero — a result of the singularity of the coefficient matrix. For the corresponding LDU decomposition, we get

$$
\begin{pmatrix} 1.0 & 0.0 & 0.0 \\ -0.5 & 1.0 & 0.0 \\ -0.5 & -1.0 & 1.0 \end{pmatrix}
\begin{pmatrix} -2.0 & 0.0 & 0.0 \\ 0.0 & -2.5 & 0.0 \\ 0.0 & 0.0 & 0.0 \end{pmatrix}
\begin{pmatrix} 1.0 & -0.5 & -1.0 \\ 0.0 & 1.0 & -0.4 \\ 0.0 & 0.0 & 1.0 \end{pmatrix}.
$$

Notice how the zero diagonal element is handled.

Let us return now to the general case. For a nonhomogeneous system of equations with nonsingular coefficient matrix, once an LU decomposition has been determined, a forward substitution step followed by a backward substitution yields the solution. However, in the case of the numerical solution of irreducible Markov chains, the system of equations is homogeneous and the coefficient matrix is singular. From the irreducibility property it is known that Q has a one-dimensional null space. In this case, the final row of U (if the Doolittle decomposition has been performed) is zero. Proceeding as for the nonhomogeneous case, we have

$$Q^T x = 0 \quad \text{or} \quad (LU)x = 0.$$

If we now set $Ux = z$ and attempt to solve $Lz = 0$ we find that, since L is nonsingular, we must have $z = 0$. Let us now proceed to the back-substitution on $Ux = z = 0$, in which the last row of U is identically zero. It is evident that we may assign any nonzero value, say η, to x_n, for this element will always be multiplied by one of the zero elements in the last row of U. We may then use back-substitution to determine the remaining elements of the vector x in terms of η. We get $x_i = c_i \eta$ for some constants c_i, $i = 1, 2, \ldots, n$, and $c_n = 1$. Thus, the solution obtained depends on the value of η. There still remains one equation that the elements of a probability vector must satisfy, namely that the sum of the probabilities must be 1. Consequently, normalizing the vector obtained from the solution of $Ux = 0$, so that its elements sum to 1, yields the desired unique stationary probability vector π of the infinitesimal generator Q.

Consider the (3×3) example again. Substituting for L and U into $LUx = 0$, we have

$$
\begin{pmatrix} 1.0 & 0.0 & 0.0 \\ -0.5 & 1.0 & 0.0 \\ -0.5 & -1.0 & 1.0 \end{pmatrix}
\begin{pmatrix} -2.0 & 1.0 & 2.0 \\ 0.0 & -2.5 & 1.0 \\ 0.0 & 0.0 & 0.0 \end{pmatrix}
\begin{pmatrix} x_1 \\ x_2 \\ x_3 \end{pmatrix} = 0.
$$

Replacing Ux with z yields

$$
\begin{pmatrix} 1.0 & 0.0 & 0.0 \\ -0.5 & 1.0 & 0.0 \\ -0.5 & -1.0 & 1.0 \end{pmatrix}
\begin{pmatrix} z_1 \\ z_2 \\ z_3 \end{pmatrix} =
\begin{pmatrix} 0.0 \\ 0.0 \\ 0.0 \end{pmatrix},
$$

and obviously $z_1 = z_2 = z_3 = 0$. By arbitrarily setting x_3 equal to η, ignoring the last equation, and performing the back-substitution on

$$
\begin{pmatrix} -2.0 & 1.0 & 2.0 \\ 0.0 & -2.5 & 1.0 \end{pmatrix} \begin{pmatrix} x_1 \\ x_2 \\ \eta \end{pmatrix} = \begin{pmatrix} 0.0 \\ 0.0 \end{pmatrix}
$$

to obtain x_2 and x_1 yields

$$
x = \eta \begin{pmatrix} 1.2 \\ 0.4 \\ 1.0 \end{pmatrix}.
$$

When normalized, this gives the correct stationary probability vector: $\pi = \left(\frac{6}{13}, \frac{2}{13}, \frac{5}{13} \right)$.

2.3.1.2 The "Replace an Equation" Approach

An alternative approach to this use of the normalization equation is to replace the last equation of the original system with $e^T x = \pi e = 1$. If the Markov chain is irreducible, this will ensure that the coefficient matrix is nonsingular. Furthermore, the system of equations will no longer be homogeneous (since the right-hand side is now e_n, where[1] $e_n = (0, 0, \ldots, 0, 1)^T$), and so the solution may be computed directly. We shall write this modified system of equations as

$$
\pi \bar{Q} = e_n^T.
$$

Notice that π thus computed will be the last column of $(\bar{Q}^T)^{-1}$.

In the example, if we replace the last equation by $e^T x = 1$, we obtain

$$
\bar{Q}^T = \begin{pmatrix} -2.0 & 1.0 & 2.0 \\ 1.0 & -3.0 & 0.0 \\ 1.0 & 1.0 & 1.0 \end{pmatrix}.
$$

For this matrix, the LU decomposition $\bar{L}\bar{U} = \bar{Q}^T$ is given by

$$
\begin{pmatrix} 1.0 & 0.0 & 0.0 \\ -0.5 & 1.0 & 0.0 \\ -0.5 & -0.6 & 1.0 \end{pmatrix} \begin{pmatrix} -2.0 & 1.0 & 2.0 \\ 0.0 & -2.5 & 1.0 \\ 0.0 & 0.0 & 2.6 \end{pmatrix} = \begin{pmatrix} -2.0 & 1.0 & 2.0 \\ 1.0 & -3.0 & 0.0 \\ 1.0 & 1.0 & 1.0 \end{pmatrix}.
$$

Setting $z = \bar{U}\pi^T$ in $\bar{L}\bar{U}\pi^T = e_n$, we obtain $\bar{L}z = e_n$, i.e.,

$$
\begin{pmatrix} 1.0 & 0.0 & 0.0 \\ -0.5 & 1.0 & 0.0 \\ -0.5 & -0.6 & 1.0 \end{pmatrix} \begin{pmatrix} z_1 \\ z_2 \\ z_3 \end{pmatrix} = \begin{pmatrix} 0.0 \\ 0.0 \\ 1.0 \end{pmatrix},
$$

---

[1] We shall adopt the convention that e is the (column) vector whose components are all equal to 1 and e_i is the ith column of the identity matrix, i.e., the vector whose components are all equal to 0 except the ith, which is equal to 1.

which implies, from forward substitution, that $z_1 = 0$, $z_2 = 0$, $z_3 = 1$. Note that, due to the special form of the right-hand side of $\bar{L}z = e_n$, we can always write z immediately as $z = e_n$. To determine the stationary probability vector, we apply back-substitution to

$$
\begin{pmatrix}
-2.0 & 1.0 & 2.0 \\
0.0 & -2.5 & 1.0 \\
0.0 & 0.0 & 2.6
\end{pmatrix}
\begin{pmatrix}
\pi_1 \\
\pi_2 \\
\pi_3
\end{pmatrix}
=
\begin{pmatrix}
0.0 \\
0.0 \\
1.0
\end{pmatrix}
$$

and obtain $\pi = \left(\frac{6}{13}, \frac{2}{13}, \frac{5}{13} \right)$.

Of course, it is not necessary to replace the *last* equation of the system by the normalization equation. Indeed, any equation could be replaced. However, this is generally undesirable, for it will entail more numerical computation. For example, if the first equation is replaced, the first row of the coefficient matrix will contain all ones and the right-hand side will be e_1. This implies that during the forward substitution stage, the entire sequence of operations must be performed to obtain the vector z; whereas if the last equation is replaced, it is possible to simply read off the solution immediately as $z = e_n$. In addition, substantial fill-in will probably occur, since a multiple of the first row, which contains all ones, must be added to all remaining rows, and a cascading effect will undoubtedly occur in all subsequent reduction steps. The problem of handling fill-in, which plagues direct methods, is considered later in this section.

This approach is likely to yield a less accurate result than the previous one. When the last equation is replaced with the normalization equation, the elimination procedure is exactly the same as for Method 1, except for the decomposition of the final row. The decomposition of this last row yields u_{nn}, which is used to compute the last component of the solution. Thus, when x_n is computed from u_{nn} in Method 2, it will be contaminated with the round-off errors from *all* of the previous elimination steps, and this will propagate throughout the back-substitution. In Method 1, an uncontaminated constant multiple is used, which then disappears after normalization.

2.3.1.3   The "Remove an Equation" Approach

An alternative approach to that of replacing one of the equations with the normalization equation is to remove one of the equations. This may be justified on the grounds that since Q is of rank $(n-1)$, any equation may be written as a linear combination of the others. Removing an equation gives $n-1$ equations in n unknowns, which allows us to assign a value for one of the unknowns and then solve for the others. Consider $Q^T x = 0$, and partition Q^T as

$$
Q^T = \begin{pmatrix} B & d \\ c^T & f \end{pmatrix},
$$

where B is of order $(n-1)$. For an irreducible Markov chain, $d \neq 0$, and so B is nonsingular. Removing the last equation and assigning the unknown x_n the value 1 is equivalent to solving

$$\begin{pmatrix} B & d \\ c^T & f \end{pmatrix} \begin{pmatrix} \hat{x} \\ 1 \end{pmatrix} = \begin{pmatrix} 0 \\ 0 \end{pmatrix}$$

by solving $B\hat{x} = -d$. The stationary probability vector is obtained by normalizing $(\hat{x}, 1)$. Essentially we remove one of the degrees of freedom by setting one of the elements of π equal to unity and later normalize to get the correct stationary probability vector. The above formulation indicates that it is the last element that is set to unity, but this could equally well have been any other. For example, writing

$$\begin{pmatrix} \alpha & \beta^T \\ \gamma & D \end{pmatrix} \begin{pmatrix} 1 \\ \tilde{x} \end{pmatrix} = \begin{pmatrix} 0 \\ 0 \end{pmatrix},$$

the solution is obtained by solving $D\tilde{x} = -\gamma$ and normalizing $(1, \tilde{x})$.

2.3.1.4   The "Inverse Iteration" Approach

In Section 2.2.4 we saw that inverse iteration can be used to obtain an eigenvector when an approximation to its corresponding eigenvalue is known. In our particular case we wish to determine the right-hand eigenvector corresponding to the zero eigenvalue of Q^T. Therefore, letting $\mu = 0$ in the iteration formula (2.6) and substituting Q^T for A, we get

$$(Q^T - 0I)x^{(k)} = Q^T x^{(k)} = x^{(k-1)}$$

and thus we are simply required to solve $Q^T x^{(1)} = x^{(0)}$.

An LU decomposition of Q^T must be found, and as explained in Section 2.2.4, the zero pivot is replaced by machine epsilon, ϵ. Note that choosing $x^{(0)} = e_n$ reduces the amount of computation involved. The iteration simply reduces to the back-substitution step, $Ux^{(1)} = \epsilon^{-1}e_n$ (assuming that the Doolittle decomposition of Q^T is available). An appropriate normalization of $x^{(1)}$ will yield the stationary probability vector π.

2.3.1.5   Summary

We have now considered four approaches for the direct solution of $\pi Q = 0$:

1. - Determine an LU decomposition of Q^T (e.g., the Doolittle decomposition).
    - Set $u_{nn} = 1$ and $x_n = \eta \neq 0$ and solve $Ux = \eta e_n$ for x.
    - Normalize x^T to obtain π.

2.   • Replace the last equation of $Q^T x = 0$ with $e^T x = 1$ to obtain $\bar{Q}^T x = e_n$.

      • Determine an LU decomposition of $\bar{Q}^T$, i.e., $\bar{Q}^T = \bar{L}\bar{U}$.

      • Perform backward substitution on $\bar{U} x = e_n$ to determine $x = \pi^T$.

3.   • Partition $Q^T = \begin{pmatrix} B & d \\ c^T & f \end{pmatrix}$.

      • Determine an LU decomposition of B, i.e., $B = \hat{L}\hat{U}$.

      • Perform forward substitution on $\hat{L}\hat{z} = -d$ to determine $\hat{z}$.

      • Perform backward substitution on $\hat{U}\hat{x} = \hat{z}$ to determine $\hat{x}$.

      • Normalize $x = (\hat{x}, 1)^T$ to determine π.

4. Inverse Iteration:

      • Determine an LU decomposition of Q^T (e.g., Doolittle decomposition).

      • Set $u_{nn} = \epsilon$ and solve $Ux = e_n$, where ϵ is machine epsilon.

      • Normalize x^T to obtain π.

Obviously, with inverse iteration we have come full circle, for approach 1 reduces to inverse iteration when η is chosen to be the reciprocal of machine epsilon. Furthermore, Method 3 corresponds exactly to the case of $\eta = 1$, since forward substitution on $\hat{L}\hat{z} = -d$ to determine $\hat{z}$ is in fact equivalent to reducing the last row of Q^T.

These three, essentially identical, approaches are to be preferred to Method 2. For example, with inverse iteration an indication of the buildup of rounding error may be obtained. Theoretically, it is known that we should obtain a zero pivot during the reduction of the final row. However, due to rounding error, this will hardly ever be exactly zero; its nonzero value will yield an indication of the magnitude of the buildup of rounding error. This is important when very large matrices are being handled on computers with a small word size, or in cases in which the Markov chain is almost reducible, for it is known that occasionally the rounding error becomes so large that it swamps the correct solution vector.

Additionally, we have already seen that Method 2 requires more numerical operations. In Method 2, during the decomposition stage, all of the elements of the normalization equation are nonzero and must be reduced by adding a nonzero multiple of each of the preceding equations into this equation. When the coefficient matrix is large and sparse, the equation that is replaced by the normalization equation will normally have mostly zero elements, and these operations could otherwise be avoided. If one of the initial equations is substituted for the normalization equation, then substantial fill-in and a subsequent large increase in the amount of computation will occur.

We also indicated that Method 2 is likely to be less accurate than the others. This topic will be taken up again in Section 2.7.5, when we introduce the concept of the *condition* of a matrix.

2.3.2 To Transpose or Not to Transpose

In the preceding discussions we first converted the original system of equations $\pi Q = 0$ into the standard form $Q^T x = 0$ and considered the application of direct methods to Q^T. We may also consider applying the methods directly to Q by forming an LU decomposition of Q rather than of Q^T. Writing

$$L_2 U_2 = Q,$$

where L_2 is unit lower triangular and U_2 is upper triangular, we obtain the solution of $\pi L_2 U_2 = 0$ by first setting $\pi L_2 = z$ and considering $z U_2 = 0$. It is obvious that $z = \eta e_n^T$ satisfies this equation for any nonzero value of η, so our task is the computation of π from

$$\pi L_2 = \eta e_n^T \tag{2.8}$$

by back-substitution followed by an appropriate normalization. Notice that when we choose $\eta = 1$, the (unnormalized) solution is given by the last row of L_2^{-1}, namely $e_n^T L_2^{-1}$.

The matrix Q^T possesses the property that its *column* sums are equal to zero and the off-diagonal elements in any *column* are less than or equal to the absolute value of the diagonal element in that column. This means that in applying Gaussian elimination to Q^T we are guaranteed to have pivots that are less than or equal to 1. This is not necessarily the case when Gaussian elimination is applied to Q, for there is no reason to assume that a diagonal element of Q is the largest in any column, so the resulting pivots may be greater than 1. This should only be considered as a minor inconvenience, rather than a major disadvantage of applying direct methods to Q, for it may be shown (see Section 2.7.4.2) that the growth of matrix elements in both cases, as measured by the growth factor, is equal to 1, so both approaches may be considered to be equally stable.

A more important drawback of applying direct methods to Q is that the unit lower triangular matrix L_2 is needed in the final back-substitution stage (Equation 2.8). The multipliers cannot be disposed of at the end of each step, as they can when Gaussian elimination is applied to Q^T. Nor can this be compensated for by discarding the rows of U_2, since these rows are needed in the reduction of all rows up until the final step of the reduction has been completed. In cases when computer memory is not scarce and both L_2 and U_2 may be readily accommodated, this is not a problem. However, when available memory is limited, this need to store both L_2 and U_2 during the reduction stage is a serious disadvantage. This leads us directly to a consideration of storage mechanisms and other important implementation details that relate to the solution of Markov chains by direct methods.

2.4   Implementation Considerations

2.4.1   Implementation in Two-Dimensional Storage Arrays

The most convenient storage mechanism for implementing variants of Gaussian elimination is a two-dimensional storage array. Given a transition rate matrix $Q \in \Re^{n \times n}$ stored by rows or by columns in some two-dimensional array structure, it is possible to perform the complete reduction in this array and to overwrite the original matrix by its LU factors. When an element is eliminated, the multiplier used to effect this elimination should be negated and stored in the position previously occupied by the eliminated element since, as shown in Section 2.2.1, this gives the matrix L. The unit diagonal elements of L are not stored explicitly. The matrix U may overwrite the upper triangular part of Q. The original matrix is always available, if and when needed, albeit in decomposed form. Since access is equally available to rows and to columns, an LU reduction of Q^T is as easily implemented as an LU reduction of Q.

2.4.2   Compact Storage Schemes for Direct Methods

Frequently the matrices generated from Markov models are too large to permit regular two-dimensional arrays to be used to store them in computer memory. Since these matrices are usually very sparse, it is economical, and indeed necessary, to use some sort of packing scheme, whereby only the nonzero elements and their positions in the matrix are stored. When a direct equation-solving method is to be applied, provision usually must be made to include elements that become nonzero during the reduction and, somewhat less important, to remove elements that have been eliminated. If memory locations are not urgently required, the easiest way of removing an element is to set it to zero without trying to recuperate the words that were used to store it. To include an element into the storage scheme, either some means of appending this element to the end of the storage arrays must be provided, or else sufficient space must be left throughout the arrays so that fill-in can be accommodated as and when it occurs. The first usually requires the use of link pointers and is most useful if the nonzero elements are randomly dispersed throughout the matrix, while the second is more useful if the pattern of nonzero elements is rather regular. Throughout this section, the reader should take care not to confuse the manner in which the coefficient matrix is stored (by rows or by columns) with the particular reduction being applied, viz., to Q^T (to obtain LU) or to Q (to obtain $L_2 U_2$). All combinations are in fact possible.

2.4.2.1   Regular Pattern Storage Schemes

Regular pattern storage schemes may be used when the nonzero elements of the matrix occur in a well-defined manner; the pattern of the nonzero elements will dictate the particular storage scheme that should be used. In queueing networks the nonzero elements often lie relatively close to the diagonal, so that a *fixed-bandwidth scheme* may be used. In other cases the nonzero elements lie along lines that run parallel to the diagonal, which implies that probably there will be considerable fill-in between these parallel lines. Again, a fixed-bandwidth scheme is appropriate.

As an example, we show below how a (6×6) matrix is stored using a fixed-bandwidth scheme of size 3.

$$
\begin{pmatrix}
a_{11} & a_{12} & 0 & 0 & 0 & 0 \\
a_{21} & a_{22} & 0 & 0 & 0 & 0 \\
0 & 0 & a_{33} & 0 & 0 & 0 \\
0 & 0 & a_{43} & a_{44} & a_{45} & 0 \\
0 & 0 & 0 & 0 & a_{55} & a_{56} \\
0 & 0 & 0 & 0 & a_{65} & a_{66}
\end{pmatrix}
\Rightarrow
\begin{pmatrix}
0 & a_{11} & a_{12} \\
a_{21} & a_{22} & 0 \\
0 & a_{33} & 0 \\
a_{43} & a_{44} & a_{45} \\
0 & a_{55} & a_{56} \\
a_{65} & a_{66} & 0
\end{pmatrix}
$$

Matrix operations in general, and equation solving in particular, can be programmed with virtually the same ease using these regular pattern storage schemes as they can using standard two-dimensional matrix storage. It is obvious that any fill-in that occurs is restricted to this band. Furthermore, there is no storage requirement for secondary arrays such as link pointers and row and column indicators, so no computation time is used in the processing of such arrays. It is advantageous to adopt these types of storage scheme where possible. Even when some of the elements within the regular pattern are zero, as is the case in the above example, it may be more economical and convenient to use a regular pattern scheme than a scheme that attempts to minimize the storage used. In fact, as the size of the matrix increases, it rapidly becomes evident that unless such a regular pattern storage scheme can be implemented, the feasibility of incorporating a direct method rapidly diminishes.

2.4.2.2 Address Links

The following is an example of a (4×4) matrix stored in compact form using address links:

$$
A = \begin{pmatrix}
-2.1 & 0.8 & 0.2 & 0.0 \\
1.7 & -0.8 & 0.0 & 0.3 \\
0.0 & 0.0 & -1.7 & 0.2 \\
0.4 & 0.0 & 0.0 & -0.5
\end{pmatrix}
$$

Real array	A :	-2.1	-0.8	-1.7	-0.5	0.8	0.2	1.7	0.3	0.2	0.4
Row array	RA :	1	2	3	4	1	1	2	2	3	4
Column array	CA :	1	2	3	4	2	3	1	4	4	1
Link array	LA :	5	8	9	0	6	7	2	3	10	4

The nonzero elements of A may be stored in any order in a real array A. Their corresponding row and column positions are stored in the integer arrays RA and

CA, respectively. The value of the ith nonzero element is stored in memory location $A(i)$, its row position is given by $RA(i)$, and its column position by $CA(i)$. In this particular example the links have been constructed so that the nonzero elements can be accessed in a row-wise sense. Thus, the value that is stored in $LA(i)$ indicates the position in the real array A at which the next nonzero element in row $RA(i)$ may be found. If the element stored in $A(i)$ is the last nonzero element in row $RA(i)$, then $LA(i)$ points to the first nonzero element in row $RA(i) + 1$.

In the example provided above, $A(6)$ contains the value 0.2, the (1,3) element of the matrix and the last (nonzero) element of row 1. The link pointer indicates that the value of the next element (in a row-wise sense) may be found in location $A(7)$. This new location contains 1.7, the (2,1) element, and its link pointer indicates that the next nonzero value is to be found in location $A(2)$.

Normally, it is useful to be able to enter the chain at several points; this is achieved in the above representation by listing the diagonal elements first in the array: the kth diagonal element is stored in location $A(k)$; the next nonzero element in row k is at location $A(LA(k))$, and so on. Once again, if the diagonal element is the final nonzero element in its row, the link pointer will indicate the position of the first nonzero in the following row.

To see how an element may be included, consider the elimination of the element in position (2,1). This causes .162 to be added into position (2,3), which was previously empty. It is handled in this storage scheme by appending

$$.162 \text{ to } A; \qquad 2 \text{ to } RA; \qquad 3 \text{ to } CA; \qquad 8 \text{ to } LA.$$

The links must also be updated so that the link that previously indicated 8 (the second element of LA) must now point to 11. This updating, in fact, constitutes a major disadvantage of this type of storage scheme, since it is not unusual for it to require more computation time than the actual operations involved in the reduction. A second disadvantage is the fact that three integer arrays are required in addition to the array that contains the nonzero elements.

2.4.2.3 Semisystematic Packing Schemes

This type of packing scheme is used most often with matrix iterative methods in which the matrix itself is not altered. However, it sometimes can be a convenient way to store LU factors. One commonly used scheme is the following. A one-dimensional real array (which we shall denote as aa) is used to hold the nonzero elements of the matrix. The elements are stored by rows; elements of row i come before those of row $i+1$, but the elements within a row need not be in order (this is why the term *semisystematic* is used). Two integer arrays are used to specify the location of the elements in the matrix. The first, ja, contains the column position of each element. Thus, $ja(k)$ gives the column position of the element stored in the kth position of aa. The second integer array, ia, is a pointer array whose l^{th} element indicates the position in aa and ja at which the elements of the lth row begin. A final element, which is set equal to 1 more than the number of nonzero elements, is appended to ia. In this case $ia(l+1) - ia(l)$ always gives the number of nonzero elements in row l, $l = 1, 2, \ldots, n$.

Consider the matrix discussed previously:

$$A = \begin{pmatrix} -2.1 & 0.8 & 0.2 & 0.0 \\ 1.7 & -0.8 & 0.0 & 0.3 \\ 0.0 & 0.0 & -1.7 & 0.2 \\ 0.4 & 0.0 & 0.0 & -0.5 \end{pmatrix}$$

One possible way to store this in the semisystematic format is

Real array	aa :	0.8	−2.1	0.2	1.7	−0.8	0.3	0.2	−1.7	−0.5	0.4
Column array	ja :	2	1	3	1	2	4	4	3	4	1
Pointer array	ia :	1	4	7	9	11					

2.4.3 Simultaneous Row Generation and Reduction

When applying direct equation-solving methods such as Gaussian elimination, it is usually assumed that the complete set of linear equations has already been generated and that the entire coefficient matrix is stored somewhere in the computer memory, albeit in a compact form. The reduction phase begins by using the first equation to eliminate all nonzero elements in the first column of the coefficient matrix from position 2 through n. More generally, during the ith reduction step the ith equation is used to eliminate all nonzero elements in the ith column from positions $(i + 1)$ through n. Naturally, it is assumed that the pivot elements are always nonzero; otherwise the reduction breaks down. Partial pivoting is almost invariably used to ensure a stable reduction.

However, in Markov chain problems it is possible to envisage an alternative approach that has several advantages over the traditional method just outlined. Assume that Q^T is available and perhaps even generated, row by row. Then, immediately after the second row has been obtained, it is possible to eliminate the element in position (2,1) by adding a multiple of the first row to it. No further changes are made to the second row, so it may now be stored in any convenient compact form. The third row may now be chosen and immediately reduced, by adding a suitable multiple of the first row to it, followed by adding a multiple of the second row. The third row is now fully reduced and may be stored in compact form. This process may be continued recursively, so that when the ith row of the coefficient matrix is considered, rows 1 through $(i-1)$ have been treated and are already reduced to upper triangular form. The first $(i-1)$

rows may then be used to eliminate all nonzero elements in row i from column positions 1 through $i - 1$, thus putting the matrix into the desired triangular form. Figure 2.3 presents the state of the matrix just after the generation of the ith row but before the reduction step has begun to alter it.

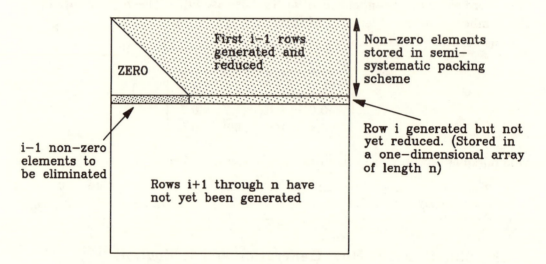

Figure 2.3: Partially reduced matrix at beginning of step i.

The proposed method has a distinct advantage in that once a row has been treated in this fashion, no more fill-in will occur into this row. It is suggested that a separate storage area be reserved to hold a single unreduced row temporarily. The reduction is performed in this storage area. Once completed, the reduced row may be compacted into any convenient form and appended to the rows that have already been reduced. In this way no storage space is wasted in holding subdiagonal elements that, due to elimination, have become zero or in reserving space for the inclusion of additional elements. The storage scheme should be chosen bearing in mind the fact that these rows will be used in the reduction of further rows and also later in the algorithm during the back-substitution phase. Since the form of the matrix will no longer be altered, the efficient semisystematic storage schemes that are used with many iterative methods can be adopted.

If this approach is applied to the matrix Q (and it is more common for Markov chain generators to obtain the matrix Q in a row-wise fashion), then an additional array must be set aside to store the multipliers, for in this case it is the matrix L that is used to compute the solution. Notice also that this approach cannot be used for solving general systems of linear equations, because it inhibits a pivoting strategy from being implemented. It is valid when solving irreducible Markov chains, since pivoting is not required for the LU decomposition of Q^T to be computed in a stable manner.

The most successful direct methods are those that combine the benefits of simultaneous row generation and reduction with a fixed-bandwidth storage scheme. Even when the matrix is extremely large (in excess of 100,000, for example) — so large in fact that some form of backing store may be needed — this approach is still viable. Suppose, for example, that the matrix is band-shaped and very large. When memory is exhausted, it is convenient to put a large section of the reduced matrix onto backing store, and this will need to be returned to memory only once, for the final back-substitution phase. Consider the matrix shown in Figure 2.4 and assume that available memory can only hold l reduced lines. Let h denote the maximum number of nonzero elements to the left of the diagonal element. Suppose that once the lth row has been generated, reduced, and appended to the rows already reduced, no memory locations remain for the reduced form of lines $(l+1)$ through n. However, since rows 1 through $(l-h)$ are no longer required for the reduction of rows below the lth, they may be put onto backing store and the memory they used made available to store a further $(l-h)$ reduced rows.

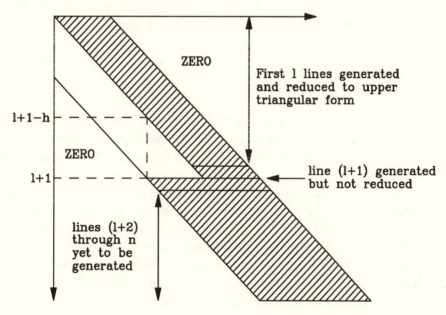

Figure 2.4: First $(l-h)$ rows are ready to be sent to backing store.

Depending on the size of the matrix and available memory, this process may have to be performed a number of times. Even when backing store is not explicitly requested — for example, in paging systems — this implementation will still be efficient, for in that case it minimizes page faults.

2.4.4   Back-Substitution and Normalization

Depending on the particular model being solved, it is possible that successive values determined during the back-substitution phase become either larger or smaller, with a net result that some sort of an adaptive normalization procedure becomes necessary. The back-substitution procedure is as follows:

$$x_n = \epsilon^{-1}$$

and

$$x_i = \left(- \sum_{k=i+1}^{n} u_{ik} x_k \right) / u_{ii}, \qquad i = n-1, \ldots, 2, 1.$$

Suppose that while we are generating successive components of the solution $x_n, x_{n-1}, x_{n-2}, \ldots, x_k$, we find that the elements start to become so large that the danger of overflow arises. In this case, an appropriate procedure is to normalize $x_n, \ldots, x_k$ by dividing all the elements by the largest and then continuing with the back-substitution. This renormalization may have to be performed on several occasions. On the other hand, if successive components become smaller, so small in fact that the danger of underflow occurs, those components may be replaced by zero.

2.5   The GTH Advantage

It is appropriate at this point to mention a version of Gaussian elimination that has attributes that appear to make it even more stable than the usual version. This procedure is commonly referred to as the GTH (Grassmann–Taksar–Heyman) algorithm [59, 146]. In GTH the diagonal elements are obtained by summing the off-diagonal elements rather than performing a subtraction; it is known that subtractions can sometimes lead to loss of significance in the representation of real numbers. The concept evolved from probabilistic arguments, and the originally suggested implementation is a backward (rather than forward) elimination procedure. However, other implementations are possible.

The key to the GTH algorithm is that the properties that characterize the infinitesimal generator of an irreducible Markov chain,

$$q_{ii} < 0, \quad q_{ij} \geq 0, \quad \text{and} \quad \sum_{j=1}^{n} q_{ij} = 0 \quad \text{for all } i$$

are invariant under the elementary row operations carried out when Gaussian elimination is applied to Q.[2] This may be seen by considering the first two

---

[2] The reader should carefully note that it is the matrix Q and *not* Q^T that is the subject of discussion in this paragraph.

rows of Q and the elimination of q_{21}, which for obvious reasons we assume to be nonzero.

$$\text{Row 1}: \qquad q_{11} \quad q_{12} \quad q_{13} \quad \cdots \quad q_{1n} \qquad \text{with} \quad q_{11} = \sum_{k=2}^{n} q_{1k}$$

$$\text{Row 2}: \qquad q_{21} \quad q_{22} \quad q_{23} \quad \cdots \quad q_{2n} \qquad \text{with} \quad q_{22} = \sum_{k \neq 2}^{n} q_{2k}$$

To eliminate $q_{21} \neq 0$, we add a multiple of row 1 into row 2. This multiple is given by $-q_{21}/q_{11} = q_{21}/\sum_{k=2}^{n} q_{1k}$, which is a positive quantity. Since q_{12}, q_{13}, $\ldots$, q_{1n} are positive, this elementary row operation causes a positive quantity to be added into each element q_{22} through q_{2n} (q_{21} is *set* to zero). The elements in row 2 become

$$q_{2j} \leftarrow q_{2j} + q_{21} \frac{q_{1j}}{\sum_{k=2}^{n} q_{1k}} = q_{2j} + q_{21} w_{1j}, \quad \text{for } j = 2, \ldots n, \qquad (2.9)$$

which shows that the quantity q_{21} is distributed over the other elements of the second row according to their weight w_{1j} in row 1. Consequently it follows that the sum of the elements in row 2 remains zero. Also, the off-diagonal elements in row 2 do not decrease (since they were greater than or equal to zero and nonnegative quantities are added), and the diagonal element, which had been strictly less than zero, can only move closer to zero — it cannot exceed zero, for otherwise the sum across the row could not be zero. The only operation in (2.9) that involves a negative number occurs with $j = 2$. We have

$$q_{22} \leftarrow q_{22} + q_{21} \frac{q_{12}}{\sum_{k=2}^{n} q_{1k}} \qquad (2.10)$$

and q_{22} is negative. However, instead of using (2.10), q_{22} may be found by summing the new off-diagonal elements and then negating this sum. We get

$$q_{22} = -\sum_{j=3}^{n} \left(q_{2j} + q_{21} \frac{q_{1j}}{\sum_{k=2}^{n} q_{1k}} \right). \qquad (2.11)$$

This procedure may be extended in the obvious manner to cover the entire Gaussian elimination algorithm. Thus, as the Gaussian elimination algorithm is unfolding, a diagonal pivot may be computed either in the usual fashion, by subtracting a multiple of the element immediately above it from its current value, or by summing the off-diagonal elements in the row once it has been reduced, and negating this sum. The latter approach involves no subtractions and yields a more stable algorithm. Notice that in an implementation, it would be redundant to actually insert a minus sign before the computed diagonal element.

The GTH implementation requires more numerical operations than the standard implementation (compare equations 2.10, and 2.11), but this may be offset by a gain in precision when the matrix Q is ill-conditioned. The extra additions are not very costly when compared with the overall cost of the elimination procedure, which leads to the conclusion that the GTH advantage should be exploited where possible in elimination procedures. It is also possible to apply the GTH approach to Q^T. If the transition rate matrix is stored in a two-dimensional or band storage structure, access is easily available to both the rows and columns of Q, and there is no difficulty in applying GTH to obtain an LU decomposition of either Q or Q^T.

Unfortunately, difficulties arise in implementing GTH when computer memory is at a premium and sparse compact storage schemes, such as those described in Section 2.4.2.3, must be used [36]. Suppose first that Q is stored by rows and an LU decomposition of Q is sought. Both L and U need to be kept during the GTH reduction stage: the rows of U are needed to eliminate nonzero elements in the unreduced part of the coefficient matrix, while L is needed to compute the solution from $\pi L = e_n$ by forward substitution. This is no different from an implementation of regular Gaussian elimination applied to Q. Nor does it change if we store Q by columns. An LU reduction of Q always requires storage for L. Let us now consider the possibilities when an LU decomposition of Q^T is sought, since in this case only U need be kept and the multipliers may be discarded immediately after they have been used. If Q^T is stored by rows, then in order to obtain the diagonal elements by adding off-diagonal terms, access is also needed to the columns of Q^T. If the storage scheme does not provide convenient access to both the rows and columns, then this approach cannot be easily used. Incorporating link pointers to provide such access can have a serious detrimental effect on computation time. That leaves us only with the case in which Q^T is stored by columns. The following algorithm, requiring access only to the rows of Q, has been proposed for implementing GTH to obtain an LU decomposition of Q^T. To accomplish this goal, it keeps a running sum of the elements that contribute to the pivot. During the ith step of the elimination, the sum for the $(i+1)^{st}$ column is accumulated and taken as the pivot element for the $(i+1)^{st}$ elimination step. Thus, diagonal elements are computed just before they are needed.

Algorithm: GTH Applied to Q^T

1. Compute $q_{11} = q_{12} + q_{13} + \ldots + q_{1n}$.

2. For $i = 1, 2, \ldots, n-1$ do

   - Set $\sigma = 0$.
   - For $j = i+1, i+2, \ldots, n$ do
      - Compute $\mu = q(j,i)/q(i,i)$.
      - For $k = i+1, i+2, \ldots, n$ do
         * $q(j,k) = q(j,k) + \mu \times q(i,k)$.
      - If $j > i+1$, then compute $\sigma = \sigma + q(j, i+1)$.
   - Set $q(i+1, i+1) = \sigma$.

A careful examination of this algorithm will reveal that although it implements the GTH approach on Q^T, it requires access only to the *rows* of Q! However, it suffers from the drawback that *all* rows below the ith are modified during the ith step, so that a process of expansion and recompaction of the unreduced portion must be performed continuously throughout the algorithm.

We may well ask ourselves where this leaves us with respect to GTH. We may conclude that if two-dimensional or band storage is implemented, then GTH should be used, because the additional time required to obtain the multipliers by adding off-diagonal elements is likely to be an insignificant part of the total computation. On the other hand, if compact storage schemes are used, then compared with the best possible implementation of Gaussian elimination (an LU decomposition of Q^T with Q^T stored by rows), GTH is likely to require either significantly more memory (if an LU decomposition of Q is computed), significantly more time (if an LU decomposition of Q^T is formed), or both. Since Gaussian elimination is known to be stable, it may be felt that the only real need for GTH occurs when the problem is very ill-conditioned.

2.6 Sample Experiments with Direct Methods

In this section we shall consider the results of some numerical experiments that were conducted using direct methods. We describe two different models. The first we solve using the implementation procedures suggested in Section 2.4. Our concern is with the amount of time and memory required as the number of states grows. Additionally we shall examine the amount of fill-in as a function of the ordering imposed on the state space. The second model yields a very highly structured coefficient matrix, and we show how advantage can be taken of this. The moral is that careful examination of a model can be beneficial.

2.6.1   An Interactive Computer System Model

A Fortran subroutine called GE (Gaussian Elimination) was implemented according to the guidelines given in Section 2.4, incorporated into the software package MARCA (for MARkov Chain Analyzer) [157], and used to obtain the results presented in this section. GE assumes that the transpose of the transition rate matrix is available in sparse row-wise semisystematic form as presented in Section 2.4.2.3. A double-precision array (aa) holds the nonzero elements; an integer array (ja) stores the column positions of each nonzero element; and a second integer array (ia) stores the position in aa and ja of the first nonzero element of each row. The subroutine GE extracts the rows of the matrix one by one from this compact form, expands each row into a full vector by inserting zeros where appropriate, and then performs the reduction on this full vector. Once the reduction of a row is complete, it is compacted and appended to previously reduced rows in a separate sparse storage facility. These rows are used later to reduce other rows, but notice that it is not necessary (and in fact inefficient) to re-expand them at that point. Also, the multipliers are *not* kept, so at the end of the reduction step only the upper triangular part U is in compact storage. The original matrix is left unaltered.

The Markov chain used in this first set of experiments was derived from the queueing network model of an interactive computer system shown in Figure 2.5. It consists of a set of N terminals from which N users generate commands; a central processing unit (CPU); a secondary memory device (SM) and a filing device, (FD).

A queue of requests is associated with each device, and the scheduling is assumed to be FCFS (first come, first served). When a command is generated, the user at the terminal remains inactive until the system responds. Symbolically, a user, having generated a command, enters the CPU queue. The behavior of the process at the CPU is characterized by a compute time followed either by a page fault, after which the process enters the SM queue, or an input/output (file request), in which case the process enters the FD queue. Processes that terminate their service at the SM or FD queue return to the CPU queue. Symbolically, the completion of a command is represented by a departure of the process from the CPU to the terminals.

A state of this system is of the form $(\eta_1, \eta_2, \eta_3, \eta_4)$ with $\sum_{i=1}^{4} \eta_i = N$, where η_1 denotes the number of users active at terminals, η_2 the number of processes at the CPU, and η_3 and η_4 the number of processes at the SM and FD, respectively.

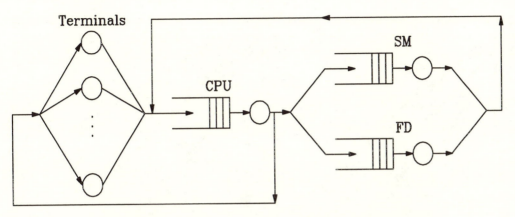

Figure 2.5: An interactive computer system model.

The state space and the infinitesimal generator matrix were generated by the software package MARCA as follows. An initial state is chosen and inserted at the top of a state list. This initial state is examined to determine which states it can reach in a single transition. These destination states are entered into the list of states in the order in which they are generated. The next state in the list of states is now examined to determine the states it can reach in a single transition. Destination states that are not already in the list of states are appended to the list. This process is continued until all states are examined and their destination states incorporated into the list of states. This obviously imposes an ordering on the states, and one of the things we wish to examine is the effect of choosing different initial states on the amount of fill-in generated by the GE algorithm.

Figures 2.6 through 2.9 show the structure of the infinitesimal generator matrix for the cases $N = 5$ and $N = 15$ using two different initial starting states for each case. The order of the matrix in the first case is $n = 56$, while in the second case it is $n = 816$. The number of nonzero elements, nz, is 266 and 4896, respectively. In these figures each dot represents a nonzero element. Notice that the matrix has a band structure in all cases, but that when the starting state is $(0, 0, 0, N)$, the band is considerably narrower than when the starting state is $(N, 0, 0, 0)$. Less fill-in will be generated in the case of the narrower bandwidth, with a resulting decrease in the computation time.

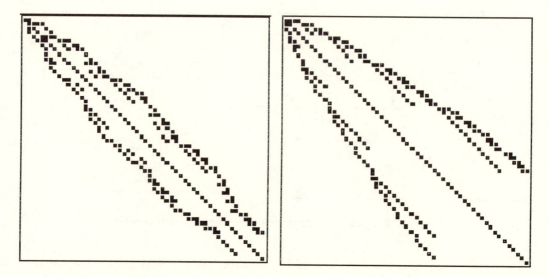

Figure 2.6: Starting state: $(0, 0, 0, 5)$ **Figure 2.8**: Starting state: $(5, 0, 0, 0)$

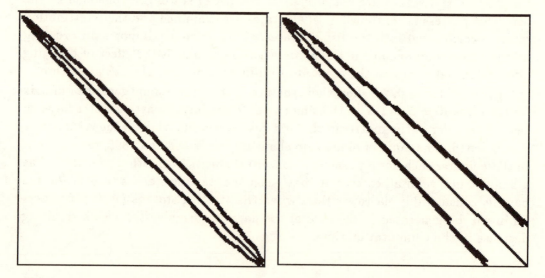

Figure 2.7: Starting state: $(0, 0, 0,$ **Figure 2.9**: Starting state: $(15, 0, 0,$
$15)$ $0)$

Table 2.1: Solution Times for Interactive Computer System Example

N	n	nz	$nzalu$	t
5	56	266	428	.01
6	84	420	809	.03
7	120	624	1423	.06
8	165	885	2364	.10
9	220	1210	3749	.17
10	286	1606	5720	.29
11	346	2080	8447	.48
12	455	2639	12129	.72
13	560	3290	16994	1.10
14	680	4040	23304	1.61
15	816	4896	31363	2.27
16	969	5865	41477	3.12
17	1140	6954	54021	4.35
18	1330	8170	69424	5.78
19	1540	9520	88149	7.72
20	1771	11011	110684	9.66

Table 2.1 presents the values of $N, n, nz, nzalu$ (the number of nonzeros in the reduced factor U), and t (the number of seconds of CPU time needed to solve this system on a SUN SPARCstation 2). The starting state in each of these experiments was taken as $(0, 0, 0, N)$. Note that quite large systems may be solved efficiently using a direct method.

The difference in the amount of fill-in generated when different initial states are used is shown in Table 2.2, which lists the number of nonzero elements in the reduced upper triangular factor U for some different starting states. Naturally these results are for a particular example. Other examples may have very different behavior. However, it is suspected that many Markov chains have the characteristics illustrated by this example.

2.6.2 Two Coxian Queues: A Highly Structured Example

It sometimes happens that the infinitesimal generator matrix of a given Markov chain is so highly structured that it is more efficient to write a specific solution procedure for that problem than to use existing software such as the GE subroutine just described. One such problem that we now consider is that of a queueing network consisting of two Coxian servers.

Table 2.2: Fill-In for Interactive Computer System Example

Initial State	nzalu	Initial State	nzalu
(0,0,12,0)	12113	(2,6,2,2)	16871
(0,0,0,12)	12129	(0,6,3,3)	19373
(0,12,0,0)	12130	(3,3,3,3)	22477
(4,0,0,8)	12590	(9,1,1,1)	24280
(6,0,6,0)	14311	(12,0,0,0)	24939

2.6.2.1 The Model and Its State Space

Consider a closed queueing network consisting of two single-server stations. We assume that the service time distribution at each server is modelled as a Coxian distribution. The queueing discipline is first-come first-served, and a fixed number of customers, N, circulates within the network. This system is illustrated in Figure 2.10.

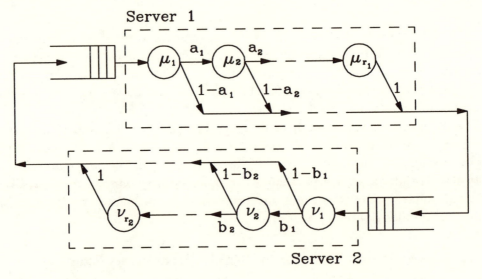

Figure 2.10: Network consisting of two Coxian servers.

Server 1 (respectively, server 2) is represented by a law of Cox of order r_1 (r_2), in which the mean service rate at stage i is μ_i (ν_i). Upon completion of service at stage i, the customer proceeds to stage $i+1$ with probability a_i (b_i), or leaves the server with probability $1 - a_i\, (1 - b_i)$. By convention a_{r_1} and b_{r_2} are chosen to be zero.

The state of the system at any time t is completely specified by the quadruple (η_1, i, η_2, j). Here η_1 and η_2 denote respectively the number of customers at server 1 and server 2, and since the total number of customers in the network is N, we must have $\eta_1 + \eta_2 = N$. The parameters i and j denote the current phase of service at server 1 and server 2, respectively. Obviously $0 \le i \le r_1$ and $0 \le j \le r_2$ and are equal to zero only if the corresponding server is idle.

2.6.2.2 The Ordering of the States of the Network

It is convenient to arrange the states of the system, first in decreasing order of the number of customers at server 1, second according to the service phase of server 1, and finally according to the service phase of server 2. This is illustrated in Table 2.3.

There is a total of $(N-1)$ blocks of states, in which the number of customers at the first station is between 1 and $N-1$, and it may be observed from the ordering diagram that each of these blocks contains $r_1 r_2$ states. In addition, there are r_1 states in which all customers are at the first station ($\eta_1 = N$) and r_2 states in which all N are at the second station ($\eta_1 = 0$). Consequently, the total number of states of the system is given by

$$n = (N - 1)r_1 r_2 + r_1 + r_2.$$

In particular, if both servers are second-order Coxians, the number of states is given by $4N$. The position of any arbitrary state $(\eta, i, N - \eta, j)$ in this list of states may be obtained from the relation

$$\text{posn}(\eta, i, N - \eta, j) = \begin{cases} i, & \text{if } \eta = N; \\ r_1 + (N - \eta - 1)r_1 r_2 + (i - 1)r_2 + j & \text{if } 0 < \eta < N; \\ r_1 + (N - 1)r_1 r_2 + j, & \text{if } \eta = 0. \end{cases}$$

The transitions that can occur from an arbitrary state of the system are shown in Table 2.4.

Table 2.3: State Space Ordering for Two-Coxian-Server Example

η_1	i	η_2	j
N	1	0	0
N	2	0	0
$\vdots$	$\vdots$	$\vdots$	$\vdots$
$\vdots$	$\vdots$	$\vdots$	$\vdots$
$\eta+1$	r_1	N-η-1	r_2-1
$\eta+1$	r_1	N-η-1	r_2
η	1	N-η	1
η	1	N-η	2
$\vdots$	$\vdots$	$\vdots$	$\vdots$
$\vdots$	$\vdots$	$\vdots$	$\vdots$
η	1	N-η	r_2-1
η	1	N-η	r_2
η	2	N-η	1
η	2	N-η	2
$\vdots$	$\vdots$	$\vdots$	$\vdots$
$\vdots$	$\vdots$	$\vdots$	$\vdots$
η	2	N-η	r_2-1
η	2	N-η	r_2
η	3	N-η	1
η	3	N-η	2
$\vdots$	$\vdots$	$\vdots$	$\vdots$
$\vdots$	$\vdots$	$\vdots$	$\vdots$
η	r_1	N-η	r_2-1
η	r_1	N-η	r_2
η-1	1	N-η+1	1
η-1	1	N-η+1	2
$\vdots$	$\vdots$	$\vdots$	$\vdots$
$\vdots$	$\vdots$	$\vdots$	$\vdots$
0	0	N	r_2-1
0	0	N	r_2

Table 2.4: Possible Transitions for Two-Coxian-Server Example

Transition	Rate of Transition
A customer leaves stage i, $(i \neq r_1)$ of server 1 and proceeds to stage $i + 1$.	$\mu_i a_i$
A customer leaves station 1 from stage i.	$\mu_i(1 - a_i)$
A customer leaves stage j, $(j \neq r_2)$ of server 2 and proceeds to stage $j + 1$.	$\nu_j b_j$
A customer leaves station 2 from stage j.	$\nu_j(1 - b_j)$

2.6.2.3   The Structure of the Infinitesimal Generator

The above information is sufficient to generate any row of the infinitesimal generator, Q. This matrix has the following block tridiagonal structure:

$$Q^T = \begin{pmatrix} D_0 & U_0 & & & & & \\ L_0 & D & U & & & & \\ & L & D & U & & & \\ & & \ddots & \ddots & \ddots & & \\ & & & L & D & U & \\ & & & & L & D & U_N \\ & & & & & L_N & D_N \end{pmatrix}.$$

The submatrices are dimensioned as follows: L, D, and U are of size $r_1 r_2 \times r_1 r_2$,

$$D_0 \in \Re^{r_1 \times r_1}; \qquad L_0 \in \Re^{r_1 r_2 \times r_1}; \qquad U_0 \in \Re^{r_1 \times r_1 r_2};$$

$$D_N \in \Re^{r_2 \times r_2}; \qquad L_N \in \Re^{r_2 \times r_1 r_2}; \qquad U_N \in \Re^{r_1 r_2 \times r_2}.$$

Blocks that are not specifically labeled must contain only zero elements, since the contrary would imply that two transitions from one station to the second could occur in a single time interval. Also, the interior blocks are identical, since the rates of transition are independent of the number of customers in the queue. The structure of this matrix is the single most critical factor that affects the efficiency of the numerical algorithm. Consequently, it is important to study the nonzero blocks of Q^T in some detail. We will consider the blocks L, D and U, first in the particular case where $r_1 = 4$ and $r_2 = 3$, and then in the general case. We shall not explicitly examine the boundary blocks, D_0, L_0, U_0 D_N, L_N, and U_N, but leave this for the reader.

Table 2.5: Diagonal Block Structure ($r_1 = 4$ $r_2 = 3$)

η_1	·	·	η_1	·	·	η_1	∘	∘	η_1	∘	∘
1	1	1	2	2	2	3	3	3	4	4	4
η_2	·	∘	η_2	∘	·	η_2	∘	∘	η_2	∘	·
1	2	3	1	2	3	1	2	3	1	2	3
$*$											
$b_1\nu_1$	$*$										
	$b_2\nu_2$	$*$									
$a_1\mu_1$			$*$								
	$a_1\mu_1$		$b_1\nu_1$	$*$							
		$a_1\mu_1$		$b_2\nu_2$	$*$						
			$a_2\mu_2$			$*$					
				$a_2\mu_2$		$b_1\nu_1$	$*$				
					$a_2\mu_2$		$b_2\nu_2$	$*$			
						$a_3\mu_3$			$*$		
							$a_3\mu_3$		$b_1\nu_1$	$*$	
								$a_3\mu_3$		$b_2\nu_2$	$*$

Diagonal Blocks: Table 2.5 shows the structure of diagonal blocks D when $r_1 = 4$ and $r_2 = 3$. Recall that the blocks have been transposed, so transitions out of a given state are to be found in the columns (and not the rows) corresponding to the state. This particular case ($r_1 = 4$, $r_2 = 3$) may be generalized to arbitrary r_1 and r_2. Keeping one eye on the particular case, the reader may readily observe that the structure of the diagonal blocks D for arbitrary r_1 and r_2 is given by

$$D = \begin{pmatrix} D_{11} & & & & \\ D_{21} & D_{22} & & & \\ & D_{32} & D_{33} & & \\ & & & \ddots & \ddots & \\ & & & & D_{r_1,r_1-1} & D_{r_1,r_1} \end{pmatrix} \in \Re^{r_1 r_2 \times r_1 r_2}$$

and has the following structural properties. All subblocks D_{ij} are of order r_2. The diagonal subblocks D_{ii} $(i = 1, 2, \ldots, r_1)$ contain nonzero elements only along the diagonal and subdiagonal. Each diagonal element of D_{ii} is also the diagonal element of the overall transition rate matrix and equals the negated sum of the off-diagonal elements (of Q^T) in the column in which it occurs. In general, the diagonal elements of D will differ. The remaining nonzero elements of the subblocks D_{ii} represent possible transitions among states which differ only in the service phase of the second server. The only transitions possible among these states are those which take server 2 from one service phase to the next. Consequently, the subdiagonal elements $(j, j-1)$ of any of the subblocks, D_{ii}, are given by $b_{j-1}\nu_{j-1}$, $j = 2, 3, \ldots, r_2$.

Nonzero elements of subdiagonal blocks, $D_{i,i-1}$, represent transitions from states in which server 1 is in phase $i - 1$ to states in which it is in phase i. Therefore, taking into account the ordering that is imposed on the states, it follows that the subdiagonal blocks $D_{i,i-1}$ of D must be defined by

$$D_{i,i-1} = a_{i-1}\mu_{i-1}I, \quad i = 2, 3, \ldots, r_1.$$

Subdiagonal Blocks: Table 2.6 shows the structure of subdiagonal blocks L when $r_1 = 4$ and $r_2 = 3$, where $\alpha_1 = (1 - a_1)\mu_1$, $\alpha_2 = (1 - a_2)\mu_2$; $\alpha_3 = (1 - a_3)\mu_3 = \mu_3$. The structure of subdiagonal blocks L for arbitrary r_1 and r_2 is

$$L = \begin{pmatrix} L_{11} & L_{12} & L_{13} & \cdots & L_{1r_1} \\ 0 & 0 & 0 & \cdots & 0 \\ \vdots & \vdots & \vdots & \ddots & \vdots \\ 0 & 0 & 0 & \cdots & 0 \end{pmatrix} \in \Re^{r_1 r_2 \times r_1 r_2}$$

All subblocks L_{1j} are of order r_2. The nonzero elements of L denote transitions that occur when a customer finishes service at any phase of server 1 and joins the queue at server 2. The next customer at server 1 immediately enters service phase 1, and hence all subblocks L_{ij} for $i > 1$ must contain only zero elements. In addition, the arrival of a customer to server 2 does not alter the service phase of that server; thus, only the diagonal elements of the blocks L_{1j} are nonzero. These blocks are given by

$$L_{1j} = (1 - a_j)\mu_j I, \quad j = 1, 2, \ldots, r_1.$$

Table 2.6: Subdiagonal Block Structure ($r_1 = 4$, $r_2 = 3$)

η_1+1	·	·	η_1+1	·	·	η_1+1	·	·	η_1+1	·	·
1	1	1	2	2	2	3	3	3	4	4	4
η_2-1	·	·	η_2-1	·	·	η_2-1	∘	∘	η_2-1	∘	∘
1	2	3	1	2	3	1	2	3	1	2	3

Superdiagonal Blocks: Table 2.7 shows the structure of superdiagonal blocks U when $r_1 = 4$ and $r_2 = 3$, where $\beta_1 = (1 - b_1)\nu_1$, $\beta_2 = (1 - b_2)\nu_2$; $\beta_3 = (1 - b_3)\nu_3 = \nu_3$. The structure of superdiagonal blocks U for arbitrary r_1 and r_2 is

$$U = \begin{pmatrix} U' & & & & \\ & U' & & & \\ & & \ddots & & \\ & & & U' & \\ & & & & U' \end{pmatrix} \in \Re^{r_1 r_2 \times r_1 r_2}$$

in which all subblocks are of order r_2. The ij^{th} element of each of the diagonal submatrices is given by

$$U'_{ij} = (1 - b_j)\nu_j\delta_{i1}$$

where δ_{ij} is the Kronecker delta. This structure is a result of the fact that nonzero elements of U indicate transitions that occur when a customer finishes service at any phase of server 2 and proceeds to join the queue at server 1. The particular pattern of the nonzero elements arises because server 2 must recommence at service phase 1 while the service phase of server 1 remains unchanged. In conclusion, the transition rate matrix for the two-station network has a unique block tridiagonal structure and is easy to generate.

Table 2.7: Superdiagonal Block Structure ($r_1 = 4$, $r_2 = 3$)

η_1-1	·	·	η_1-1	·	·	η_1-1	·	·	η_1-1	·	·
1	1	1	2	2	2	3	3	3	4	4	4
η_2+1	·	·	η_2+1	·	·	η_2+1	·	·	η_2+1	·	·
1	2	3	1	2	3	1	2	3	1	2	3
β_1	β_2	β_3									
			β_1	β_2	β_3						
						β_1	β_2	β_3			
									β_1	β_2	β_3

2.6.2.4   Implementation and Operation Count

We have seen that the coefficient matrix is such that any row can be generated
quickly and efficiently. Our task now is to determine an LU factorization of
this matrix. It is necessary to obtain and store in memory only the reduced
upper triangular portion, and this may be generated in a row-by-row fashion as
described previously. The structure of the matrix dictates that a two-dimensional
storage array is most suitable. The number of rows in this array should be equal
to the number of states in the queueing network ($n = (N-1)r_1r_2 + r_1 + r_2$),
and the number of columns should be sufficiently large to allow the storage of
any elements to the right of the diagonal element that might become nonzero.
From the structure of the matrix, this is equal to $r_1r_2 + r_2$. The diagonal element
of a reduced row will always be the first element stored in the corresponding
row of the array. No additional storage is required for the solution vector, since
the components, obtained in reverse order, may simply overwrite corresponding
components in one column of the array used to store the reduced matrix.

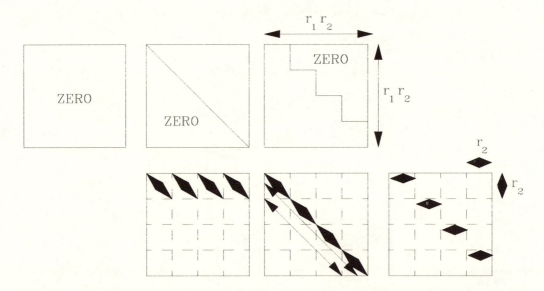

Figure 2.11: Block-by-block reduction.

Figure 2.11 is useful in determining the number of operations involved in reducing a block of rows to triangular form. It is assumed that all rows prior to
the designated block have already been reduced. It is apparent that the first
r_2 rows of the block will require considerably more operations than the remaining $r_2(r_1 - 1)$ rows. It is also apparent that once the reduction in any row
begins, all zero elements between the element being reduced and the diagonal
element will experience fill-in. Consequently, in each of the first r_2 rows, r_1r_2

elements will eventually have to be reduced before the row conforms to the upper triangular form. The reduction of each of these elements requires one division and at most $(r_1r_2 + r_2 - 1)$ multiplications and additions. Therefore a total of $r_2 \times r_1r_2 \times (r_1r_2 + r_2)$ multiplications and divisions is required to bring the first r_2 rows to upper triangular form. Each of the remaining $(r_1r_2 - r_2)$ rows may be reduced in $r_2 \times (r_1r_2 + r_2)$ multiplications and divisions. The total number of multiplications and divisions for each section of the reduction is then strictly less than

$$r_2 \times r_1r_2 \times (r_1r_2 + r_2) + (r_1r_2 - r_2) \times r_2 \times (r_1r_2 + r_2).$$

In the back-substitution stage each element of the solution vector may be obtained with a maximum of $(r_1r_2 + r_2 - 1)$ multiplications and additions and one division. A total of $r_1r_2(r_1r_2 + r_2)$ multiplications and divisions is therefore sufficient to obtain all components of the solution vector corresponding to any section. An upper bound for the operation count for the entire algorithm is then given by

$$(N + 1)\{2r_1^2r_2^3 + r_1^2r_2^2 + r_1r_2^3 + r_1r_2^2 - r_2^3\}$$

In the special case when both servers are Coxians of order 2 ($r_1 = r_2 = 2$), the operation count is strictly less than $96(N + 1)$.

2.7 Stability, Conditioning, and Error Analysis

2.7.1 Stable Algorithms and Well-Conditioned Problems

We begin by examining the difference between "problems" and "algorithms" in the setting of numerical computations. The figures that follow are inspired by those of G.W. Stewart [150]. When we are given a *problem*, we are essentially given data that must be manipulated in such a way that an "answer" is found. The answer is also data. It is assumed that the manipulation is such that no inaccuracies or faults arise. The set of all possible inputs is called a *data set*, which we denote by D, and the set of all outputs is called a *result set*, denoted R. We may view a problem as a transformation or function that operates on input to produce output. We shall therefore denote a problem by the symbol $\mathcal{F}$; the input data shall be denoted x, a vector quantity, and the result is then $\mathcal{F}(x)$, a second vector quantity, not necessarily the same length as x. For certain types of problems it turns out that small differences in the input data may result in large differences in the results. Such problems are called *ill-conditioned* (see Figure 2.12). When small differences in the data lead to small differences in the results, the problem is said to be *well-conditioned*. Note that ill-conditioning and well-conditioning are characteristics of a *problem* rather than of an algorithm.

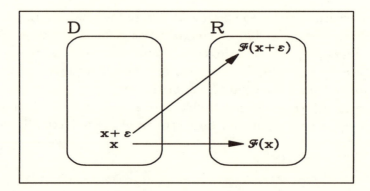

Figure 2.12: An ill-conditioned problem. $\mathcal{F}(x + \epsilon)$ is not close to $\mathcal{F}(x)$, even though $x + \epsilon \approx x$.

An *algorithm*, on the other hand, is a computer-based procedure that purports to calculate the solution to a problem — or an approximation to the solution. However, even if the problem involves only the basic operations of adding and multiplying, an algorithm will not exactly mimic the operations of the problem. Addition and multiplication on a computer are *not* the same as regular (mathematical) addition and multiplication. The computer-based implementations of the basic arithmetic operations usually generate an error, so that only an approximation to the result is obtained. Additionally, the implementations of multiplication and division are neither associative nor distributive, and thus the usual rules of algebra do not apply. We shall use the notation F to denote the algorithm that corresponds to the problem $\mathcal{F}$. We have now two concerns about numerical problem solving on a computer:

1. The problem $\mathcal{F}$ may be well-conditioned or ill-conditioned.

2. The algorithm F may or may not be a "good" algorithm.

In the second case, we use the terms *stable* and *unstable* to describe algorithms.

It is important to keep the concept of the *stability of an algorithm* distinct from the concept of conditioning (which depends only on the problem). This means that we cannot simply say that an algorithm is stable only if it always produces an accurate solution. There may be instances when a perfectly good algorithm is used to obtain the solution to an ill-conditioned problem; its failure to compute an accurate result should not be taken to imply that the algorithm is unstable. The definition of stability must be independent of the conditioning of the problem. An algorithm should not be expected to compute an accurate answer to an ill-conditioned problem. However, it should not introduce unacceptable inaccuracies of its own. When applied to a well-conditioned problem, a stable algorithm will always give an accurate result.

The usual definition is that an algorithm is termed stable if the algorithm and its implementation on a computer yield a solution that is near the exact solution of a slightly perturbed problem (i.e., slightly perturbed input data), as shown in Figure 2.13. In other words, F is a stable algorithm if for any data $x \in D$ there exists an $x + \xi \in D$ with $x + \xi \approx x$ such that $\mathcal{F}(x + \xi) \approx \mathsf{F}(x)$.

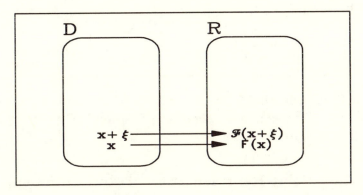

Figure 2.13: A stable algorithm. There exists an $x + \xi$ close to x such that $\mathcal{F}(x + \xi) \approx \mathsf{F}(x)$.

It follows that if $\mathcal{F}$ is a well-conditioned problem and F is a stable algorithm, then starting from approximate data $x + \epsilon$, the computed result will be $\mathsf{F}(x + \epsilon)$, and this will be near $\mathcal{F}(x + \xi)$ for some $x + \xi \approx x + \epsilon$. Since $x + \xi \approx x + \epsilon \approx x$ and $\mathcal{F}$ is well-conditioned, it follows that $\mathcal{F}(x + \xi) \approx \mathcal{F}(x)$. Therefore we should expect that $\mathsf{F}(x + \epsilon) \approx \mathcal{F}(x)$. Indeed, writing

$$\|\mathsf{F}(x + \epsilon) - \mathcal{F}(x)\| \leq \|\mathsf{F}(x + \epsilon) - \mathsf{F}(x)\| + \|\mathsf{F}(x) - \mathcal{F}(x + \xi)\| + \|\mathcal{F}(x + \xi) - \mathcal{F}(x)\|$$

it may be seen that the three norms on the right-hand side are all small — the first because F is a continuous algorithm, the second because it is stable, and the third because the problem $\mathcal{F}$ is well-conditioned. We may conclude that a stable algorithm will always produce an accurate result to a well-conditioned problem. Indeed, given that $x \approx x + \epsilon \approx x + \xi$, then $\mathsf{F}(x + \epsilon) \approx \mathcal{F}(x + \xi)$, since the algorithm is stable, and $\mathcal{F}(x + \xi) \approx \mathcal{F}(x)$, since the algorithm is well-conditioned. Therefore, $\mathsf{F}(x + \epsilon) \approx \mathcal{F}(x)$.

When $\mathcal{F}$ is an ill-conditioned problem, we cannot guarantee that $\mathcal{F}(x + \xi) \approx \mathcal{F}(x)$ even though $x + \xi \approx x$ (see Figure 2.14). However, we may still assert that the computed result is the exact solution of a slightly perturbed problem, i.e., that $\mathsf{F}(x + \epsilon) \approx \mathcal{F}(x + \xi)$.

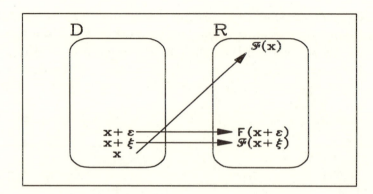

Figure 2.14: Stable algorithm and ill-conditioned problem. Given $x \approx x + \epsilon \approx x + \xi$, $\mathsf{F}(x + \epsilon) \approx \mathcal{F}(x + \xi)$, since algorithm is stable; $\mathcal{F}(x + \xi) \not\approx \mathcal{F}(x)$, since algorithm is ill-conditioned; therefore, $\mathsf{F}(x + \epsilon) \not\approx \mathcal{F}(x)$.

2.7.2   Floating-Point Representation of Real Numbers

To progress further, we need more information concerning the manner in which real numbers are stored in a computer. Floating-point representation is the usual way in which this is accomplished. Any real number x may be written in floating-point form as $x = a\beta^\gamma$, where a is the *mantissa*, β is the number *base*, and γ is an integer *exponent*. Thus, the following are all valid representations of the decimal number 1.234:

$$1234.0 \times 10^{-3}, \quad 123.4 \times 10^{-2}, \quad 12.34 \times 10^{-1}, \quad 1.234 \times 10^0,$$

$$.1234 \times 10^1, \quad .01234 \times 10^2, \quad .001234 \times 10^3, \quad \text{etc.}$$

To have a unique computer representation of a real number x, we insist that the mantissa satisfy $1/\beta \le a < 1$ when $x \neq 0$. The floating-point numbers thus obtained are said to be normalized. Hence $.1234 \times 10^1$ is the unique normalized floating-point representation of the decimal number 1.234. In what follows we shall consider only normalized floating-point numbers.

Both the mantissa and the exponent of x are stored in the computer; the base β is, of course, available implicitly. Because of the finite nature of the computer's ability to represent numbers, only a certain number of digits of the mantissa can be stored. This number is referred to as the *precision* of the representation. We shall denote it by t. To be able to refer to the digits of the mantissa explicitly, we write $a = \pm c_1 c_2 \ldots c_t$. Similarly, the range of values that the exponent can assume is also limited. We shall denote the lower and upper bounds of the exponent by MIN and MAX respectively, so that permissible exponent values satisfy $MIN \le \gamma \le MAX$. The sizes of t, MIN, and MAX depend on the characteristics of the particular computer being used and on whether single-precision or double-precision representation is used. Some typical values are shown in Table 2.8.

Table 2.8: Floating-Point Representations

Computer		β	t	MIN	MAX
IBM 370, 308x	single-precision	16	6	-64	63
	double-precision	16	14	-64	63
DEC VAX 11/780, 11/750		2	24	-128	127
Cray-1		2	48	-16384	8191

A given set of values $\{\beta, t, MIN, MAX\}$ defines a finite set of floating-point numbers. We shall denote this set of floating-point values, together with the number zero, by F. Thus $F = \{a\beta^\gamma \mid a = \pm c_1 c_2 \ldots, c_t, \ 0 \le c_i < \beta, \ c_1 \ne 0, \ MIN \le \gamma \le MAX\} \cup \{0\}$. The number of elements in this set is given by

$$2(\beta - 1)\beta^{t-1}(MAX - MIN + 1) + 1.$$

These are the only numbers the computer has for representing the infinite set of real numbers. Each floating-point number must therefore represent an interval of real numbers. A number in the interval is either rounded or chopped to obtain the floating-point number that is used to represent it. Let fl(x) denote the floating-point number that is used to represent the real number x. Obviously fl(x) $\in F$. We have

$$\text{fl}(x) = \begin{cases} \text{nearest } \xi \in F \text{ to } x \text{ if rounding is used. In the case of a tie, round away from zero.} \\ \text{nearest } \xi \in F \text{ to } x \text{ satisfying } |\xi| \le |x| \text{ if chopping is used.} \end{cases}$$

The spacing between adjacent floating-point numbers is not the same for all pairs, but instead depends on the exponent of adjacent values; the larger the exponent value, the greater the spacing.

When the real number x is greater than the largest floating-point number that can be represented, i.e.,

$$|x| > (1 - \beta^{-t})\beta^{MAX},$$

then *overflow* is said to have occurred, and the computer will generally stop executing. On the other hand, when

$$0 < |x| < (1/\beta)\beta^{MIN},$$

underflow occurs. Often, at this point fl(x) is taken to be zero and execution continues. However, the user should be wary in these instances, for underflow often indicates that something somewhere is amiss and should be studied carefully. It is worthwhile stressing at this point that overflow and underflow conditions depend uniquely on the values of the exponent and do not concern the *precision* of the representation.

Generally, there is an error made when using fl(x) to represent the real number x. It is sometimes surprising to realize that apparently "nice" numbers like 0.1 have nonterminating binary representations and cannot be exactly represented. We conclude by providing bounds on these errors.

Lemma 2.1 *The relative error made in representing the real number x by* fl(x) *satisfies*

$$\left|\frac{\text{fl}(x) - x}{x}\right| \leq .5\beta^{1-t} \text{ when rounding is used, and}$$

$$\left|\frac{\text{fl}(x) - x}{x}\right| \leq \beta^{1-t} \text{ when chopping is used.}$$

We shall assume that rounding is used. The quantity $.5\beta^{1-t}$ is called the *unit round-off error*. Let ρ be the relative error, i.e.,

$$\rho = \frac{\text{fl}(x) - x}{x}.$$

It then follows that fl(x) = $x(1 + \rho)$, with $|\rho| \leq .5\beta^{1-t}$, and we may claim that the error made in representing x by fl(x) is small.

2.7.3 Backward Error Analysis

Armed with the definition of stability for numerical algorithms given in Section 2.7.1 and the representation of floating-point numbers of Section 2.7.2, we now proceed to examine a technique that is often used to determine whether a particular algorithm satisfies the stability condition. The particular technique we shall examine was pioneered by Wilkinson [178] and is referred to as a *backward error analysis*. It is not the only possibility, but it is perhaps the simplest. We illustrate it first to show that simple arithmetic operations are stable operations, and then we demonstrate its application to the computation of inner products. In the next section, we shall use a backward error analysis to examine the stability of Gaussian elimination.

Let us begin by considering the product of two floating-point numbers y_1, $y_2 \in F$. We denote the computed result by fl($y_1 y_2$) $\in F$. This operation is carried out by adding the exponents, multiplying the two mantissas in a double-precision accumulator (the mantissa of the product may need to be shifted one place to the left if its first digit becomes 0; in this case the exponent of the product must be decreased by 1), and finally rounding the mantissa to single-precision. Since the only error involved in this operation is the final representation error (incurred by rounding the result in the double-precision accumulator to single-precision), we must have

$$\mathrm{fl}(y_1 y_2) = y_1 y_2 (1 + \rho) \quad \text{where } |\rho| \leq .5\beta^{1-t}.$$

We may therefore claim that floating-point multiplication is a stable algorithm for computing the product of two floating-point numbers, because the computed result $y_1 y_2 (1 + \rho)$ is the exact product of a slightly perturbed problem (viz., the exact product of y_1 and $y_2(1 + \rho)$ or of $y_1(1 + \rho)$ and y_2, or $y_1(1 + \rho)^{.5}$ and $y_2(1 + \rho)^{.5}$, etc.). Similarly we may show that

$$\mathrm{fl}(y_1 / y_2) = y_1 / y_2 (1 + \rho) \quad \text{where } |\rho| \leq .5\beta^{1-t}.$$

The bounds for floating-point addition and subtraction are slightly different, since in general only a single-precision accumulator is used. It may be shown [177] that

$$\mathrm{fl}(a \pm b) = (a \pm b)(1 + \rho'), \quad \text{where } |\rho'| \leq .55\beta^{1-t}.$$

To be able to use the same bound for all the basic arithmetic operations, we need to alter the notation slightly. Furthermore, since all our numerical examples are in decimal, we shall take $\beta = 10$. Introducing μ, a quantity of order unity, we have

$$\left. \begin{array}{rcl} \mathrm{fl}(xy) &=& xy(1 + \rho) \\ \mathrm{fl}(x/y) &=& x/y(1 + \rho) \\ \mathrm{fl}(x \pm y) &=& (x \pm y)(1 + \rho) \end{array} \right\} \quad \text{where } |\rho| \leq \mu 10^{-t} \text{ in all cases.}$$

As for floating-point multiplication, we therefore see that floating-point division, addition, and subtraction are stable algorithms for performing these operations on floating-point numbers.

We now consider a backward error analysis applied to the computation of the inner product $x^T y$ of two vectors x and y. This operation is often required in numerical computations. The algorithm usually adopted is

$$\begin{array}{ll} \text{Set} & s_1 = x_1 y_1. \\ \text{For} & i = 2, 3, \ldots, n \text{ do} \\ \bullet & s_i = s_{i-1} + x_i y_i. \end{array}$$

At the end of these operations we have $s_n = x^T y$, if the operations are carried out in exact arithmetic. On a computer, what we calculate is

Set $s_1 = \mathrm{fl}(x_1 y_1)$.
For $i = 2, 3, \ldots, n$ do
 • $s_i = \mathrm{fl}(s_{i-1} + \mathrm{fl}(x_i y_i))$.

Let us assume that the components of x and y are already stored in the computer, i.e., $x_i, y_i \in F$ for $i = 1, 2, \ldots, n$. Then

$$
\begin{aligned}
s_1 &= \mathrm{fl}(x_1 y_1) = x_1 y_1 (1 + \rho_1) \quad \text{where } |\rho_1| \le \mu 10^{-t} \\
s_2 &= \mathrm{fl}(x_1 y_1 (1 + \rho_1) + x_2 y_2 (1 + \rho_2)) \\
&= x_1 y_1 (1 + \rho_1)(1 + \gamma_2) + x_2 y_2 (1 + \rho_2)(1 + \gamma_2) \\
&\quad \text{where } |\rho_1|, |\rho_2|, |\gamma_2| \le \mu 10^{-t}.
\end{aligned}
$$

Continuing in this manner, we eventually obtain

$$
\begin{aligned}
\mathrm{fl}(x^T y) &= \mathrm{fl}\left(\sum_{i=1}^{n} x_i y_i \right) \\
&= x_1 y_1 (1 + \rho_1)(1 + \gamma_2)(1 + \gamma_3) \ldots (1 + \gamma_n) \\
&+ x_2 y_2 (1 + \rho_2)(1 + \gamma_2)(1 + \gamma_3) \ldots (1 + \gamma_n) \\
&+ x_3 y_3 (1 + \rho_3)(1 + \gamma_3)(1 + \gamma_4) \ldots (1 + \gamma_n) \\
&+ \cdots \\
&+ x_{n-1} y_{n-1} (1 + \rho_{n-1})(1 + \gamma_{n-1})(1 + \gamma_n) \\
&+ x_n y_n (1 + \rho_n)(1 + \gamma_n),
\end{aligned}
$$

where $|\rho_i|, |\gamma_i| \le \mu 10^{-t}$ for $i = 1, 2, \ldots, n$. Thus $\mathrm{fl}(x^T y) = \sum_{i=1}^{n} x_i y_i (1 + \delta_i)$, where

$$
(1 + \delta_i) = (1 + \rho_i) \prod_{k=i}^{n} (1 + \gamma_k)
$$

and $\gamma_1 = 0$. To proceed further, we need the following two lemmas:

Lemma 2.2 (Stewart [150]) *If $|\alpha_k| \le \mu 10^{-t}$ for $k = 1, 2, \ldots, n$, then there exists an α such that*

$$
(1 + \alpha)^n = \prod_{k=1}^{n} (1 + \alpha_k)
$$

and

$$
|\alpha| \le \mu 10^{-t}.
$$

Lemma 2.3 (Stewart [150]) *Suppose $n\mu 10^{-t} < .1$ (a reasonable assumption). If $|\alpha| \le \mu 10^{-t}$, then*

$$
(1 + \alpha)^n = 1 + n\alpha'
$$

where

$$
|\alpha'| \le \mu' 10^{-t} \quad \text{and} \quad \mu' = 1.06\mu.
$$

It follows from these lemmas that

$$\text{fl}(x^T y) = \sum_{i=1}^{n} x_i y_i (1 + \delta_i), \quad \text{where } |\delta_i| \leq (n - i + 2)\mu' 10^{-t}.$$

To complete the backward error analysis, we have

$$\text{fl}(x^T y) = \sum_{i=1}^{n} x_i (y_i + y_i \delta_i) = \sum_{i=1}^{n} x_i (y_i + \epsilon_i) = x^T (y + \epsilon),$$

where the ith component of ϵ ($\epsilon_i = y_i \delta_i$) is small compared to the ith component of y. Therefore, we may claim that the computed result is the exact result of a slightly perturbed problem, and hence, the algorithm is a stable algorithm for computing the inner product of two vectors.

We are now in a position to apply a backward error analysis to investigate the stability of Gaussian elimination for computing the stationary probability vector of irreducible Markov chains.

2.7.4 Error Analysis for Gaussian Elimination

In examining the stability of Gaussian elimination, we shall proceed by steps. We shall first consider the reduction phase and show that the computed error bound for this part of the algorithm depends on a quantity called the *growth factor*. In a second part we present a number of results that bound this growth factor in specific cases, including the case of infinitesimal generator matrices. Finally, we incorporate the back-substitution phase into the analysis to determine complete bounds for the errors in Gaussian elimination. We shall see that Gaussian elimination without pivoting is a stable way to compute the stationary probability vector of an irreducible Markov chain.

2.7.4.1 The Elimination Phase

As was shown previously, the reduction phase of Gaussian elimination applied to the matrix A produces a sequence of matrices $A = A^{(0)}, A^{(1)}, A^{(2)}, \ldots, A^{(n-1)} \equiv U$, the last of which is upper triangular. During the kth step of this reduction, the following operations are performed on $A^{(k-1)}$ to obtain $A^{(k)}$:

- Compute the multipliers, $m_{ik} = a_{ik}^{(k-1)} / a_{kk}^{(k-1)} : \quad i = k+1, k+2, \ldots, n$. We get

$$\begin{aligned} \text{fl}(m_{ik}) &= \text{fl}\left(a_{ik}^{(k-1)} / a_{kk}^{(k-1)}\right) \\ &= \left(a_{ik}^{(k-1)} / a_{kk}^{(k-1)}\right)(1 + \delta_1) \quad \text{with } |\delta_1| \leq \mu 10^{-t}. \end{aligned}$$

- Compute the elements of $A^{(k)}$: $a_{ij}^{(k)}$ for $i > k$, $j \geq k$. This gives

$$
\begin{aligned}
a_{ij}^{(k)} &= \text{fl}\left(a_{ij}^{(k-1)} - m_{ik}a_{kj}^{(k-1)}\right) \\
&= \left[a_{ij}^{(k-1)} - m_{ik}a_{kj}^{(k-1)}(1+\delta_2)\right](1+\delta_3) \\
&= \left[a_{ij}^{(k-1)} - \left(a_{ik}^{(k-1)}/a_{kk}^{(k-1)}\right)(1+\delta_1)a_{kj}^{(k-1)}(1+\delta_2)\right](1+\delta_3) \\
&= a_{ij}^{(k-1)}(1+\delta_3) - \left(a_{ik}^{(k-1)}/a_{kk}^{(k-1)}\right)a_{kj}^{(k-1)}(1+\delta_1)(1+\delta_2)(1+\delta_3),
\end{aligned}
$$

with $|\delta_1|, |\delta_2|, |\delta_3| \leq \mu 10^{-t}$.

This may be written as

$$
\begin{aligned}
a_{ij}^{(k)} &= a_{ij}^{(k-1)} - \left(a_{ik}^{(k-1)}/a_{kk}^{(k-1)}\right)a_{kj}^{(k-1)} + a_{ij}^{(k-1)}\delta_3 \\
&\quad - \left(a_{ik}^{(k-1)}/a_{kk}^{(k-1)}\right)a_{kj}^{(k-1)}\left[(1+\delta_1)(1+\delta_2)(1+\delta_3)-1\right].
\end{aligned}
$$

Rearranging, we find

$$
\begin{aligned}
a_{ij}^{(k-1)} - \left(a_{ik}^{(k-1)}/a_{kk}^{(k-1)}\right)a_{kj}^{(k-1)} &= a_{ij}^{(k)} - a_{ij}^{(k-1)}\delta_3 + \left(a_{ik}^{(k-1)}/a_{kk}^{(k-1)}\right)a_{kj}^{(k-1)} \times \\
&\quad \left[(1+\delta_1)(1+\delta_2)(1+\delta_3)-1\right].
\end{aligned}
$$

i.e.,

$$
a_{ij}^{(k-1)} - \left(a_{ik}^{(k-1)}/a_{kk}^{(k-1)}\right)a_{kj}^{(k-1)} = a_{ij}^{(k)} - \epsilon_{ij}^{(k)}, \tag{2.12}
$$

where

$$
\epsilon_{ij}^{(k)} = a_{ij}^{(k-1)}\delta_3 - \left(a_{ik}^{(k-1)}/a_{kk}^{(k-1)}\right)a_{kj}^{(k-1)}\left[(1+\delta_1)(1+\delta_2)(1+\delta_3)-1\right].
$$

By applying Lemma 2.2 followed by Lemma 2.3, the term inside the brackets simplifies to 3δ. Since the multipliers are not greater than 1 in absolute value, we obtain the following bound:

$$
|\epsilon_{ij}^{(k)}| \leq |a_{ij}^{(k-1)}|\,\mu 10^{-t} + (1)|a_{kj}^{(k-1)}|\,3 \times 1.06\mu 10^{-t}.
$$

Let $\beta_k = \max_k\{|a_{ij}^{(k)}|;\ i,j = 1,2,\ldots,n\}$. Then

$$
|\epsilon_{ij}^{(k)}| \leq 4.18(\mu 10^{-t})\beta_{k-1}: \qquad i > k,\ j \geq k.
$$

Thus the perturbation incurred in moving from $A^{(k-1)}$ to $A^{(k)}$ is given by the matrix $E^{(k)}$ whose first k rows and first $(k-1)$ columns are zero and whose remaining elements are given by $\epsilon_{ij}^{(k)}$ $(i > k,\ j \geq k)$.

We have thus been able to determine bounds on the magnitude of the perturbation in step k of the decomposition stage. Note that this bound is given in terms of $\max_k |a_{ij}^{(k)}|$. We shall return to this point momentarily. What we must do now is bound the total perturbation made during the decomposition. Specifically, we need bounds for the matrix E given by

$$\hat{L}\hat{U} = A + E,$$

where $\hat{L}\hat{U}$ are the approximate lower (unit) and upper triangular matrices computed by the algorithm. In the current notation, this becomes

$$\hat{L}A^{(n-1)} = A^{(0)} + E.$$

Referring back to equation (2.12),

$$a_{ij}^{(k-1)} - m_{ik}a_{kj}^{(k-1)} = a_{ij}^{(k)} - \epsilon_{ij}^{(k)} \quad i > k, \; j \geq k$$
$$a_{ij}^{(k-1)} = a_{ij}^{(k)} \quad \text{otherwise.}$$

Recalling the definition of M_k from equation (2.1), we may write this in matrix form as

$$M_k A^{(k-1)} = A^{(k)} - E^{(k)}, \qquad (2.13)$$
$$\text{where} \quad \left(M_k A^{(k-1)}\right)_{ij} = a_{ij}^{(k-1)} - m_{ik}a_{kj}^{(k-1)}.$$

From (2.13) we have

$$A^{(n-1)} = M_{n-1}A^{(n-2)} + E^{(n-1)},$$
$$\text{i.e.,} \quad M_{n-1}^{-1}A^{(n-1)} = A^{(n-2)} + M_{n-1}^{-1}E^{(n-1)}$$
$$= A^{(n-2)} + E^{(n-1)}$$

since $M_{n-1}^{-1}E^{(n-1)} = E^{(n-1)}$. In fact, since the first k rows of $E^{(k)}$ are zero,

$$M_l^{-1}E^{(k)} = E^{(k)} \quad \text{for } l = 1, 2, \ldots, k.$$

Similarly

$$A^{(n-2)} = M_{n-2}A^{(n-3)} + E^{(n-2)},$$

and substituting into the previous equation, we have

$$M_{n-1}^{-1}A^{(n-1)} = M_{n-2}A^{(n-3)} + E^{(n-2)} + E^{(n-1)}$$
$$M_{n-2}^{-1}M_{n-1}^{-1}A^{(n-1)} = A^{(n-3)} + M_{n-2}^{-1}E^{(n-1)} + M_{n-2}^{-1}E^{(n-1)}$$
$$M_{n-2}^{-1}M_{n-1}^{-1}A^{(n-1)} = A^{(n-3)} + E^{(n-2)} + E^{(n-1)}.$$

Continuing in this fashion, at the next step we have

$$M_{n-3}^{-1}M_{n-2}^{-1}M_{n-1}^{-1}A^{(n-1)} = A^{(n-4)} + E^{(n-3)} + E^{(n-2)} + E^{(n-1)},$$

and so on until

$$M_1^{-1} \ldots M_{n-2}^{-1} M_{n-1}^{-1} A^{(n-1)} = A^{(0)} + E^{(1)} + E^{(2)} + \cdots + E^{(n-1)}.$$

Letting

$$\hat{L} = M_1^{-1} M_2^{-1} \cdots M_{n-1}^{-1}$$

and

$$E = E^{(1)} + E^{(2)} + \cdots + E^{(n-1)}$$

we have the required form

$$\hat{L} A^{(n-1)} = A^{(0)} + E.$$

The elements e_{ij} of $E = \sum_{k=1}^{n-1} E^{(k)}$ satisfy the bounds

$$|e_{ij}| \leq 4.18(i-1) \max_{ijk} |a_{ij}^{(k)}| (\mu 10^{-t}).$$

It only remains to bound the term $\max_{ijk} |a_{ij}^{(k)}|$. Let us introduce the *growth factor* g_A, defined by:

$$g_A = \max_{ijk} |a_{ij}^{(k)}| / \max_{ij} |a_{ij}|.$$

Then $|e_{ij}| \leq 4.18(i-1) \max_{ij} |a_{ij}| g_A (\mu 10^{-t})$. The factor $(i-1)$ is an overestimation for most (i, j). It is due to the number of arithmetic operations that take place at position (i, j). When the matrix is large and sparse, or has some other special structure such as banded, the number of arithmetic operations may be significantly less.

To summarize, after the elimination stage the computed decomposition $\hat{L}\hat{U}$ is equal to the exact solution of the perturbed system $A + E$. The magnitude of these perturbations depends on the quantity g_A; if g_A is small, the perturbation will be small and the LU decomposition will be stable. In other words, a rounding error analysis for Gaussian elimination reduces to estimating the growth factor g_A.

2.7.4.2 Growth Factors

We now seek bounds for the growth factors, particularly in the case of infinitesimal generators. The following general bounds on growth factors were obtained by Wilkinson [177] and others.

Gaussian Elimination with Complete Pivoting:

$$g_A \leq \left[n 2^1 3^{1/2} 4^{1/3} \cdots n^{1/n-1} \right]^{1/2}, \quad \text{for all nonsingular } A.$$

Note that the right-hand side increases rather slowly with n, and therefore we may conclude that Gaussian elimination with full pivoting is unconditionally stable.

Gaussian Elimination with Partial Pivoting:

$$g_A \leq 2^{n-1}, \quad \text{for all nonsingular } A.$$

Note that here the right-hand side may be large, and therefore we cannot say that, with only partial pivoting, Gaussian elimination is unconditionally stable. However, in practice, partial pivoting is usually adequate.

Symmetric, Positive Definite Matrices:

$$g_A = 1$$

This is for sequential elimination (without pivoting). Gaussian elimination is unconditionally stable for any symmetric, positive definite matrix.

Strictly Diagonally Dominant Matrices:

$$g_A \leq 2$$

A is strictly (column) diagonally dominant if

$$|a_{jj}| > \sum_{i, i \neq j} |a_{ij}|, \qquad 1 \leq j \leq n.$$

Gaussian elimination is therefore unconditionally stable for the class of strictly diagonally dominant matrices. Note also that symmetric positive definite and strictly diagonally dominant matrices preserve these properties under symmetric permutations of A, and consequently, elimination in any order will also be stable.

Diagonally Dominant M-Matrices:

$$g_A = 1$$

There are many possible characterizations of M-matrices [10]. One commonly used definition is that a matrix is an M-matrix if all of its off-diagonal elements are nonpositive and all eigenvalues have nonnegative real parts. Another and equivalent definition is that a matrix whose diagonal elements are nonnegative and whose off-diagonal elements are nonpositive is an M-matrix. The matrix $-Q$ is an M-matrix, although not strictly a diagonally dominant M-matrix. Funderlic, Neumann and Plemmons [48] show that for diagonally dominant M-matrices, $g_A = 1$.

Infinitesimal Generator Matrices:

$$g_A = 1$$

When applied to infinitesimal generator matrices obtained from irreducible Markov chains, each step of Gaussian elimination forces the negative diagonal elements closer to zero. This is sufficient to show that the growth factor cannot exceed 1. We need only consider the effect of the elementary operation that adds a multiple of row 1 into row 2 to eliminate the element in position (2,1) of the matrix Q. The properties observed in this case extend in an obvious fashion to the elimination of any element at any step of the decomposition. Initially we have

$$
\begin{array}{lllllll}
\text{Row 1}: & q_{11} & q_{12} & q_{13} & \cdots & q_{1n} & \sum_{j=1}^{n} q_{1j} = 0 \\
\text{Row 2}: & q_{21} & q_{22} & q_{23} & \cdots & q_{2n} & \sum_{j=1}^{n} q_{2j} = 0
\end{array}
$$

The element q_{11} is strictly less than zero (if it is equal to zero, the Markov chain is reducible) and is equal to the negated sum of the off-diagonal elements in row 1; none of the off-diagonal elements can be less than zero. After the elimination of q_{21}, the second row is

$$
\text{Row 2}: \quad 0, \quad q_{22} - \frac{q_{21}}{q_{11}}q_{12}, \quad q_{23} - \frac{q_{21}}{q_{11}}q_{13}, \quad \ldots, \quad q_{2n} - \frac{q_{21}}{q_{11}}q_{1n}
$$

We have previously seen (Section 2.5) that the zero row-sum property is unaffected by this operation. Observe that the negative diagonal element q_{22} is modified by subtracting a *negative* quantity from it. This element therefore increases in algebraic value. Also, since

$$
\frac{|q_{12}|}{|q_{11}|} \leq 1 \quad \text{and} \quad |q_{22}| \geq |q_{21}|,
$$

the new diagonal element, $q_{22} - \frac{q_{21}}{q_{11}}q_{12}$, cannot become positive. Nor can it be equal to zero (unless $n = 2$), for then the sum of the off-diagonal elements would also be zero and the matrix would be reducible. It follows that the absolute values of the diagonal elements decrease during the LU decomposition. Now consider the off-diagonal elements. These positive elements are modified by the addition of a zero or positive quantity. They therefore do not decrease in magnitude. However, because of the zero row-sum property, none of the off-diagonal elements can exceed the absolute value of the diagonal element, which in turn has decreased in magnitude. Recalling that the growth factor is defined as the maximum (absolute) size attained by any element during the reduction divided by the maximum element in the initial matrix, it necessarily follows that the growth factor for infinitesimal generator matrices is 1.

We have argued on the basis of Gaussian elimination applied to the matrix Q. We might well ask whether the same result is true if applied to the matrix Q^T.

The answer is yes, as we now show. We consider only the first step, since all following steps behave in a similar manner. Let

$$Q = \begin{pmatrix} q_{11} & q_1^T \\ q_2 & Q_2 \end{pmatrix},$$

where q_1^T is the row vector containing elements 2 through n of the first row of Q; q_2 is the column vector consisting of elements 2 through n of the first column of Q; and Q_2 is the right lower $(n-1) \times (n-1)$ submatrix of Q. Applying the standard version of Gaussian elimination to Q results in q_2 becoming zero, while applying Gaussian elimination to Q^T causes q_1^T to disappear. In *both* cases, however, the submatrix Q_2 is given by

$$Q_2 - \frac{q_2 q_1^T}{q_{11}}.$$

We may view q_2/q_{11} to be a vector of multipliers when Gaussian elimination is applied to Q, and q_1^T/q_{11} to be the multipliers when applied to Q^T. Naturally, the same analysis can be performed on Q_2 to carry the reduction process one step further. Thus, no matter whether Gaussian elimination is performed on Q or on Q^T, the result as concerns the bounds on the growth factor will be the same.

2.7.4.3 The Computed Solution

To conclude our stability analysis and tie all the loose ends together, we need to show that the solution computed by Gaussian elimination to the system of equations, $\pi Q = 0$, is the exact solution of a slightly perturbed system and to bound the magnitude of that perturbation. Precisely, we shall show that $\hat{\pi}$, the computed solution to $\pi Q = 0$, satisfies the perturbed system $\hat{\pi}(Q + F) = 0$ exactly [69], where for $1 \leq i, j \leq n$,

$$|f_{ij}| \leq (5.19 + 1.01n)i \max_{ij} |q_{ij}| \mu 10^{-t}.$$

Using the results of Section 2.7.4.1 and the fact that $g_A = 1$ for infinitesimal generator matrices, the computed decomposition of Q^T is seen to satisfy

$$\hat{L}\hat{U} = Q^T + E,$$

where

$$|e_{ij}| \leq 4.18(i-1)\mu 10^{-t} \max_j |q_{jj}|.$$

When the decomposition phase has finished, we have

$$
\hat{U} = \begin{pmatrix}
\hat{u}_{11} & \hat{u}_{12} & \cdots & \hat{u}_{1n} \\
0 & \hat{u}_{22} & \cdots & \hat{u}_{2n} \\
\vdots & \vdots & \ddots & \vdots \\
0 & 0 & \cdots & \hat{u}_{n-1,n} \\
0 & 0 & \cdots & 0
\end{pmatrix}.
$$

Let

$$
\tilde{U} = \begin{pmatrix}
\hat{u}_{11} & \hat{u}_{12} & \cdots & \hat{u}_{1n} \\
0 & \hat{u}_{22} & \cdots & \hat{u}_{2n} \\
\vdots & \vdots & \ddots & \vdots \\
0 & 0 & \cdots & \hat{u}_{n-1,n} \\
0 & 0 & \cdots & 1
\end{pmatrix}.
$$

The solution $\hat{\pi}$ is computed from $\tilde{U}$ by back-substitution. We could at this point begin to carry out a backward error analysis of this back-substitution phase. It is much less involved than the analysis of the LU factorization phase, but somewhat more complex than the analysis of the inner product computation considered in Section 2.7.3. However, we shall be content to refer the reader to [150], in which it is shown that there exists a perturbation matrix G such that the computed solution $\hat{\pi}$ to $\tilde{U}\pi = e_n$ satisfies the system

$$(\tilde{U} + G)\hat{\pi} = e_n \tag{2.14}$$

where the elements of G satisfy

$$|g_{ij}| \leq 1.01(n+1)|\tilde{u}_{ij}|\mu 10^{-t} \leq 1.01\mu 10^{-t}(n+1)\max_l |q_{ll}|.$$

Multiplying equation (2.14) on the left by $\hat{L}$, we have

$$(\hat{L}\tilde{U} + \hat{L}G)\hat{\pi} = \hat{L}e_n = e_n,$$

and since $\hat{L}\tilde{U} = \hat{L}\hat{U} + e_n e^T$, it follows that

$$(\hat{L}\hat{U} + e_n e^T + \hat{L}G)\hat{\pi} = e_n$$

$$(\hat{L}\hat{U} + \hat{L}G)\hat{\pi} + e_n e_n^T \hat{\pi} = e_n.$$

Since $\hat{\pi}_n = 1$, $e_n^T \hat{\pi} = 1$, and it follows that

$$(\hat{L}\hat{U} + \hat{L}G)\hat{\pi} = (Q^T + E + \hat{L}G)\hat{\pi} = 0.$$

Thus, the computed solution $\hat{\pi}$ is the exact solution of $(Q^T + F)\hat{\pi} = 0$, where $F = E + \hat{L}G$. The elements of F satisfy

$$|f_{ij}| \leq |e_{ij}| + \sum_{k=1}^{i} |\hat{l}_{ik}||g_{kj}| \leq |e_{ij}| + \sum_{k=1}^{i} |g_{ik}|,$$

which implies that

$$|f_{ij}| \leq 4.18(i-1)\mu 10^{-t} \max_j |q_{jj}| + 1.01(i)\mu 10^{-t}(n+1) \max_l |q_{ll}|.$$

Finally,

$$|f_{ij}| \leq i(5.19 + 1.01n)\mu 10^{-t} \max_l |q_{ll}|.$$

This then completes the error analysis of Gaussian elimination applied to Markov chain models.

2.7.5 Condition Numbers, Residuals, and the Group Inverse

The condition of a matrix A is frequently measured in terms of its *condition number*, defined as

$$\text{cond}(A) = \|A\| \, \|A^{-1}\|,$$

where $\|.\|$ is any induced matrix norm. The larger the condition number, the more ill-conditioned the matrix. Notice that

$$1 = \|AA^{-1}\| \leq \|A\| \, \|A^{-1}\| = \text{cond}(A),$$

so the condition number must be at least equal to 1.

When we compute the solution of a nonhomogeneous system of linear equations $Ax = b$ with nonsingular coefficient matrix A, it is tempting to verify its accuracy by substituting the computed solution into the system of equations and checking to see how close the result matches the right-hand side vector b. What we are doing is using the *residual* as an accuracy check. Let $\hat{x}$ be a computed solution and define an *error vector*, $\hat{e}$, as $\hat{e} = x - \hat{x}$, and a *residual vector*, r, as $r = b - A\hat{x}$. It is then easy to show that

$$\frac{1}{\text{cond}(A)} \frac{\|r\|}{\|b\|} \leq \frac{\|\hat{e}\|}{\|x\|} \leq \text{cond}(A) \frac{\|r\|}{\|b\|}. \tag{2.15}$$

The term $\|\hat{e}\|/\|x\| = \|x - \hat{x}\|/\|x\|$ is the relative error in the computed solution, while $\|r\|/\|b\| = \|b - A\hat{x}\|/\|b\|$ is the relative residual. Equation (2.15) says that the error in the computed solution can be as large as $\text{cond}(A)$ times the error in the residual; thus, if the condition number is large, the residual is not a good indicator of the accuracy of the solution. On the other hand, if $\text{cond}(A)$ is small, the relative error in the solution is "squeezed" between a small magnification and a small decrease of the relative residual, and in this case the magnitude of the residual is a good estimator of the accuracy of the solution.

In Markov chain problems, the infinitesimal generator matrix Q is singular; hence, its inverse does not exist. In such cases the concept of a *group inverse* or *generalized inverse* plays a role analogous to that of the matrix inverse. The group inverse of Q is denoted by $Q^{\#}$ and is characterized as the unique matrix satisfying

$$QQ^{\#}Q = Q, \qquad Q^{\#}QQ^{\#} = Q^{\#}, \quad \text{and} \quad QQ^{\#} = Q^{\#}Q.$$

Many inportant properties of the group inverse may be found in the text of Campbell and Meyer [17]. In particular, it is shown that $Q^{\#}$ satisfies the important relationship

$$I - QQ^{\#} = e\pi \tag{2.16}$$

Also, we note in passing that Golub and Meyer [55] show that $Q^{\#}$ is a fundamental quantity governing the sensitivity of the stationary distribution of an ergodic Markov chain.

Before delving into an algorithm for computing a group inverse, recall that the usual procedure for computing the inverse of a nonsingular matrix A is to compute an LU decomposition of A and then determine the columns of A^{-1} one at a time as

$$z_i = U^{-1}L^{-1}e_i, \qquad i = 1, 2, \ldots, n.$$

It is easy to verify that

$$AA^{-1} = A[z_1, z_2, \ldots, z_n] = [e_1, e_2, \ldots, e_n] = I.$$

The algorithm presented below for the computation of the group inverse of an infinitesimal generator is based on this approach and on the fact that the group inverse of Q is the unique matrix that satisfies the equations

$$QX = I - e\pi \tag{2.17}$$
$$\pi X = 0. \tag{2.18}$$

That the group inverse satisfies these equations may be seen by postmultiplying equation (2.17) by Q and using the fact that $\pi Q = 0$ to get $QXQ = Q$. To show that $QX = XQ$, observe that equations (2.16) and (2.17) imply that $QX = QQ^{\#}$. Furthermore, substituting (2.16) into $e\pi X = 0$ yields

$$X = QQ^{\#}X \tag{2.19}$$

and therefore

$$XQ = QQ^{\#}XQ = Q^{\#}QXQ = Q^{\#}Q = QQ^{\#}, \tag{2.20}$$

and the desired result follows. $X = XQX$ follows by substituting from (2.20) into (2.19).

Algorithm: Computation of the Group Inverse [47]

1. Compute an LU decomposition of Q.

2. Compute the stationary probability vector π.

3. For $i = 1, 2, \ldots, n$

   - Compute $z_i = U^{-1}L^{-1}(e_i - \pi_i e)$. Note that each z_i satisfies equation (2.17).

   - *Modify z_i to get i^{th} column of $Q^{\#}$:* Compute $q_i^{\#} = z_i - (\pi z_i)e$. The vector $q_i^{\#}$ now satisfies both (2.17) and (2.18).

4. *Form the group inverse:* Set $Q^{\#} = \left[q_1^{\#}, q_2^{\#}, \ldots, q_n^{\#} \right]$.

Although we provide this algorithm for the computation of the group inverse of Q, this inverse, like the regular inverse, is seldom needed explicitly.

In Section 2.3, we discussed four approaches to applying direct methods to Markov chain problems and concluded that three of them were essentially identical but that approach 2 was inherently different. We alluded to the fact that this approach, in addition to requiring more numerical operations than the others, had other undesirable properties. It is to this aspect that we now return. At one point or another, in all of the implementations, we need to determine an LU factorization; the real difference stems from the matrix to which this factorization is applied. In approach 2, the matrix Q is replaced by a nonsingular matrix $\bar{Q}$, derived from Q by replacing one of the equations by a normalization equation. However, although Q is singular, it can be well-conditioned in the sense that $\|Q\| \, \|Q^{\#}\|$ may be small, whereas the modified matrix $\bar{Q}$, although nonsingular, can be very ill-conditioned, in that $\|\bar{Q}\| \, \|\bar{Q}^{\#}\| = \|\bar{Q}\| \, \|\bar{Q}^{-1}\|$ can be very large.

Meyer [98] gives the following matrix as an example.

$$Q = \begin{pmatrix} 1 - \epsilon & \epsilon/(n-1) & \epsilon/(n-1) & \cdots & \epsilon/(n-1) \\ \epsilon/(n-1) & 1 - \epsilon & \epsilon/(n-1) & \cdots & \epsilon/(n-1) \\ \vdots & \vdots & \vdots & \ddots & \vdots \\ \epsilon/(n-1) & \epsilon/(n-1) & \epsilon/(n-1) & \cdots & 1 - \epsilon \end{pmatrix} \quad 0 < \epsilon < 1,$$

and shows that

$$\|Q\|_2 \, \|Q^{\#}\|_2 = 1$$

(i.e., independent of n and ϵ), whereas

$$\|\bar{Q}\|_2 \, \|\bar{Q}^{\#}\|_2 = (n-1)/\epsilon$$

regardless of which equation is replaced by the normalization equation. This latter can be made arbitrarily large by choosing ϵ small or n large.

We know of no formal proof that allows us to claim that Q will always be better conditioned than $\bar{Q}$. Extensive numerical investigations by Meyer and others seem to indicate that this will generally be true. Nor are there counterexamples to show that $\bar{Q}$ is sometimes better conditioned than Q. Nevertheless, the above example, along with the arguments presented in Section 2.3.1.2, should convince the reader that the "Replace an Equation" approach to the direct solution of Markov chains should be avoided.

Chapter 3

Iterative Methods

3.1 The Power Method

3.1.1 Introduction

Consider a discrete-time Markov chain whose matrix of transition probabilities is

$$P = \begin{pmatrix} .0 & .8 & .2 \\ .0 & .1 & .9 \\ .6 & .0 & .4 \end{pmatrix}. \tag{3.1}$$

If the system starts in state 1, the initial probability vector is given by

$$\pi^{(0)} = (1, \ 0, \ 0).$$

Immediately after the first transition, the system will be either in state 2, with probability .8, or in state 3, with probability .2. The vector $\pi^{(1)}$, which denotes the probability distribution after one transition (or one step) is thus

$$\pi^{(1)} = (0, \ .8, \ .2).$$

Notice that this result may be obtained by forming the product $\pi^{(0)}P$.

The probability of being in state 1 after two time steps is obtained by summing (over all i) the probability of being in state i after 1 step, given by $\pi_i^{(1)}$, multiplied by the probability of making a transition from state i to state 1. We have

$$\sum_{i=1}^{3} \pi_i^{(1)} p_{i1} = \pi_1^{(1)} \times .0 + \pi_2^{(1)} \times .0 + \pi_3^{(1)} \times .6 = .12.$$

Likewise, the system will be in state 2 after two steps with probability .08 ($=$ .0×.8+.8×.1+.2×.0), and in state 3 with probability .8 ($=$ .0×.2+.8×.9+.2×.4).

121

Thus, given that the system begins in state 1, we have the following probability distribution after two steps:

$$\pi^{(2)} = (.12, \; .08, \; .8).$$

Notice once again that $\pi^{(2)}$ may be obtained by forming the product $\pi^{(1)}P$.

$$\pi^{(2)} = (.12, \; .08, \; .8) = (0.0, \; 0.8, \; 0.2) \begin{pmatrix} .0 & .8 & .2 \\ .0 & .1 & .9 \\ .6 & .0 & .4 \end{pmatrix} = \pi^{(1)}P.$$

We may continue in this fashion, computing the probability distribution after each transition step. For any integer k, the state of the system after k transitions is obtained by multiplying the probability vector obtained after $(k-1)$ transitions by P. Thus

$$\pi^{(k)} = \pi^{(k-1)}P = \pi^{(k-2)}P^2 = \ldots = \pi^{(0)}P^k.$$

When the Markov chain is finite, aperiodic, and irreducible (as in the current example), the vectors $\pi^{(k)}$ converge to the stationary probability vector π regardless of the choice of initial vector. We have

$$\lim_{k \to \infty} \pi^{(k)} = \pi.$$

Table 3.1 presents the probability distribution of the states of this example at specific steps, for each of three different starting configurations. After 25 iterations, no further changes are observed in the first four digits for any of the starting configurations.

For an ergodic Markov chain, the stationary probability vector is referred to as the *long-run* (also *steady-state, equilibrium*) probability vector, signifying simply that sufficient transitions have occurred as to erase all influence of the starting state. This method of determining the stationary probability vector is referred to as the *power method* or *power iteration*.

3.1.2   Application to an Arbitrary Matrix, A

The power method is well known in the context of determining the right-hand eigenvector corresponding to a dominant eigenvalue of a matrix. Let A be a square matrix of order n. The power method is described by the iterative procedure

$$z^{(k+1)} = \frac{1}{\xi_k} A z^{(k)}, \qquad (3.2)$$

where ξ_k is a normalizing factor, typically $\xi_k = \|Az^{(k)}\|_\infty$, and $z^{(0)}$ is an arbitrary starting vector. To examine the rate of convergence of this method, let A have eigensolution

$$Ax_i = \lambda_i x_i, \qquad i = 1, 2, \ldots, n,$$

Table 3.1: Convergence in Power Method

Step	Initial State			Initial State			Initial State		
	1.0	.0	.0	.0	1.0	.0	.0	.0	1.0
1:	.0000	.8000	.2000	.0000	.1000	.9000	.6000	.0000	.4000
2:	.1200	.0800	.8000	.5400	.0100	.4500	.2400	.4800	.2800
3:	.4800	.1040	.4160	.2700	.4330	.2970	.1680	.2400	.5920
4:	.2496	.3944	.3560	.1782	.2593	.5626	.3552	.1584	.4864
$\vdots$	$\vdots$	$\vdots$	$\vdots$	$\vdots$	$\vdots$	$\vdots$	$\vdots$	$\vdots$	$\vdots$
10:	.2860	.2555	.4584	.2731	.2573	.4696	.2827	.2428	.4745
$\vdots$	$\vdots$	$\vdots$	$\vdots$	$\vdots$	$\vdots$	$\vdots$	$\vdots$	$\vdots$	$\vdots$
25:	.2813	.2500	.4688	.2813	.2500	.4688	.2813	.2500	.4688

and suppose that

$$|\lambda_1| > |\lambda_2| \geq |\lambda_3| \geq \cdots \geq |\lambda_n|.$$

Let us further assume that the initial vector may be written as a linear combination of the eigenvectors of A, i.e.,

$$z^{(0)} = \sum_{i=1}^{n} \alpha_i x_i.$$

The rate of convergence of the power method may then be determined from the relationship

$$z^{(k)} = A^k z^{(0)} = \sum_{i=1}^{n} \alpha_i \lambda_i^k x_i = \lambda_1^k \left\{ \alpha_1 x_1 + \sum_{i=2}^{n} \alpha_i \left(\frac{\lambda_i}{\lambda_1} \right)^k x_i \right\}. \qquad (3.3)$$

It may be observed that the process converges to the dominant eigenvector x_1. The rate of convergence depends on the ratios $|\lambda_i|/|\lambda_1|$ for $i = 2, 3, \ldots, n$. The smaller these ratios, the quicker the summation on the right-hand side tends to zero. It is, in particular, the magnitude of the subdominant eigenvalue, λ_2, that determines the convergence rate. Thus, the power method will not perform satisfactorily when $|\lambda_2| \approx |\lambda_1|$. Obviously major difficulties arise when $|\lambda_2| = |\lambda_1|$. Note that, although omitted from equation (3.3), it is generally necessary to normalize successive iterates, since otherwise the term λ_1^k may cause successive approximations to become too large (if $\lambda_1 > 1$) or too small (if $\lambda_1 < 1$) and may result in overflow or underflow. Additionally, this normalization is required to provide a standardized vector with which to implement convergence testing.

For the 3×3 example given by equation (3.1), the eigenvalues of P are $\lambda_1 = 1$ and $\lambda_{2,3} = -.25 \pm .5979i$. Thus $|\lambda_2| \approx .65$. Notice that $.65^{10} \approx .01$, $.65^{25} \approx 2 \times 10^{-5}$ and $.65^{100} = 2 \times 10^{-19}$. Observe from the table given in Section 3.1.1 that approximately two decimal places of accuracy have been obtained after 10 iterations and four places after 25 iterations, as predicted by the magnitude of the subdominant eigenvalue.

3.1.3   Application to a Stochastic Matrix, P

When the power method is applied to stochastic transition probability matrices, it is the left-hand eigenvector corresponding to a unit eigenvalue that is required. The matrix to which the method is applied is thus P^T. Notice also that since the matrix has 1 as a dominant eigenvalue ($\lambda_1 = 1$), the requirement for periodic normalization of iterates disappears. Indeed, if the initial starting approximation is a probability vector, all successive approximations will also be probability vectors. In this case, the above iteration, equation (3.2), takes the form

$$z^{(k+1)} = P^T z^{(k)} \tag{3.4}$$

It is known that the unit eigenvalue of a stochastic matrix is a dominant eigenvalue and that if the matrix is irreducible, there are no other unit eigenvalues. When the matrix is cyclic, however, there exist other eigenvalues on the unit circle, which are different from 1 but whose modulus is equal to 1. A straightforward application of the power method in this case will fail. This situation may be circumvented by a slight modification that leaves the unit eigenvalue and its corresponding eigenvector unchanged. The matrix P is usually obtained from the infinitesimal generator by means of the relationship

$$P = (Q\Delta t + I),$$

where $\Delta t \leq 1/\max_i |q_{ii}|$. If Δt is chosen so that $\Delta t < 1/\max_i |q_{ii}|$, the resulting stochastic matrix has diagonal elements $p_{ii} > 0$ and therefore cannot be cyclic. As previously mentioned in Chapter 1, Wallace and Rosenberg, in their Recursive Queue Analyzer [174], chose a value $\Delta t = .99/\max_i |q_{ii}|$. Under these conditions (irreducible and acyclic), the power method can be guaranteed to converge. Its rate of convergence is governed by the ratio $|\lambda_2|/|\lambda_1|$, i.e., by $|\lambda_2|$.

Unfortunately, the difference between theoretical conditions for the convergence of an iterative method and its observed behavior in practical situations can be quite drastic. What in theory will converge may take such a large number of iterations that for all practical purposes the method should be considered unworkable. This occurs in the power method when the modulus of the subdominant eigenvalue, $|\lambda_2|$, is close to unity. For example, stochastic matrices that are nearly completely decomposable (NCD) arise frequently in modelling physical and mathematical systems; such matrices have subdominant eigenvalues that are necessarily close to unity. In these cases the power method will converge extremely slowly.

3.1.4 Comparison with Matrix Powering

Given the fact that it often takes many iterations to achieve convergence, it may be thought that a more economical approach is to repeatedly square the matrix P. Let $k = 2^m$ for some integer m. Using the basic iterative formula $\pi^{(k)} = \pi^{(k-1)}P$ requires k iterations to obtain $\pi^{(k)}$; each iteration includes a matrix-vector product. Repeatedly squaring the matrix requires only m matrix products to determine P^{2^m}, from which $\pi^{(k)} = \pi^{(0)}P^k$ is quickly computed. It may be further speculated that, since a matrix-vector product requires n^2 multiplications and a matrix-matrix product requires n^3, that the squaring approach is to be recommended when $mn^3 < 2^m n^2$, i.e., when $nm < 2^m$. Unfortunately, this analysis completely omits the fact that the matrix P is usually large and sparse. Thus, a matrix-vector product requires only n_z multiplications, where n_z is the number of nonzero elements in P. The matrix-squaring operation will increase the number of nonzero elements in the matrix (in fact, for an irreducible matrix, convergence will not be attained before all the elements have become nonzero), thereby increasing not only the number of multiplications needed but also the amount of memory needed. It is perhaps memory requirements more than time constraints that limit the applicability of matrix powering.

3.2 Jacobi, Gauss–Seidel, SOR, and Symmetric SOR

We have just seen how the power method may be applied to a stochastic transition probability matrix P and, under certain conditions, used to determine the left-hand eigenvector π corresponding to its dominant eigenvalue. Similarly, iterative methods may be applied to the infinitesimal generator matrix, the matrix of transition rates, Q. In the first case, the power method is used to obtain the solution of an eigenproblem ($\pi P = \pi$), while in the second the iterative methods that we are about to consider are used to obtain the solution of the homogeneous system of linear equations

$$\pi Q = 0.$$

Although these methods are usually applied to nonhomogeneous systems of equations with nonsingular coefficient matrices, they may also be successfully applied to singular homogeneous systems that satisfy certain conditions.

3.2.1   The Nonhomogeneous Case

The standard and well-known iterative methods for the solution of systems of linear equations are the methods of Jacobi, Gauss–Seidel, and successive overrelaxation (SOR). These methods derive from a nonhomogeneous system of linear equations

$$Ax = b,$$

an iterative formula of the form

$$x^{(k+1)} = Hx^{(k)} + c, \qquad k = 0, 1, \ldots \tag{3.5}$$

This is accomplished by splitting the coefficient matrix A. Given a splitting

$$A = M - N$$

with nonsingular M, we have

$$(M - N)x = b,$$

or

$$Mx = Nx + b,$$

which leads to the iterative procedure

$$x^{(k+1)} = M^{-1}Nx^{(k)} + M^{-1}b = Hx^{(k)} + c, \qquad k = 0, 1, \ldots.$$

The matrix $H = M^{-1}N$ is called the *iteration* matrix.

If x is the true solution and $x^{(k)}$ the approximation obtained after k applications of the iterative process (3.5), then

$$e^{(k)} = x^{(k)} - x$$

is a measure of the error after k iterations. It follows that $e^{(k)} = H^k e^{(0)}$. For convergence, it is required that $e^{(k)} \to 0$ as $k \to \infty$, i.e., that $H^k \to 0$ as $k \to \infty$. This will be true if and only if all the eigenvalues of H are strictly less than 1 in modulus, i.e., iff $\rho(H) < 1$. When this condition holds, the method will converge for any starting vector $x^{(0)}$.

Now consider the application of these methods to Markov chain problems. In the previous chapter, we saw that by partitioning Q^T as

$$Q^T = \begin{pmatrix} B & d \\ c^T & f \end{pmatrix}$$

where B is of order $(n-1)$ and nonsingular, then the solution vector π may be obtained by solving the nonhomogeneous system of equations

$$B\hat{x} = -d \qquad (3.6)$$

and then normalizing $(\hat{x}, 1)$. In that chapter we were concerned with direct methods for finding $\hat{x}$, but it is equally possible to apply iterative methods to $B\hat{x} = -d$. This formulation fits precisely into the assumptions just given (non-homogeneous system of linear equations with nonsingular coefficient matrix). In Section 3.6, it is shown that because B is a principal submatrix of an irreducible stochastic matrix, the spectral radius of the iteration matrices for the methods of Jacobi, Gauss–Seidel, and SOR (for $0 < \omega \leq 1$) are all strictly less than 1, thereby guaranteeing convergence. In general, however, this approach is not to be recommended, because of the extremely slow rate of convergence that often accompanies it. Instead, it is usually more advantageous to apply the standard iterative methods to the homogeneous system of equations $\pi Q = 0$.

3.2.2 The Method of Jacobi

We wish to solve

$$Q^T \pi^T = 0$$

For notational convenience, set $x = \pi^T$ and let

$$Q^T = D - (L + U) \qquad (3.7)$$

where D is a diagonal matrix and L and U are respectively strictly lower and strictly upper triangular matrices[1]. The method of Jacobi corresponds to the splitting $M = D$ and $N = (L + U)$. Its iteration matrix is given by

$$H_J = D^{-1}(L + U).$$

Note that D^{-1} exists, since $d_{jj} \neq 0$ for all j. Once the kth approximation, $x^{(k)}$, to the solution vector x has been formed, the next approximation is obtained by solving the system of equations

$$Dx^{(k+1)} = (L + U)x^{(k)}.$$

This gives

$$x^{(k+1)} = D^{-1}(L + U)x^{(k)},$$

which in scalar form is simply

$$x_i^{(k+1)} = \frac{1}{d_{ii}} \left\{ \sum_{j \neq i} (l_{ij} + u_{ij}) x_j^{(k)} \right\}, \qquad i = 1, 2, \ldots, n. \qquad (3.8)$$

---

[1]The matrices L and U should not be confused with the LU factors obtained from direct methods such as Gaussian elimination.

Notice that substituting $Q^T = D - (L + U)$ into $Q^T x = 0$ gives $(L + U)x = Dx$, and since D is nonsingular, this yields the eigenvalue equation

$$D^{-1}(L + U)x = x \qquad (3.9)$$

in which x is seen to be the right-hand eigenvector corresponding to a unit eigenvalue of the matrix $D^{-1}(L + U)$. This matrix will immediately be recognized as the iteration matrix for the method of Jacobi, H_J. That H_J has a unit eigenvalue is obvious from equation (3.9). Furthermore, from the zero column-sum property of Q^T, we have

$$d_{jj} = \sum_{i=1,\ i\neq j}^{n} (l_{ij} + u_{ij}), \qquad j = 1, 2, \ldots$$

with l_{ij}, $u_{ij} \leq 0$ for all i, j; $i \neq j$, and it follows directly from the theorem of Gerschgorin that no eigenvalue of H_J can have modulus greater than unity. The stationary probability vector π is therefore the eigenvector corresponding to a dominant eigenvalue of H_J, and the method of Jacobi is identical to the power method applied to the iteration matrix H_J.

3.2.3   The Method of Gauss–Seidel

Usually the computations specified by equation (3.8) are carried out sequentially; the components of the vector $x^{(k+1)}$ are obtained one after the other as $x_1^{(k+1)}, x_2^{(k+1)}, \ldots, x_n^{(k+1)}$. When evaluating $x_i^{(k+1)}$ for $i \geq 2$, only components from the previous iteration $x_j^{(k)}$, $j = 1, 2, \ldots, n$ are used, even though elements from the current iteration $x_j^{(k+1)}$ for $j < i$ are available and are (we hope) more accurate. The Gauss–Seidel method makes use of these most recently available component approximations. This may be accomplished by simply overwriting elements as soon as a new approximation is determined. Mathematically, the Gauss–Seidel iteration formula is given by

$$x_i^{(k+1)} = \frac{1}{d_{ii}} \left(\sum_{j=1}^{i-1} l_{ij} x_j^{(k+1)} + \sum_{j=i+1}^{n} u_{ij} x_j^{(k)} \right), \qquad i = 1, 2, \ldots, n. \qquad (3.10)$$

In matrix terms, given an approximation $x^{(k)}$, the next approximation is obtained by solving

$$(D - L)x^{(k+1)} = U x^{(k)}. \qquad (3.11)$$

Rearranging equation (3.11) into the standard form of equation (3.5), we find

$$x^{(k+1)} = (D - L)^{-1} U x^{(k)} \qquad (3.12)$$

which shows that the iteration matrix for the method of Gauss–Seidel is

$$H_{GS} = (D - L)^{-1} U. \qquad (3.13)$$

Notice that since $d_{jj} \neq 0$ for all j, the inverse $(D - L)^{-1}$ exists. This iterative method corresponds to the splitting $M = (D - L)$ and $N = U$.

In some cases, the matrix Q^T is scaled by postmultiplying it with D^{-1} before applying the Gauss–Seidel method. This has the effect of making all the diagonal elements equal to 1. In this case the iteration matrix is given by

$$(I - LD^{-1})^{-1}UD^{-1} \;=\; D\left((D - L)^{-1}U\right)D^{-1}$$

and is therefore related to the iteration matrix of equation (3.13) through a similarity transformation. This scaling has therefore no effect on the rate of convergence of Gauss–Seidel. Since the system solved in this case is $Q^T D^{-1} x = 0$, the stationary probability vector is given by $\pi = \left(D^{-1}x\right)^T$.

The stationary probability vector $\pi = x^T$ obviously satisfies $H_{GS}x = x$, which shows that x is the right-hand eigenvector corresponding to a unit eigenvalue of H_{GS}. As a consequence of the Stein–Rosenberg theorem ([170], p. 70) and the fact that the corresponding Jacobi iteration matrix H_J possesses a dominant unit eigenvalue, the unit eigenvalue of the matrix H_{GS} is a dominant eigenvalue. The method of Gauss–Seidel is therefore identical to the power method applied to H_{GS}.

As indicated in equation (3.8), the method of Gauss–Seidel corresponds to computing the ith component of the current approximation from $i = 1$ through $i = n$, i.e., from top to bottom. To denote specifically the direction of solution, this is sometimes referred to as *forward* Gauss–Seidel. A *backward* Gauss–Seidel iteration takes the form

$$(D - U)x^{(k+1)} = Lx^{(k)}, \qquad k = 0, 1, \ldots$$

and corresponds to computing the components from bottom to top. Forward and backward iterations in a Jacobi setting are meaningless, since in Jacobi only components of the previous iteration are used in the updating procedure.

As a general rule of thumb, a forward iterative method is usually recommended when the preponderance of the elemental mass is to be found below the diagonal, for in this case the iterative method essentially works with the inverse of the lower triangular portion of the matrix, $(D-L)^{-1}$, and intuitively, the closer this is to the inverse of the entire matrix, the faster the convergence. In fact, the case in which $D - L = Q^T$ (and $U = 0$) requires only one iteration. The method at this point has been reduced to a back-substitution. Ideally, in a general context, a splitting should be such that M is chosen as close to Q^T as possible, subject only to the constraint that M^{-1} be easy to find. On the other hand, a backward iterative scheme works with the inverse of the upper triangular portion, $(D - U)^{-1}$, and is generally recommended when most of the nonzero mass lies above the diagonal. However, some examples that run counter to this "intuition" are known to exist [80]. Also, in one of the examples we examine below, no benefit is gained from

a backward as opposed to a forward procedure, regardless of the values assigned
to the nonzero elements.

Little information is available on the effect of the ordering of the state space
on the convergence of Gauss–Seidel. Examples are available in which it works
extremely well for one ordering but not at all for an opposing ordering [100].
In these examples the magnitude of the nonzero elements appears to have little
effect on the speed of convergence. It seems that an ordering that in some sense
preserves the direction of probability flow works best. This topic is considered
more fully in Section 3.5.6.

The following MATLAB code performs a fixed number of iterations of the basic
forward Gauss–Seidel method on the system of equation $Ax = b$. It accepts an
input matrix A (which may be set equal to an infinitesimal generator Q^T), an
initial approximation x_0, a right-hand side vector b (which may be set to zero),
and the number of iterations to be carried out, *itmax*. It returns the computed
solution, *soln*, and a vector containing the residual computed at each iteration,
resid. This code will be used in conjunction with other methods for demonstration
purposes in this and in later chapters. It is not designed to be a "production"
code, but rather simply a way to generate some results. In several chapters we
shall make use of MATLAB programs to explicitly show the sequence of steps
that must be followed in an algorithm as well as to get some intuition as to
performance characteristics.

Algorithm: Gauss–Seidel.

```
function [soln,resid] = gs(A,x0,b,itmax)
%    Performs ''itmax'' iterations of Gauss--Seidel on Ax = b

[n,n] = size(A); L = zeros(n,n); U = L;   D = diag(diag(A));
for i = 1:n,
    for j = 1:n,
        if i<j, U(i,j) = -A(i,j); end
        if i>j, L(i,j) = -A(i,j); end
    end
end
B = inv(D-L)*U;          % B is the iteration matrix

for iter = 1:itmax,
    soln = B*x0+inv(D-L)*b;
    resid(iter) = norm(A*soln-b,2);
    x0 = soln;
end
resid = resid';
```

3.2.4   The SOR Method

The SOR method with relaxation parameter ω, applied to the homogeneous system $Q^T x = (D - L - U)x = 0$, may be written as

$$x_i^{(k+1)} = (1-\omega)x_i^{(k)} + \omega \left\{ \frac{1}{d_{ii}} \left(\sum_{j=1}^{i-1} l_{ij} x_j^{(k+1)} + \sum_{j=i+1}^{n} u_{ij} x_j^{(k)} \right) \right\}, \qquad i = 1, 2, \ldots, n,$$

or in matrix form as

$$x^{(k+1)} = (1 - \omega)x^{(k)} + \omega \left\{ D^{-1}(Lx^{(k+1)} + Ux^{(k)}) \right\}. \tag{3.14}$$

A *backward* SOR relaxation may also be written. Rewriting equation (3.14) we find

$$(D - \omega L)x^{(k+1)} = [(1 - \omega)D + \omega U]x^{(k)},$$

or

$$x^{(k+1)} = (D - \omega L)^{-1}[(1 - \omega)D + \omega U]x^{(k)}, \tag{3.15}$$

and thus the iteration matrix for the SOR method is

$$H_\omega = (D - \omega L)^{-1}[(1 - \omega)D + \omega U].$$

It corresponds to the splitting $M = \omega^{-1}[D - \omega L]$ and $N = \omega^{-1}[(1 - \omega)D + \omega U]$. When the system of equations is scaled, the iteration matrix takes the form

$$(I - \omega L D^{-1})^{-1}[(1 - \omega)I + \omega U D^{-1}].$$

Notice that in equation (3.14), the expression within the braces is exactly the result that is obtained when Gauss–Seidel is used and that the method reduces to Gauss–Seidel when ω is set equal to 1. For $\omega > 1$, the process is said to be one of *overrelaxation*; for $\omega < 1$ it is said to be *underrelaxation*.

From equation (3.15), it is evident that the stationary probability vector is the eigenvector corresponding to a unit eigenvalue of the SOR iteration matrix. It is not necessarily true that this eigenvalue is the dominant eigenvalue, because the magnitude of the dominant eigenvalue depends on the choice of the relaxation parameter ω. It is known that when ω is chosen within a certain range, the unit eigenvalue is a dominant eigenvalue, and in this case the SOR method is identical to the power method applied to H_ω.

The SOR method converges only if $0 < \omega < 2$ [184]. The value of ω that maximizes the difference between the unit eigenvalue and the subdominant eigenvalue of H_ω is the optimal choice for the relaxation parameter, and the convergence rate achieved with this value of ω can be a considerable improvement over that of Gauss–Seidel. The choice of an optimal, or even a reasonable, value for ω has been the subject of much study, especially for problems arising in the numerical solution of partial differential equations. Some results have been obtained for certain classes of matrices. Unfortunately, little is known at present for arbitrary nonsymmetric linear systems.

3.2.5   The Symmetric SOR Method: SSOR

The symmetric successive overrelaxation method (SSOR) consists of a relaxation sweep from top to bottom followed by a relaxation sweep from bottom to top. Thus, the case $\omega = 1$ corresponding to a SGS (symmetric Gauss–Seidel) scheme is as follows:

$$(D - L)y^{(k+1/2)} = Uy^{(k)}$$
$$(D - U)y^{(k+1)} = Ly^{(k+1/2)}$$

Notice that the forward and backward sweeps can be carried out for the price of a single forward (or backward) sweep. Successive steps may be written as

$$Dy^{(k+1/2)} - Ly^{(k+1/2)} = Uy^{(k)}$$
$$Dy^{(k+1)} - Uy^{(k+1)} = Ly^{(k+1/2)}$$

$$Dy^{(k+3/2)} - Ly^{(k+3/2)} = Uy^{(k+1)}$$
$$Dy^{(k+2)} - Uy^{(k+2)} = Ly^{(k+3/2)}$$

and in all cases the right-hand side is already available from the previous equation.

For arbitrary ω, the forward and backward sweeps are written as

$$(D - \omega L)y^{(k+1/2)} = [(1 - \omega)D + \omega U]y^{(k)}$$
$$(D - \omega U)y^{(k+1)} = [(1 - \omega)D + \omega L]y^{(k+1/2)}$$

The main attraction of SSOR schemes is that the iteration matrix is similar to a symmetric matrix when the original matrix is symmetric. This situation rarely occurs in Markov chain models. However, SSOR may help to reduce poor convergence behavior that results from a badly ordered state space.

To conclude this section, we have now seen that the power method may be used to obtain π from one of four sources: P^T, H_J, H_{GS}, and H_ω. The eigenvalues (with the exception of the unit eigenvalue) will not be the same from one matrix to the next, and sometimes a considerable difference in the number of iterations required to obtain convergence may be observed. Since the computational effort to perform an iteration step is the same in all four cases, it is desirable to apply

Table 3.2: Eigenvalues of Iteration Matrices for Courtois Problem

P^T	H_J	H_{GSf}	H_{GSb}
1.0	1.0	1.0	1.0
0.99980000000000	0.99938353640096	0.99878460378460	0.99874765491570
0.99849479689134	0.99790895882791	0.99206152274985	0.99102834317366
0.75002623150850	0.99579335294684	0.25325763435349	0.13871647028072
0.55006663986827	0.58325797025592	0.10423733661151	0.00000712680448
0.40003338516447	0.50277001092246	0.00000000025144	0.00000712680448
0.30071431880876	0.50277001092246	0.00000000004771	0.0
0.14953537224133	0.41640735625114	0.0	0.0

the power method to the matrix that yields convergence in the smallest number of iterations, i.e., to the matrix whose subdominant eigenvalues are, in modulus, furthest from unity.

3.2.6   Examples

Throughout this chapter we shall use two small examples for purposes of illustration. The first is an (8×8) stochastic matrix adapted from Courtois [28], and is given as

$$
P^T = \begin{pmatrix}
0.85000 & 0.10000 & 0.10000 & 0 & 0.00050 & 0 & 0.00003 & 0 \\
0 & 0.65000 & 0.80000 & 0.00040 & 0 & 0.00005 & 0 & 0.00005 \\
0.14900 & 0.24900 & 0.09960 & 0 & 0.00040 & 0 & 0.00003 & 0 \\
0.00090 & 0 & 0.00030 & 0.70000 & 0.39900 & 0 & 0.00004 & 0 \\
0 & 0.00090 & 0 & 0.29950 & 0.60000 & 0.00005 & 0 & 0.00005 \\
0.00005 & 0.00005 & 0 & 0 & 0.00010 & 0.60000 & 0.10000 & 0.19990 \\
0 & 0 & 0.00010 & 0.00010 & 0 & 0.24990 & 0.80000 & 0.25000 \\
0.00005 & 0.00005 & 0 & 0 & 0 & 0.15000 & 0.09990 & 0.55000
\end{pmatrix}
$$

$$(3.16)$$

We wish to examine the consequences of applying the power method to P^T and the other iterative methods to the homogeneous system $(I - P^T)x = 0$. Table 3.2 displays the *absolute* value of the eigenvalues of P^T and the iteration matrices H_J, of Jacobi; H_{GSf}, of forward Gauss–Seidel; and H_{GSb}, of backward Gauss–Seidel.

These are very discouraging results, for they show that the power method will need approximately 69,000 iterations to achieve an accuracy of 10^{-6} (since $.99980^{69,000} \approx 10^{-6}$) and Jacobi will need approximately 23,000. Even Gauss–Seidel will need in excess of 11,000 iterations to achieve this accuracy, and very little is gained by using a backward rather than a forward approach.

Figure 3.1 plots the magnitude of the subdominant eigenvalue of H_ω as ω varies. The optimal value of the relaxation parameter occurs at $\omega = 1.532$, and for this value the magnitude of the subdominant eigenvalue is 0.996115. Thus, even with the optimal value of ω, more than 3,500 iterations are needed to compute the stationary probability vector of this small example, correct to 6 decimal places.

An examination of this matrix will reveal that it is nearly completely decomposable (NCD) into a block of order 3, a second of order 2, and a third of order 3. None of the iterative methods described above is effective in handling this situation; in Chapter 6 we shall examine iterative methods that exploit this NCD property. There are, however, many examples that are well suited to the basic Gauss–Seidel and SOR iterative methods. Reliability models often fall into this category, and it is from this type of model that we draw our second small example.

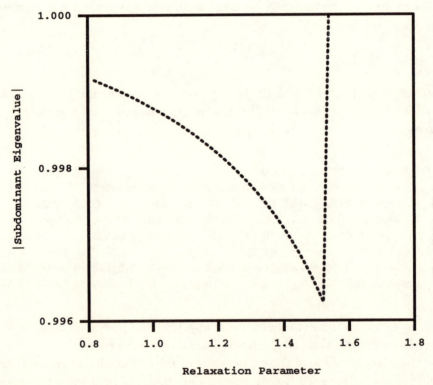

Figure 3.1: SOR convergence characteristics for Courtois problem.

Next we consider a reliability model that consists of two different classes of components. We shall assume that there are only two components per class in order to keep the model small. The components are subject to failure and subsequent repair. A state may be completely specified by the ordered pair (η_1, η_2), where η_1 (respectively, η_2) denotes the number of components of class 1 (respectively class 2) that are operational. We shall order the states from 1 to 9 as follows:

$$(2,2) \prec (2,1) \prec (2,0) \prec (1,2) \prec (1,1) \prec (1,0) \prec (0,2) \prec (0,1) \prec (0,0)$$

The failure rate of the first class is taken as λ_1, and that of the second class λ_2. The repair rates for the two classes are given by μ_1 for class 1 and μ_2 for class 2. The infinitesimal generator matrix is therefore

$$Q = \begin{pmatrix} * & 2\lambda_2 & & 2\lambda_1 & & & & & \\ \mu_2 & * & \lambda_2 & & 2\lambda_1 & & & & \\ & 2\mu_2 & * & & & 2\lambda_1 & & & \\ \mu_1 & & & * & 2\lambda_2 & & \lambda_1 & & \\ & \mu_1 & & \mu_2 & * & \lambda_2 & & \lambda_1 & \\ & & \mu_1 & & 2\mu_2 & * & & & \lambda_1 \\ & & & 2\mu_1 & & & * & 2\lambda_2 & \\ & & & & 2\mu_1 & & \mu_2 & * & \lambda_2 \\ & & & & & 2\mu_1 & & 2\mu_2 & * \end{pmatrix}. \qquad (3.17)$$

The diagonal elements are equal to the negated sum of the off-diagonal elements and are denoted by an asterisk. Notice that nonzero elements above the diagonal specify failure rates, while nonzero elements below the diagonal specify repair rates. It might be expected that when failures occur infrequently and failed components are quickly repaired, we might observe some difference in the backward as opposed to forward Gauss–Seidel iterations. In this case, elements above the diagonal will be small compared to those below the diagonal. However, as we shall see, this difference fails to materialize in this example. We begin by presenting the results obtained using the values:

$$\lambda_1 = 0.2; \qquad \lambda_2 = 0.3; \qquad \mu_1 = 5.0; \quad \text{and} \quad \mu_2 = 6.0.$$

The absolute values of the eigenvalues of the different iteration matrices used in solving $\pi(Q\Delta t + I) = \pi$ by the power method or $\pi Q\Delta t = 0$ by iterative equation-solving methods, where $\Delta t = (\max_i |q_{ii}|)^{-1}$, are given in the table below. They show that the power method will yield six decimal places accuracy in 52 iterations, while Gauss–Seidel will give the same accuracy in only six iterations! Note however, that the Jacobi iteration matrix has two eigenvalues of modulus equal to 1, which implies that this method will fail to converge. It turns out that the iteration matrix for the method of Jacobi is cyclic with index 2, and hence, for each nonzero eigenvalue λ of H_J there is another eigenvalue equal to $-\lambda$. Table 3.3 presents only the *absolute* values of the eigenvalues.

Table 3.3: Eigenvalues of Iteration Matrices for Reliability Problem (Similar Failure/Repair Rates)

P^T	H_J	H_{GSf}	H_{GSb}
1.0	1.0	1.0	1.0
0.76363636363636	1.0	0.07217321571772	0.07217321571772
0.71363636363636	0.26865073183917	0.03662668253823	0.03662668253823
0.52727272727273	0.26865073183917	0.00354119404637	0.00354119404637
0.47727272727273	0.19138098792260	0.00000000050525	0.00000000176264
0.42727272727273	0.19138098792260	0.00000000050525	0.00000000176264
0.24090909090909	0.05950793263396	0.0	0.0
0.19090909090909	0.05950793263396	0.0	0.0
0.04545454545455	0.0	0.0	0.0

Table 3.4: Eigenvalues of Iteration Matrices for Reliability Problem (Dissimilar Failure/Repair Rates)

P^T	H_J	H_{GSf}	H_{GSb}
1.0	1.0	1.0	1.0
0.99421487603306	1.0	0.98273525248789	0.98273525248789
0.98842975206612	0.99133004215947	0.96569082611991	0.96569082611991
0.49917355371901	0.99133004215947	0.00004862394243	0.00004862393889
0.49338842975207	0.98269569354908	0.00000000019127	0.00000001262423
0.48760330578512	0.98269569354908	0.00000000019127	0.00000001262069
0.25619834710744	0.00697308700858	0.0	0.0
0.25041322314050	0.00697308700858	0.0	0.0
0.24462809917355	0.0	0.0	0.0

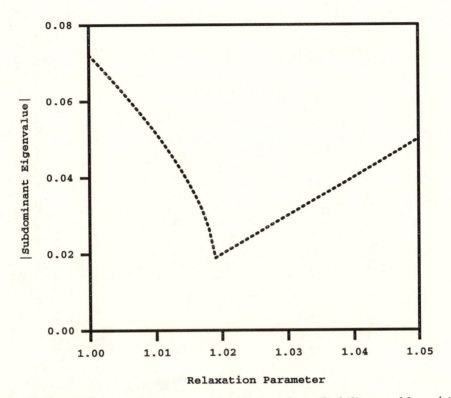

Figure 3.2: SOR convergence characteristics for reliability problem (similar failure/repair rates).

Figure 3.2 plots the magnitude of the subdominant eigenvalue of H_ω as ω varies. The optimal value of the relaxation parameter occurs at $\omega = 1.0187$, and at this point the magnitude of the subdominant eigenvalue is 0.0187. With this value, six decimal places of accuracy are attained in four iterations.

Let us now alter the parameters so that the failure and repair rates of the second type of component are much larger than those of the first type. We shall take

$$\lambda_1 = 0.2; \qquad \lambda_2 = 30.0; \qquad \mu_1 = 0.5; \quad \text{and} \quad \mu_2 = 60.0.$$

Table 3.4 shows the modulus of the eigenvalues. Now the power method will require 2,375 iterations for an accuracy of six decimal digits, while Gauss–Seidel will need approximately 790. As before, the method of Jacobi will fail to converge. With these values of the parameters the model is almost completely decomposable. The state space may be partitioned into three groups, each group distinguishable by the number of class 2 components that are operational.

With SOR the optimal value of ω is now 1.7677 (see Figure 3.3), at which point

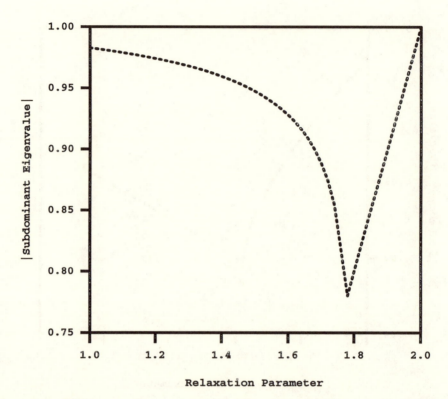

Figure 3.3: SOR convergence characteristics for reliability problem (dissimilar failure/repair rates).

the subdominant eigenvalue is equal to 0.7677. A total of 53 iterations is now needed to achieve six-digit accuracy.

3.3   Block Iterative Methods

In Markov chain problems it is frequently the case that the state space can be meaningfully partitioned into subsets. Perhaps the states of a subset interact only infrequently with the states of other subsets, or perhaps the states possess some property that merits special consideration. In these cases it is possible to partition the transition rate matrix accordingly and to develop iterative methods that are based on this partition. In general such block iterative methods require more computation per iteration, but this is offset by a faster rate of convergence.

Let us partition the defining homogeneous system of equations $\pi Q = 0$ as

$$(\pi_1, \ \pi_2, \ \ldots, \ \pi_N) \begin{pmatrix} Q_{11} & Q_{12} & \cdots & Q_{1N} \\ Q_{21} & Q_{22} & \cdots & Q_{2N} \\ \vdots & \vdots & \ddots & \vdots \\ Q_{N1} & Q_{N2} & \cdots & Q_{NN} \end{pmatrix} = 0.$$

We now introduce the *block* splitting:

$$Q^T = D_N - (L_N + U_N),$$

where D_N is a *block* diagonal matrix and L_N and U_N are respectively strictly lower and upper *block* triangular matrices. We have

$$D_N = \begin{pmatrix} D_{11} & 0 & \cdots & 0 \\ 0 & D_{22} & \cdots & 0 \\ \vdots & \vdots & \ddots & \vdots \\ 0 & 0 & \cdots & D_{NN} \end{pmatrix}$$

$$L_N = \begin{pmatrix} 0 & 0 & \cdots & 0 \\ L_{21} & 0 & \cdots & 0 \\ \vdots & \vdots & \ddots & \vdots \\ L_{N1} & L_{N2} & \cdots & 0 \end{pmatrix}, \qquad U_N = \begin{pmatrix} 0 & U_{12} & \cdots & U_{1N} \\ 0 & 0 & \cdots & U_{2N} \\ \vdots & \vdots & \ddots & \vdots \\ 0 & 0 & \cdots & 0 \end{pmatrix}.$$

In analogy with equation (3.11), the block Gauss–Seidel method is given by

$$(D_N - L_N)x^{(k+1)} = U_N x^{(k)}.$$

If we write this out in full, we get

$$D_{ii}x_i^{(k+1)} = \left(\sum_{j=1}^{i-1} L_{ij}x_j^{(k+1)} + \sum_{j=i+1}^{N} U_{ij}x_j^{(k)} \right), \qquad i = 1, 2, \ldots, N$$

where the subvectors x_i are partitioned conformally with D_{ii}, $i = 1, 2, \ldots, N$. This implies that at each iteration we must now solve N systems of linear equations

$$D_{ii}x_i^{(k+1)} = z_i, \qquad i = 1, 2, \ldots, N, \tag{3.18}$$

where

$$z_i = \left(\sum_{j=1}^{i-1} L_{ij}x_j^{(k+1)} + \sum_{j=i+1}^{N} U_{ij}x_j^{(k)} \right), \qquad i = 1, 2, \ldots, N.$$

The right-hand side, z_i, may always be computed before the ith system has to be solved.

In a similar vein to block Gauss–Seidel, we may also define a *block* Jacobi method

$$D_{ii}x_i^{(k+1)} = \left(\sum_{j=1}^{i-1} L_{ij}x_j^{(k)} + \sum_{j=1+1}^{N} U_{ij}x_j^{(k)} \right), \qquad i = 1, 2, \ldots, N,$$

and a *block* SOR method

$$x_i^{(k+1)} = (1-\omega)x_i^{(k)} + \omega \left\{ D_{ii}^{-1} \left(\sum_{j=1}^{i-1} L_{ij}x_j^{(k+1)} + \sum_{j=i+1}^{N} U_{ij}x_j^{(k)} \right) \right\}, \quad i = 1, 2, \ldots, N.$$

If the matrix Q is irreducible, the N systems of equations (3.18) are nonhomogeneous and have nonsingular coefficient matrices. We may use either direct or iterative methods to solve them. Naturally, there is no requirement to use the same method to solve all the diagonal blocks. Instead, it is possible to tailor methods to the particular block structures.

If a direct method is to be used, then an LU decomposition of each block D_{ii} may be formed once and for all before beginning the iteration, so that solving $D_{ii}x_i^{(k+1)} = z_i$, $i = 1, \ldots, N$ in each iteration simplifies to a forward and backward substitution. The nonzero structure of the blocks may be such that this is a particularly attractive approach. For example, if the diagonal blocks are themselves diagonal matrices, or if they are upper or lower triangular matrices or even tridiagonal matrices, then it is very easy to obtain their LU decomposition, and a block iterative method becomes very attractive. The LU factors may overwrite the storage arrays used to hold the diagonal blocks.

If the diagonal blocks do not possess such a structure, and when they are of large dimension, it may be appropriate to use an iterative method to solve each of the block systems. In this case, we have many inner iterative methods (one per block) within an outer (or global) iteration. A number of tricks may be used to speed up this process. First, the solution computed for any block D_{ii} at iteration k should be used as the initial approximation to the solution of this same block at iteration $k + 1$. Second, it is hardly worthwhile computing a highly accurate solution in early (outer) iterations. We should require only a small number of digits of accuracy until the global process begins to converge. One convenient way to achieve this is to carry out only a fixed, small number of iterations for each inner solution. Initially, this will not give much accuracy, but when combined with the first suggestion, the accuracy achieved will increase from one outer iteration to the next.

Intuitively, it is expected that for a given transition rate matrix Q, the larger the block sizes (and thus the smaller the number of blocks), the fewer the (outer) iterations needed to achieve convergence. This has been shown to be true under fairly general assumptions on the coefficient matrix for general systems of equations (see [170]). In the special case of only one block, the method degenerates to a standard direct method and we compute the solution in a single "iteration."

The reduction in the number of iterations that usually accompanies larger blocks is offset to a certain degree by an increase in the number of operations that must be performed at each iteration. However, in some important cases it may be shown that there is no increase. For example, when the matrix is block tridiagonal (as arises in quasi-birth-death processes) and the diagonal blocks are also tridiagonal, it may be shown that the computational effort per iteration is the same for both point and block iterative methods. In this case the reduction in the number of iterations makes the block methods very attractive indeed.

3.3.1 MATLAB Code and Examples

We wish to provide some examples to show the performance of block iterative methods. The experiments were conducted using MATLAB and the following code. The *Block Gauss–Seidel* program accepts a stochastic matrix, P; a partitioning vector, ni, whose i^{th} component stores the length of the ith block; and two integers: *itmax1*, which denotes the number of outer iterations to perform, and *itmax2*, which denotes the number of iterations to use to solve each of the blocks *if* Gauss–Seidel is used to solve these blocks. The program returns the solution vector π and a vector of residuals. It calls the Gauss–Seidel MATLAB program given in Section 3.2.3. The vector of residuals obtained when this program is supplied with the Courtois (8×8) NCD matrix is given in Table 3.5. The block method is apparently very effective in this example.

If the partitioning provided to the algorithm does not match the decomposability characteristics of the matrix, the convergence behavior may be much less satisfactory. For instance, in the foregoing example, the vector ni was taken to be $(3, 2, 3)$. If instead we use $(2, 2, 2, 2)$ or $(2, 4, 2)$, the procedure converges extremely slowly. On the other hand, if we use $(4, 4)$ or $(2, 3, 3)$, convergence is almost as rapid as with $(3, 2, 3)$.

Table 3.5: Residuals from Block Gauss–Seidel Method in Courtois Problem

Iter	Block-GS Residuals
	$1.0e-05\times$
1	0.94805408435419
2	0.01093707688215
3	0.00046904081241
4	0 .00002012500900
5	0.00000086349742
6	0.00000003705098
7	0.00000000158929
8	0.00000000006641
9	0.00000000000596
10	0.00000000000395

Algorithm: Block Gauss–Seidel.

```
function [x,res] = bgs(P,ni,itmax1,itmax2)
[n,n] = size(P); [na,nb] = size(ni);

%          BLOCK GAUSS-SEIDEL FOR P^T x = x

bl(1) = 1;                                  % Get beginning and end
for k = 1:nb, bl(k+1) = bl(k)+ni(k); end    % points of each block
x = ones(n,1)/n;                            % Initial approximation

%%%%%%%%%%%%%%%%%%%%%%%%%% BEGIN OUTER LOOP %%%%%%%%%%%%%%%%%%%%%%%%%%%%
for iter = 1:itmax1,
    for m = 1:nb,                           % All diagonal blocks
        A = P(bl(m):bl(m+1)-1,bl(m):bl(m+1)-1)'; % Get A_mm
        b = -P(1:n,bl(m):bl(m+1)-1)'*x+A*x(bl(m):bl(m+1)-1); % RHS
        z = inv(A-eye(ni(m)))*b;            % Solve for z

%          *** To solve the blocks using Gauss-Seidel    ***
%          *** instead of a direct method, substitute    ***
%          *** the next two lines for the previous one.   ***
%**        x0 = x(bl(m):bl(m+1)-1);        % Get starting vector
%**        [z,r] = gs(A-eye(ni(m)),x0,b,itmax2); % Solve for z

   x(bl(m):bl(m+1)-1) = z;                  % Update x
        end
    res(iter) = norm((P'-eye(n))*x,2);      % Compute residual
end
x = x/norm(x,1); res = res';
```

3.4   Preconditioned Power Iterations

We have already seen that the power method can be extremely slow to converge when the subdominant eigenvalue of P is close to 1. The relaxation schemes described in Section 3.2 typically have better convergence rates. This means that the iteration matrices corresponding to these schemes have a subdominant eigenvalue whose modulus is farther from 1 than the subdominant eigenvalue of P. This provides the basic idea behind preconditioning. The original system of equations is modified in such a way that the solution is unchanged, but the distribution of eigenvalues is better suited for iterative methods.

In a general context, preconditioning consists of replacing a system of equations $Ax = b$ with a modified system

$$M^{-1}Ax = M^{-1}b. \tag{3.19}$$

Here M^{-1} is a preconditioning matrix; M should be chosen so that its LU factorization is inexpensive to compute. If M^{-1} approximates A^{-1}, then it is expected that the application of iterative methods to (3.19) will yield the solution in very few iterations.

When the coefficient matrix is singular and the right-hand side is zero, preconditioning with M is identified as the application of the power method to the matrix

$$(I - M^{-1}A).$$

For Markov chain problems, the power method may be written as

$$x^{(k+1)} = (I - (I - P^T))x^{(k)},$$

and preconditioning involves premultiplying the matrix $(I - P^T)$ with a matrix M^{-1} chosen to approximate the group inverse $(I - P^T)^{\#}$. In this case the iteration matrix

$$(I - M^{-1}(I - P^T)) \tag{3.20}$$

has one unit eigenvalue and the remaining eigenvalues are close to zero, leading to a rapidly converging iterative procedure. We refer to such methods as *preconditioned power iterations* or *fixed-point iterations*.

Other arrangements are possible. Given an LU factorization, $M = \tilde{L}\tilde{U}$,

$$M^{-1} = \tilde{U}^{-1}\tilde{L}^{-1} \approx (I - P^T)^{\#},$$

we may also use

$$x^{(k+1)} = (I - (I - P^T)M^{-1})x^{(k)}$$

or

$$x^{(k+1)} = (I - \tilde{U}^{-1}(I - P^T)\tilde{L}^{-1})x^{(k)}.$$

3.4.1   Gauss–Seidel, SOR, and SSOR Preconditionings

A look at the defining equation for the method of Gauss–Seidel (Equation 3.11) reveals an interesting connection with the power method. We have

$$
\begin{aligned}
x^{(k+1)} &= (D-L)^{-1}Ux^{(k)} \\
&= (D-L)^{-1}\left\{(D-L)-Q^T\right\}x^{(k)} \\
&= x^{(k)} - (D-L)^{-1}Q^Tx^{(k)},
\end{aligned}
$$

which shows Gauss–Seidel to be the power method applied to the matrix

$$
I - (D-L)^{-1}Q^T. \tag{3.21}
$$

Thus $(D-L)^{-1}$ performs the role of the preconditioning matrix M^{-1}, and we may view the Gauss–Seidel method as a preconditioned power iteration. Similarly, we may define an SOR preconditioning and an SSOR preconditioning.

3.4.2   *ILU* Preconditioning

Most preconditioners are obtained from incomplete LU factorizations. The resulting fixed-point iterations are sometimes referred to as *combined direct–iterative* methods. They are composed of two phases. First, an LU decomposition of $(I - P^T)$ is initiated. At various points during the decomposition, nonzero elements are omitted according to various rules. Some possibilities are discussed in the following paragraphs. In all cases, instead of arriving at an exact LU decomposition, what we obtain is of the form

$$
(I - P^T) = \tilde{L}\tilde{U} - E \tag{3.22}
$$

where E, called the *remainder*, is expected to be small in some sense. When this has been achieved, the "direct" phase of the computation is complete. In the second phase, the incomplete factorization is incorporated into an iterative procedure by writing

$$
(I - P^T)x = (\tilde{L}\tilde{U} - E)x = 0
$$

and then using

$$
\tilde{L}\tilde{U}x^{(k+1)} = Ex^{(k)}
$$

or equivalently

$$
x^{(k+1)} = x^{(k)} - (\tilde{L}\tilde{U})^{-1}(I - P^T)x^{(k)}
$$

as the iteration scheme. Note that this is the same as solving the preconditioned (from the left) system of equations

$$
\tilde{U}^{-1}\tilde{L}^{-1}(I - P^T)x = 0
$$

by the power method.

One possibility that comes to mind when the matrix is block-structured is to let $\tilde{L}\tilde{U}$ consist of the diagonal blocks and E be the matrix of off-diagonal blocks. In this case, the incomplete factorization method is just the block Jacobi method and is likely to be less satisfactory than block Gauss–Seidel.

We shall consider three types of incomplete factorizations that have appeared in the literature. The first forces the computed factors to have the same nonzero structure as the original matrix and is called *ILU(0)*; the second is a threshold-based approach that is called *ILUTH*; the third permits the computed factors to have only a fixed number of nonzero elements per row and is called *ILUK*. There has been only limited study of what constitutes good incomplete factorizations for Markov chain models. The concept is still in its infancy in this domain, and much more extensive studies are still needed. Nevertheless, the results that are currently available are encouraging.

3.4.2.1   The *ILU(0)* Incomplete Factorization

The first type of incomplete factorization has been widely used and found to be successful when applied to systems of equations that arise in the solution of elliptic partial differential equations. Given the matrix $(I - P^T)$, this *ILU* factorization consists of performing the usual Gaussian elimination factorization and discarding any fill-in during the process. In other words, the factorization is given by

$$(I - P^T) = \tilde{L}\tilde{U} - E$$

where $\tilde{L}$ is unit lower triangular, $\tilde{U}$ is upper triangular, and $\tilde{L} + \tilde{U}$ has the same zero structure as the matrix $(I - P^T)$. This is referred to as *ILU(0)*, or *IC(0)* for incomplete Cholesky, in the symmetric case. The *ILU(0)* factorization is known to exist for nonsingular *M*-matrices [95]. It may also be shown to exist for the matrices $(I - P)$ and $(I - P^T)$ by trivially extending the results in ([7], p. 42).

3.4.2.2   The *ILUTH* Incomplete Factorization

In *ILUTH*, the decomposition proceeds in a row-by-row manner similar to that described for the GE method of Chapter 2. However, after a row of the matrix has been reduced and before that row is recompacted and stored, each nonzero element is examined. If its absolute value is less than a prespecified threshold, then it is replaced by zero. Similarly, if any of the multipliers formed during the reduction are less than the threshold, they are dropped from further consideration. The only exceptions to this drop threshold are the diagonal elements, which are kept no matter how small they become.

3.4.2.3   The *ILUK* Incomplete Factorization

The final type of incomplete factorization that we consider is based on a realization that only a fixed amount of memory may be available to store the incomplete factors $\tilde{L}$ and $\tilde{U}$, so only a fixed number of nonzero elements is kept in each row. These are usually chosen to be the largest in magnitude. The algorithm proceeds as for *ILUTH*. When a row has been reduced, a search is conducted to find the K largest elements in absolute value. This search is conducted over both the multipliers and the nonzero elements to the right of the diagonal element. As for *ILUTH*, the diagonal elements are kept regardless of their magnitude.

3.4.3   Examples

The following MATLAB code was used to generate *ilu(0)* factors used in the experiments of this section. Given an infinitesimal generator Q, the program constructs a matrix R, whose upper triangular section contains the upper triangular factor $\tilde{U}$ and whose strictly lower triangular part stores $\tilde{L}$ (the unit diagonal elements of $\tilde{L}$ are known implicitly). The rows to be reduced are taken one at a time; for each nonzero element to the left of the diagonal, a multiplier is computed and overwrites the nonzero element. The product of the multiplier and the appropriate preceding row is subtracted from the row undergoing the reduction. So far this is what normally happens in Gaussian elimination. The difference comes in the final operation on each row, for only elements (those belonging to the reduced upper triangular part as well as multipliers) that occupy positions that store nonzero elements in the original matrix are actually kept. This means that the nonzero structure of the original matrix and the nonzero structure of the incomplete factors are identical. The benefit of this is not apparent in the following code. Only when Q is large and sparse and compact storage schemes are used do the benefits become important. As mentioned before, the purpose of these sections of MATLAB code is to clearly show the sequence of operations that must be followed and to provide some feel for the applicability of the algorithms. They are *not* meant to represent production versions of the algorithms. The reader may wish to verify that when the last *for* loop is commented out, an exact *LU* factorization is obtained.

Algorithm: Incomplete Factorization, *ILU(0)*.

```
function [R] = ilu0(Q)
[n,n] = size(Q);

%     ***    ILU(0) --- FOR INFINITESIMAL GENERATOR, Q  ***
```

```
        R = Q;
        for k = 2:n,                      %  Rows 2 thro' n
            for i = 1:k-1,                %  Get elements to reduce
                if R(k,i) > 0,            %  Nonzero element?
                    R(k,i) = R(k,i)/R(i,i);   %  Multiplier
                    for j = i+1:n,
                        R(k,j) = R(k,j) - R(k,i)*R(i,j); %  Now eliminate
                    end
                end
            end
            for i = 1:n,
                if Q(k,i)==0, R(k,i)=0; end %  Keep original structure
            end
        end
```

To implement a threshold-based incomplete factorization, the lines of code

```
        for i = 1:n,
            if Q(k,i)==0, R(k,i)=0; end %  Keep original structure
        end
```

must be replaced by

```
        for i = 1:n,
            if abs(R(k,i)) < threshold & i~=k, R(k,i)=0; end
        end
```

Similarly, in an *ILUK* factorization these lines will be replaced with code that keeps only the k largest elements. When using an *ILUTH* or *ILUK* incomplete factorization, it is appropriate to scale the matrix first so that the diagonal elements are equal to one. This simply means dividing the ith equation by the ith diagonal element.

The matrix chosen for the experiments was the (9×9) reliability model with parameters

$$\lambda_1 = 0.2, \qquad \lambda_2 = 30.0, \qquad \mu_1 = 0.5, \qquad \mu_2 = 60.0.$$

The infinitesimal generator matrix is given by

$$
\begin{pmatrix}
-60.40 & 60.00 & 0 & 0.40 & 0 & 0 & 0 & 0 & 0 \\
60.00 & -90.40 & 30.00 & 0 & 0.40 & 0 & 0 & 0 & 0 \\
0 & 120.00 & -120.40 & 0 & 0 & 0.40 & 0 & 0 & 0 \\
0.50 & 0 & 0 & -60.70 & 60.00 & 0 & 0.20 & 0 & 0 \\
0 & 0.50 & 0 & 60.00 & -90.70 & 30.00 & 0 & 0.20 & 0 \\
0 & 0 & 0.50 & 0 & 120.00 & -120.70 & 0 & 0 & 0.20 \\
0 & 0 & 0 & 1.00 & 0 & 0 & -61.00 & 60.00 & 0 \\
0 & 0 & 0 & 0 & 1.00 & 0 & 60.00 & -91.00 & 30.00 \\
0 & 0 & 0 & 0 & 0 & 1.00 & 0 & 120.00 & -121.00
\end{pmatrix}
$$

and, when divided by the largest diagonal element, becomes

$$
\begin{pmatrix}
-0.4992 & 0.4959 & 0 & 0.0033 & 0 & 0 & 0 & 0 & 0 \\
0.4959 & -0.7471 & 0.2479 & 0 & 0.0033 & 0 & 0 & 0 & 0 \\
0 & 0.9917 & -0.9950 & 0 & 0 & 0.0033 & 0 & 0 & 0 \\
0.0041 & 0 & 0 & -0.5017 & 0.4959 & 0 & 0.0017 & 0 & 0 \\
0 & 0.0041 & 0 & 0.4959 & -0.7496 & 0.2479 & 0 & 0.0017 & 0 \\
0 & 0 & 0.0041 & 0 & 0.9917 & -0.9975 & 0 & 0 & 0.0017 \\
0 & 0 & 0 & 0.0083 & 0 & 0 & -0.5041 & 0.4959 & 0 \\
0 & 0 & 0 & 0 & 0.0083 & 0 & 0.4959 & -0.7521 & 0.2479 \\
0 & 0 & 0 & 0 & 0 & 0.0083 & 0 & 0.9917 & -1.0000
\end{pmatrix}
$$

We shall present all the steps in the computation of the *ILU(0)* factorization of this matrix and leave the other incomplete factorizations as an exercise for the reader. Obviously the first row is unaltered. Consider the second row of Q:

$$
\begin{pmatrix} 0.4959 & -0.7471 & 0.2479 & 0 & 0.0033 & 0 & 0 & 0 & 0 \end{pmatrix}
$$

The first element is eliminated by subtracting a multiple of the first row from the second; the multiplier takes the place of this element:

$$
\begin{pmatrix} -0.9934 & -0.2545 & 0.2479 & 0.0033 & 0.0033 & 0 & 0 & 0 & 0 \end{pmatrix}
$$

Before we finish with this second row, we must set the fourth element to zero, since the fourth element of the second row of Q is zero. The final values of row 2 are

$$
\begin{pmatrix} -0.9934 & -0.2545 & 0.2479 & 0 & 0.0033 & 0 & 0 & 0 & 0 \end{pmatrix}
$$

This procedure must be performed for each row. The complete sequence of operations is given in Table 3.6, which shows the initial value of each row, the values after each reduction of a nonzero component, and the final values before the row is stored. The matrix R that holds the *ILU(0)* factors is finally obtained as

$$
\begin{pmatrix}
-0.4992 & 0.4959 & 0 & 0.0033 & 0 & 0 & 0 & 0 & 0 \\
-0.9934 & -0.2545 & 0.2479 & 0 & 0.0033 & 0 & 0 & 0 & 0 \\
0 & -3.8964 & -0.0290 & 0 & 0 & 0.0033 & 0 & 0 & 0 \\
-0.0083 & 0 & 0 & -0.5016 & 0.4959 & 0 & 0.0017 & 0 & 0 \\
0 & -0.0162 & 0 & -0.9885 & -0.2593 & 0.2484 & 0 & 0.0017 & 0 \\
0 & 0 & -0.1426 & 0 & -3.8246 & -0.0470 & 0 & 0 & 0.0017 \\
0 & 0 & 0 & -0.0165 & 0 & 0 & -0.5041 & 0.4959 & 0 \\
0 & 0 & 0 & 0 & -0.0319 & 0 & -0.9837 & -0.2642 & 0.2482 \\
0 & 0 & 0 & 0 & 0 & -0.1757 & 0 & -3.7538 & -0.0680
\end{pmatrix}
$$

Table 3.6: Consecutive Operations of *ILU(0)* on Each Row of Q

Row 2:

0.4959	-0.7471	0.2479	0	0.0033	0	0	0	0
-0.9934	-0.2545	0.2479	0.0033	0.0033	0	0	0	0
-0.9934	-0.2545	0.2479	0	0.0033	0	0	0	0

Row 3:

0	0.9917	-0.9 50	0	0	0.0033	0	0	0
0	-3.8964	-0.0290	0	0.0129	0.0033	0	0	0
0	-3.8964	-0.0290	0	0	0.0033	0	0	0

Row 4:

0.0041	0	0	-0.5017	0.4959	0	0.0017	0	0
-0.0083	0.0041	0	-0.5016	0.4959	0	0.0017	0	0
-0.0083	-0.0161	0.0040	-0.5016	0.4959	0	0.0017	0	0
-0.0083	-0.0161	-0.1380	-0.5016	0.4959	0.0005	0.0017	0	0
-0.0083	0	0	-0.5016	0.4959	0	0.0017	0	0

Row 5:

0	0.0041	0	0.4959	-0.7496	0.2479	0	0.0017	0
0	-0.0162	0.0040	0.4959	-0.7495	0.2479	0	0.0017	0
0	-0.0162	-0.1389	0.4959	-0.7495	0.2484	0	0.0017	0
0	-0.0162	-0.1389	-0.9885	-0.2593	0.2484	0.0016	0.0017	0
0	-0.0162	0	-0.9885	-0.2593	0.2484	0	0.0017	0

Row 6:

0	0	0.0041	0	0.9917	-0.9975	0	0	0.0017
0	0	-0.1426	0	0.9917	-0.9970	0	0	0.0017
0	0	-0.1426	0	-3.8246	-0.0470	0	0.0063	0.0017
0	0	-0.1426	0	-3.8246	-0.0470	0	0	0.0017

Row 7:

0	0	0	0.0083	0	0	-0.5041	0.4959	0
0	0	0	-0.0165	0.0082	0	-0.5041	0.4959	0
0	0	0	-0.0165	-0.0315	0.0078	-0.5041	0.4959	0
0	0	0	-0.0165	-0.0315	-0.1664	-0.5041	0.4959	0.0003
0	0	0	-0.0165	0	0	-0.5041	0.4959	0

Row 8:

0	0	0	0	0.0083	0	0.4959	-0.7521	0.2479
0	0	0	0	-0.0319	0.0079	0.4959	-0.7520	0.2479
0	0	0	0	-0.0319	-0.1683	0.4959	-0.7520	0.2482
0	0	0	0	-0.0319	-0.1683	-0.9837	-0.2642	0.2482
0	0	0	0	-0.0319	0	-0.9837	-0.2642	0.2482

Row 9:

0	0	0	0	0	0.0083	0	0.9917	-1.0000
0	0	0	0	0	-0.1757	0	0.9917	-0.9997
0	0	0	0	0	-0.1757	0	-3.7538	-0.0680
0	0	0	0	0	-0.1757	0	-3.7538	-0.0680

Table 3.7: Eigenvalues of Iteration Matrices With and Without Preconditioners

$I + Q \, (= P)$	$I - \tilde{U}^{-1}\tilde{L}^{-1}Q$	$I - \hat{U}^{-1}\hat{L}^{-1}Q$	$I - U^{-1}L^{-1}Q$
1.0000000000	1.0000000000	1.0000000000	1.0000000000
0.9942148760	−0.8172637875	−0.0437860838	0.0000000000
0.9884297520	−0.0652755760	0.0073502785 + 0.0229713363i	0.0000000000
−0.4991735537	0.0462325061	0.0073502785 − 0.0229713363i	0.0000000000
−0.4876033058	−0.0051050607	−0.0035033159	0.0000000000
−0.4933884298	0.0051670576	0.0035093086	0.0000000000
0.2561983471	0.0000000000	0.0000000000	0.0000000000
0.2446280992	0.0000000000	0.0000000000	0.0000000000
0.2504132231	0.0000000000	0.0000000000	0.0000000000

To investigate the benefits of the preconditioners we need to examine the eigenvalues of the iteration matrices. Table 3.7 contains the eigenvalues of

1. The original stochastic matrix, $P = I + Q$,

2. The iteration matrix when the *ILU(0)* factorization is used, $\tilde{U}^{-1}\tilde{L}^{-1}Q$;

3. The iteration matrix when the *ILUTH* factorization is used, $\hat{U}^{-1}\hat{L}^{-1}Q$ (in this particular case the threshold was taken to be 0.1);

4. Finally, just for comparison purposes, the fourth column contains the values obtained when an exact *LU* factorization is used: $U^{-1}L^{-1}Q$.

Note in this last case that all the eigenvalues are equal to 0 except the first, which is equal to 1. The number of iterations needed to achieve convergence may be obtained from an examination of the subdominant eigenvalue.

3.4.4 Summary

Experiments with different incomplete factorizations applied to Markov chain models tend to show that *ILUTH* is most effective, provided a threshold of 10^{-3} or smaller can be accommodated; *ILU(0)* appears to be the least effective, despite the success that it has achieved in other domains.

We wish to point out that these three ILU factorizations are not the only ones possible. One that has been proposed in the numerical analysis literature makes use of the *symmetric nonzero structure* of a matrix; $\tilde{L}\tilde{U}$ is the exact decomposition of the symmetric nonzero portion of the matrix. $M = \tilde{L}\tilde{U}$ is chosen as

$$m_{ij} = a_{ij} \quad \text{if } a_{ij}a_{ji} \neq 0, \quad \text{and}$$
$$m_{ij} = 0 \quad \text{otherwise,}$$

and now standard symmetric ordering schemes, such as those available in SPARS-PAK, can be modified and used quite effectively. However, it is not known how this or other incomplete factorizations perform on Markov models.

A final word of warning. When searching for suitable preconditioners of the ILU type, $(I - P^T) = \tilde{L}\tilde{U} - E$, a small norm for E does not necessarily imply that $(\tilde{L}\tilde{U})^{-1}$ is close to the inverse of $(I - P^T)$!

3.5 Implementation Considerations

We focus now on various aspects that must be taken into account when producing software that implements iterative methods for the solution of large-scale Markov chains. Although in certain parts this discussion evolves primarily around the SOR method, the general guidelines are applicable to all iterative methods.

3.5.1 Sparse Storage Schemes

One of the major advantages that iterative methods have over direct methods is that no modification of the elements of the coefficient matrix occurs during the execution of the algorithm. Thus, the matrix may be stored once and for all in some convenient compact form without the need to provide mechanisms to handle insertions (due to zero elements becoming nonzero) and deletions (due to the elimination of nonzero elements). All that is usually required is the ability to compute the product of the matrix and a given vector. Naturally, the semisystematic packing schemes and the linking techniques described in Chapter 2 in the context of direct methods may also be used for iterative methods. However, the linking scheme is unnecessarily complex for iterative applications, and more efficient approaches can be developed and implemented.

One simple approach is to use a real (double-precision) one-dimensional array aa to store the nonzero elements of the matrix and two integer arrays ia and ja to indicate, respectively, the row and column positions of these elements. Thus, if the nonzero element a_{ij} is stored in position k of aa, i.e., $aa(k) = a_{ij}$, we have $ia(k) = i$ and $ja(k) = j$. For example, the (4×4) matrix A given by

$$A = \begin{pmatrix} -2.1 & 0.0 & 1.7 & 0.4 \\ 0.8 & -0.8 & 0.0 & 0.0 \\ 0.2 & 1.5 & -1.7 & 0.0 \\ 0.0 & 0.3 & 0.2 & -0.5 \end{pmatrix}$$

may be stored as

$$
\begin{array}{lccccccccccc}
aa: & -2.1 & 1.7 & 0.4 & -0.8 & 0.8 & -1.7 & 0.2 & 1.5 & -0.5 & 0.3 & 0.2 \\
ia: & 1 & 1 & 1 & 2 & 2 & 3 & 3 & 3 & 4 & 4 & 4 \\
ja: & 1 & 3 & 4 & 2 & 1 & 3 & 1 & 2 & 4 & 2 & 3
\end{array}
$$

To form the product $z = Ax$, when A is stored in this form, we proceed as follows:

Algorithm: Sparse Matrix-Vector Multiply I.

1. Set $z_i = 0$ for all i.

2. For $next = 1$ to n_z do

   - Set $nrow = ia(next)$.
   - Set $ncol = ja(next)$.
   - Compute $z(nrow) = z(nrow) + aa(next) \times x(ncol)$.

Here n_z denotes the number of nonzero elements in the matrix A. To perform the product $z = A^T x$ it suffices simply to interchange the arrays ia and ja.

In this algorithm it is necessary to begin by setting the elements of the vector z to zero, because the inner products are computed in a piecemeal fashion. If some ordering is imposed on the positions of the nonzero elements in the array aa, this can be avoided. In the example given above, the elements are stored by rows; elements of row i precede those of row $i + 1$. The elements within a row may or may not be in order. In this case, the product $z = Ax$ may be formed as follows:

Algorithm: Sparse Matrix-Vector Multiply II.

1. Set $next = 1$.

2. For $i = 1$ to n do

   - Set $sum = 0$.
   - While $ia(next) = i$ do
     - Set $ncol = ja(next)$.
     - Compute $sum = sum + aa(next) \times x(ncol)$.
     - Compute $next = next + 1$.
   - Set $z(i) = sum$.

However, to form the product $z = A^T x$, it is still necessary to first initialize the vector z to zero.

When the nonzero elements are stored by rows, it is possible to dispense with the integer array ia and to replace it with a smaller array. The most commonly used compact storage scheme uses the elements of ia as pointers into the arrays aa and ja. The kth element of ia denotes the position in aa and ja at which the first element of row k is stored. Thus, we always have $ia(1) = 1$. Additionally, it is usual to store the first empty position of aa and ja in position $(n + 1)$ of ia. Most often this means that $ia(n + 1) = n_z + 1$. The number of nonzero elements in row i is then given by $ia(i + 1) - ia(i)$. The 4×4 example matrix is stored in this form as

aa :	-2.1	1.7	0.4	-0.8	0.8	-1.7	0.2	1.5	-0.5	0.3	0.2
ja :	1	3	4	2	1	3	1	2	4	2	3
ia :	1	4	6	9	12						

It is not necessary for the elements in any row to be in order; it suffices that all the nonzero elements of row i come before those of row $i + 1$ and after those of row $i - 1$. Using this storage scheme, the matrix-vector product $z = Ax$ may be computed by

Algorithm: Sparse Matrix-Vector Multiply III.

1. For $i = 1$ to n do

   - Set $sum = 0$.
   - Set $initial = ia(i)$.
   - Set $last = ia(i + 1) - 1$.
   - For $j = initial$ to $last$ do
     - Compute $sum = sum + aa(j) \times x(ja(j))$.
   - Set $z(i) = sum$.

When programming SOR, we use the formula

$$x_i^{(k+1)} = (1 - \omega)x_i^{(k)} + \frac{\omega}{a_{ii}} \left(b_i - \sum_{j=1}^{i-1} a_{ij}x_j^{(k+1)} - \sum_{j=i+1}^{n} a_{ij}x_j^{(k)} \right).$$

By scaling the matrix so that $a_{ii} = 1$ for all i and setting $b_i = 0$, for all i, this reduces to

$$x_i^{(k+1)} = (1 - \omega)x_i^{(k)} - \omega \left(\sum_{j=1}^{i-1} a_{ij}x_j^{(k+1)} + \sum_{j=i+1}^{n} a_{ij}x_j^{(k)} \right)$$

$$= x_i^{(k)} - \omega \left(\sum_{j=1}^{i-1} a_{ij} x_j^{(k+1)} + a_{ii} x_i^{(k)} + \sum_{j=i+1}^{n} a_{ij} x_j^{(k)} \right).$$

At iteration k, the program may be written (assuming A is stored in the compact form just described) simply as

Algorithm: Sparse SOR.

1. For $i = 1$ to n do

   - Set $sum = 0$.
   - Set $initial = ia(i)$.
   - Compute $last = ia(i + 1) - 1$.
   - For $j = initial$ to $last$ do
     - Compute $sum = sum + aa(j) \times x(ja(j))$.
   - Compute $x(i) = x(i) - \omega \times sum$.

It is possible here to overwrite the elements of x with their new values once they are obtained, simply because they are determined sequentially. The computation of element $x(i + 1)$ does not begin until the new value of $x(i)$ has already been computed.

In all cases in which the matrix is stored by rows, it is necessary to initialize the array z to zero before the product $z = A^T x$ can be computed. Thus, when implementing SOR to solve systems of the form $Q^T x = 0$, it is necessary to carry an extra working array into which the product $Q^T x$ can be accumulated. Also, regardless of whether Qx or $Q^T x$ is being computed, a one-dimensional array of length n must be used to keep a previous vector against which to test for convergence. When only consecutive iterates are tested, this vector may serve to accumulate the product $Q^T x$. We shall see later that it is often necessary to compare iterates that are not consecutive but are spaced farther apart. In this case, if Q is stored by rows and SOR is used to determine a solution of $Q^T x = 0$, then three one-dimensional arrays of length n are needed; one to hold the current iterate $x^{(k)}$, one into which to accumulate the new iterate $x^{(k+1)}$, and finally one to hold the iterate $x^{(k-m)}$ against which convergence testing is conducted. The moral is obviously to try to store the matrix Q by columns. Unfortunately, in many Markov chain applications, it is much more convenient to determine all destination states that occur from a given source state (row-wise generation) than to determine all source states which lead to a given destination state (columnwise generation). Given current memory/processor trade-offs, in which processing time is the more precious resource, it is usual to adopt a row-wise storage scheme and to use the three-array approach. An alternative is to implement a procedure to transpose the matrix in this compact form right at the beginning and iterate with $A = Q^T$ instead.

3.5.2 Choice of an Initial Iteration Vector

When choosing an initial starting vector for an iterative method, it is tempting to choose something simple, such as a vector of all ones or a vector whose components are all zero except for one or two entries. If such a choice is made, care must be taken to ensure that the initial vector is not deficient in some component of the basis of the solution vector. Consider, for example, the (2×2) system

$$Q = \begin{pmatrix} -\lambda & \lambda \\ \mu & -\mu \end{pmatrix}.$$

Applying Gauss–Seidel with $\pi^{(0)} = (1, \ 0)$ yields $\pi^{(k)} = (0, \ 0)$ for all $k \geq 1$, since

$$x^{(1)} = (D - L)^{-1} U x^{(0)} = (D - L)^{-1} \begin{pmatrix} 0 & -\mu \\ 0 & 0 \end{pmatrix} \begin{pmatrix} 1 \\ 0 \end{pmatrix} = (D - L)^{-1} \begin{pmatrix} 0 \\ 0 \end{pmatrix} = 0.$$

If some approximation to the solution is known, then it should be used. However, if this is not possible, the elements of the initial iterate should be assigned random numbers uniformly distributed between 0 and 1 and then normalized to produce a probability vector.

Sometimes it is possible to compute an approximation to the solution vector. Consider, for example, the case of dependability models, which consist of components that break down and are subsequently repaired. The states may be ordered so that those with j failed components precede those in which $j + 1$ have failed. With this ordering, state 1 corresponds to the case in which no components have failed; generally this is the state in which the system spends most of its time. Thus, the numerical value of the first component of the solution vector will be much larger than the others. However, rather than setting the first component of the initial iterate equal to one and all others equal to zero, it is better to assign small, randomly distributed, positive numbers to $\pi_i^{(0)}$, $i \neq 1$ and assign the remaining probability mass to $\pi_1^{(0)}$. If the failure and repair rates of the different components are identical, this idea may be taken a step further. As an initial approximation, elements representing states in which j components have failed may be assigned the value 10^{-j} or 10^{-mj} for some small positive integer m (say, $m = 2$ or 3). The vector $\pi^{(0)}$ should then be normalized so that $||\pi^{(0)}||_1 = 1$. If we consider the (9×9) reliability model whose parameters are given by

$$\lambda_1 = 0.2; \qquad \lambda_2 = 0.3; \qquad \mu_1 = 5.0; \quad \text{and} \quad \mu_2 = 6.0,$$

we find that the stationary probabilities of the states are

$(2,2)$	$(2,1)$	$(2,0)$	$(1,2)$	$(1,1)$	$(1,0)$	$(0,2)$	$(0,1)$	$(0,0)$
.838600	.083860	.067088	.002096	.006709	.001342	.000168	.000134	.000003

The magnitude of these probabilities is proportional to the number of type 1 components that are operational; the number of type 2 operational components plays a subsidary role.

3.5.3   Normalization of Successive Approximations

It is sometimes unnecessary to constrain successive approximations to be probability vectors. Removing the normalization procedure will reduce the amount of computation required per iteration. Once the convergence criteria have been satisfied, a final normalization will give the desired probability vector. Let us examine the question of overflow and underflow of the elements of the successive approximations.

Underflow: While this is not fatal, it often gives rise to an undesirable error message. This may be avoided by periodically checking the magnitude of the elements and by setting those that are less than a certain threshold (e.g., 10^{-25}) to zero. Note that the underflow problem can arise even when the approximations are normalized at each iteration, since some elements of the solution vector, although strictly positive, may be extremely small. The concern when normalization is not performed is that *all* of the elements will become smaller with each iteration until they are all set to zero. This problem can be avoided completely by a periodic scan of the approximation to ensure that at least one element exceeds a certain minimum threshold and to initiate a normalization when this test fails.

Overflow: With a reasonable starting vector, overflow is unlikely to occur, since the eigenvalues of the iteration matrices should not exceed 1. All doubt can be eliminated by keeping a check on the magnitude of the largest element and normalizing the iterate if this element exceeds a certain maximum threshold, say 10^{10}.

3.5.4   Testing for Convergence

The number of iterations k needed to satisfy a tolerance criterion ϵ may be obtained approximately from the relationship

$$\rho^k = \epsilon, \quad \text{i.e., } k = \frac{\log \epsilon}{\log \rho},$$

where ρ is the spectral radius of the iteration matrix. In Markov chain problems, the magnitude of the subdominant eigenvalue is used in place of ρ. Thus, when we wish to have six decimal places of accuracy, we set $\epsilon = 10^{-6}$ and find that the number of iterations needed for different spectral radii are as follows:

ρ	.1	.5	.6	.7	.8	.9	.95	.99	.995	.999
k	6	20	27	39	62	131	269	1,375	2,756	13,809

The usual method of testing for convergence is to examine some norm of the difference of successive iterates; when this becomes less than a certain prespecified tolerance, the iterative procedure is stopped. This is a satisfactory approach when the procedure converges relatively rapidly and the magnitude of the largest element of the vector is of order unity. However, when the procedure is converging slowly, it is possible that the difference in successive iterates is smaller than the tolerance specified, even though the vector may be far from the solution. This problem may be overcome by testing not successive iterates,

$$\|\pi^{(k)} - \pi^{(k-1)}\| < \epsilon,$$

but rather iterates spaced further apart, e.g.,

$$\|\pi^{(k)} - \pi^{(k-m)}\| < \epsilon.$$

Ideally, m should be determined as a function of the convergence rate. A simple but less desirable alternative is to allow m to assume different values during the iteration procedure. For example, when the iteration number k is

$$
\begin{aligned}
k < 100 &\qquad \text{let } m = 5 \\
100 \le k < 500 &\qquad \text{let } m = 10 \\
500 \le k < 1,000 &\qquad \text{let } m = 20 \\
k \ge 1,000 &\qquad \text{let } m = 50
\end{aligned}
$$

A second problem arises when the approximations converge to a vector in which the elements are all small. Suppose the problem has 100,000 states, and all are approximately equally probable. Then each element of the solution is approximately equal to 10^{-5}. If the tolerance criterion is set at 10^{-3} and the initial vector is chosen such that $\pi_i^{(0)} = 1/n$ for all i, then the process will probably "converge" after one iteration!

This same problem can arise in a more subtle context if the approximations are not normalized before convergence testing *and* the choice of initial vector results in the iterative method converging to a vector in which all components are small. This may happen even though some elements of the solution may be large relative to others. For example, it happens in the (2×2) example given previously when the initial approximation is chosen as $(\xi,\ 10^{-6})$ for any value of ξ. (Note that the opposite effect will occur if the approximations converge to a vector with all elements relatively large, e.g., let the initial vector be $(\xi,\ 10^6)$ in the (2×2) example.)

A solution to this latter aspect of the problem (when vectors are not normalized) is, of course, to normalize the iterates before testing for convergence. If this normalization is such that it produces a probability vector, the original problem (all of the components may be small) still remains. A better choice of normalization in this instance is $\|\pi^{(k)}\|_\infty = 1$ (i.e., normalize so that the largest element of $\pi^{(k)}$ is equal to 1).

A better solution, however, and the one that is recommended, is to use a relative measure; e.g.,

$$\max_i \left(\frac{|\pi_i^{(k)} - \pi_i^{(k-m)}|}{|\pi_i^{(k)}|} \right) < \epsilon.$$

This effectively removes the *exponent* from consideration in the convergence test and hence gives a better estimate of the *precision* that has been achieved.

Other criteria that have been used for convergence testing include residuals (the magnitude of $\|\pi^{(k)}Q\|$, which should be small) and the use of subdominant eigenvalue approximations. Residuals will work fine in many and perhaps even most modelling problems. Unfortunately, a small residual does not always imply that the error in the solution vector is also small. In ill-conditioned systems the residual may be very small indeed, yet the computed solution may be hopelessly inaccurate. A small residual is a necessary condition for the error in the solution vector to be small — but it is not a sufficient condition. The most suitable approach is to check the residual after the relative convergence test indicates that convergence has been achieved. In fact, it is best to envisage a battery of convergence tests, all of which must be satisfied before the approximation is accepted as being sufficiently accurate.

We now turn to the frequency with which the convergence test should be administered. Normally it is performed during each iteration. This may be wasteful, especially when the matrix is very large and the iterative method is converging slowly. Sometimes it is possible to estimate the rate of convergence and to determine from this rate the approximate numbers of iterations that must be performed. It is now possible to proceed "full steam ahead" and carry out this number of iterations without testing for convergence or normalizing. For rapidly converging problems this may result in more iterations being performed than is strictly necessary, thereby achieving a more accurate result than the user actually needs. When the matrix is large and an iteration costly, this may be undesirable. One possibility is to carry out only a proportion (say 80 to 90%) of the estimated number of iterations before beginning to implement the battery of convergence tests.

An alternative approach is to implement a relatively inexpensive convergence test (e.g., using the relative difference in the first nonzero component of successive approximations) at each iteration. When this simple test is satisfied, more rigorous convergence tests may be initiated.

Finally, for reasons largely unknown, experience shows that iterative methods sometimes have a habit of doing strange things during early iterations, such as appearing to "diverge" wildly, to apparently "converge" while far from the solution, etc. It is suggested that in all cases some minimum number of iterations be performed, e.g., 10, 15 or 20. Again, we acknowledge that in certain cases this may be wasteful.

3.5.5 Choosing a Relaxation Parameter for SOR

The choice of a value for the relaxation parameter will obviously affect the eigenvalues of the SOR iteration matrix. The optimum value is the one that minimizes the magnitude of the subdominant eigenvalue. Unfortunately there are very few results that permit the localization of an optimum ω_b.

If a series of related experiments is to be conducted, it is worthwhile carrying out some numerical experiments to try to determine a suitable value; some sort of adaptive procedure might be incorporated into the algorithm. For example, it is possible to begin iterating with a value of $\omega = 1$ and, after some iterations have been carried out, to estimate the rate of convergence from the computed approximations. The value of ω may now be augmented to 1.1, say, and after some further iterations a new estimate of the rate of convergence computed. If this is better than before, ω should again be augmented, to 1.2, say, and the same procedure used. If the rate of convergence is not as good, the value of ω should be diminished.

Aside from using an adaptive scheme, some results due to Young [184] are available for the solution of nonhomogeneous systems of equations $Ax = b$ with nonsingular, *consistently ordered*, coefficient matrices, and they may be helpful in the Markov chain case.

Definition 3.1 (Consistently Ordered Matrix) *The matrix A of order n is consistently ordered if for some t there exist disjoint subsets $S_1, S_2, \ldots, S_t$ of $W = \{1, 2, \ldots, n\}$ such that $\cup_{k=1}^{t} S_k = W$ and such that if $a_{ij} \neq 0$ or $a_{ji} \neq 0$, then $j \in S_{k+1}$ if $j > i$ and $j \in S_{k-1}$ if $j < i$, where S_k is the subset containing i.*

Young uses the term *T-matrix* to refer to a matrix that is block tridiagonal and whose diagonal blocks are square, diagonal matrices. It is immediately evident that if A is a *T-matrix*, then A is consistently ordered. The states of the reliability example of Section 3.2.6 may be ordered in such a way that its infinitesimal generator exhibits this consistently ordered property. In fact, if the nine states of this model are ordered as

$$(2,2) \prec (1,0) \prec (0,1) \prec (2,1) \prec (1,2) \prec (0,0) \prec (2,0) \prec (1,1) \prec (0,2),$$

the infinitesimal generator matrix becomes

$$
Q = \begin{pmatrix}
* & & & 2\lambda_2 & 2\lambda_1 & & & & \\
 & * & & & \lambda_1 & \mu_1 & 2\mu_2 & & \\
 & & * & & \lambda_2 & & 2\mu_1 & \mu_2 & \\
\mu_2 & & & * & & \lambda_2 & 2\lambda_1 & & \\
\mu_1 & & & & * & 2\lambda_2 & & \lambda_1 & \\
 & 2\mu_1 & 2\mu_2 & & & * & & & \\
 & & 2\lambda_1 & & 2\mu_2 & & * & & \\
 & \lambda_2 & \lambda_1 & \mu_1 & \mu_2 & & & * & \\
 & & 2\lambda_2 & & 2\mu_1 & & & & *
\end{pmatrix}
\tag{3.23}
$$

and is seen to be a *T-matrix*. This matrix, however, is not nonsingular.

Theorem 3.1 (Young) *Let A be a consistently ordered matrix with nonvanishing diagonal elements, such that $H_J = I - (\operatorname{diag} A)^{-1} A$ has real eigenvalues, and whose spectral radius satisfies $\bar{\mu} \equiv \rho(H_J) < 1$. If ω_b is defined as*

$$\omega_b = \frac{2}{1 + (1 - \bar{\mu}^2)^{1/2}},\tag{3.24}$$

then

$$\rho(H_{\omega_b}) = \omega_b - 1$$

and if $\omega \neq \omega_b$, then

$$\rho(H_\omega) > \rho(H_{\omega_b}).$$

Moreover, for any ω in the range $0 < \omega < 2$, we have

$$\rho(H_\omega) = \begin{cases} \left[\frac{\omega\bar{\mu} + (\omega^2\bar{\mu}^2 - 4(\omega-1))^{1/2}}{2} \right]^2 & \text{if } 0 < \omega \leq \omega_b \\ \omega - 1 & \text{if } \omega_b \leq \omega < 2 \end{cases}$$

Finally, if $0 < \omega < \omega_b$, then $\rho(H_\omega)$ is a strictly decreasing function of ω.

The value $\bar{\mu}$ may be estimated by observing successive iterates $\pi^{(k)}$ and $\pi^{(k-1)}$. Specifically if

$$\delta^{(k)} = \|\pi^{(k)} - \pi^{(k-1)}\|_\infty$$

then

$$\bar{\mu} \approx \frac{\bar{\theta} + \omega - 1}{\omega\bar{\theta}^{1/2}}\tag{3.25}$$

where

$$\bar{\theta} = \lim_{k \to \infty} \{ \frac{\delta^{(k+1)}}{\delta^{(k)}} \},$$

and ω is the value of the relaxation parameter used when the iterates $\pi^{(k)}$, $k = 1, 2, \ldots$ were generated.

When estimating ω_b from iterates obtained using an $\omega > \omega_b$, problems will arise, since values of $\theta^{(k)} \equiv \delta^{(k+1)}/\delta^{(k)}$ oscillate. This is because the subdominant eigenvalue of H_ω is complex when $\omega \geq \omega_b$. The simplest approach to implement is to carry out a certain number of iterations using $\omega = 1$ (Gauss–Seidel) and to observe the behavior of $\theta^{(k)}$. When successive differences in $\theta^{(k)}$ are small, an estimate of ω_b can be calculated using equations (3.25) and (3.24), and this value may be used until convergence. In a slowly converging problem, it is probably worthwhile applying this procedure two or three times until it is estimated that a value very close to ω_b has been determined. Alternatively, during the iteration procedure, the maximum relative errors in successive approximations may be computed and the sign patterns examined. If the signs change, this implies that the subdominant eigenvalue is complex, so the value of ω is too large and should be reduced. This approach is sometimes incorporated into adaptive procedures.

If some of the eigenvalues of $(L + U)$ are complex, then to guarantee convergence, ω must be chosen such that $0 < \omega < 2/(1+\zeta)$. The variable ζ is related to the size of the imaginary part of the complex eigenvalues of $(L+U)$. Specifically, if $\alpha + i\beta$ (where $i = \sqrt{-1}$) is an eigenvalue of $(L + U)$, then ζ must be chosen sufficiently large so that (α, β) lies in the interior of the ellipse

$$E(1, \zeta), \quad x^2 + y^2(\zeta^2)^{-1} = 1.$$

In experimenting with reliability models, such as the one described in Section 3.2.6, it is found that many of the aforementioned properties hold, even though the infinitesimal generator is singular. In the first set of experiments provided in Section 3.2.6, using the parameters

$$\lambda_1 = 0.2, \qquad \lambda_2 = 0.3, \qquad \mu_1 = 5.0, \quad \text{and} \quad \mu_2 = 6.0,$$

it was seen that the matrix H_J contains two eigenvalues of modulus equal to 1. The next largest eigenvalues have modulus equal to .26865, and again there are two of them. Using this value in the formula for the optimum relaxation parameter, equation (3.24), the value we compute is 1.0187, which is exactly the observed value of the optimum ω parameter. In the second set of experiments, we choose

$$\lambda_1 = 0.2, \qquad \lambda_2 = 30.0, \qquad \mu_1 = 0.5, \quad \text{and} \quad \mu_2 = 60.0.$$

In this case, the subdominant eigenvalue of H_J is .99133, and using this value in equation (3.24) gives an optimum value of 1.7677, again exactly the same as the observed value.

Notice also that Figures 3.1 through 3.3 in Section 3.2.6, which plot the subdominant eigenvalue of H_ω for values of ω for $1 \le \omega < 2$, are identical to those described by Young: ρ decreases monotonically as ω varies from 1 to ω_b; the rate at which it decreases, increases sharply close to ω_b and tends to a limit that is perpendicular to the x-axis. We have

$$\lim_{\omega \to \omega_b^+} \frac{d\rho(\omega)}{d\omega} = \infty \quad \text{and} \quad \lim_{\omega \to \omega_b^-} \frac{d\rho(\omega)}{d\omega} = 1$$

For values of $\omega > \omega_b$, ρ increases linearly; in fact, $\rho = \omega - 1$ for $\omega > \omega_b$. Also, although not apparent from the graph, at $\omega = \omega_b$ *all* the eigenvalues of H_ω have modulus $\omega - 1$ (i.e., they all lie on the circumference of a circle of radius $\omega - 1$ and center 0) — a further property predicted by the results of Young.

As far as the estimation procedure, equation (3.25), is concerned, the results of the experiment are also favorable. We consider the case when

$$\lambda_1 = 0.2, \qquad \lambda_2 = 0.3, \qquad \mu_1 = 5.0, \quad \text{and} \quad \mu_2 = 6.0.$$

The results obtained for the various quantities needed for the estimation are given in Table 3.8. The first column contains the iteration number (the iterations were performed with $\omega = 1$); column 2 contains $\delta^{(k)}$, the difference in consecutive approximations to the solution; the next column displays the difference between the exact value of $\bar{\theta}$ $(= 0.07217321571772)$ and the approximation computed at each iteration; column 4 shows the difference between the exact value of $\bar{\mu}$ $(=0.26865073183917)$ and its estimated value at iteration k. Finally, the last column has the approximation ω_k to the optimal relaxation parameter, which in this case is $\omega_b = 1.01877$. Notice that the process performs well until approximately iteration 15. At this point the difference between the vector approximations becomes too small and the approximation ceases to be as good. Of course, at this point the algorithm has converged and we have computed the desired solution.

The same type of result is obtained when the parameters of the model are changed to

$$\lambda_1 = 0.2, \qquad \lambda_2 = 30.0, \qquad \mu_1 = 0.5, \quad \text{and} \quad \mu_2 = 60.0.$$

The basic iterative procedure converges much more slowly in this case, and so more advantage can be taken of the estimation procedure. As a general rule, it was found that the estimation procedure worked well in this type of reliability model. However, it was also found that if similar experiments are conducted using iterates obtained from applying SOR with a relaxation parameter greater than 1, the estimation procedure did not work.

Table 3.8: Quantities Used in Relaxation Parameter Selection

Iter	$\|\pi^{(k)} - \pi^{(k-1)}\|_\infty$	$\theta_k - \bar\theta$	$\mu_k - \bar\mu$	ω_k
2	0.88888888888889	0.81671567317117	0.67415830974289	1.50000000000000
3	0.16530293297449	0.11379258387858	0.16258678925136	1.05139290597431
4	0.04373443356897	0.19239821554488	0.24571434401902	1.07667467653066
5	0.03091613821171	0.63473289209677	0.57212635384674	1.29753733489625
6	0.00163854497888	-0.01917355037594	-0.03843417000710	1.01361311709476
7	0.00011321346606	-0.00307931379685	-0.00579354262634	1.01789730575269
8	0.00000803660140	-0.00118694813687	-0.00221824962116	1.01840586067648
9	0.00000057532852	-0.00058468024336	-0.00109039196354	1.01856792967394
10	0.00000004135228	-0.00029726836588	-0.00055383265561	1.01864530831097
11	0.00000000297827	-0.00015135210921	-0.00028183720250	1.01868460178937
12	0.00000000021472	-0.00007696026717	-0.00014327300320	1.01870463696610
13	0.00000000001549	-0.00003905847168	-0.00007270361265	1.01871484527215
14	0.00000000000112	-0.00002010909950	-0.00003742870952	1.01871994916918
15	0.00000000000008	-0.00001287596434	-0.00002396520304	1.01872189739695
16	0.00000000000000	-0.00006858656611	-0.00012768040483	1.01870689226707
17	0.00000000000000	0.00181246447321	0.00335235672956	1.01921403007900
18	0.00000000000000	-0.00765708668546	-0.01465047783879	1.01667129887010

3.5.6 The Effect of State Space Orderings on Convergence

We now consider the effect on convergence of the ordering that is imposed on the states. Obviously, for methods like Jacobi and the power method, the ordering is unimportant. However, it has long been known that for methods like Gauss–Seidel the role of the order assigned to the states can be extremely important. This has been graphically demonstrated by the analysis of Tsoucas and Mitra [100]. As an example, we consider a tandem queueing model consisting of m finite-buffer, single-server, exponential service centers. The capacity of queue i is N_i. The arrival process is Poisson with rate μ_0, except that an arrival to the system is lost if it finds the first buffer full. The ith server is blocked if the $(i+1)^{th}$ queue is full, and server i does no work on the current job (*blocking before service*). The service times are independent and exponentially distributed with rates μ_i, $1 \le i \le m$. A state of the system may be completely described by the vector

$$\eta = (\eta_1,\ \eta_2,\ \ldots,\ \eta_m), \qquad 0 \le \eta_i \le N_i, \qquad 1 \le i \le m$$

in which η_i denotes the number of customers at the ith service center. We shall let index(η) denote the position of this state in the list of states.

Let us introduce the following splitting of the infinitesimal generator

$$Q = \hat{L} + \hat{U} - \hat{D}$$

where $\hat{U}$ is the strictly upper triangular part of Q, $\hat{L}$ is its strictly lower triangular part, and $\hat{D}$ its negated diagonal. Notice that this splitting is somewhat different from the one introduced previously in that the previous splitting was applied to the transpose of the infinitesimal generator Q^T, while this one is applied to the matrix Q itself. Once an ordering has been assigned to the states, the kth iteration of Gauss–Seidel is

$$\pi^{(k+1)}\hat{D} = \pi^{(k+1)}\hat{U} + \pi^{(k)}\hat{L}.$$

The matrix $\hat{L}$ reflects *down-stepping* or *low-stepping* transitions, which means that this matrix contains the transition rates from each state i to states with a lower index j (lower meaning that state j comes before state i in the list of states), while $\hat{U}$ reflects *high-stepping* transitions (transitions to states of higher index).

Two natural orderings for this example were considered by Tsoucas and Mitra. Define coefficients C_r and c_r as

$$C_1 = 1; \qquad C_r = \prod_{i=1}^{r-1}(N_i + 1), \qquad 2 \leq r \leq m,$$

$$c_m = 1; \qquad c_r = \prod_{i=r+1}^{m}(N_i + 1), \qquad 1 \leq r \leq m - 1.$$

The *lexicographic* ordering for the state space is defined as follows. Given a state $\eta = (\eta_1, \eta_2, \ldots, \eta_m)$, the index assigned to this state (i.e., its position in the list of states) is

$$\text{index}_{lex}(\eta) = \prod_{i=1}^{m} c_i \eta_i.$$

This definition causes the states to be ordered as

$$(0, 0, \ldots, 0, 0) \prec (0, 0, \ldots, 0, 1) \prec \cdots \prec (0, 0, \ldots, 0, N_m) \prec (0, 0, \ldots, 1, 0) \prec$$

$$(0, 0, \ldots, 1, 1) \prec \cdots \prec (N_1, N_2, \ldots, N_m).$$

The *antilexicographical* ordering is defined as

$$\text{index}_{alex}(\eta) = \prod_{i=1}^{m} C_i \eta_i$$

and the states are now ordered as

$$(0, 0, \ldots, 0) \prec (1, 0, \ldots, 0) \prec \cdots \prec (N_1, 0, \ldots, 0) \prec (0, 1, \ldots, 0) \prec \cdots \prec$$

$$(1, 1, \ldots, 0) \prec \cdots \prec (N_1, N_2, \ldots, N_m).$$

In the antilexicographical ordering, as a customer enters the system and moves from one queue to the next, the system performs high-stepping transitions. In this ordering the only low-stepping transitions are due to departures of customers from the system. Tsoucas and Mitra show that Gauss–Seidel is guaranteed to converge in this case. Indeed, when SOR is used, the fastest convergence is achieved with $\omega = 1$.

However, in the lexicographical ordering, customer movements from queue to queue, as well as departures from the system, correspond to down-stepping transitions; high-stepping transitions are caused only by new arrivals to the system. In this case it may be shown that Gauss–Seidel is divergent. SOR can be made to converge when ω is chosen to be less than 1.

To gain some insight into these results, we consider random walks defined on the states of the Markov chain. Let i_0 be a state that is occupied after a low-stepping transition, i.e., after the system moves from some state i_{-1} to the state i_0 with $\text{index}(i_{-1}) > \text{index}(i_0)$. We define a *high-stepping random walk* to be the sequence of states $(i_0, i_1, \ldots, i_\kappa)$ visited in sequence such that

$$\text{index}(i_0) < \text{index}(i_1) < \cdots < \text{index}(i_\kappa),$$

i.e., always to a higher index, until the transition from state i_κ to state $i_{\kappa+1}$, which is a *low-stepping* one: $\text{index}(i_\kappa) > \text{index}(i_{\kappa+1})$. In other words, this random walk is the sequence of states visited after a low-stepping transition up to the next low-stepping transition. Tsoucas and Mitra show that the Gauss–Seidel iterations have the property of "leaping" over the individual Markov chain steps so as to cover entire high-stepping random walks in one iteration. So, orderings that make high-stepping transitions likely (as the antilexicographical ordering of the example), and hence make high-stepping random walks longer, will make the Gauss–Seidel method visit more states in each iteration and may be expected to converge faster.

As an example, we consider the special case of two service centers, each of which can hold at most two customers. This gives a total of nine states, which are, in lexicographical order,

$$(0,0) \prec (0,1) \prec (0,2) \prec (1,0) \prec (1,1) \prec (1,2) \prec (2,0) \prec (2,1) \prec (2,2)$$

and in antilexicographical order,

$$(0,0) \prec (1,0) \prec (2,0) \prec (0,1) \prec (1,1) \prec (2,1) \prec (0,2) \prec (1,2) \prec (2,2)$$

The infinitesimal generators, Q_{lex} and Q_{alex}, in these two cases are given in Figures 3.4 and 3.5. The diagonal elements, which are equal to the negated sum of the off-diagonal elements in each row, are represented by asterisks.

$$\begin{pmatrix}
* & & & \lambda & & & & & \\
\mu_2 & * & & & \lambda & & & & \\
& \mu_2 & * & & & \lambda & & & \\
& \mu_1 & & * & & & \lambda & & \\
& & \mu_1 & \mu_2 & * & & & \lambda & \\
& & & & \mu_2 & * & & & \lambda \\
& & & & \mu_1 & & * & & \\
& & & & & \mu_1 & \mu_2 & * & \\
& & & & & & & \mu_2 & *
\end{pmatrix}$$

Figure 3.4: Infinitesimal generator: Lexicographical ordering.

$$\begin{pmatrix}
* & \lambda & & & & & & & \\
& * & \lambda & \mu_1 & & & & & \\
& & * & & \mu_1 & & & & \\
\mu_2 & & & * & \lambda & & & & \\
& \mu_2 & & & * & \lambda & \mu_1 & & \\
& & \mu_2 & & & * & & \mu_1 & \\
& & & \mu_2 & & & * & \lambda & \\
& & & & \mu_2 & & & * & \lambda \\
& & & & & \mu_2 & & & *
\end{pmatrix}$$

Figure 3.5: Infinitesimal generator: Antilexicographical ordering.

In this example we also consider the behavior of orderings other than those just described. In many models it is often convenient to generate the states from a given initial state, since the number of states may not be known in advance, much less the position of any state in the list. The initial state is given the index 1 and is appended to an empty list of states. States that can be reached in a single step from the initial state are appended to this list and assigned an index according to their place in the list. When all the transitions from the first state have been considered, the next state in the list is chosen, and any states that it can reach in a single transition that are not already in the list are appended to the list. If the Markov chain is irreducible, then all the states will eventually be generated and a unique index assigned to each. This is the ordering incorporated in the software package MARCA, and we shall refer to it as the *Marca ordering*.

This transitionlike ordering tends to assign state indices in a way that "follows" the natural flow in the network and makes high-stepping transitions probable. It

$$
\begin{pmatrix}
* & \lambda & \mu_1 & \mu_2 & & & & & \\
 & * & & & \mu_1 & \mu_2 & & & \\
 & & * & \lambda & & & \mu_2 & & \\
 & & & * & \lambda & & \mu_1 & & \\
\mu_2 & & & & * & & \lambda & & \\
\mu_1 & & & & & * & & & \\
\lambda & & & & & & * & \mu_2 & \\
 & \mu_2 & & & & & & * & \\
 & & & \lambda & & & & & *
\end{pmatrix}
$$

Figure 3.6: Infinitesimal generator, "Marca" ordering.

would seem, however, that the choice of the initial state is crucial. Consider the specific example of this section. If one specifies the "empty" state $(0,0)$ as the initial state, then an almost antilexicographical ordering may be expected. If, on the other hand, one chooses the "full" state $(2,2)$ to be the initial state, one may expect an "oscillation" in the assignment of indices that would tend to make high-stepping walks rather short. In the results given below, the state $(1,1)$ was chosen as the initial state. For each state, transitions due to possible arrivals were considered first, followed by transitions from queue 1 to queue 2, and finally transitions due to departures. The resulting ordering is

$$(1,1) \prec (2,1) \prec (0,2) \prec (1,0) \prec (1,2) \prec (2,0) \prec (0,1) \prec (2,2) \prec (0,0),$$

and the infinitesimal generator matrix Q_{Marca} is shown in Figure 3.6.

The absolute values of the eigenvalues for the forward and backward Gauss–Seidel iteration matrices are given in Table 3.9 for each of the three state space orderings. These results indicate that forward Gauss–Seidel applied to the lexicographically ordered matrix Q_{lex} and backward Gauss–Seidel applied to either Q_{alex} or Q_{Marca} will diverge. In these cases the iteration matrices possess multiple eigenvalues of modulus 1. In the other cases, forward Gauss–Seidel applied to Q_{lex} or Q_{Marca} or backward Gauss–Seidel applied to Q_{lex}, the method will converge. Furthermore, the convergence will be extremely rapid, due to the large separation of dominant and subdominant eigenvalues. The reader will hardly fail to observe that the eigenvalues are identical in those instances in which Gauss–Seidel converges, and in those in which it diverges.

Some experiments were conducted to assess the effect of varying the magnitude of the parameters λ, μ_1, and μ_2. These were chosen in such a way as to make the iteration matrices upper or lower triangular–dominant, in the sense that the most of the probability mass is concentrated in either the upper triangular or

Table 3.9: Eigenvalues of Gauss–Seidel Iteration Matrices for Different State Orderings

Lexicographical		Antilexicographical		Marca	
Forward GS	Backward GS	Forward GS	Backward GS	Forward GS	Backward GS
0	0	0	0	0	0
1.00000000	0	0	1.00000000	0	0
1.00000000	0	0	1.00000000	0	0
0.38352139	1.00000000	1.00000000	0.38352139	0	1.00000000
0.38352139	0.14708866	0.14708866	0.38352139	1.00000000	1.00000000
0.21224477	0.04504784	0.04504784	0.21224477	0.00000000	0.38352139
0.21224477	0.00000000	0.00000000	0.21224477	0.14708866	0.38352139
0.00000000	0.00000000	0.00000000	0.00000000	0.04504784	0.21224477
0.00000000	0.00000000	0.00000000	0.00000000	0.00000000	0.21224477

lower triangular part of the matrix. Although the actual values of the nonzero eigenvalues with modulus different from 1 changed somewhat, no noticable difference was observed in the convergence characteristics. In particular, in cases of convergence the rate of that convergence was always extremely fast.

When SOR rather than Gauss–Seidel is used, again we find somewhat similar results. It is, however, always possible to find values of ω for which SOR converges, regardless of the ordering imposed or the direction of solution (forward or backward). Our concern here is primarily with the value of the optimum relaxation parameter. We consider only forward SOR. For all tested values of the model parameters, λ, μ_1, and μ_2, the best SOR parameter, using antilexicographic ordering, was practically 1 (never farther from 1 than 0.005). The best SOR parameter for the lexicographic ordering was found to lie in the range $[0.70, 0.85]$ (underrelaxed SOR). The Gauss–Seidel with the antilexicographical ordering needed only 16% to 50% of the iterations required for the optimally tuned SOR with the lexicographic ordering.

The behavior of the Marca ordering was similar to that of the antilexicographical ordering. Additional experiments with the Marca orderings showed that the choice of the initial state from which the entire state space is generated was not a factor. In all cases, the best SOR parameter was found to be 1, and the differences in performance between the orderings derived from different initial states were insignificant.

One final note: a series of experiments on *feed-forward* networks exhibited the same convergence characteristics as the tandem network just considered.

3.6 Convergence Properties

We now turn our attention to convergence properties of iterative methods that are based on splittings, such as those described in Section 3.2. The books of Berman and Plemmons [10] and of Varga [170] provide invaluable reference material on this subject. Additional papers of interest include [109, 130, 141]. In a general context, given a splitting $A = M - N$ in which M is nonsingular, we seek a solution of $Ax = b$ by way of the iterative procedure

$$x^{(k+1)} = M^{-1}Nx^{(k)} + M^{-1}b = Hx^{(k)} + c$$

for $k = 0, 1, \ldots$ and some given initial starting vector $x^{(0)}$. Of particular interest to us is the case in which $b = 0$ and $A = I - P^T$ is singular, where naturally P is a stochastic transition probability matrix.

Throughout this section we shall use the notation

$$\begin{aligned} A \geq B \quad &\text{if } a_{ij} \geq b_{ij} & &\text{for all } i \text{ and } j \\ A > B \quad &\text{if } A \geq B \text{ and } A \neq B \\ A \gg B \quad &\text{if } a_{ij} > b_{ij} & &\text{for all } i \text{ and } j \end{aligned}$$

Also, recall that the spectrum, or set of eigenvalues, of A is denoted by $\sigma(A)$ and the spectral radius of A by $\rho(A)$, i.e., $\rho(A) = \max\{|\lambda|, \ \lambda \in \sigma(A)\}$. Additionally we shall denote the maximum magnitude over all elements in $\sigma(A)\backslash\{1\}$ by $\gamma(A)$, i.e.,

$$\gamma(A) = \max\{|\lambda|, \ \lambda \in \sigma(A), \lambda \neq 1\}.$$

Clearly, $\gamma(A) = \rho(A)$ if $1 \notin \sigma(A)$. A matrix A is said to be *primitive* if it possesses only one eigenvalue with modulus $\rho(A)$. Finally, the *index* of a square matrix A, denoted index(A), is the smallest nonnegative integer k such that rank A^{k+1} = rank A^k. We begin by giving a number of important definitions.

3.6.1 Definitions

Definition 3.2 (Convergent Matrices) *A matrix H is said to be convergent whenever $\lim_{k\to\infty} H^k$ exists and is the zero matrix.*

Definition 3.3 (Semiconvergent Matrices) *A matrix H is said to be semiconvergent whenever $\lim_{k\to\infty} H^k$ exists. This limit need not be zero.*

Definition 3.4 (Regular and Weak Regular Splittings) *A splitting $A = M - N$ is called a regular splitting if $M^{-1} \geq 0$ and $N \geq 0$. It is called a weak regular splitting if $M^{-1} \geq 0$ and $M^{-1}N \geq 0$.*

Definition 3.5 (Convergent and Semiconvergent Splittings) *A splitting $A = M - N$ is said to be convergent if $M^{-1}N$ is convergent. It is said to be semiconvergent if $M^{-1}N$ is semiconvergent.*

Definition 3.6 (Convergent Iterative Method) *An iterative method is said to converge to the solution of a given linear system if the iteration*

$$x^{(k+1)} = Hx^{(k)} + c$$

associated with that method converges to the solution for every starting vector $x^{(0)}$.

Definition 3.7 (Asymptotic Convergence Rate) *The number $R_\infty(H) = -\ln \gamma(H)$ is called the asymptotic convergence rate of $x^{(k+1)} = Hx^{(k)} + c$ and applies whether H is convergent or semiconvergent.*

Definition 3.8 (M-Matrix) *A finite matrix A with nonpositive off-diagonal elements and nonnegative diagonal elements can be expressed in the form*

$$A = sI - G, \qquad s > 0, \qquad G \geq 0.$$

A matrix A of this form for which $s \geq \rho(G)$ is called an M-matrix.

Definition 3.9 (Property C) *An M-matrix is said to have property C if it can be split into $A = sI - G$, $s > 0$, $G \geq 0$ and the matrix $H \equiv (1/s)G$ is semiconvergent.*

3.6.2 Convergence Theorems

The following theorems, together with their proofs, may be found in the texts of Berman and Plemmons [10] and of Varga [170].

Theorem 3.2

1. *H is convergent iff $\rho(H) < 1$.*

2. *H is convergent iff $(I - H)$ is nonsingular and $(I - H)^{-1} = \sum_{k=0}^{\infty} H^k$.*

Theorem 3.3 *H is semiconvergent iff all of the following conditions hold*

1. $\rho(H) \le 1$.

2. *If $\rho(H) = 1$, then all the elementary divisors associated with the unit eigen-value of H are linear.*

3. *If $\rho(H) = 1$, then $\lambda \in \sigma(H)$ with $|\lambda| = 1$ implies that $\lambda = 1$.*

Theorem 3.4 *Let $A = M - N$ with M nonsingular and let b be in the range of A. Then, with $H = M^{-1}N$ and $c = M^{-1}b$, the iterative method*

$$x^{(k+1)} = Hx^{(k)} + c$$

converges to a solution of $Ax = b$ for each $x^{(0)}$ if and only if H is semiconvergent.

Note that Theorem 3.4 is valid when A is singular and $b = 0$.

Theorem 3.5 *If A is an M-matrix, then*

1. $A^{-1} \ge 0$.

2. *If $\tilde{A}$ is obtained from A by setting any off-diagonal element to zero, then $\tilde{A}$ is also an M-matrix.*

This theorem implies that a Gauss–Seidel splitting of an M-matrix A, whether A is singular or not, is a regular splitting, since given that $A = D - (L + U)$, it follows that $D - L$ must be an M-matrix and hence $(D - L)^{-1} \ge 0$. This, along with the fact that $U \ge 0$, satisfies the conditions for a regular splitting.

Theorem 3.6 *For singular and nonsingular M-matrices, the methods of Jacobi and Gauss–Seidel are based on regular splittings. For $0 < \omega \le 1$, SOR is based upon a regular splitting.*

We next present a number of theorems that are applicable when the coefficient matrix is *nonsingular*. Although infinitesimal generator matrices are singular, these theorems are still important in Markov chain applications. It was seen in Chapter 2 that a possible approach for computing the stationary probability vector of an irreducible Markov chain was to partition the matrix as

$$Q^T = \begin{pmatrix} B & d \\ c^T & f \end{pmatrix}$$

and to compute the solution of the nonhomogeneous system $B\hat{x} = -d$ in which B is nonsingular. The stationary probability vector is then obtained by normalizing $(\hat{x}, 1)$. When Q^T is irreducible, the matrix B is in fact a nonsingular M-matrix (see Theorem 3.16).

Theorem 3.7 *For a nonsingular matrix A,*

1. *Every regular splitting of A is convergent,*

2. *Every weak regular splitting of A is convergent.*

It therefore follows that Gauss–Seidel applied to a nonsingular M-matrix is convergent. Unfortunately it does not imply the same when A is singular, for it is not true that a regular splitting of a singular M-matrix is always convergent or even semiconvergent. We return to this point after we present some additional theorems.

Theorem 3.8 *Let A be a nonsingular M-matrix and assume, without loss of generality, that $D = \text{diag}\{A\} = I$. The Jacobi iteration matrix is therefore $H_J = (L + U)$, and the SOR iteration matrix for A is $H_\omega = (I - \omega L)^{-1}[(1 - \omega)I + \omega U]$. Then, for $0 < \omega \le 1$:*

1. $\rho(H_J) < 1$ *if and only if* $\rho(H_\omega) < 1$,

2. $\rho(H_J) < 1$ *(and $\rho(H_\omega) < 1$) if and only if A is a nonsingular M-matrix, in which case $\rho(H_\omega) \le 1 - \omega + \omega\rho(H_J)$,*

3. *If $\rho(H_J) \ge 1$, then $\rho(H_\omega) \ge 1 - \omega + \omega\rho(H_J) \ge 1$.*

Theorem 3.9 *Let A be a nonsingular M-matrix and let $0 < \omega_1 \le \omega_2 \le 1$. Then $\rho(H_{\omega_2}) \le \rho(H_{\omega_1}) < 1$ and hence $R_\infty(H_{\omega_2}) \ge R_\infty(H_{\omega_1})$.*

Theorem 3.10 *If A is a nonsingular M-matrix, then $\rho(H_\omega) < 1$ for all ω satisfying*

$$0 < \omega < \frac{2}{1 + \rho(H_J)}.$$

Before directing our attention to theorems that relate specifically to singular coefficient matrices, we present three theorems that concern M-matrices with property C.

Theorem 3.11 *All nonsingular M-matrices (but not all M-matrices) possess property C.*

Theorem 3.12 *A is an M-matrix with property C if and only if every regular splitting of A into $A = M - N$ satisfies $\rho(M^{-1}N) \leq 1$ and $\text{index}(I - M^{-1}N) \leq 1$.*

Theorem 3.13 *A is an M-matrix with property C if and only if $\text{index}(A) \leq 1$.*

The remaining theorems relate directly to stochastic probability matrices, P. The first in fact applies to any nonnegative matrix A. In particular it applies to stochastic matrices P and to the iteration matrices derived from regular and weak regular splittings.

Theorem 3.14 *If $A \geq 0$, then the spectral radius of A, $\rho(A)$ is an eigenvalue of A. Furthermore,*

- *If $A \geq 0$ is irreducible, $\rho(A)$ is a simple eigenvalue, and any eigenvalue of A of the same modulus is also simple.*

- *If $A \geq 0$ is irreducible, A has eigenvalues of the same modulus as $\rho(A)$ but different from $\rho(A)$ if and only if A is periodic.*

Theorem 3.15 *If P is a stochastic matrix, then $I - P^T$ is a singular M-matrix with property C.*

It follows from Theorems 3.12 and 3.15 that the iteration matrix H obtained from any regular splitting of $I - P^T$ satisfies condition 1 of semiconvergence (item 1 of Theorem 3.3). Furthermore, if P is irreducible it follows from Theorems 3.14 and 3.15 that H satisfies condition 2 of semiconvergence.

Theorem 3.16 *If A is an irreducible singular M-matrix, then*

1. *$\text{index}(A) = 1$.*

2. *A has rank $n - 1$.*

3. *A has property C.*

4. *Each principal submatrix of A other than A itself is a nonsingular M-matrix.*

A regular (or weak regular) splitting of a *singular* M-matrix is not necessarily a convergent or a semiconvergent splitting. This explains why Gauss–Seidel does not necessarily converge when applied to the matrix $I - P^T$. For the iteration matrix to be semiconvergent, it must satisfy all three conditions of Theorem 3.3. Condition 3 requires that the only eigenvalue of the iteration matrix with modulus equal to 1 is the unit eigenvalue itself. Even though the matrix $I - P^T$ may be irreducible and aperiodic, iteration matrices obtained by means of regular splittings may be periodic and possess multiple eigenvalues equally distributed around the unit circle. For example, the irreducible, aperiodic stochastic matrix

$$P = \begin{pmatrix} p_{11} & 0 & 0 & \cdots & 0 & p_{1n} \\ p_{21} & p_{22} & 0 & \cdots & 0 & 0 \\ 0 & p_{32} & p_{33} & \cdots & 0 & 0 \\ \vdots & \vdots & \vdots & \ddots & \vdots & \vdots \\ 0 & 0 & 0 & \cdots & p_{n-1,n-1} & 0 \\ 0 & 0 & 0 & \cdots & p_{n,n-1} & p_{nn} \end{pmatrix}$$

gives rise to Jacobi and Gauss–Seidel iteration matrices that are periodic. The following three theorems propose alterations that force the diagonal elements of the iteration matrices to be nonzero in order to eliminate any possibility of periodicity, so that the final condition of semiconvergence can be satisfied.

Theorem 3.17 *Let A be a singular irreducible M-matrix and let $\epsilon > 0$. Then the splitting*

$$A = (D - L - U) = (D - L + \epsilon I) - (U + \epsilon I) = M - N$$

yields an iteration matrix $H_\epsilon = (D - L + \epsilon I)^{-1}(U + \epsilon I)$, which is semiconvergent for every $\epsilon > 0$.

In fact the splitting is a regular splitting, and since A is irreducible, it follows that H_ϵ has spectral radius 1 and the elementary divisors associated with the eigenvalue 1 are linear. The semiconvergence of H_ϵ follows from the fact that since both $(D - L + \epsilon I)^{-1}$ and $(U + \epsilon I)$ have a strictly positive diagonal, so also does H_ϵ. Note that the iteration matrix H_ϵ reduces to the usual SOR iteration matrix

$$H_\omega = (I - \omega D^{-1}L)^{-1}[(1 - \omega)I + \omega D^{-1}U]$$

when $\epsilon = 1/\omega - 1$, or $\omega = 1/(1+\epsilon)$. We may thus conclude that SOR is convergent for all ω such that $1/\omega - 1 > 0$, i.e., $0 < \omega < 1$.

The previous theorem pertains to Gauss–Seidel–like iteration matrices. However, the concept may be extended to the iteration matrix derived from any (weak) regular splitting. We have the following theorem.

Theorem 3.18 *If H is the iteration matrix arising from a (weak) regular splitting of an irreducible singular M-matrix A, then the transformed matrix*

$$H_\alpha = (1 - \alpha)I + \alpha H$$

is semiconvergent for all $\alpha \in (0,1)$.

Since H is obtained from a weak regular splitting of an irreducible singular M-matrix, it has one simple unit eigenvalue; other eigenvalues of the same modulus (if any) are also simple. Now if λ is an eigenvalue of H, then $1 - \alpha + \alpha\lambda$ is an eigenvalue of H_α. Therefore, with $0 < \alpha < 1$, any eigenvalue $\lambda \neq 1$ of H gives rise to an eigenvalue in H_α with modulus strictly less than 1. It follows that H_α has only one eigenvalue on the unit circle: the unit eigenvalue.

If H has at least one nonzero diagonal element, it is already semiconvergent and it is not necessary to perform this suggested modification. On the other hand, if H has only zero diagonal elements, then this procedure forces all the diagonal elements equal to $1 - \alpha$. The modification may also be applied to an irreducible stochastic matrix P. The matrix $P_\alpha = (1 - \alpha)I + \alpha P$ for $\alpha \in (0,1)$ is easily verified to be an aperiodic, irreducible stochastic matrix, and the power method when applied to P_α can be guaranteed to converge.

Theorem 3.19 *If A is an M-matrix and $A = sI - G$, $G \geq 0$ with $s > \max_i(a_{ii})$ then, with $H = (1/s)G$, condition 3 for semiconvergence (Theorem 3.3) is satisfied.*

Using this theorem, the jth diagonal element of H is equal to $(s - a_{jj})/s$, which is strictly greater than 0 for all $j = 1, 2, \ldots, n$. Below, as an example, we give an M-matrix A and two iteration matrices H_1 and H_s,

$$A = \begin{pmatrix} 1 & -1 \\ -1 & 1 \end{pmatrix}; \qquad H_1 = \begin{pmatrix} 0 & 1 \\ 1 & 0 \end{pmatrix}; \qquad H_s = \begin{pmatrix} 1 - 1/s & 1/s \\ 1/s & 1 - 1/s \end{pmatrix}.$$

The first iteration matrix, H_1, is obtained when we take $s = 1$, and it is not semiconvergent, $(\sigma(H_1) = \{\pm 1\})$. The second, H_s, is semiconvergent for any $s > 1 = \max_i(a_{ii})$, $(\sigma(H_s) = \{1, 1 - 2/s\})$.

It is worth noting that a nonzero diagonal term is a sufficient but not a necessary condition for convergence, as the following example shows. Let

$$I - P^T = \begin{pmatrix} 1 & 0 & -1 & 0 & 0 \\ -1 & 1 & 0 & 0 & 0 \\ 0 & 0 & 1 & 0 & -1 \\ 0 & -0.5 & 0 & 1 & 0 \\ 0 & -0.5 & 0 & -1 & 1 \end{pmatrix}.$$

The iteration matrix for forward Gauss–Seidel, H_{GSf}, is not semiconvergent, but that for backward Gauss–Seidel, H_{GSb}, is convergent, even though in both cases all the diagonal elements of the iteration matrices are zero. We have

$$H_{GSf} = \begin{pmatrix} 0 & 0 & 1 & 0 & 0 \\ 0 & 0 & 1 & 0 & 0 \\ 0 & 0 & 0 & 0 & 1 \\ 0 & 0 & 0.5 & 0 & 0 \\ 0 & 0 & 1 & 0 & 0 \end{pmatrix}, \quad H_{GSb} = \begin{pmatrix} 0 & 0.5 & 0 & 1 & 0 \\ 1 & 0 & 0 & 0 & 0 \\ 0 & 0.5 & 0 & 1 & 0 \\ 0 & 0.5 & 0 & 0 & 0 \\ 0 & 0.5 & 0 & 1 & 0 \end{pmatrix}$$

with spectra $\sigma(H_{GSf}) = \{0, 0, 0, -1, 1\}$ and $\sigma(H_{GSb}) = \{0, 0, (1 \pm i)/2, 1\}$, respectively.

3.6.3   Application to Markov Chains

We conclude this section and chapter by summarizing the convergence results as they relate to iterative methods for computing solutions of irreducible Markov chains. Consider first the case of solving the $(n-1) \times (n-1)$ system $B\hat{x} = -d$ given in equation (3.6). The matrix $-B$ is a nonsingular M-matrix, since it is a principal submatrix of an irreducible singular M-matrix. It follows that iterative methods based on regular splittings are guaranteed to converge when applied to $B\hat{x} = -d$. In particular, we can guarantee the convergence of Gauss–Seidel. However, numerical experiments with this approach often turn out to be very disappointing, because the rate of convergence is frequently extremely slow — usually much slower than that attained with the same method applied to $\pi Q = 0$. This is a direct consequence of the closeness of the spectral radius $\rho(H) = \gamma(H)$ to unity. When the infinitesimal generator matrix is large, the effect of the removal of a single row and column is often to shift the zero eigenvalue only slightly from zero, and this undesirable trait seems to be maintained in the iteration matrix. However, there are currently no theoretical results that allow us to claim that iterative methods applied to $\pi Q = 0$ will always outperform the same methods applied to $B\hat{x} = -d$.

Consider now the application of iterative methods to $\pi Q = 0$. The matrix $-\Delta t Q \ (= I - P)$ is a singular M-matrix, and the iterative scheme $Mx^{(k+1)} = Nx^{(k)}$, $k = 0, 1, \ldots$ converges if and only if its iteration matrix $H = M^{-1}N$ is semiconvergent, i.e., if and only if H satisfies the three conditions of Theorem 3.3. It follows from the theorems of Section 3.6.2 that for regular splittings (e.g., Gauss–Seidel) of $I - P^T$, the first condition is satisfied for all stochastic matrices P, the second is satisfied when P is irreducible, and the third when H is aperiodic. The aperiodicity of P is not sufficient to guarantee the aperiodicity of H; so one of the devices offered by Theorems 3.17 through 3.19 may need to be used when the aperiodicity of H cannot be guaranteed, as is the case when all its diagonal elements are zero.

Chapter 4

Projection Methods

4.1 Introduction

4.1.1 Preliminaries

We begin this chapter by introducing some notation and definitions. We shall consider only the real domain $\Re$, and as usual, we shall denote the space of all real vectors of length n by $\Re^n$. The vectors $\{u_1, u_2, \ldots, u_m\}$ of $\Re^n$ are said to be *linearly independent* if no vector can be written as a linear combination of the others — in other words, $\sum_{i=1}^{m} \alpha_i u_i = 0$ implies that $\alpha_i = 0$ for all i. The *space spanned* by a subset of vectors $\mathcal{S} = \{u_1, u_2, \ldots, u_m\}$ of $\Re^n$ is called the *span* of that subset and consists of all possible linear combinations of these vectors. The span of a subset of vectors of $\Re^n$ is called a *subspace* of $\Re^n$. A set of vectors forms a *basis* for a subspace if they are linearly independent and span that subspace. Thus the set of vectors e_i, $i = 1, 2, \ldots, n$ that are the columns of the identity matrix of order n form a basis for $\Re^n$. All bases for a subspace $\mathcal{S}$ contain the same number of vectors. This number is referred to as the *dimension* of the subspace and is denoted by $\dim(\mathcal{S})$. Two important subspaces of $\Re^n$ that are associated with an $(n \times m)$ matrix A are the *range space*, range(A), and the *null space*, null(A), defined as

$$
\begin{aligned}
\text{range}(A) &= \{v \in \Re^n \mid v = Au, \ \forall \ u \in \Re^m\}, \\
\text{null}(A) &= \{u \in \Re^n \mid Au = 0\}.
\end{aligned}
$$

Notice that the range of A is the subspace consisting of all linear combinations of the *columns* of A. A subspace $\mathcal{S} \subseteq \Re^n$ that has the property that

$$
u \in \mathcal{S} \Rightarrow Au \in \mathcal{S}
$$

is said to be an *invariant subspace* for A. This implies that each eigenvector of A constitutes an invariant subspace of dimension 1 for A.

Let u and w be two nonzero vectors in $\Re^n$. It follows that $u^T w = w^T u$. If $u^T w = 0$, then u and w are said to be *orthogonal*. More generally, if $\mathcal{S} = \text{span}\{u_1, u_2, \ldots, u_m\}$ is a subspace of $\Re^n$, then $\mathcal{S}$ is said to be *orthogonal* if $u_i^T u_j = 0$ for $i \neq j$. $\mathcal{S}$ is said to be *orthonormal* if $u_i^T u_j = \delta_{ij}$, where δ_{ij} is the Kronecker delta. Two subspaces $\mathcal{S}_i$ and $\mathcal{S}_j$ are said to be *mutually orthogonal* if for any $u \in \mathcal{S}_i$ and $w \in \mathcal{S}_j$, we have $u^T w = 0$ when $i \neq j$. For any subspace $\mathcal{S}$, we can define its *orthogonal complement*:

$$\mathcal{S}^\perp = \{w \in \Re^n \mid w^T u = 0, \ \forall \ u \in \mathcal{S}\}.$$

The vectors $\{u_1, u_2, \ldots, u_m\}$ form an *orthonormal basis* for the subspace $\mathcal{S}$ if they are orthonormal and span $\mathcal{S}$. A matrix $Q \in \Re^{n \times n}$ is said to be *orthogonal* if $Q^T Q = I$.[1] In this case the columns of Q form an orthonormal basis for $\Re^n$. An important property that results from this definition is that the vector 2-norm is invariant under an orthogonal transformation, i.e., we have

$$\|Qv\|_2^2 = v^T Q^T Q v = v^T v = \|v\|_2^2.$$

A set of nonzero vectors $\{u_1, u_2, \ldots, u_m\}$ is said to be *A-orthogonal* if $u_j^T A u_i = 0$ when $i \neq j$ and A is a symmetric positive-definite matrix. It is said to be *A-orthonormal* when $u_j^T A u_i = \delta_{ij}$.

A matrix $R \in \Re^{n \times n}$ is said to be an *orthogonal projector* onto a subspace $\mathcal{S} \subseteq \Re^n$ if $\text{range}(R) = \mathcal{S}$ and $R = R^T = R^2$. It then follows that for any vector $u \in \Re^n$, $Ru \in \mathcal{S}$ and $(I - R)u \in \mathcal{S}^\perp$. The orthogonal projector for a subspace $\mathcal{S}$ is unique. If the matrix $U = [u_1, u_2, \ldots, u_m]$ has orthonormal columns that provide a basis for a subspace $\mathcal{S}$, then $R = UU^T \in \Re^{n \times n}$ is the unique orthogonal projector onto the subspace $\mathcal{S}$.

Finally, we consider a process for extracting eigenvalue and eigenvector approximations from a given subspace, the so-called *Rayleigh–Ritz* procedure. Given some real matrix $A \in \Re^{n \times n}$ and a subspace $\mathcal{S} \subseteq \Re^n$, we proceed as follows:

Rayleigh–Ritz Procedure

1. Compute an orthonormal basis $Q = \{q_1, q_2, \ldots, q_m\}$ for the subspace $\mathcal{S}$.

2. Form the Rayleigh quotient matrix, $B = Q^T A Q$ $(\in \Re^{m \times m})$.

3. Compute eigensolution of B: $Bx_i = \theta_i x_i, \quad \|x_i\|_2 = 1, \quad i = 1, 2, \ldots, m.$

4. Form the Ritz approximations $y_i = Qx_i, \quad i = 1, 2, \ldots, m.$

---

[1] In the numerical analysis literature, the use of the letter Q to denote an orthogonal matrix is pervasive. At the risk of creating some confusion, we also shall use this notation. It should be clear from the context whether the Q in question refers to an orthogonal matrix or to an infinitesimal generator.

The approximations to the eigenvalues and eigenvectors of A are taken to be θ_i and y_i, respectively. Let $\Theta = \text{diag}\{\theta_1, \theta_2, \ldots, \theta_m\}$ be a diagonal matrix of the eigenvalues of B, and let $Y = [y_1, y_2, \ldots, y_m]$. It may be shown that the Rayleigh–Ritz approximations are optimal in the sense that $\|AY - Y\Theta\|_F$ is less than the norm of the residual matrix from any other set of orthonormal vectors in $\mathcal{S}$ with any approximate eigenvalues [116].

4.1.2 General Projection Processes

The basic idea behind projection techniques is quite simple. They consist of approximating an exact solution from a sequence of approximations taken from small-dimension subspaces. These projection steps are repeated until convergence is reached. In the sections that follow, we shall use the letter m to denote the dimension of the subspace from which the approximate solutions are extracted, and in general we shall have $m \ll n$. The various projection methods differ in the way in which the subspaces are selected and on how the approximations are extracted from them. Therefore, a basic projection step is defined formally with two objects: a subspace $\mathcal{K}$ of dimension m, from which the approximation is selected, and another subspace $\mathcal{L}$ that sets constraints necessary for extracting the approximation from $\mathcal{K}$. Orthogonality-type constraints are the most commonly used.

4.1.3 Projection Processes for Linear Systems

Consider a linear system

$$Ax = b$$

and let $\mathcal{K}$ be spanned by a set of m vectors $V \equiv [v_1, \ldots, v_m]$. We can write an approximate solution as a linear combination of these vectors, $x = Vy$, where y is a vector of m unknown components. This gives m degrees of freedom. In order to extract a unique y, one possibility is to require that the residual vector, $b - Ax$, be orthogonal to m linearly independent vectors $w_1, \ldots, w_m$, i.e.,

$$b - AVy \perp w_i, \qquad i = 1, 2, \ldots, m.$$

The set of vectors $W \equiv [w_1, \ldots, w_m]$ is a basis of the subspace $\mathcal{L}$. The above constraints mean that the residual vector is required to be orthogonal to $\mathcal{L}$, so we may rewrite the constraints as

$$W^T(b - AVy) = 0,$$

which yields, assuming the matrix $W^T AV$ is nonsingular,

$$y = [W^T AV]^{-1} W^T b. \tag{4.1}$$

If an initial guess x_0 to the solution of the system is known, then we may seek a correction δ to x_0 such that $x_0 + \delta$ is a solution, i.e., $A(x_0 + \delta) = b$. As a result, if we set $r_0 \equiv b - Ax_0$, the projection step must be applied to the system

$$A\delta = r_0,$$

to compute the unknown vector δ.

Notice that what we have just described is a basic projection step. As was already mentioned, most methods require a sequence of such steps, in which the most recent approximation is used as an initial guess. Therefore, a general projection algorithm is as follows:

Algorithm: Prototype Projection Method

1. Do until convergence

   - Select a pair of subspaces $\mathcal{K}$ and $\mathcal{L}$ and choose two bases $V = [v_1, \ldots, v_m]$ and $W = [w_1, \ldots, w_m]$ for these subspaces.
   - Compute $r = b - Ax$.
   - Compute $y = (W^T A V)^{-1} W^T r$.
   - Compute $x = x + Vy$.

Early projection methods proposed for linear systems involved mostly one-dimensional projection processes. For example, for symmetric positive-definite matrices the well-known steepest descent algorithm consists of taking $\mathcal{L} = \mathcal{K}$ at every step and choosing $\mathcal{K}$ to be the (one-dimensional) subspace spanned by the current residual vector, $r = b - Ax$. From the previous expressions, we immediately see that the corresponding iteration takes the form

$$x \leftarrow x + \alpha r \quad \text{where } \alpha = (r^T r)/(r^T A r).$$

Denoting the exact solution by x^*, an important property of the steepest descent algorithm relates to the function $f(x) = \frac{1}{2}(x - x^*)^T A(x - x^*)$, which is by definition the square of the A-norm of the error $x - x^*$. The residual vector r for any given x lies in the direction opposite to that of the gradient of f, and it is well known that this represents the direction of greatest rate of reduction for f. It may be shown that each iteration of the algorithm takes a step in this favorable direction and minimizes the above A-norm of the error in this direction (hence the name "steepest descent").

This optimality property is shared by many of the projection methods in the symmetric positive-definite case. For example, the conjugate gradient algorithm minimizes the A-norm of the error in the Krylov subspace, span$\{r, Ar, \ldots, A^{m-1}r\}$. Generally, for matrices that are symmetric positive-definite, it can be shown that the optimality property just mentioned is true whenever $\mathcal{L} = \mathcal{K}$; the A-norm of the error is minimized in the subspace $\mathcal{K}$.

For nonsymmetric problems, taking $\mathcal{L} = \mathcal{K}$ does not entail any such optimality property. In fact, the A-norm may be degenerate if A is not positive-definite. In this case a common alternative is to define projection methods that attempt to minimize the 2-norm of the residual vector, $b - Ax$, instead of the A-norm of the error. One such method is the GMRES algorithm [137], which will be described in Section 4.4.

4.1.4 Projection Processes for Eigenvalue Problems

Projection methods for the eigenvalue problem

$$Ax = \lambda x$$

may be defined similarly. Here, we seek an approximate eigenvalue λ and an approximate eigenvector $x \in \mathcal{K}$ such that the residual vector $Ax - \lambda x$ is orthogonal to a subspace $\mathcal{L}$. Writing $x = Vy$ and translating the orthogonality constraints, we get

$$W^T(AVy - \lambda Vy) = 0$$

or

$$W^T AVy = \lambda W^T Vy \qquad (4.2)$$

which is a generalized eigenvalue problem of dimension m. In many algorithms, the two bases W and V are chosen to ensure that the matrix $W^T V$ is the identity matrix, in which case the approximate eigenvalues of A are the eigenvalues of the $m \times m$ matrix $C = W^T AV$. The corresponding approximate eigenvectors are the vectors $x_i = Vy_i$, where y_i are the eigenvectors of C. This provides m approximate eigenpairs λ_i, x_i. The most common situation is when $\mathcal{K} = \mathcal{L}$ and $V (= W)$ is an orthogonal basis of $\mathcal{K}$, in which case the projection process is referred to as an *orthogonal projection process*.

4.1.5 Application to Markov Chains

When we apply projection methods to Markov chain problems, we are free to take either of the two viewpoints: eigenvalue problem or linear systems approach. The only potential problem is that for the linear systems approach the original linear system is homogeneous ($b = 0$), and this may cause some difficulties when defining the projected problem. The minimum assumptions that must be made in order for these projection processes to be feasible are that $W^T AV$ be nonsingular for linear systems and that $W^T V$ be nonsingular for eigenvalue problems.

Throughout this chapter, we shall demonstrate the behavior of various projection-type algorithms on the same two small matrices used in Chapter 3. The first is the stochastic (8×8) Courtois NCD matrix, given by

$$P^T = \begin{pmatrix} 0.85000 & 0.10000 & 0.10000 & 0 & 0.00050 & 0 & 0.00003 & 0 \\ 0 & 0.65000 & 0.80000 & 0.00040 & 0 & 0.00005 & 0 & 0.00005 \\ 0.14900 & 0.24900 & 0.09960 & 0 & 0.00040 & 0 & 0.00003 & 0 \\ 0.00090 & 0 & 0.00030 & 0.70000 & 0.39900 & 0 & 0.00004 & 0 \\ 0 & 0.00090 & 0 & 0.29950 & 0.60000 & 0.00005 & 0 & 0.00005 \\ 0.00005 & 0.00005 & 0 & 0 & 0.00010 & 0.60000 & 0.10000 & 0.19990 \\ 0 & 0 & 0.00010 & 0.00010 & 0 & 0.24990 & 0.80000 & 0.25000 \\ 0.00005 & 0.00005 & 0 & 0 & 0 & 0.15000 & 0.09990 & 0.55000 \end{pmatrix}$$

$$(4.3)$$

The eigenvalues of this matrix, in descending order, are

1.0   0.999800   0.998495   0.750026   0.550067   0.400033   0.300714 -0.149535

The second is the (9×9) infinitesimal generator matrix derived from a reliability model:

$$\begin{pmatrix} * & 2\lambda_2 & & 2\lambda_1 & & & & & \\ \mu_2 & * & \lambda_2 & & 2\lambda_1 & & & & \\ & 2\mu_2 & * & & & 2\lambda_1 & & & \\ \mu_1 & & & * & 2\lambda_2 & & \lambda_1 & & \\ & \mu_1 & & \mu_2 & * & \lambda_2 & & \lambda_1 & \\ & & \mu_1 & & 2\mu_2 & * & & & \lambda_1 \\ & & & 2\mu_1 & & & * & 2\lambda_2 & \\ & & & & 2\mu_1 & & \mu_2 & * & \lambda_2 \\ & & & & & 2\mu_1 & & 2\mu_2 & * \end{pmatrix}.$$

In this second example, the diagonal elements are equal to the negated sum of the off-diagonal elements and are simply denoted by an asterisk. Two cases are considered corresponding to the parameter values

$\lambda_1 = 0.2$; $\lambda_2 = 0.30$; $\mu_1 = 5.0$;   and $\mu_2 = 6.00$,   (non$-$NCD case) and
$\lambda_1 = 0.2$; $\lambda_2 = 30.0$; $\mu_1 = 0.5$;   and $\mu_2 = 60.0$,   (NCD case).

The eigenvalues, in descending order, of the stochastic matrices corresponding to both these cases are given in Table 4.1.

4.2   Simultaneous Iteration

4.2.1   A Generic Subspace Iteration Algorithm

Among the simplest methods for computing invariant subspaces are the *subspace iteration methods* or *simultaneous iteration methods*. In its simplest form, sub-

Table 4.1: Eigenvalues of Reliability Model Matrices

Case I (non-NCD)	Case II (NCD)
1.00000000000000	1.00000000000000
0.76363636363636	0.99421487603306
0.71363636363636	0.98842975206612
0.52727272727273	-0.49917355371901
0.47727272727273	-0.49338842975207
0.42727272727273	-0.48760330578512
0.24090909090909	0.25619834710744
0.19090909090909	0.25041322314050
-0.04545454545455	0.24462809917355

space iteration for the computation of approximations to the dominant eigenvalues and corresponding right-hand eigenvectors of P^T can be described as follows (see [78, 151] for details):

Algorithm: Subspace Iteration

1. Choose an initial orthonormal system $U \equiv [u_1, u_2, \ldots, u_m]$ and an integer k.

2. Compute $V = (P^T)^k U$ and orthonormalize V to get Q.

3. Perform an (orthogonal) projection process onto span$\{Q\}$, i.e., compute the eigenvalues $\theta_1, \theta_2, \ldots, \theta_m$ and eigenvectors $Z = [z_1, z_2, \ldots, z_m]$ of $B = Q^T P^T Q$.

4. Test for convergence. If satisfied, then exit; else continue.

5. Compute $U = QZ$ (the system of approximate eigenvectors), choose a value for k, and go to Step 2.

This algorithm utilizes the matrix P^T only to compute successive matrix-by-vector products $P^T u$, so sparsity can be exploited. However, it can sometimes be a slow method — slower than some of the alternatives to be described later in this chapter. In fact, a more satisfactory alternative is to use a *Chebyshev-subspace iteration*: Step 2 is replaced by $V = T_k(P^T)U$, where T_k is obtained from the Chebyshev polynomial of the first kind of degree k by a linear change of variables.

The three-term recurrence relation for Chebyshev polynomials allows us to compute the vector $T_k(P^T)u$ at almost the same cost as $(P^T)^k u$, and performance can be dramatically improved. Details on implementation and some experiments are described in [134].

4.2.2 "LOPSI": A Lopsided Simultaneous Iteration Algorithm

4.2.2.1 Lopsided Iteration

We may view simultaneous iteration methods as extensions of the power method whereby iteration is carried out with a number of trial vectors that converge onto the eigenvectors corresponding to the dominant eigenvalues. The first simultaneous iteration methods are due to Bauer [9] and have been discussed by Wilkinson [178]. Subsequent methods are related to Bauer's *bi-iteration* technique but improve the convergence rate by including an *interaction analysis* in the iteration cycle. These methods have been most highly developed for the real symmetric eigenvalue problem, in which the left- and right-hand eigenvector sets coincide. Here we shall briefly describe a method, called *Lopsided iteration*, that predicts only one set of eigenvectors. It has been found to be simpler than bi-iteration and to be more efficient in computer time and memory space.

4.2.2.2 The Basic Iteration Cycle

Each iteration cycle of the Lopsided iteration algorithm involves a premultiplication and a reorientation, followed by a normalization and a tolerance test. Of these, the normalization and tolerance test are trivial as far as the amount of computation is concerned.

Premultiplication: Let P^T be a real unsymmetric matrix of order n for which the dominant right-hand eigenvector set is required. The premultiplication phase involves premultiplying a set of m normalized vectors, $U = [u_1, u_2, \ldots, u_m]$, by P^T. If the resulting set of vectors is designated $V = [v_1, v_2, \ldots, v_m]$, then

$$V = P^T U. \tag{4.4}$$

Let $\Lambda = \text{diag}(\Lambda_a \,|\, \Lambda_b) = \text{diag}(\lambda_1, \ldots, \lambda_m, \,|\, \lambda_{m+1}, \ldots, \lambda_n)$ be a diagonal matrix of the eigenvalues of P^T arranged in descending order of absolute magnitude, and let $Y = [Y_a \,|\, Y_b] = [y_1, \ldots, y_m \,|\, y_{m+1}, \ldots, y_n]$ be a matrix of the corresponding right-hand eigenvectors. It follows from the eigenvalue properties of P that

$$P^T Y_a = Y_a \Lambda_a \quad \text{and} \quad P^T Y_b = Y_b \Lambda_b. \tag{4.5}$$

Let us assume that the set of trial vectors U may be written as a linear combination of the set of eigenvectors; thus

$$U = Y_a C_a + Y_b C_b, \tag{4.6}$$

where C_a and C_b are coefficient matrices of size $m \times m$ and $(n-m) \times m$ respectively. It follows from equations (4.4) through (4.6) that

$$V = Y_a \Lambda_a C_a + Y_b \Lambda_b C_b. \tag{4.7}$$

It may be noticed that the lower eigenvectors contribute relatively less to V than they do to U. After a number of iterations have taken place, it may be assumed that due to the washing-out process of earlier iterations the coefficients of C_b are small compared to those of C_a.

Reorientation: The reorientation process involves the complete eigensolution of the $m \times m$ *interaction matrix* B obtained as the solution of

$$(U^T U) B = (U^T V). \tag{4.8}$$

By substituting the values of U and V given in equations (4.6) and (4.7) into equation (4.8), and assuming that the C_b coefficients are negligible compared with the C_a coefficients, it follows that

$$U^T Y_a C_a B \approx U^T Y_a \Lambda_a C_a.$$

Hence, if $U^T Y_a$ is nonsingular,

$$C_a B \approx \Lambda_a C_a,$$

which shows that the matrix of left-hand eigenvectors of B is an approximation to C_a and that the eigenvalues of B give an approximation to Λ_a. If Z is the $m \times m$ matrix of right-hand eigenvectors of B, then

$$Z \approx C_a^{-1}$$

and hence the set of vectors obtained by the multiplication,

$$W = VZ \approx Y_a \Lambda_a + Y_b \Lambda_b C_b C_a^{-1},$$

gives an improved set of right-hand vectors of P^T.

4.2.2.3 Convergence Rate

As in all simultaneous iteration algorithms, the rate of convergence onto the dominant set of eigenvectors is first-order, with the error in the ith eigenvector prediction reducing (approximately) by the factor $|\lambda_{m+1}/\lambda_i|$ at each iteration, where m is the number of trial vectors used, and λ_i and λ_{m+1} are the ith and $(m+1)^{th}$ largest eigenvalues in modulus, respectively. If the r dominant eigenvalues and corresponding eigenvectors are required, then it is usual to iterate with

m ($m > r$) trial vectors; the additional vectors are called *guard vectors*. The slowest convergence rate will be for the rth eigenvector, as given by the factor $|\lambda_{m+1}/\lambda_r|$. When only the dominant eigenvector is required, the convergence factor is given by $|\lambda_{m+1}/\lambda_1|$. For stochastic matrices this is simply $|\lambda_{m+1}|$. In Lopsided iteration, the errors in the eigenvalue predictions tend to be of the same relative order of magnitude as those in the eigenvector predictions.

Because the convergence factor behaves like $|\lambda_{m+1}/\lambda_i|$, there is no advantage when the distribution of the modulus of the eigenvalues is uniform. Simultaneous iteration methods profit from gaps in the eigenvalue distribution to accelerate convergence. For example, for stochastic matrices, the rate of convergence of the power method is $|\lambda_2|$, whereas that of simultaneous iteration with m trial vectors is $|\lambda_{m+1}|$. If the eigenvalues are uniformly distributed, then a simultaneous iteration method will need only $1/m$ times the number of iterations needed by the power method. However, the amount of work involved in each iteration of simultaneous iteration is more than m times that of the power method.

4.2.2.4 Examples

Elementary MATLAB programs were written to implement basic versions of many of the algorithms presented in this chapter. Indeed, all the experiments discussed in this chapter were performed in MATLAB. These programs have been collected together and are reproduced at the end of this chapter. However, we caution our readers that these implementations were not designed to be production versions of the algorithms. The MATLAB implementations gloss over many of the implementation details that would need to be incorporated in a production version.

Consider the results obtained with the Lopsided MATLAB program (function *lopsi.m* in Section 4.6), when it is applied to the Courtois matrix, equation (4.3), with three trial vectors ($m = 3$). Figure 4.1 shows the residual as a function of iteration number. The theoretical asymptotic convergence factor is also plotted. Since the fourth eigenvalue is given as $\lambda_4 = 0.750026$ (the set of eigenvalues is given in Section 4.1.5), we should expect the eigenvector corresponding to the dominant unit eigenvalue to have an accuracy of approximately 0.75^k at the kth iteration.

Figure 4.2 shows the residual when m, the size of the subspace, is varied. It clearly demonstrates the benefit that can be derived from the gap in the eigenvalue distribution between λ_3 and λ_4. The rate of convergence using only two trial vectors is extremely slow. Iterating with three or more yields a much faster convergence rate. These results clearly demonstrate the fact that simultaneous iteration methods should be used only when advantage can be taken of such gaps in the eigenvalue distribution.

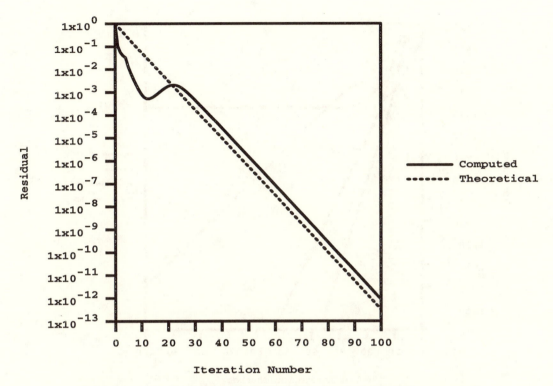

Figure 4.1: Convergence behavior: theoretical and computed.

4.3 Krylov Subspaces

We have just seen that simultaneous iteration methods iterate with a certain *fixed* number of vectors, which under suitable conditions converge onto the right-hand eigenvectors corresponding to the dominant eigenvalues. Other projection methods — in fact, the majority of projection methods — begin with a single vector and construct a subspace one vector at a time. One subspace of particular importance that is built in this fashion is the *Krylov subspace*, given by

$$\mathcal{K}_m(A, v) = \text{span}\{v, Av, A^2v, \ldots, A^{m-1}v\}.$$

Notice that this subspace is spanned by consecutive iterates of the power method. Many of the methods that we shall consider in this chapter require that an orthonormal basis be found for this Krylov subspace. This leads us to consider the Gram–Schmidt orthogonalization procedures.

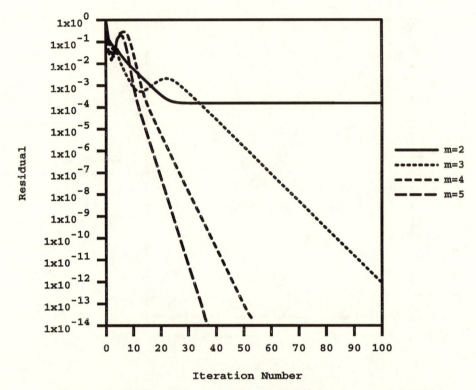

Figure 4.2: Residuals for different numbers of trial vectors.

4.3.1   Gram–Schmidt Orthogonalization Procedures

The Gram–Schmidt algorithm is often used to orthogonalize a set of vectors. Let X be an $(n \times m)$ matrix of rank m $(\leq n)$. The Gram–Schmidt algorithm in fact computes a QR factorization of X. In other words, it constructs an $(n \times m)$ matrix Q, with orthonormal columns, and an $(m \times m)$ upper triangular matrix R for which

$$X = QR.$$

Thus, Q is an orthonormal basis for X. The columns of X, $[x_1, x_2, \ldots, x_m]$, are given as linear combinations (defined by the elements of R) of the columns of Q. The classical Gram–Schmidt algorithm is as follows:

Algorithm: Classical Gram–Schmidt

1. *Initialize:*   Compute $r_{11} = \|x_1\|_2$ and $q_1 = x_1/r_{11}$.

2. *Iterate:*   For $j = 2, 3, \ldots, m$ do

- Compute $r_{ij} = q_i^T x_j$, $\quad i = 1, 2, \ldots, j-1$.
- Compute $q_j = x_j - \sum_{i=1}^{j-1} r_{ij} q_i$.
- Compute $r_{jj} = \|q_j\|_2$ and $q_j = q_j / r_{jj}$.

Observe that q_j, the vector computed at step j, has 2-norm equal to 1 and, for $j > 1$, is orthogonal to all previously computed vectors q_i, $i = 1, \ldots, j-1$. In fact, we have

$$q_j = x_j - \sum_{i=1}^{j-1} (q_i^T x_j) q_i$$

and for any k such that $1 \le k \le j-1$,

$$
\begin{aligned}
q_k^T q_j &= q_k^T x_j - \sum_{i=1}^{j-1} q_k^T (q_i^T x_j) q_i \\
&= q_k^T x_j - \sum_{i=1}^{j-1} q_k^T q_i (q_i^T x_j) = q_k^T x_j - q_k^T x_j = 0.
\end{aligned}
$$

Unfortunately, the classical Gram–Schmidt algorithm has poor numerical properties, for there is usually a loss of orthogonality among the columns of Q. A modified Gram–Schmidt algorithm, which turns out to be simply a rearrangement of the manner in which the computations are performed, has much better numerical properties. In the classical Gram–Schmidt, it suffices to replace the first two statements of Step 2,

- Compute $r_{ij} = q_i^T x_j$, $\quad i = 1, 2, \ldots, j-1$.
- Compute $q_j = x_j - \sum_{i=1}^{j-1} r_{ij} q_i$.

with

- Set $q_j = x_j$.
- For $i = 1, 2, \ldots, j-1$ do
  - Compute $r_{ij} = q_i^T q_j$.
  - Compute $q_j = q_j - r_{ij} q_i$.

When the vectors to be orthonormalized are the successive vectors in a Krylov subspace, the Gram–Schmidt process may be incorporated with the generation of the vectors. In this case, we refer to the procedure as the *Arnoldi process*.

4.4   GMRES and the Method of Arnoldi

We shall begin by showing how GMRES and the method of Arnoldi are used in the case of a matrix $A \in \Re^{n \times n}$, either to compute approximate eigenvalues and corresponding right-hand eigenvectors of A or to obtain an approximate solution to the system of linear equations $Ax = b$. In Section 4.4.4, we consider what changes, if any, must be incorporated when A is derived from a Markov chain problem. Indeed, it suffices to replace A with P^T, when the method is geared toward the determination of approximate eigensolutions, and with Q^T (or $I - P^T$), when computing solutions of linear systems, where P is the stochastic transition probability matrix of the Markov chain and Q the infinitesimal generator.

The method of Arnoldi [4, 132] has two variants, depending on whether the problem at hand is that of solving a system of linear equations or that of determining eigenvalues and corresponding eigenvectors of A. In the case of solving a system of equations, it is often called the *full orthogonalization method (FOM)*. In both cases, the method of Arnoldi is an orthogonal projection process onto the Krylov subspace

$$\mathcal{K}_m = \operatorname{span}\{v_1, Av_1, \ldots, A^{m-1}v_1\}.$$

As in simultaneous iteration, the only operations that must be performed on A are matrix-vector multiplications of the form $w = Av$, which makes these methods attractive for large sparse problems.

4.4.1   Arnoldi for Eigensolutions

The Arnoldi algorithm may be used to compute approximations to the largest eigenvalues and the corresponding eigenvectors of an unsymmetric matrix A. Unlike simultaneous iteration, these eigenvalues are not necessarily the largest in modulus, but the largest in an algebraic sense. The method starts with some nonzero approximation v_1, normalized so that $\|v_1\|_2 = 1$, and uses the modified Gram–Schmidt orthogonalization procedure to construct an orthonormal basis of the Krylov subspace

$$\mathcal{K}_m = \operatorname{span}\{v_1, Av_1, \ldots, A^{m-1}v_1\}.$$

This is what we have previously referred to as the *Arnoldi process*. The sequence of orthonormal vectors generated, v_i, is computed as follows.

Arnoldi Process

1. For $j = 1, 2, \ldots, m$ do

   - Compute $w = Av_j$.
   - For $i = 1, 2, \ldots, j$ do
     - Compute $h_{ij} = v_i^T w \ \ (= v_i^T Av_j)$.
     - Compute $w = w - h_{ij}v_i$.
   - Compute $h_{j+1,j} = \|w\|_2$ and $v_{j+1} = w/h_{j+1,j}$.

Let H_m be the $(m \times m)$ upper Hessenberg matrix whose nonzero elements are the coefficients h_{ij}, $1 \le i \le \min(j+1, m)$, $1 \le j \le m$. Then H_m represents the restriction of the linear transformation A to the subspace $\mathcal{K}_m$ with respect to the basis V_m, i.e., we have $H_m = V_m^T A V_m$, where $V_m = [v_1, v_2, \ldots, v_m]$ is the computed orthonormal basis of the Krylov subspace.

Observe that at the completion of the jth step of the Arnoldi process we have

$$h_{j+1,j}v_{j+1} = Av_j - h_{1j}v_1 - h_{2j}v_2 - \cdots - h_{jj}v_j,$$

or

$$Av_j = v_1 h_{1j} + v_2 h_{2j} + \cdots + v_j h_{jj} + v_{j+1} h_{j+1,j},$$

so that by setting $V_{m+1} = [v_1, v_2, \ldots, v_{m+1}]$, we find the important relationship

$$AV_m = V_{m+1}\bar{H}_m, \tag{4.9}$$

where $\bar{H}$ is the $(m+1) \times m$ matrix obtained by appending the row

$$[0, 0, \ldots, 0, h_{m+1,m}]$$

to H_m.

Approximations to some of the eigenvalues of A can be obtained from the eigenvalues of H_m. As m increases, the eigenvalues of H_m that are located in the outermost part of the spectrum start converging toward corresponding eigenvalues of A. If λ_i is an eigenvalue of H_m and y_i the corresponding eigenvector, i.e.,

$$H_m y_i = \lambda_i y_i, \tag{4.10}$$

then λ_i is taken as an approximation to an eigenvalue of A, and $V_m y_i$ as an approximation to its corresponding right-hand eigenvector. This is Arnoldi's method in its simplest form. Notice that if λ_i is any eigenvalue of H_m and y_i the corresponding eigenvector, then, for all $v \in \mathcal{K}_m$,

$$v^T \left(A(V_m y_i) - \lambda_i(V_m y_i) \right) = 0, \quad \text{for all } i = 1, \ldots, m.$$

This condition is called a *Galerkin condition*. The Arnoldi algorithm is completely specified by the following:

Algorithm: Arnoldi for Eigensolutions

1. *Initialize:* Choose a size m for the Krylov subspace and an initial vector v_1 such that $\|v_1\|_2 = 1$.

2. *Arnoldi process:* For $j = 1, 2, \ldots, m$ do

   - Compute $w = Av_j$.
   - For $i = 1, 2, \ldots, j$ do
     - Compute $h_{ij} = v_i^T w$.
     - Compute $w = w - h_{ij} v_i$.
   - Compute $h_{j+1,j} = \|w\|_2$ and $v_{j+1} = w/h_{j+1,j}$.

3. *Compute approximations to eigenvalues and corresponding eigenvectors:*

   - Compute eigensolution: $H_m y_i = \lambda_i y_i, \quad i = 1, 2, \ldots, m$.
   - Compute approximate eigenvectors $V_m y_i$.

Note that in this algorithm it is not strictly necessary to compute the element $h_{m+1,m}$ or the vector v_{m+1}. Also, a number of possibilities exists for computing the eigenvalues and corresponding eigenvectors of H_m. For example, in EISPACK the subroutine *HQR* may be used to find the eigenvalues and then *INVIT* used to compute selected eigenvectors by inverse iteration.

4.4.2   FOM — The Full Orthogonalization Method

Like the version of Arnoldi just given, the *full orthogonalization method (FOM)* and the *generalized minimum residual (GMRES)* algorithms begin by constructing an orthonormal basis for a Krylov subspace. However, the initial vector with which to begin the construction of the subspace is not an approximation to the solution but rather the residual produced by an initial approximation.

Let x_0 be an initial approximation to the solution of the linear system of equations

$$Ax = b. \tag{4.11}$$

The initial residual, r_0, is given by $r_0 = b - Ax_0$, and the Krylov subspace that is generated is

$$\mathcal{K}_m = \text{span}\{v_1, Av_1, \ldots, A^{m-1}v_1\},$$

where $v_1 = r_0/\|r_0\|_2$. The FOM method and GMRES both compute an approximation to the solution of (4.11) as

$$x_m = x_0 + z_m, \quad \text{with } z_m \in \mathcal{K}_m. \tag{4.12}$$

For the FOM approach, the computed solution is forced to satisfy the Galerkin condition

$$v^T r_m = 0, \quad \text{for all } v \in \mathcal{K}_m. \tag{4.13}$$

Since we have chosen $v_1 = r_0/\|r_0\|_2$, the Arnoldi vectors $V_m = [v_1, v_2, \ldots, v_m]$ span $\mathcal{K}_m$, so for some vector $y_m \in \Re^m$ equation (4.12) may be written as

$$x_m = x_0 + V_m y_m. \tag{4.14}$$

From $Ax_0 = b - r_0$, $Ax_m = b - r_m$ and $v_1 = r_0/\|r_0\|_2$, it follows that by multiplying equation (4.14) by A and substituting, we obtain

$$r_m = \|r_0\|_2 v_1 - A V_m y_m.$$

Since the Arnoldi vectors are orthonormal, we have

$$
\begin{aligned}
V_m^T r_m &= V_m^T(\|r_0\|_2 v_1 - A V_m y_m) \\
&= \|r_0\|_2 V_m^T v_1 - V_m^T A V_m y_m \\
&= \|r_0\|_2 e_1 - H_m y_m
\end{aligned}
$$

where e_1 is the first column of the identity matrix of order m. Therefore, the Galerkin condition, equation (4.13), is imposed by setting

$$y_m = H_m^{-1} \|r_0\|_2 e_1.$$

The FOM is completely specified by the following algorithm:

Algorithm: FOM — Full Orthogonalization Method

1. *Initialize:*

   - Choose a size m for the Krylov subspace and an initial approximate solution x_0.
   - Compute $r_0 = b - Ax_0$, $\beta = \|r_0\|_2$, and $v_1 = r_0/\beta$.

2. *Arnoldi process:* For $j = 1, 2, \ldots, m$ do

   - Compute $w = Av_j$.
   - For $i = 1, 2, \ldots, j$ do
     - Compute $h_{ij} = v_i^T w$.
     - Compute $w = w - h_{ij}v_i$.
   - Compute $h_{j+1,j} = \|w\|_2$ and $v_{j+1} = w/h_{j+1,j}$.

3. *Form the approximate solution:*

   - Solve the $(m \times m)$ Hessenberg system $H_m y_m = \|r_0\|_2 e_1$ for y_m.
   - Compute $x_m = x_0 + V_m y_m$.

Like the previous algorithm, it is not strictly necessary to form $h_{m+1,m}$ or the vector v_{m+1}. The system of equations in Step 3 may be solved by means of a standard LU factorization. However, it is more usual to compute a QR factorization of H_m, since this may easily be computed at the same time that the orthonormal basis V_m is generated. We shall return to this point during our discussion of the GMRES method.

The full orthogonalization method is theoretically equivalent to the conjugate gradient method, when A is symmetric and positive-definite, and to the Lanczos method, when A is symmetric [133]. When A is nonsingular, it is guaranteed to converge to the exact solution of (4.11) in at most n steps using exact arithmetic.

4.4.3 GMRES — The Generalized Minimal Residual Method

A disadvantage of the full orthogonalization method is that it does not satisfy an optimality property. GMRES does not suffer from this disadvantage. Whereas in the FOM algorithm the computed solution,

$$x_m = x_0 + z_m,$$

is forced to satisfy a Galerkin condition, in GMRES the vector z_m is chosen from the Krylov subspace in such a way as to minimize $\|b - Ax_m\|_2$.

Again, let $\beta = \|r_0\|_2$, $v_1 = r_0/\beta$, and let $\bar{H}$ be the $(m+1 \times m)$ matrix obtained by appending the row $[0, 0, \ldots, 0, h_{m+1,m}]$ to H_m. In GMRES we seek an approximation of the form $x_m = x_0 + z_m$, $z_m \in \mathcal{K}_m$ that minimizes the residual norm over $\mathcal{K}_m$. Writing $z_m = V_m y_m$, we see that y_m must minimize the following function of y,

$$
\begin{aligned}
J(y) &= \|b - A(x_0 + V_m y)\|_2 \\
&= \|r_0 - AV_m y\|_2 \\
&= \|\beta v_1 - AV_m y\|_2
\end{aligned}
\tag{4.15}
$$

Using the relation (4.9), $AV_m = V_{m+1}\bar{H}_m$, this becomes

$$
J(y) = \|V_{m+1}\left(\beta e_1 - \bar{H}_m y\right)\|_2 = \|\beta e_1 - \bar{H}_m y\|_2
\tag{4.16}
$$

by the orthogonality of V_{m+1}. As a result the vector y_m can be obtained inexpensively by solving an $(m+1) \times m$ least squares problem; the solution computed by GMRES is given by $x_0 + V_m y_m$, where y_m is the solution to the upper Hessenberg least squares problem

$$
\min_{y \in \Re^m} \|\beta e_1 - \bar{H}_m y\|_2.
$$

The GMRES algorithm is therefore identical to the FOM algorithm, with the sole exception that the first statement of Step 3 of FOM,

- Solve the $(m \times m)$ Hessenberg system $H_m y_m = \|r_0\|_2 e_1$ for y_m.

is replaced with

- Find the vector $y_m \in \Re^m$ that minimizes the function $J(y) = \|\beta e_1 - \bar{H}_m y\|_2$.

We should point out that this procedure is also a projection process. More precisely, it may be shown that the minimization of $J(y)$ is equivalent to imposing the Gram condition that

$$
r_0 - AV_m y \perp v \quad \forall\, v \in \text{span}\{AV_m\},
$$

which means that we are solving $Az = r_0$ with a projection process with

$$
\mathcal{K} = \text{span}\{r_0, Ar_0, \ldots, A^{m-1}r_0\}
$$

and

$$
\mathcal{L} = A\mathcal{K} = \text{span}\{Ar_0, A^2 r_0, \ldots, A^m r_0\}.
$$

(Saad and Schultz [137]). It may be shown that its convergence is monotonic, but not necessarily strictly monotonic. We have

$$\|r_{j+1}\|_2 \leq \|r_j\|_2.$$

GMRES is a generalization of the MINRES algorithm presented by Paige and Saunders [115]. For arbitrary nonsingular matrices it is mathematically equivalent to the ORTHODIR algorithm of Hageman and Young [66]. For matrices with positive-definite symmetric part, it is equivalent to GCR, the generalized conjugate residual method [39].

4.4.4   Application to Markov Chains

The Arnoldi procedure for computing eigensolutions may be applied directly to a stochastic transition probability matrix, P. In Section 4.4.1 the matrix A is set equal to P^T and the operations performed exactly as indicated. The methods FOM and GMRES are oriented toward nonhomogeneous systems of equations of the form $Ax = b$. In cases where the right-hand side b is zero and the matrix A is singular (e.g., $A = Q^{T\ 2}$ or $A = I - P^T$), these algorithms may also be used to compute approximate solutions. The only condition needed is to choose an $x_0 \neq 0$ in such a manner as to avoid a breakdown in the first step. From what was said earlier, GMRES will compute an approximate solution by attempting to minimize $\|Q^T x\|_2$ over the affine subspace $x_0 + \mathrm{span}\{r_0, Q^T r_0, \ldots, (Q^T)^{m-1} r_0\}$, which is typically of dimension m when $r_0 = Q^T x_0$ is nonzero. Thus, whenever $x_0 \neq 0$, one can expect the method to work without any difference from the nonhomogeneous case. It is subject to the same conditions of breakdown as the usual GMRES algorithm for general linear systems: that is, the only possible case of breakdown is when the initial vector r_0 has minimal degree not exceeding $m - 1$ with respect to Q^T. In this case $\mathcal{K}_m$ becomes invariant under Q^T, and the algorithm stops prematurely, delivering the exact solution. However, this happens very rarely in practice.

4.4.5   Examples

Before proceeding to consider these projection methods' more usual implementations, which include preconditioning and iterative versions, we shall present some results obtained with the basic algorithms applied to some simple examples. We shall consider the 8×8 Courtois matrix described previously. Table 4.2 presents the eigenvalues of the Hessenberg matrix H_m given in equation (4.10).

---

[2]Here Q *is* an infinitesimal generator matrix rather than the unitary part of the QR factorization.

Table 4.2: Eigenvalues of Hessenberg Matrix

$m = 1$	$m = 2$	$m = 3$	$m = 4$
1.000000000000	1.000000000000	1.000000000000	1.000000000000
	−0.000926468497	0.540409782347	0.584441888366
		−0.134822595267	0.409634960609
			−0.154427193158

$m = 5$	$m = 6$	$m = 7$	$m = 8$
1.000000000000	1.000000000000	1.000000000000	0.9999999999
0.735223766274	0.751133734732	0.998481371985	0.999800000076
0.549189634632	0.548977723488	0.750027552603	0.998494796881
0.326744563643	0.433904138005	0.550071148856	0.750026231509
−0.149280864797	0.301739584006	0.400028763006	0.550066639868
	−0.149620139804	0.300712817863	0.400033385164
		−0.149535360949	0.300714318809
			−0.149535372241

Observe that the second and third eigenvalues do not appear until the penultimate and final steps. This may appear to be a contradiction to the more usually observed phenomenon, whereby the dominant eigenvalues appear first. The reason in this case is that the second and third eigenvalues are difficult to obtain, since they are almost coincident with the first one. Choosing a different starting vector from the vector $e/\|e\|_2$ used in these experiments will not force these eigenvalues to appear sooner.

Table 4.3 contains the residuals obtained when the Arnoldi and GMRES methods (MATLAB programs *arnoldi.m* and *gmres.m* of Section 4.6) are applied to the Courtois example.

4.4.6 Iterative, Incomplete, and Preconditioned Methods

4.4.6.1 Iterative and Incomplete Methods

In practice one difficulty with the foregoing algorithms is that as m increases, cost and storage increase rapidly. One solution is to use the methods iteratively. A fixed value, which may depend on the amount of computer memory available, is assigned to m. If, when the Krylov subspace reaches this maximum size, the residual does not satisfy the tolerance requested, the process is begun all over again, this time using $x_m = x_0 + V_m y_m$ as the initial approximation.

Table 4.3: Residuals in Courtois Problem

	Arnoldi	GMRES
$m = 1$	0.09822373530937	0.02943574503629
$m = 2$	0.03085379797675	0.00514655070532
$m = 3$	0.00522706407017	0.00165837821929
$m = 4$	0.00175181833849	$4.564001667262711e{-}04$
$m = 5$	$4.747321883146810e{-}04$	$1.313235690891818e{-}04$
$m = 6$	$1.371225888578969e{-}04$	$1.585827790962993e{-}05$
$m = 7$	$1.597518327446166e{-}05$	$3.535102781248365e{-}17$
$m = 8$	$5.240184483061559e{-}11$	$5.133724873074499e{-}17$

Algorithm: Iterative GMRES

1. *Start:* Choose x_0 and a dimension m for the Krylov subspaces.

2. *Arnoldi process:*

   - Compute $r_0 = b - Ax_0$, $\beta = \|r_0\|_2$ and $v_1 = r_0/\beta$.
   - Use the Arnoldi procedure to generate $v_2, \ldots, v_{m+1}$ and the matrix $\bar{H}_m$.

3. *Form the approximate solution:*

   - Find the vector $y_m \in \Re^m$ that minimizes the function $J(y) = \|\beta e_1 - \bar{H}_m y\|_2$.
   - Compute $x_m = x_0 + V_m y_m$.

4. *Restart:* If satisfied stop; else set $x_0 = x_m$ and go to 2.

Saad and Schultz [137] show how to compute r_m without first explicitly form-ing x_m. The algorithm is restarted as often as is needed until the residual falls below the tolerance requested or the maximum number of iterations is completed. The algorithm above presents the procedure for iterative GMRES. Similar algo-rithms may be written for iterative FOM and iterative Arnoldi. If the method of Arnoldi for eigensolutions is used, the new initial approximation is chosen as a linear combination of the approximate eigenvectors. For stochastic matrices the eigenvector closest to the unit eigenvalue should be used.

A second approach is to orthogonalize v_{j+1} not with respect to *all* the previous vectors, $v_1, v_2, \ldots, v_j$, but only with respect to the last l of these, v_{j-l+1}, v_{j-l+2},

$\ldots, v_j$. These methods are referred to as *incomplete orthogonalization methods (IOM)* and *incomplete GMRES (IGMRES)*. They may be combined with iterative versions when $l < m$.

4.4.6.2 Preconditioned Arnoldi and GMRES Algorithms

Preconditioning techniques, such as those discussed in Chapter 3, can also be used to improve the convergence rates of Arnoldi's method and GMRES. This typically amounts to replacing the original equations by the system

$$M^{-1}Ax = M^{-1}b, \tag{4.17}$$

where M^{-1} approximates A^{-1} and $M^{-1}w$ is inexpensive to compute for any vector w. Thus, in both Arnoldi and GMRES it suffices to replace the original matrix A with the preconditioned matrix $M^{-1}A$. We may also precondition from the right, i.e., we may replace A with AM^{-1}, thus solving a system of the form $AM^{-1}z = b$. We should note that in this case the solution to the original problem is $M^{-1}z$, which requires a solve with M at the end of a projection loop. On the other hand, the computation of the initial residual $Ax_0 = AM^{-1}Mx_0$ does not require such a solve, in contrast with left preconditioning. Thus, both versions require the same number of matrix-vector products and M-solves. If $M = LU$, then the preconditioning can also be split between left and right, by replacing the original matrix with the preconditioned matrix $L^{-1}AU^{-1}$.

4.4.7 Examples, continued

Table 4.4 shows the results obtained with preconditioned iterative variants of Arnoldi and GMRES (MATLAB programs *arniter.m* and *gmriter.m* of Section 4.6 respectively) when applied to the Courtois matrix. Results obtained from other examples may be found in [119].

The leftmost column displays the absolute value of the eigenvalues of the matrix $I - M^{-1}Q$, where M^{-1} is the preconditioner obtained from the threshold-based, incomplete LU factorization $ILUTH$, with the threshold chosen as $th = 0.10$. The eigenvalues for the unpreconditioned case were given previously in Section 4.1.5. The two rightmost columns of this table show the residuals for iterative GMRES, using a subspace of size 3, and iterative Arnoldi, using a subspace of size 4. Observe that the convergence is extremely rapid.

For purposes of comparison, we also present results for the 9×9 reliability model. We examine only the case with parameters

$$\lambda_1 = 0.2; \quad \lambda_2 = 30.0; \quad \mu_1 = 0.5, \quad \text{and} \quad \mu_2 = 60.0,$$

and consider both preconditioned and unpreconditioned variants. Table 4.5 shows the differences in the eigenvalue distribution without this preconditioning and

Table 4.4: Results for the Courtois (8×8) NCD Matrix

$\|Eigenvalues\|$ $I - M^{-1}Q$	Residuals	
	$Gmriter(Q, 3, 10)$	$Arniter(P, 4, 10)$
	$1.0e{-}03$ *	$1.0e{-}03$ *
1.00000000000000	0.17382452506826	0.34550505311762
0.76421158980663	0.00047850149078	0.00003413429162
0.00531651289873	0.00000045765750	0.00000005318678
0.00531651289873	0.00000000125034	0.00000000000529
0.00080065735515	0.00000000000119	0.00000000000014
0.00080065735515	0.00000000000003	0.00000000000002
0.00003564383643	0.00000000000006	0.00000000000003
0.00000000000000	0.00000000000002	0.00000000000013
	0.00000000000006	0.00000000000005
	0.00000000000003	0.00000000000010

Table 4.5: Eigenvalues for the (9×9) Reliability Model

No Preconditioning	With $ILUTH$, $th = .1$
1.00000000000000	1.00000000000000
0.99421487603306	0.79158480243233
0.98842975206612	0.02784924420615
0.49917355371901	0.00671654679881
0.49338842975207	0.00406424204919
0.48760330578512	0.00168400344546
0.25619834710744	0.00000000000014
0.25041322314050	0.00000000000000
0.24462809917355	0.00000000000000

Table 4.6: Residuals for the (9×9) Reliability Model

No Preconditioning		$ILUTH$ Preconditioning, $th = .1$	
$gmriter(Q, 4, 10)$	$arniter(P, 4, 10)$	$gmriter(Q, 4, 10)$	$arniter(P, 4, 10)$
		$1.0e{-}07$ *	$1.0e{-}03$ *
0.001505985427	0.004580644138	0.451676030603	0.218609371328
0.000620909184	0.001558403500	0.000000288010	0.000103375112
0.000265238077	0.000105441919	0.000000000294	0.000000046690
0.000029361131	0.000044319228	0.000000000432	0.000000000004
0.000016882231	0.000026228516	0.000000000391	0.000000000002
0.000002536337	0.000010700481	0.000000000383	0.000000000002
0.000001057168	0.000010681357	0.000000000512	0.000000000002
0.000000181748	0.000017222891	0.000000000552	0.000000000002
0.000000027484	0.000012291051	0.000000000340	0.000000000002
0.000000012844	0.000001541936	0.000000000363	0.000000000002

with threshold preconditioning using $th = 0.10$. The next table (Table 4.6) shows the residuals obtained from the first 10 iterations of the methods of Arnoldi and GMRES with and without the threshold-based ILU preconditioners. In all cases, the subspace generated was of size 4. Notice that GMRES outperforms Arnoldi in this example.

4.4.8 Implementation

4.4.8.1 QR Factorizations

The recommended approach for solving the linear least squares problem

$$\min_{y \in \Re^m} \|\beta e_1 - \bar{H}_m y\|_2, \qquad (4.18)$$

in this context is that of the QR factorization. We begin by showing how a QR factorization of $\bar{H}_m$, $(\bar{H}_m = Q_m R_m)$, may be used to compute the least squares solution, and then later we describe how the QR factorization may best be found. It turns out that solving (4.18) is equivalent to solving the triangular system of equations

$$R_m y = \beta Q_m^T e_1. \qquad (4.19)$$

In a general context, we wish to find the least squares solution

$$y_{LS} = \min_{y \in \Re^m} \|Fy - f\|_2,$$

where $F \in \Re^{n \times m}$, with $n \geq m$ and $f \in \Re^n$, by way of a QR factorization. Let $Q \in \Re^{n \times n}$ be an orthogonal matrix for which

$$Q^T F = R = \begin{pmatrix} R_1 \\ 0 \end{pmatrix} \begin{array}{c} m \\ n - m \end{array}$$

where $R_1 \in \Re^{m \times m}$ is upper triangular, and let

$$Q^T f = \begin{pmatrix} c \\ d \end{pmatrix} \begin{array}{c} m \\ n - m \end{array}$$

Then, for any $y \in \Re^m$,

$$\|Fy - f\|_2^2 = \|Q^T Fy - Q^T f\|_2^2 = \|R_1 y - c\|_2^2 + \|d\|_2^2.$$

It is clear that when F has full column rank (i.e., $\operatorname{rank}(F) = \operatorname{rank}(R_1) = m$), then the least squares solution is unique and is defined as

$$y_{LS} = R_1^{-1} c. \tag{4.20}$$

Furthermore, if we define the *minimum residual* as

$$\rho_{LS} = \|Fy_{LS} - f\|_2,$$

then

$$\rho_{LS} = \|d\|_2.$$

Thus, the full-rank least squares solution can readily be obtained from the QR factorization of F by solving $R_1 y_{LS} = c$. For the problem at hand, we have $F = \bar{H}_m$ and $f = \beta e_1$, so that the solution is obtained by solving the triangular system (4.19).

4.4.8.2   Givens Rotations

We now turn our attention to determining the QR factorization of $\bar{H}_m$. Since this matrix is upper Hessenberg, the method of choice is that of Givens rotations.

A Givens rotation is given by

$$
G_{ik} = \begin{pmatrix}
1 & & & & & & & & & \\
& \ddots & & & & & & & & \circ \\
& & 1 & & & & & & & \\
& & & c & & s & & & & i \\
& & & & \ddots & & & & & \cdot \\
& & & -s & & c & & & & k \\
& & & & & & 1 & & & \cdot \\
& & & & & & & \ddots & & \\
& & & & & & & & 1 &
\end{pmatrix}
$$

where $c = \cos\theta$ and $s = \sin\theta$ for some angle θ. Since $\sin^2\theta + \cos^2\theta = 1$,

$$
G_{ik}^T G_{ik} = I,
$$

and so G_{ik} is an orthogonal matrix. The usefulness of Givens rotations is their ability to zero out a selected element of a vector (or matrix). If $z = G_{ik}x$, then

$$
z_j = \begin{cases}
cx_i + sx_k, & \text{if } j = i \\
-sx_i + cx_k, & \text{if } j = k \\
x_j, & \text{if } j \neq i,\ k
\end{cases}
$$

and so, by setting

$$
c = \frac{x_i}{\sqrt{x_i^2 + x_k^2}}; \qquad s = \frac{x_k}{\sqrt{x_i^2 + x_k^2}}, \tag{4.21}
$$

we force the element z_k to be zero. There are better ways than a direct implementation of (4.21) to determine c and s. For example, when $|x_k| > |x_i|$, c and s may be formed as

$$
\xi = x_i/x_k; \qquad s = 1/\sqrt{1 + \xi^2}; \qquad c = s\xi.
$$

When $|x_k| \leq |x_i|$, similar formulae may be used.

4.4.8.3 The QR Factorization in GMRES

In GMRES (and FOM) we may compute the QR factorization of the matrix $\bar{H}_j$ progressively at each step of the Arnoldi process. A single element must be eliminated at each step: $h_{j+1,j}$ is eliminated at step j, $j = 1, 2, \ldots, m$, so a single rotation per step is needed. However, it is important to remember that as new columns are appended, all previous rotations must be applied to the new column.

Let us begin by assuming that rotations $G_{i+1,i}, i = 1, 2, \ldots, j$ have already been applied to $\bar{H}_j$ to produce the upper triangular matrix of dimension $(j + 1) \times j$:

$$R_j = \begin{pmatrix} \times & \times & \times & \times & \times & \times \\ 0 & \times & \times & \times & \times & \times \\ 0 & 0 & \times & \times & \times & \times \\ 0 & 0 & 0 & \times & \times & \times \\ 0 & 0 & 0 & 0 & \times & \times \\ 0 & 0 & 0 & 0 & 0 & \times \\ 0 & 0 & 0 & 0 & 0 & 0 \end{pmatrix}.$$

Here $\times$ indicates a nonzero element. At the next step, a new vector is added to the Krylov subspace. This means that the last column and row of $\bar{H}_{j+1}$ appear and are appended to R_j. We must now modify this new $(j + 2) \times (j + 1)$ matrix to obtain R_{j+1}. We begin by premultiplying the new column by all previous rotations. This yields a $(j + 2) \times (j + 1)$ matrix of the form

$$\begin{pmatrix} \times & \times & \times & \times & \times & \times & \vdots & \times \\ 0 & \times & \times & \times & \times & \times & \vdots & \times \\ 0 & 0 & \times & \times & \times & \times & \vdots & \times \\ 0 & 0 & 0 & \times & \times & \times & \vdots & \times \\ 0 & 0 & 0 & 0 & \times & \times & \vdots & \times \\ 0 & 0 & 0 & 0 & 0 & \times & \vdots & \times \\ 0 & 0 & 0 & 0 & 0 & 0 & \vdots & r \\ 0 & 0 & 0 & 0 & 0 & 0 & \vdots & h \end{pmatrix},$$

the principal upper $(j + 1) \times j$ submatrix of which is simply R_j. Notice that the final element $h = h_{j+2,j+1}$, is not affected by the previous rotations. We now apply the Givens rotation $G_{j+2,j+1}$ to eliminate h. From (4.21) it follows that $G_{j+2,j+1}$ is defined by

$$\begin{aligned} c_{j+1} &= r/(r^2 + h^2)^{1/2} \\ s_{j+1} &= h/(r^2 + h^2)^{1/2}. \end{aligned}$$

Before proceeding to the next step, we must remember to apply the rotation to the right-hand side so that after the final rotation has been applied, we have $Q^T b = Q^T \beta e_1$. Continuing in this fashion, we eventually obtain the decomposition

$$Q_m^T \bar{H}_m = R_m,$$

where $Q_m^T \in \Re^{(m+1) \times (m+1)}$ and is the accumulated product of the rotation matrices G, and R_m is an upper triangular matrix of dimension $(m+1) \times m$, whose last row is zero. From the fact that Q_m is unitary, it follows that

$$J(y) = \|\beta e_1 - \bar{H}_m y\|_2 = \|Q_m^T[\beta e_1 - \bar{H}_m y]\|_2 = \|g_m - R_m y\|_2, \qquad (4.22)$$

where $g_m = Q_m^T \beta e_1$ is the transformed right-hand side. Therefore, the minimization of (4.22) is achieved by solving the linear system whose $m \times m$ coefficient matrix is the upper triangular matrix that results from removing the last row of R_m and whose right-hand side contains the first m elements of the vector g_m. This solution provides y_m and, in turn, the approximate solution $x_m = x_0 + V_m y_m$.

This manner of computing the QR factorization of $\bar{H}_m$ allows the residual norm of the corresponding approximate solution x_k to be evaluated at every step without having to compute it explicitly from $r_m = b - A x_m$ [136]. From the definition of $J(y)$, the residual norm is equal to $\|g_m - R_m y_m\|$, and because y_m is formed as the minimum least squares solution, it follows that this norm is the absolute value of the last component of g_m. Since g_m is updated at each step, the residual norm is available at every step of the QR factorization at no extra cost. As a result, one can stop as soon as the desired accuracy is achieved, even when this occurs in the middle of constructing the Krylov subspace.

4.4.9 The Complete Iterative GMRES Algorithm with Preconditioning

Because of the importance of this algorithm we now present a complete specification.

Algorithm: Iterative GMRES with Preconditioning

1. *Initialize:* Choose a size m for the Krylov subspace and an initial approximate solution x_0.

2. *Begin outer loop:*

   - Compute $r_0 = M^{-1}(b - A x_0)$ and $\beta = \|r_0\|_2$. If $\beta < \epsilon$, stop.
   - Set $v_1 = r_0/\beta$ and initialize first element in right-hand side, $g_1 = \beta$.

3. *Inner loop:* For $j = 1, 2, \ldots, m$ do

- *Arnoldi process:*
    - Compute $w = M^{-1}Av_j$.
    - For $i = 1, 2, \ldots, j$ do
        * Compute $h_{ij} = v_i^T w$.
        * Compute $w = w - h_{ij}v_i$.
    - Compute $h_{j+1,j} = \|w\|_2$, and if $h_{j+1,j} \neq 0$, set $v_{j+1} = w/h_{j+1,j}$.
- *Update factorization of H:*
    - *Perform previous transformations on j^{th} column of H:*
      For $i = 1, 2, \ldots, j-1$ do
        * Compute $h_{i,j} = c_i \times h_{i,j} + s_i \times h_{i+1,j}$.
        * Compute $h_{i+1,j} = -s_i \times h_{i,j} + c_i \times h_{i+1,j}$.
    - *Determine next plane rotation:*
        * Compute $\gamma = \sqrt{h_{jj}^2 + h_{j+1,j}^2}$; $c_j = h_{jj}/\gamma$; and $s_j = h_{j+1,j}/\gamma$.
        * Compute $g_{j+1} = -s_j \times g_j$ and $g_j = c_j \times g_j$.
        * Compute $h_{jj} = c_j \times h_{jj} + s_i \times h_{j+1,j}$.
- *Convergence test:*
    - If $\|g_{j+1}\| < \epsilon$, perform Step 4, then stop.

4. *Form the approximate solution:*

    - Solve $(m \times m)$ triangular system: $H_m y_m = g$.
    - Compute $x_m = x_0 + V_m y_m$.

5. *End outer loop:*   If satisfied stop, else set $x_0 = x_m$ and go to Step 2.

The following are some important points to note in this algorithm.

- M^{-1} is the preconditioning matrix.

- The upper Hessenberg matrix H obtained from the modified Gram–Schmidt algorithm is overwritten with the triangular matrix R belonging to the QR decomposition of H.

- The parameters $\cos\theta$ and $\sin\theta$ of the Givens rotations, which define the unitary matrix Q of the QR decomposition, are stored in two one-dimensional arrays $c(\cdot)$ and $s(\cdot)$, respectively.

- ϵ is the tolerance requested.

- The vector g is used to store the computed right-hand side, $\beta Q^T e_1$.

4.5 Lanczos and Conjugate Gradients

We now turn our attention to the methods of Lanczos, conjugate gradients, and related ideas. These algorithms generate approximations by means of simple three-term recurrence relations and thus require much less computation per iteration than the methods of the preceding section. Unfortunately, however, they do not satisfy an optimality property. The Lanczos algorithms are used to compute eigensolutions; hence, when Markov chain problems are to be solved, the matrix A should be replaced by a transposed stochastic transition probability matrix, P^T. The various versions of conjugate gradients are applicable to systems of linear equations, so in these cases the matrix A needs to be replaced by an infinitesimal generator matrix, Q^T or $I - P^T$.

4.5.1 The Symmetric Lanczos Algorithm

When the coefficient matrix A is symmetric, Arnoldi's method is identical to the method of Lanczos. The upper Hessenberg matrix H that is generated by Arnoldi's method becomes a real symmetric tridiagonal matrix. It is usual to set $h_{jj} = \alpha_j$ and $h_{j-1,j} = \beta_j$, which gives

$$
T_m = \begin{pmatrix}
\alpha_1 & \beta_2 & & & & \\
\beta_2 & \alpha_2 & \beta_3 & & & \\
& \beta_3 & \alpha_3 & \beta_4 & & \\
& & \ddots & \ddots & \ddots & \\
& & & \beta_{m-1} & \alpha_{m-1} & \beta_m \\
& & & & \beta_m & \alpha_m
\end{pmatrix}.
$$

The method of Lanczos generates a sequence of tridiagonal matrices, T_j, $j = 1, 2 \ldots$ whose eigenvalues are taken as approximations to a subset of the eigenvalues of A. If the computations are performed in exact arithmetic and if no breakdowns occur, the algorithm terminates in at most n steps, and the tridiagonal matrix generated represents the restriction of A to an invariant subspace of $\Re^n$. Thus, all eigenvalues of T are also eigenvalues of A, and a set of basis vectors for the A-invariant subspace is generated.

If carried out to completion, the objective of the Lanczos algorithm is to compute a symmetric tridiagonal matrix T such that $T = V^T A V$, where $V = [v_1, v_2, \ldots, v_n]$ is orthonormal. We shall let T_m denote the leading principal submatrix of size $(m \times m)$ of T. Writing $T = V^T A V$ as $AV = VT$, and equating the jth column of each side, yields

$$
Av_j = \beta_j v_{j-1} + \alpha_j v_j + \beta_{j+1} v_{j+1} \tag{4.23}
$$

for $j = 1, 2, \ldots, n$, with the understanding that $\beta_1 v_0 = 0$ and $\beta_{n+1} v_{n+1} = 0$. From equation (4.23) and the orthonormality of V it immediately follows that

$$v_j^T A v_j = \alpha_j. \tag{4.24}$$

Furthermore, writing

$$\beta_{j+1} v_{j+1} = (A - \alpha_j I) v_j - \beta_j v_{j-1} \equiv \hat{v}_{j+1} \tag{4.25}$$

it follows that, so long as $\hat{v}_{j+1} \neq 0$,

$$v_{j+1} = \hat{v}_{j+1} / \beta_{j+1} \quad \text{with} \quad \beta_{j+1} = \|\hat{v}_{j+1}\|_2. \tag{4.26}$$

The vectors v_1, v_2, etc., are called the *Lanczos vectors*. Equations (4.24) through (4.26) suggest the following iterative algorithm.

Algorithm: Lanczos for Hermitian Matrices

1. *Initialize:*

   - Choose a size m for the Krylov subspace and an initial vector v_1 such that $\|v_1\|_2 = 1$.
   - Set $\beta_1 = 0$ and $v_0 = 0$.

2. For $j = 1, 2, \ldots, m$ do

   - Compute $\alpha_j = v_j^T A v_j$.
   - Compute $\hat{v}_{j+1} = (A - \alpha_j I) v_j - \beta_j v_{j-1}$.
   - Compute $\beta_{j+1} = \|\hat{v}_{j+1}\|_2$ and $v_{j+1} = \hat{v}_{j+1} / \beta_{j+1}$.

3. *Compute approximations to eigenvalues and corresponding eigenvectors:*

   - Compute eigensolution: $T_m y_i = \lambda_i y_i, \quad i = 1, 2, \ldots, m$.
   - Compute approximate eigenvectors $V_m y_i$.

As pointed out in Golub and Van Loan [54], with careful overwriting, Step 2 of this algorithm may be implemented with only two vectors of length n plus storage for the α_i and β_i. The matrix A is used only in the matrix-vector multiplication $A v_j$, only one of which must be performed at each step. Also, since T is symmetric tridiagonal, many efficient methods are available for the computation of its eigenvalues and eigenvectors. In particular, we cite the bisection method, based on the Sturm sequence property, and the symmetric QR algorithm as possible candidates. To obtain approximations to the eigenvectors of A, all of the Lanczos vectors must be kept.

If carried out in exact arithmetic, the vectors v_j, $j = 1, 2, \ldots, m$ are orthogonal. Unfortunately, on a computer this theoretical orthogonality is observed only in the first few steps and is quickly lost thereafter. The work of Paige [114] is critical to an understanding of the numerical difficulties that gave the Lanczos method a poor reputation. It follows from equation (4.23) that at step j of the algorithm, we have computed a matrix $V_j = [v_1, v_2, \ldots, v_j]$ and a tridiagonal matrix T_j such that

$$AV_j = V_j T_j + \beta_{j+1} v_{j+1} e_j^T. \qquad (4.27)$$

In the analysis of Paige it is shown that although at step j the computed Lanczos vectors and tridiagonal matrix satisfy equation (4.27) to working accuracy, the Lanczos vectors need not be orthogonal. There may be a substantial departure from orthogonality when the β_i are small, and this loss of orthogonality may result in a loss of accuracy in the eigenvalue predictions; there may be a large difference between the eigenvalues of T_j and the eigenvalues of A, even though the Lanczos vectors are correct to machine precision! This observed behavior is one of the primary reasons why, after much initial excitement, the Lanczos algorithm fell from grace.

A possible solution that comes immediately to mind is to reorthogonalize new vectors against all previously generated Lanczos vectors as and when these new vectors are generated from the three-term recurrence relation. This is called *Lanczos with complete reorthogonalization*, and in some cases it results in a viable algorithm. However, it is expensive except when the number of eigenvalues needed is small and the subspace that must be generated is also small.

A better alternative, called *Lanczos with selective orthogonalization*, has been suggested by Parlett [116]. The analysis of Paige shows that, instead of being orthogonal to the previously generated Lanczos vectors, newly generated vectors have a component that is nonnegligible in the direction of any *converged* Ritz vector. This suggests that it suffices to ensure that newly generated Lanczos vectors be made orthogonal to all converged Ritz vectors, and since this will generally be a much smaller number than the number of previously generated Lanczos vectors, a considerable savings ensues. The set of converged Ritz vectors is initially empty. Computationally simple formulae are available that permit the loss of orthogonality to be tracked during the Lanczos procedure. When this loss is deemed to be unacceptable, the tridiagonal matrix must be diagonalized and the converged Ritz vectors computed. In the steps that follow, all newly generated Lanczos vectors are orthogonalized against these Ritz vectors until once again an unacceptable loss of orthogonality is detected. The diagonalization of the tridiagonal matrix must then be updated and the set of converged Ritz vectors enlarged. For more details, the interested reader should consult [116].

4.5.2   The Unsymmetric Lanczos Algorithm

As we stated at the beginning of this section, the Lanczos algorithm may be considered as a symmetric version of the method of Arnoldi. Conversely, we may view Arnoldi as a generalization of Lanczos to nonsymmetric matrices. However, this is not the only possible extension of Lanczos to the nonsymmetric case. We now present the algorithm that is more generally referred to as the *unsymmetric Lanczos method.*

We have seen that for symmetric matrices, an *orthogonal* similarity transformation may be applied to reduce A to tridiagonal form. We now consider the case when a *general* similarity transformation is applied to a nonsymmetric matrix A to reduce it to tridiagonal form. We should not expect this to be as stable as an orthogonal transformation, and indeed this is the case. Let

$$V = [v_1, v_2, \ldots, v_n] \quad \text{and} \quad W = [w_1, w_2, \ldots, w_n]$$

be such that $V^{-T} = W$ and

$$V^{-1}AV = T = \begin{pmatrix} \alpha_1 & \beta_2 & & & & \\ \gamma_2 & \alpha_2 & \beta_3 & & & \\ & \gamma_3 & \alpha_3 & \beta_4 & & \\ & & \ddots & \ddots & \ddots & \\ & & & \gamma_{n-1} & \alpha_{n-1} & \beta_n \\ & & & & \gamma_n & \alpha_n \end{pmatrix}.$$

By considering the jth column of $AV = VT$ and $A^TW = WT^T$ respectively, we have analagously to the symmetric case,

$$Av_j = \beta_j v_{j-1} + \alpha_j v_j + \gamma_{j+1} v_{j+1},$$

$$A^T w_j = \gamma_j w_{j-1} + \alpha_j w_j + \beta_{j+1} w_{j+1},$$

for $j = 1, 2, \ldots, n$, where we take $\beta_1 v_0 = \gamma_1 w_0 = 0$ and $\beta_{n+1} w_{n+1} = \gamma_{n+1} v_{n+1} = 0$. As before, from the biorthogonality of V and W, we may derive

$$\alpha_j = w_j^T A v_j,$$

and

$$\begin{aligned} \gamma_{j+1} v_{j+1} &= (A - \alpha_j I) v_j - \beta_j v_{j-1}, \\ \beta_{j+1} w_{j+1} &= (A - \alpha_j I)^T w_j - \gamma_j w_{j-1}. \end{aligned}$$

In essence, the unsymmetric Lanczos method constructs a pair of biorthogonal bases for the subspaces

$$\begin{aligned} \mathcal{K}_m(A, v_1) &= \text{span}\{v_1, Av_2, \ldots, A^{m-1}v_1\} \\ \mathcal{K}_m(A^T, w_1) &= \text{span}\{w_1, A^T w_1, \ldots, (A^T)^{m-1}w_1\}. \end{aligned}$$

The algorithm is as follows:

Algorithm: Lanczos for Nonsymmetric Matrices

1. *Initialize:* Choose a size m for the Krylov subspace and two initial vectors v_1, w_1 such that $\|v_1\|_2 = 1$ and $w_1^T v_1 = 1$. Set $\beta_1 = \gamma_1 = 0$, and $w_0 = v_0 = 0$.

2. For $j = 1, 2, \ldots, m$ do

   - Compute $\alpha_j = w_j^T A v_j$.
   - Compute $\hat{v}_{j+1} = (A - \alpha_j I)v_j - \beta_j v_{j-1}$ and $\beta_{j+1} = \|\hat{v}_{j+1}\|_2$.
   - Compute $\hat{w}_{j+1} = (A^T - \alpha_j I)w_j - \gamma_j w_{j-1}$ and $\gamma_{j+1} = \hat{w}_{j+1}^T \hat{v}_{j+1} / \beta_{j+1}$.
   - Compute $v_{j+1} = \hat{v}_{j+1}/\beta_{j+1}$ and $w_{j+1} = \hat{w}_{j+1}/\gamma_{j+1}$.

3. *Compute approximations to eigenvalues and corresponding eigenvectors:*

   - Compute eigensolution: $T_m y_i = \lambda_i y_i, \quad i = 1, 2, \ldots, m$.
   - Compute approximate eigenvectors $V_m y_i$.

Note that the algorithm must be halted when $\hat{w}_{j+1}^T \hat{v}_{j+1} = 0$, for otherwise a division by zero will ensue. This quantity will be zero if either $\hat{v}_{j+1} = 0$ or $\hat{w}_{j+1} = 0$. In most circumstances this is considered a happy situation, for it means that either an A-invariant or A^T-invariant subspace has been computed; it is referred to as a *regular termination*. Notice, however, that in Markov chain problems we already have at our disposal the dominant eigenvalue and the corresponding right-hand eigenvector of P. If the matrix P^T is provided to the algorithm and the initial choice of w is given as $w = e$, then a "regular termination" will occur during the first iteration! Strictly speaking, our problem is not the computation of an eigenvalue, but only the left-hand eigenvector corresponding to a unit eigenvalue. It is possible that the Lanczos algorithm will terminate after computing the right-hand vector while still a long way from finding the corresponding left-hand vector.

Furthermore, it may also happen that

$$\hat{w}_{j+1}^T \hat{v}_{j+1} = 0, \quad \text{but} \quad \hat{v}_{j+1} \neq 0 \quad \text{and} \quad \hat{w}_{j+1} \neq 0,$$

and the process halts without computing an invariant subspace. This is referred to as a *serious breakdown*. When implemented on a computer, it is unlikely that $\hat{w}_{j+1}^T \hat{v}_{j+1}$ will be exactly zero, but it is entirely possible that a *near-breakdown* will occur. In this case we find

$$\hat{w}_{j+1}^T \hat{v}_{j+1} \approx 0, \quad \text{with} \quad \hat{v}_{j+1} \not\approx 0 \quad \text{and} \quad \hat{w}_{j+1} \not\approx 0.$$

Current research and experimental results suggest that "look-ahead" versions of unsymmetric Lanczos algorithms can be successful in detecting when these situations will arise and then avoid them by essentially jumping over them. This is briefly discussed at the end of this section.

If approximations to the left-hand eigenvectors are required instead of the right-hand ones as shown in the algorithm, these may be computed as $W_m z_i$, where z_i is the left eigenvector of T_m corresponding to the eigenvalue λ_i. This is useful when estimating the condition number of a computed eigensolution, for usually both sets of vectors are needed.

Notice that β_{j+1} and γ_{j+1} are simply scaling factors, chosen so that the vectors v_{j+1} and w_{j+1} satisfy the biorthogonality condition, $w_{j+1}^T v_{j+1} = 1$. Obviously there is an infinity of ways in which this may be accomplished. It is simply necessary that

$$\gamma_{j+1}\beta_{j+1} = \hat{w}_{j+1}^T \hat{v}_{j+1}.$$

The choice made in the algorithm above, i.e.,

$$\beta_{j+1} = \|\hat{v}_{j+1}\|_2; \qquad \gamma_{j+1} = \hat{w}_{j+1}^T \hat{v}_{j+1}/\beta_{j+1},$$

is perhaps a natural choice to make. An alternative choice, suggested by Saad [135], is

$$\gamma_{j+1} = |\hat{w}_{j+1}^T \hat{v}_{j+1}|^{1/2}; \qquad \beta_{j+1} = (\hat{w}_{j+1}^T \hat{v}_{j+1})/\gamma_{j+1}.$$

This results in the two vectors being divided by two scalars having the same modulus, so that v_i has the same norm as w_i for all i, assuming that v_1 and w_1 have the same norm to begin with.

If we let V_j denote the first j columns of V and T_j the leading principal submatrix of order j of T, then it is easy to show that the following relations hold for the unsymmetric Lanczos algorithm:

$$
\begin{aligned}
AV_j &= V_j T_j + \gamma_{j+1} v_{j+1} e_j^T, \\
A^T W_j &= W_j T_j^T + \beta_{j+1} w_{j+1} e_j^T, \\
W_j^T AV_j &= T_j.
\end{aligned}
$$

Theorem 4.1 *The sets of vectors $[v_1, v_2, \dots]$ and $[w_1, w_2, \dots]$, generated by the unsymmetric Lanczos algorithm, are biorthonormal.*

Proof   The proof is by induction.

Basis Step: By construction, we have $w_1^T v_1 = 1$.

Induction Hypothesis: We assume that the vectors $v_1, v_2, \ldots, v_j$ and $w_1, w_2, \ldots, w_j$ are biorthonormal, i.e., $w_i^T v_k = \delta_{ik}$ for $1 \leq i \leq j$ and $1 \leq k \leq j$.

To show: $v_1, v_2, \ldots, v_{j+1}$ and $w_1, w_2, \ldots, w_{j+1}$ are biorthonormal. We begin by showing that under the induction hypothesis the following proposition holds:

Proposition 4.1

$$
w_i^T A v_j = \begin{cases} \alpha_j & \text{if } i = j & (a) \\ \beta_j & \text{if } i = j - 1 & (b) \\ 0 & \text{if } i < j - 1 & (c) \end{cases} \tag{4.28}
$$

Proof Obviously part *(a)* is true, since $\alpha_j = w_j^T A v_j$ by construction. To prove part *(b)*, we have

$$
\begin{aligned}
w_{j-1}^T A v_j &= (A^T w_{j-1})^T v_j \\
&= (\beta_j w_j + \alpha_{j-1} w_{j-1} + \gamma_{j-1} w_{j-2})^T v_j \\
&= \beta_j (w_j^T v_j) + \alpha_{j-1}(w_{j-1}^T v_j) + \gamma_{j-1}(w_{j-2}^T v_j) \\
&= \beta_j \qquad \text{(from the induction hypothesis)}
\end{aligned}
$$

and hence part *(b)* follows.

Finally

$$
\begin{aligned}
w_i^T A v_j &= (A^T w_i)^T v_j \\
&= \beta_{i+1}(w_{i+1}^T v_j) + \alpha_i(w_i^T v_j) + \gamma_i(w_{i-1}^T v_j).
\end{aligned}
$$

For $i < j-1$ it follows from the induction hypothesis that all of the inner products in this expression vanish, thereby proving part *(c)*. $\square$

We show that $w_i^T v_{j+1} = 0$ for $i \leq j$ by considering the three cases:

1. For $i = j$:

$$
w_j^T v_{j+1} = \gamma_{j+1}^{-1}(w_j^T A v_j - \alpha_j w_j^T v_j - \beta_j w_j^T v_{j-1}).
$$

Since $\alpha_j = w_i^T A v_j$, from part *(a)* in equation (4.28), and $w_j^T v_j = 1$, from the induction hypothesis, the first two terms cancel. Also, the last term in the parenthesis is zero, from the induction hypothesis. Hence

$$
w_j^T v_{j+1} = 0.
$$

2. For $i = j - 1$:

$$
w_{j-1}^T v_{j+1} = \gamma_{j+1}^{-1}(w_{j-1}^T A v_j - \alpha_j w_{j-1}^T v_j - \beta_j w_{j-1}^T v_{j-1}).
$$

In this case the middle term vanishes as a result of the induction hypothesis. Since $w_{j-1}^T A v_j = \beta_j$, (from part *(b)* in equation (4.28), and $w_{j-1}^T v_{j-1} = 1$, from the induction hypothesis, the remaining terms cancel, leaving the right-hand side equal to zero. Therefore

$$w_{j-1}^T v_{j+1} = 0.$$

3. For $i < j - 1$:

$$w_i^T v_{j+1} = \gamma_{j+1}^{-1}(w_i^T A v_j - \alpha_j w_i^T v_j - \beta_j w_i^T v_{j-1}).$$

From part *(c)* in equation (4.28), the first term is zero. For $i < j - 1$ it follows from the induction hypothesis that the second and third terms also vanish, leaving

$$w_i^T v_{j+1} = 0 \quad \text{for } i < j - 1.$$

In a similar manner, we may show that

$$w_{j+1}^T v_i = 0 \quad \text{for } i \le j.$$

To complete the proof, we note that, by construction

$$w_{j+1}^T v_{j+1} = 1.$$

$\square$

4.5.3 The "Look-Ahead" Lanczos Algorithm

A major disadvantage of the Lanczos algorithm is its propensity to breakdown or near-breakdown. We have observed that near-breakdowns of a serious nature can occur when

$$\hat{w}_{j+1}^T \hat{v}_{j+1} \approx 0, \quad \text{with } \hat{v}_{j+1} \not\approx 0 \quad \text{and} \quad \hat{w}_{j+1} \not\approx 0.$$

Look-ahead versions were developed to remedy this situation [117]. Recent versions developed by Freund, Gutknecht, and Nachtigal [45] appear to be particularly useful.

The look-ahead Lanczos generates a sequence of vectors $[v_1, v_2, \ldots]$ and $[w_1, w_2, \ldots]$ such that

$$\text{span}\{v_1, A v_1, \ldots, A^{m-1} v_1\} = \mathcal{K}_m(A, v_1) \quad \text{and}$$
$$\text{span}\{w_1, A^T w_1, \ldots, (A^T)^{m-1} w_1\} = \mathcal{K}_m(A^T, w_1),$$

but now, unlike the regular Lanczos, we no longer insist that the biorthogonality condition

$$w_j^T v_i = \delta_{ji},$$

be strictly enforced. This condition is relaxed whenever an exact or near-break-down situation would result. As the Lanczos vectors are generated, they are separated into groups that satisfy a group biorthogonality condition. If we denote these groups V_l and W_l, i.e.,

$$V_l = [v_{n_l}, v_{n_l+1}, \ldots, v_{n_{l+1}-1}]$$
$$W_l = [w_{n_l}, w_{n_l+1}, \ldots, w_{n_{l+1}-1}],$$

where v_{n_l} and w_{n_l} are the first vectors of the lth group, then the *blocks* must satisfy the biorthogonality relationship

$$W_j^T V_i = \begin{cases} 0 & j \neq i \\ D_j & j = i \end{cases}$$

In other words, the blocks must be biorthogonal with respect to one another, but vectors within each block need not be. This mechanism is used solely to avoid breakdown and near-breakdowns, so where possible, the original biorthogonal condition should be used. Blocks should therefore be kept as small as possible. In most cases they will be of size 1, but from time to time it may be necessary to generate some of size 2 or 3 or even greater. Numerical experiments seem to indicate that only rarely are blocks of size greater than 4 needed. More information on these methods, including implementations and code, may be obtained directly from the references.

The attractiveness of the basic Lanczos algorithm stems from the three-term recurrence relation that is used to generate successive Lanczos vectors. Both the amount of computation and the number of vectors needed is small compared to the method of Arnoldi, for example. This can be maintained in the look-ahead versions.

4.5.4 CG – The Conjugate Gradient Algorithm

The *conjugate gradient (CG)* method is well-known for solving systems of linear equations

$$Ax = b$$

when A is symmetric and positive-definite. Since we shall consider only real matrices A, this means that

$$A = A^T \quad \text{and} \quad x^T A x > 0 \quad \forall\, x \neq 0.$$

For symmetric positive-definite A, the problem of solving $Ax = b$ is equivalent to minimizing the quadratic form

$$\phi(x) = \frac{1}{2}x^T A x - b^T x. \tag{4.29}$$

If we consider $\phi(x + \alpha v)$, we find that by substituting $x + \alpha v$ for x in equation (4.29) and rearranging,

$$\phi(x + \alpha v) = \phi(x) + \alpha(Ax - b)^T v + \frac{1}{2}\alpha^2(v^T Av).$$

Since A is positive-definite, the coefficient of α^2 is strictly positive, and thus $\phi(x + \alpha v)$ has a minimum and not a maximum. This minimum is given when α has the value

$$\hat{\alpha} = \frac{(b - Ax)^T v}{v^T Av},$$

and at this value

$$\phi(x + \hat{\alpha} v) = \phi(x) - \frac{[(b - Ax)^T v]^2}{2v^T Av}. \tag{4.30}$$

Notice that the second term of (4.30) is positive, and so a reduction occurs in going from $\phi(x)$ to $\phi(x + \hat{\alpha} v)$, unless v is orthogonal to the residual. Notice also that no reduction is possible when x is a solution of $Ax = b$, for in this case the minimum has been found. Conversely, if $Ax \neq b$, then there are many vectors that are not orthogonal to the residual and thus the minimum has not yet been attained. The minimum value of the functional ϕ is given by $-b^T A^{-1} b/2$.

This leads to the development of iterative methods that seek to minimize ϕ along a set of direction vectors. A first direction vector is chosen, and the minimum in this direction is computed. At this point the next direction is chosen (in fact, the direction is often only computed at this point), and the minimum in this new direction is found. At each step, a direction vector is selected to pass through the previous minimum, x_k, and a new minimum, x_{k+1}, is computed. This may be written in the generic form

$$x_{k+1} = x_k + \alpha_k v_k,$$

where α_k is a scalar quantity and v_k is a direction vector. Methods differ in their choice of v_k; α_k is computed to attain the minimum in the chosen direction. In the *method of steepest descent*, v_k is specified as the negative gradient of ϕ at x_k, since the function ϕ decreases most rapidly in this direction. It is easy to show that the negative gradient points in the direction of the residual $r_k = b - Ax_k$. If the residual at step j is nonzero, there exists a positive α_j for which $\phi(x_j + \alpha_j r_j) < \phi(x_j)$. Therefore, in the method of steepest descent, $v_j = r_j$, and setting $\alpha_j = r_j^T r_j / r_j^T Ar_j$ minimizes $\phi(x_j + \alpha_j r_j)$.

Algorithm: Steepest Descent

1. *Initialize:*   Set $x_0 = 0$ and $r_0 = b$.

2. For $j = 1, 2, \ldots$ do

   - Compute $\alpha_j = r_{j-1}^T r_{j-1} / r_{j-1}^T A r_{j-1}$.
   - Compute $x_j = x_{j-1} + \alpha_j r_{j-1}$.
   - Compute $r_j = r_{j-1} - \alpha_j A r_{j-1}$ $(= b - A x_j)$.

However, it turns out that the method of steepest descent is rather slow at finding the solution of $Ax = b$, and other methods are more commonly used. Primarily among these is the *conjugate gradient* method (Hestenes and Stiefel [72]). The search directions chosen in the conjugate gradient method are different from, but related to, those of steepest descent. They are taken to be as close as possible to those of the steepest descent method but are subject to an orthogonality condition. Specifically, the search directions v_i must be chosen to form an A-orthogonal system (i.e., $v_i^T A v_j = \delta_{ij}$). Note that when the search direction is different from the (nonzero) residual, it follows from equation (4.30) that it must *not* be orthogonal to the residual. The algorithm is completely specified by

Algorithm: CG — Conjugate Gradient

1. *Initialize:*

   - Choose an initial approximate solution, x_0.
   - Compute $r_0 = b - A x_0$.
   - Set $v_0 = 0$ and $\beta_0 = 0$.

2. For $j = 1, 2, \ldots$ do

   - Compute $v_j = r_{j-1} + \beta_{j-1} v_{j-1}$.
   - Compute $\alpha_j = r_{j-1}^T r_{j-1} / v_j^T A v_j$.
   - Compute $x_j = x_{j-1} + \alpha_j v_j$.
   - Compute $r_j = r_{j-1} - \alpha_j A v_j$.
   - Compute $\beta_j = r_j^T r_j / r_{j-1}^T r_{j-1}$.

In CG we can see that the search directions form an A-orthogonal system. Additionally, it may be shown that the residuals themselves form an orthogonal system in the ordinary sense (i.e., $r_i^T r_j = \delta_{ij}$, where $r_i = b - A x_i$). The following theorem, adapted from Stoer and Burlirsch ([162], p. 573), provides many of the theoretical results concerning the direction vectors and the residuals.

Theorem 4.2 *Let A be a symmetric positive-definite (real) $n \times n$ matrix and $b \in \Re^n$, and let x_i, r_i and v_i be defined as in the CG algorithm. Then, for each initial vector $x_0 \in \Re^n$, there exists a smallest nonnegative integer $l \leq n$ such that $v_l = 0$. The vectors x_k, v_k and r_k, $k \leq l$ generated by the conjugate gradient method have the following properties:*

$$
\begin{aligned}
&(a) && Ax_l = b. \\
&(b) && r_i^T v_j = 0 && 0 \leq j < i \leq l. \\
&(c) && r_i^T v_i = r_i^T r_i && i \leq l \\
&(d) && v_i^T A v_j = 0 && 0 \leq i < j \leq l, \\
& && v_j^T A v_j > 0 && j < l. \\
&(e) && r_i^T r_j = 0 && 0 \leq i < j \leq l, \\
& && r_j^T r_j > 0 && j < l. \\
&(f) && r_i = b - Ax_i && i \leq l
\end{aligned}
$$

Proof The proof is by induction and may be found in Stoer and Bulirsch. □

Although theoretically the CG method will compute the exact solution of $Ax = b$ in at most n steps, this does not happen in practice. The value of CG lies in the fact that it generates good approximations to the solution in much less than n iterations. One final word: the conjugate gradient method is generally used in conjunction with a preconditioner. However, we shall delay discussion of this aspect until Section 4.5.7. By that time we shall have considered several possibilities of applying CG to nonsymmetric systems and may therefore deal with the preconditioning issue in a more general setting.

4.5.5 CGNR — Conjugate Gradient for the Normal Equations

The conjugate gradient method, particularly when preconditioned, has proven itself to be very effective at solving systems of linear equations whose coefficient matrix A is symmetric and positive-definite. Unhappily, the situation is less rosy for the nonsymmetric case.

When A is a real, nonsingular matrix, then $A^T A$ is symmetric and positive-definite, and so CG may be used to compute the solution x of $Ax = b$ from

$$A^T A x = A^T b.$$

These are called the *normal equations*. Alternatively, CG may be applied to the system

$$AA^T z = b \tag{4.31}$$

and the solution to $Ax = b$ computed as $x = A^T z$. For convenience, we shall refer to the CG method applied to equation (4.31) as *CGNR (conjugate gradient for the normal equations).* In the case when A is sparse, the product AA^T need not be explicitly formed but may be implicitly incorporated into the conjugate gradient method. In the following algorithm the matrix M^{-1} is the preconditioner.

Algorithm: CGNR — Conjugate Gradient for the Normal Equations

1. *Initialize:*

   - Choose an initial approximate solution, x_0, and tolerance criterion, ϵ.
   - Compute $r_0 = M^{-1}(b - Ax_0)$.
   - Set $v_0 = 0$; $\beta_0 = 0$; and $j = 0$.

2. While $r_j^T r_j > \epsilon$ do

   - Compute $j = j + 1$.
   - Compute $v_j = A^T M^{-T} r_{j-1} + \beta_{j-1} v_{j-1}$.
   - Compute $\alpha_j = r_{j-1}^T r_{j-1} / v_j^T v_j$.
   - Compute $x_j = x_{j-1} + \alpha_j v_j$.
   - Compute $r_j = r_{j-1} - \alpha_j M^{-1} A v_j$.
   - Compute $\beta_j = r_j^T r_j / r_{j-1}^T r_{j-1}$.

The norm of the residual after j iterations is given by $r_j^T r_j$. This must be computed at each step and is used to test for convergence. Notice that this algorithm, unlike the original CG algorithm, requires vector products with both A and A^T (and M^{-1} and M^{-T}), which means that extra numerical operations must be performed. Unfortunately, this is not the only drawback of the normal equations approach: the condition of the normal equations is worse than that of the original system. In fact, the condition number of AA^T is the square of the condition number of A. Until recently, this was sufficient grounds for dismissing this approach as a viable method for solving nonsymmetric linear systems. Nevertheless, there is currently some sentiment that there is indeed a place for CGNR. The method has a strong theoretical basis from which users may draw confidence; it has proven to be successful in numerous applications from many different domains; and extensive experiments have provided insight into the choice and development of good preconditioners.

When the matrix A is nonsingular, CGNR is guaranteed to find the correct solution in at most n steps, if the arithmetic is performed exactly. On a computer, as a result of rounding error, the method will sometimes fail to compute the correct result. When convergence does occur, then it may be shown that this convergence is strictly monotonic, i.e., if $\|r_i\| > 0$, then

$$\|r_{i+1}\| < \|r_i\|.$$

4.5.6   BCG and CGS — Conjugate Gradient for Nonsymmetric Systems

We now consider alternatives to working with the normal equations. In one of his original papers, Lanczos suggested how his method for tridiagonalizing nonsymmetric matrices might be applied to solve nonsymmetric systems of equations. The approach he suggested was slightly modified by Fletcher [43] and is known as the *BCG (biconjugate gradient)* algorithm. This method uses two residual vectors, r_j and $\tilde{r}_j$, which satisfy the biorthogonality condition $\tilde{r}_j^T r_i = 0$ if $i \neq j$, and two search direction vectors, v_j and $\tilde{v}_j$, which satisfy the biconjugacy condition $\tilde{v}_j^T A v_i = 0$ if $i \neq j$. The BCG algorithm may be viewed as being related to GMRES, for both build approximations in the same Krylov subspace. The BCG approximation is computed from simple three-term recurrence relations and thus requires much less work than GMRES, in which the amount of work and the amount of memory needed grow with the size of the subspace. However, the approximations computed by BCG do not satisfy an optimality property, which means that it cannot outperform GMRES in terms of number of iterations. The BCG algorithm is as follows:

Algorithm: BCG — Biconjugate Gradient

1. *Initialize:*

    - Choose an initial approximate solution, x_0, and tolerance criterion, ϵ.
    - Compute $r_0 = M^{-1}(b - Ax_0)$.
    - Set $v_0 = \tilde{v}_0 = \tilde{r}_0 = r_0$ and $j = 0$.

2. While $r_j^T r_j > \epsilon$ do

    - Compute $j = j + 1$.
    - Compute $\alpha_j = \tilde{r}_{j-1}^T r_{j-1} / \tilde{v}_{j-1}^T M^{-1} A v_{j-1}$.
    - Compute $r_j = r_{j-1} - \alpha_j M^{-1} A v_{j-1}$.
    - Compute $\tilde{r}_j = \tilde{r}_{j-1} - \alpha_j A^T M^{-T} \tilde{v}_{j-1}$.
    - Compute $x_j = x_{j-1} + \alpha_j v_{j-1}$.
    - Compute $\beta_j = \tilde{r}_j^T r_j / \tilde{r}_{j-1}^T r_{j-1}$.
    - Compute $v_j = r_j + \beta_j v_{j-1}$.
    - Compute $\tilde{v}_j = \tilde{r}_j + \beta_j \tilde{v}_{j-1}$.

Notice that the scalars α_j are chosen to ensure the biorthogonality conditions, while the β_j are chosen to enforce the biconjugacy conditions. Also, the norm of the residual, $r_j^T r_j$, is computed in each iteration and used to test for convergence. The algorithm will break down if either of the products $\tilde{r}_{j-1}^T r_{j-1}$ or

$\tilde{v}_{j-1}^T M^{-1} A v_{j-1}$ is zero. When this does not occur, the algorithm will converge in at most n steps in exact arithmetic.

The BCG algorithm requires multiplication by both A and A^T in each iteration. A variation due to Sonneveld [149], which is called the *CGS (conjugate gradient squared)* method, requires only one matrix-vector multiplication per iteration. Not only does this algorithm require less work per iteration, but it also converges at a faster rate than BCG. From the presentation of the BCG algorithm just given, it may be observed that the residual r_j can be expressed as a polynomial $\Phi_j(B)$ of degree j in the matrix $B = M^{-1}A$, applied to r_0, the original residual. We have

$$r_j = \Phi_j(B)(r_0).$$

Similarly, we have

$$\tilde{r}_j = \Phi_j(B^T)(\tilde{r}_0)$$

for exactly the same polynomial Φ. Now consider the scalar product $\tilde{r}_j^T r_j$. Using the relationship $(\Phi s)^T \Phi r = s^T(\Phi^T \Phi r)$, this may be computed as

$$\tilde{r}_j^T r_j = \tilde{r}_0^T[(\Phi_j(B^T))^T \Phi_j(B)(r_0)] = \tilde{r}_0^T[(\Phi_j(B))^2(r_0)].$$

This means that these scalar products may be computed *without* requiring the transpose of either A or M^{-1}. Furthermore, not only does this mean that we avoid the need to work with A^T; it also means that CGS will converge faster than BCG. After j iterations of BCG, the residual is $\Phi_j(B)(r_0)$; after the same number of iterations of CGS, the residual is $[\Phi_j(B)]^2(r_0)$. $\Phi_j(B)$ will be a contraction when BCG converges, so that $[\Phi_j(B)]^2$ will be smaller, ensuring faster convergence.

At each iteration of CGS, the new approximation x_j, the new residual r_j, and the new search direction vector v_j, are computed as

$$\begin{aligned}
x_j &= x_{j-1} + \alpha_j(r_{j-1} + \beta_j q_{j-1} + q_j) \\
r_j &= r_{j-1} - \alpha_j A(r_{j-1} + \beta_j q_{j-1} + q_j) \\
v_j &= r_{j-1} + 2\beta_j q_{j-1} + \beta_j^2 v_{j-1}
\end{aligned}$$

where

$$\alpha_j = \frac{\tilde{r}_0^T r_{j-1}}{\tilde{r}_0^T A v_j}; \qquad \beta_j = \frac{\tilde{r}_0^T r_{j-1}}{\tilde{r}_0^T r_{j-2}}; \quad \text{and} \quad q_j = r_{j-1} + \beta_j q_{j-1} - \alpha_j A v_j.$$

It is usual to set the residual $\tilde{r}_0$ equal to r_0. These may be incorporated into an algorithm as follows:

Algorithm: CGS — Conjugate Gradient Squared

1. *Initialize:*

   - Choose an initial approximate solution, x_0, and tolerance criterion, ϵ.
   - Compute $r_0 = M^{-1}(b - Ax_0)$.
   - Set $\tilde{r}_0 = r_0$; $q_0 = v_0 = 0$; $\rho_0 = 1$; and $j = 0$.

2. While $r_j^T r_j > \epsilon$ do

   - Compute $j = j + 1$.
   - Compute $\rho_j = \tilde{r}_0^T r_{j-1}$ and $\beta_j = \rho_j / \rho_{j-1}$.
   - Compute $u_j = r_{j-1} + \beta_j q_{j-1}$.
   - Compute $v_j = u_j + \beta_j(q_{j-1} + \beta_j v_{j-1})$.
   - Compute $w_j = M^{-1} A v_j$.
   - Compute $\sigma_j = \tilde{r}_0^T w_j$ and $\alpha_j = \rho_j / \sigma_j$.
   - Compute $q_j = u_j - \alpha_j w_j$.
   - Compute $r_j = r_{j-1} - \alpha_j M^{-1} A(u_j + q_j)$.
   - Compute $x_j = x_{j-1} + \alpha_j(u_j + q_j)$.

As in BCG, the norm of the residual is computed in each iteration and used to test for convergence. Also as with BCG, there is no guarantee that the algorithm will not break down before convergence has been reached. This is in contrast to the monotonicity convergence results of GMRES and CGNR. Neither the residuals nor the error in the solution decreases monotonically at each iteration for either BCG or CGS.

4.5.7 Preconditioning

The conjugate gradient method and its variants are generally used in conjunction with a preconditioner. It is usual to write the system $Ax = b$ as

$$P_L A P_R P_R^{-1} x = P_L b,$$

where P_L and P_R are nonsingular left and right preconditioners. We now apply the conjugate gradient method to

$$\tilde{A}\tilde{x} = \tilde{b},$$

where $\tilde{A} = P_L A P_R$, $\tilde{x} = P_R^{-1} x$, $\tilde{b} = P_L b$. We hope the preconditioning is such that $\tilde{A}$ has better convergence behavior than A. This usually means a reduction in the spectral condition number or an improved distribution of the smallest eigenvalues. The product $P_L P_R$ is often thought of as an approximate inverse of A. If A is sparse, this sparsity may be lost in $\tilde{A}$, so rather than forming $\tilde{A}$ explicitly, the effect of the preconditioner is implicitly incorporated into the conjugate gradient algorithms.

Let $M^{-1} = P_L P_R$. When $M^{-1} = I$, the preconditioned CG algorithm reverts to the standard CG version. On the other hand, when $M^{-1} = A^{-1}$, the exact solution is obtained in a single iteration — which includes the direct solution of the original system! Since the preconditioned CG method requires the solution of a system of the form $Mx = z$ at each iteration, it is important that this be relatively inexpensive. The incomplete factorizations discussed in the previous chapter, *ILU(0)*, *ILUTH*, and *ILUK*, as well as SOR and SSOR preconditioning, are all possible. When an incomplete factorization is performed, we compute a lower triangular matrix L and an upper triangular matrix U such that $A = LU + E$ and E is small. Some different possibilities for P_L and P_R are given by

$$P_L = U^{-1} \qquad P_R = L^{-1}$$
$$P_L = U^{-1}L^{-1} \quad P_R = I$$
$$P_L = I \qquad P_R = U^{-1}L^{-1}$$

In the descriptions of the algorithm given prior to this point, the matrix $M = LU$ and the particular choice of P_L and P_R corresponds to the third, i.e., $P_L = I$; $P_R = U^{-1}L^{-1}$. We now present a version of CGS that incorporates both left and right preconditioning.

Algorithm: CGS with Left and Right Preconditioning

1. *Initialize:*

   - Choose an initial approximate solution x_0 and tolerance criterion, ϵ.
   - Compute $r_0 = P_L(b - Ax_0)$.
   - Set $\tilde{r}_0 = r_0$; $q_0 = v_0 = 0$; $\rho_0 = 1$; and $j = 0$.

2. While $r_j^T r_j > \epsilon$ do

   - Compute $j = j + 1$.
   - Compute $\rho_j = \tilde{r}_0^T r_{j-1}$ and $\beta_j = \rho_j / \rho_{j-1}$.
   - Compute $u_j = r_{j-1} + \beta_j q_{j-1}$.
   - Compute $v_j = u_j + \beta_j(q_{j-1} + \beta_j v_{j-1})$.
   - Compute $w_j = P_L A P_R v_j$.
   - Compute $\sigma_j = \tilde{r}_0^T w_j$ and $\alpha_j = \rho_j / \sigma_j$.
   - Compute $q_j = u_j - \alpha_j w_j$.

- Compute $s_j = \alpha_j P_R(u_j + q_j)$.
- Compute $r_j = r_{j-1} - P_L s_j$.
- Compute $x_j = x_{j-1} + s_j$.

Notice that the computed solution x of this system is the unique solution of $Ax = b$. For Markov chains it suffices to take $b = 0$ and $A = Q^T$ or $A = I - P^T$.

4.5.8 Examples

We finish up this chapter by plotting the residuals for the three methods, CGNR, BCG, and CGS, when applied to the Courtois matrix (Figure 4.3) and to both cases of the reliability model (Figures 4.4 and 4.5). In these examples it appears that CGS performs best.

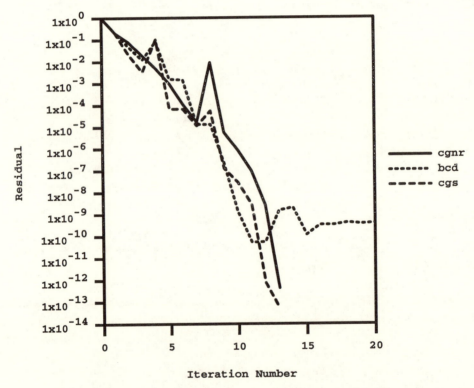

Figure 4.3: Residuals for the Courtois matrix.

Although it is likely that CGS will generally outperform BCG, the same remarks should not be made in comparisons between CGS and CGNR. In this regard, an interesting series of experiments was conducted by Nachtigal, Reddy, and Trefethen [107]. They compared GMRES, CGNR, and CGS and presented examples of matrices for which each method outperforms the others by a factor of $O(\sqrt{n})$ or $O(n)$, where n is the order of the matrix.

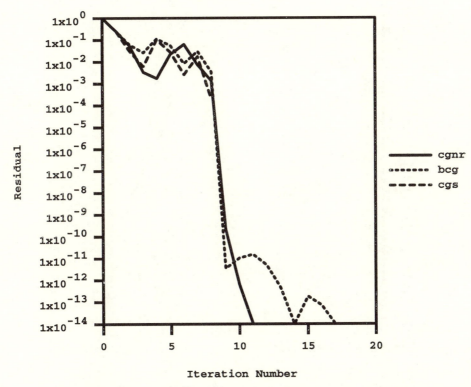

Figure 4.4: Residuals for Case I of reliability model.

4.6   MATLAB Programs

The following MATLAB programs were used to generate the experimental results presented throughout this chapter. As we previously cautioned, these implementations are not designed to be production versions of the algorithms. Their purpose is to show the exact sequence of operations that need to be performed and to give an indication of the behavior of the algorithm in some very limited cases. Indeed, they gloss over many important details that need to be incorporated in a production version. In the Lopsided algorithm, for example, details concerning the handling of complex conjugate pairs of vectors as real vectors, the choice of an optimum number of matrix-vector multiplications in each iteration, problems of consistently sorting the eigenvalues and eigenvectors of the interaction matrix B, and the development and implementation of appropriate normalization

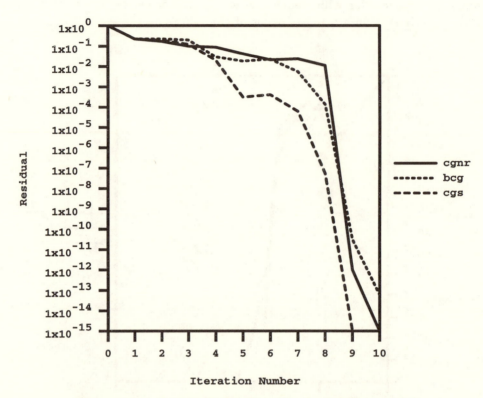

Figure 4.5: Residuals for Case II of reliability model.

and convergence-testing procedures need to be carefully considered, as shown in [160]. The original LOPSI algorithm, as well as a version of GMRES, are available through *netlib*. Instructions on obtaining these or any other programs from the *netlib* system are provided in Chapter 10.

Input parameters to the MATLAB programs include

- A real matrix A, which should be set to P^T or Q^T depending on the algorithm. The second line of each program indicates which is expected.

- An initial vector v. In the three CG programs this is generated automatically as a vector of random positive numbers and then normalized. When input to other programs, v was taken to be the vector $e/\|e\|_1$.

- A value for m, the dimension of the subspace.

- A value for *itmax*, the maximum number of iterations to perform.

Output from all the programs except *ilu.m* and *modgs.m* consists of an approximation to the solution vector, *soln*, and a vector of residuals, *resid*. In *arnoldi.m* and *gmres.m* the vector of residuals is of length 1. The output from *ilu.m* is the matrix $\tilde{U}^{-1}\tilde{L}^{-1}A$, while the output from *modgs.m* consists of the upper Hessenberg matrix $\bar{H}$ and a set of m orthonormal vectors.

Programs

```
***********************************************************************
    function [soln,resid] = lopsi(A,m,itmax)
    %   ***   Lopsided Iteration   (A = P^T)   ***

    [n,n] = size(A);   s = ones(n,1)/n;   iter = 1;   res = 1;
    rand('seed',0);    U = [s,2*rand(n,m-1)-1];

    while iter <= itmax & res > eps,
        V = A*U;   G = U'*U;   H = U'*V;        %  Premultiplications
        B = inv(G)*H;                           %  Get Interaction Matrix
        [X,Y] = eig(B);                         %  Eigensolution of B
        z = abs(Y*ones(m,1));   [zz,perm] = sort(z); %  Sort Eigenvectors
        pmat = zeros(m);
        for jj = 1:m, pmat(perm(m+1-jj),jj) = 1; end
        X = X*pmat;
        W = V*X;                                %  Multiply
        s = W(1:n,1); s = s/s(1); U = W/s(1);   %  Normalize
        res = norm((eye(n)-A)*s,2);             %  Get Residual
        resid(iter) = res; iter = iter+1;
    end

    soln = s/norm(s,1);   resid = resid';

***********************************************************************
    function [MA] = ilu(A)
    %   ***   ILU --- Incomplete Factorization   (A = Q^T)   ***

    [n,n] = size(A);   thr = .1;   R = A;
    for k = 2:n,                        %  Rows 2 thro' n
        for i = 1:k-1,                  %  Get elements to reduce
            if R(k,i) > 0,              %  Nonzero element?
                R(k,i) = R(k,i)/R(i,i); %  Multiplier
                for j = i+1:n,
                    R(k,j) = R(k,j) - R(k,i)*R(i,j); %  Now eliminate
                end
            end
        end
        for i = 1:n,
    %       if A(k,i)==0, R(k,i)=0; end                  %  ilu0
            if abs(R(k,i))<thr & i~=k , R(k,i)=0; end %  iluth
        end
    end
```

```
    L = eye(n);  U = diag(diag(R));                      %  Now form MA
    for i = 1:n,
        for j = 1:n,
            if i<j,  U(i,j) = R(i,j); end
            if i>j,  L(i,j) = R(i,j); end
        end
    end
    if U(n,n) < 10^(-16), U(n,n) = 1; end
    MA = inv(U) * (inv(L)*A);


******************************************************************************
    function [Hbar,v] = modgs(A,v,m)
    %  ***   Modified Gram-Schmidt/Arnoldi Process   (A = P^T)   ***

    [n,n] = size(A);
    for j = 1:m,
        vj = v(1:n,j);
        w = A*vj;
        for i = 1:j,
            vi = v(1:n,i);
            Hbar(i,j) = vi'*w;
            w = w-Hbar(i,j)*vi;
        end
        Hbar(j+1,j) = norm(w,2);
        v = [v,w/Hbar(j+1,j)];
    end


******************************************************************************
    function [soln,res] = arnoldi(A,v,m)
    %  ***   Noniterative Arnoldi   (A = P^T)   ***

    [n,n] = size(A); v = v/norm(v,2);

    [Hbar,v] = modgs(A,v,m);        % Modified Gram-Schmidt
    [H] = Hbar(1:m,1:m);    [v] = v(1:n,1:m);

    [evect,eval] = eig(H);          % Eigenanalysis
    x = abs(ones(m,1) - eig(H));
    [xx,perm] = sort(x);
    evect1 = real(col(evect,perm(1)));
    soln = v*evect1;      soln = soln/sum(soln);

    res = norm((A-eye(n))*soln,2);


******************************************************************************
    function [soln,res] = gmres(A,v,m)
    %  ***   Noniterative GMRES   (A = Q^T)   ***

    [n,n] = size(A); x0 = v;  r0 = -A*x0;  beta = norm(r0,2);
    v = r0/beta;        b = [beta;zeros(m,1)];
```

```
    [Hbar,v] = modgs(A,v,m);      % Modified Gram-Schmidt
    y = Hbar\b;                   % Least Squares Solution
    [v] = v(1:n,1:m);
    soln = x0 + v*y;         soln = soln/sum(soln);

    res = norm(A*soln,2);
```

****
```
    function [soln,resid] = arniter(A,m,itmax)
    %  ***   Iterative Arnoldi  (A = P^T)   ***

    [n,n] = size(A);  v = ones(n,1)/n;  iter=1;  res = 1;
    MA = eye(n)+ilu(A-eye(n));          % Precondition

    while iter <= itmax & res > eps,    % Iterate
        [soln,res] = arnoldi(MA,v,m);
        v = soln;  resid(iter) = res;  iter = iter+1;
    end

    resid = resid';
```

****
```
    function [soln,resid] = gmriter(A,m,itmax)
    %  ***   Iterative GMRES  (A = Q^T)   ***

    [n,n] = size(A);  x0 = ones(n,1)/n; iter=1; res = 1;
    MA = ilu(A);                        % Precondition

    while iter <= itmax & res > eps,    % Iterate
        [soln,res] = gmres(MA,x0,m);
        x0 = soln;  resid(iter) = res;  iter = iter+1;
    end

    resid = resid';
```

****
```
    function [soln,res] = cgnr(A,itmax)
    %  ***   CG for the Normal Equations   (A = Q^T)   ***

    [n,n] = size(A);  M = eye(n);
    rand('seed',0);   x = rand(n,1);   x = x/norm(x,1);
    r = -M*(A*x);      v = zeros(n,1);  beta = 0; iter = 1;
    res(1) = norm(A*x,2);

    while iter <= itmax & res(iter) > eps,
        v= A'*(M'*r) + beta*v;
        alpha = (r'*r)/(v'*v);
        x = x + alpha*v;
        t = r - alpha*M*(A*v);
        beta = (t'*t)/(r'*r);
```

```
          r = t;
          iter = iter+1; res(iter) = norm(A*x,2);
      end

      soln = x/norm(x,1);  res = res';

****************************************************************************
      function [soln,res] = bcg(A,itmax)
      %   ***   Biconjugate Gradient   (A = Q^T)   ***

      [n,n] = size(A); M = eye(n);
      rand('seed',0);  x = rand(n,1);        x = x/norm(x,1);
      r = -M*(A*x);    s = r;  rho = s'*r;  v = r;  w = r;  iter = 1;
      res(1) = norm(A*x,2);

      while iter <= itmax & res(iter) > eps,
          alpha = rho/(w'*(M*(A*v)));
          r = r - alpha*M*(A*v);
          s = s - alpha*A'*(M'*s);
          x = x + alpha*v;
          rho1 = s'*r; beta = rho1/rho; rho = rho1;
          v = r + beta*v;
          w = s + beta*w;
          iter = iter+1;  res(iter) = norm(A*x,2);
      end

      soln = x/norm(x,1);   res = res';

****************************************************************************
      function [soln,res] = cgs(A,itmax)
      %   ***   Conjugate Gradient Squared   (A = Q^T)   ***

      [n,n] = size(A); M = eye(n);
      rand('seed',0);  x = rand(n,1);  x = x/norm(x,1);
      r = -M*(A*x);    r0 = r; v = 0;  q = 0;  rho = 1;  iter = 1;
      res(1) = norm(A*x,2);

      while iter <= itmax & res(iter) > eps,
          rho1 = r0'*r;  beta = rho1/rho;  rho = rho1;
          u = r + beta*q;
          v = u + beta*(q + beta*v);
          w = M*(A*v);  sigma = r0'*w;  alpha = rho/sigma;
          q = u - alpha*w;
          r = r - alpha*M*A*(u + q);
          x = x + alpha*(u + q);
          iter = iter+1;    res(iter) = norm(A*x,2);
      end

      soln = x/norm(x,1);   res = res';

****************************************************************************
```

Chapter 5

Block Hessenberg Matrices and Solution by Recursion

5.1 Hessenberg Matrices

5.1.1 Definitions

A matrix H is said to be *upper Hessenberg* if $h_{ij} = 0$ for $i > j + 1$. The following is an example of an upper Hessenberg matrix of order 6:

$$
H = \begin{pmatrix}
h_{00} & h_{01} & h_{02} & h_{03} & h_{04} & h_{05} \\
h_{10} & h_{11} & h_{12} & h_{13} & h_{14} & h_{15} \\
0 & h_{21} & h_{22} & h_{23} & h_{24} & h_{25} \\
0 & 0 & h_{32} & h_{33} & h_{34} & h_{35} \\
0 & 0 & 0 & h_{43} & h_{44} & h_{45} \\
0 & 0 & 0 & 0 & h_{54} & h_{55}
\end{pmatrix}.
$$

H is said to be lower Hessenberg if $h_{ij} = 0$ for $i < j - 1$. A tridiagonal matrix is simultaneously both upper Hessenberg and lower Hessenberg. Irreducible Markov chains with infinitesimal generators that are upper Hessenberg are sometimes referred to as *skip-free to the left*; the nonzero structure permits the process to move down (toward zero) by only one state at a time (therefore free of skips), but up by any number of states. Notice that a Markov chain whose transition matrix is upper Hessenberg is irreducible only if all elements below the diagonal are strictly greater than zero.

Infinitesimal generator matrices that are of Hessenberg form (either upper or lower) arise frequently in queueing models. The paradigm for the upper Hessenberg case is the $M/G/1$ queue, and for the lower Hessenberg case, the $GI/M/1$ queue. The $M/M/1$ queue has an infinitesimal generator that is tridiagonal.

231

5.1.2   Standard Queueing Recursions as Forward Substitutions

The $M/M/1$ queue with service rate μ and arrival rate λ has the following infinite infinitesimal generator

$$
Q = \begin{pmatrix}
-\lambda & \lambda & & & \\
\mu & -(\lambda + \mu) & \lambda & & \\
& \mu & -(\lambda + \mu) & \lambda & \\
& & \mu & -(\lambda + \mu) & \lambda \\
& & & \ddots & \ddots & \ddots
\end{pmatrix}.
$$

From $\pi Q = 0$, it is obvious that $-\lambda \pi_0 + \mu \pi_1 = 0$, i.e., that $\pi_1 = (\lambda/\mu)\pi_0$. In general, we have

$$
\lambda \pi_{i-1} - (\lambda + \mu)\pi_i + \mu \pi_{i+1} = 0,
$$

from which, by induction, we may derive

$$
\pi_{i+1} = ((\lambda + \mu)/\mu)\pi_i - (\lambda/\mu)\pi_{i-1} = (\lambda/\mu)\pi_i.
$$

Thus, once π_0 is known, the remaining values, π_i, $i = 1, 2, \ldots$, may be determined recursively. For the $M/M/1$ queue it is easy to show that the probability that the queue is empty is given by $\pi_0 = (1 - \lambda/\mu)$.

This recursive approach may be extended to upper and lower Hessenberg matrices. When the transition probability matrix P of an irreducible Markov chain is upper Hessenberg, we have

$$
(\pi_0, \pi_1, \pi_2, \ldots)
\begin{pmatrix}
p_{00} & p_{01} & p_{02} & p_{03} & \cdots \\
p_{10} & p_{11} & p_{12} & p_{13} & \cdots \\
0 & p_{21} & p_{22} & p_{23} & \cdots \\
0 & 0 & p_{32} & p_{33} & \cdots \\
\vdots & \vdots & \vdots & \vdots & \ddots
\end{pmatrix}
= (\pi_0, \pi_1, \pi_2, \ldots).
$$

These equations may be written individually as

$$
\pi_j = \sum_{i=0}^{j+1} \pi_i p_{ij}, \qquad j = 0, 1, 2, \ldots
$$

or equivalently

$$
\pi_{j+1} = p_{j+1\,j}^{-1} \left[\pi_j(1 - p_{jj}) - \sum_{i=0}^{j-1} \pi_i p_{ij} \right], \qquad j = 0, 1, 2, \ldots
$$

This latter form exposes the recursive nature of these equations. If π_0 is known or may be determined, then all remaining π_i may be computed. If the matrix is finite, it is possible to assign π_0 an arbitrary value ($\pi_0 = 1$, for example) to compute the remaining π_i and then renormalize the vector π so that $\|\pi\|_1 = 1$.

This may be viewed from a matrix approach by considering the corresponding homogeneous system

$$\pi Q = 0$$

and partitioning it as

$$(\pi_0, \pi_*) \begin{pmatrix} a^T & \gamma \\ B & d \end{pmatrix} = 0,$$

in which a, d and $\pi_*^T \in \Re^{(n-1)}$, $B \in \Re^{(n-1) \times (n-1)}$; and γ and π_0 are scalars. This gives

$$\pi_0 a^T + \pi_* B = 0,$$

or

$$\pi_* B = -\pi_0 a^T.$$

If we assume that $\pi_0 = 1$, this nonhomogeneous system of $(n-1)$ equations in $(n-1)$ unknowns (the elements of π_*) may be solved very efficiently for π_*. Since the coefficient matrix B is triangular and has nonzero diagonal elements, it is simply a forward substitution procedure.

As discussed previously in Chapter 2, forward substitution is a stable procedure, and consequently, recursive algorithms based on it are also stable. However, the stability of the algorithm does not preclude problems that may arise because of overflow or underflow. A check should be kept on the size of the components of the solution vector and an adaptive normalization procedure incorporated if necessary. Also, recall that the stability of an algorithm does *not* guarantee an accurate solution but only the exact solution of a slightly perturbed problem. The solution of the perturbed problem may be far from the exact solution of the unperturbed problem. For example, in an upper triangular system of equations with large off-diagonal elements, any slight perturbation of the last equation will lead to a huge change in the computed solution. In that case the problem is ill-conditioned. We can only guarantee an accurate solution when a *stable algorithm* is applied to a *well-conditioned* problem.

5.1.3 Block Hessenberg Matrices

Block upper (or lower) Hessenberg matrices are the obvious generalization of the upper (respectively lower) Hessenberg matrices just discussed. The following infinitesimal generator Q is finite block Hessenberg:

$$Q = \begin{pmatrix} Q_{00} & Q_{01} & Q_{02} & Q_{03} & \cdots & Q_{0,N-1} & Q_{0,N} \\ Q_{10} & Q_{11} & Q_{12} & Q_{13} & \cdots & Q_{1,N-1} & Q_{1,N} \\ 0 & Q_{21} & Q_{22} & Q_{23} & \cdots & Q_{2,N-1} & Q_{2,N} \\ 0 & 0 & Q_{32} & Q_{33} & \cdots & Q_{3,N-1} & Q_{3,N} \\ \vdots & \vdots & \vdots & \vdots & \ddots & \vdots & \vdots \\ 0 & 0 & 0 & 0 & \cdots & Q_{N-1,N-1} & Q_{N-1,N} \\ 0 & 0 & 0 & 0 & \cdots & Q_{N,N-1} & Q_{N,N} \end{pmatrix}.$$

The diagonal blocks are square matrices of order n_i, for $i = 0, 1, \ldots, N$; the off-diagonal blocks Q_{ij} $i \neq j$ have dimension $(n_i \times n_j)$. All of the elements of all of the blocks of Q are nonnegative except the diagonal elements of the diagonal blocks, which are all strictly negative. (As always, the sum of elements across any row of Q is zero.)

Block Hessenberg matrices arise most commonly when the associated Markov chain is two-dimensional, $\{(X(t), Y(t)), t \geq 0\}$, on a state space $\{(\eta, k), 0 \leq \eta \leq N, 1 \leq k \leq K\}$ from which *single-step* transitions from state (i, k) to (j, l) are possible only if $i \leq j + 1$. The set of states $\{(\eta, k), 1 \leq k \leq K\}$ is sometimes referred to as *level* η, and since each level contains K states, all the blocks Q_{ij} have the same dimensions, $K \times K$. In some models, level 0 contains less than K distinct states, and in these cases the blocks Q_{0i} and Q_{i0}, $i = 0, 1, \ldots$, have correspondingly smaller dimensions.

The nonzero structure of a block upper Hessenberg infinitesimal generator permits the process to move down (toward level 0) by only one level at a time and to move up by several levels at a time. The process is said to be *skip-free to the left*. A matrix that is both block upper Hessenberg and block lower Hessenberg is block tridiagonal and is sometimes referred to as a block Jacobi matrix.

5.2   Block Recursive Methods

It is sometimes possible to develop methods to solve *block* Hessenberg matrices recursively, like their point counterparts, for the stationary probability vector (see, for example, [52], [77], [106], and [180]). These recursive procedures often arise naturally from the formulation of the Chapman–Kolmogorov equations and are easily programmed. Unfortunately, implementations based on block recursions may lead to grossly inaccurate results, as the example in Section 5.2.2 shows. This happens because the process may be numerically unstable. The analyst must be acutely aware of the conditions under which recursive solution methods will give accurate results.

5.2.1   The Recursive Procedure

Let us consider first the case

$$
Q = \begin{pmatrix}
Q_{00} & Q_{01} & Q_{02} & Q_{03} & \cdots & Q_{0,N-1} & Q_{0,N} \\
Q_{10} & Q_{11} & Q_{12} & Q_{13} & \cdots & Q_{1,N-1} & Q_{1,N} \\
0 & Q_{21} & Q_{22} & Q_{23} & \cdots & Q_{2,N-1} & Q_{2,N} \\
0 & 0 & Q_{32} & Q_{33} & \cdots & Q_{3,N-1} & Q_{3,N} \\
\vdots & \vdots & \vdots & \vdots & \ddots & \vdots & \vdots \\
0 & 0 & 0 & 0 & \cdots & Q_{N-1,N-1} & Q_{N-1,N} \\
0 & 0 & 0 & 0 & \cdots & Q_{N,N-1} & Q_{N,N}
\end{pmatrix}.
$$

To implement a *forward* recursive procedure, we require that the subdiagonal blocks $Q_{k+1,k}$, for $k = 0, 1, \ldots, N-1$, be nonsingular. Among other properties, this implies that these blocks must be square. Let the stationary probability vector π be partitioned conformally with Q: i.e.,

$$
\pi = (\pi_0, \pi_1, \pi_2, \ldots, \pi_N).
$$

The ith block equation of $\pi Q = 0$ may be written as

$$
\sum_{k=0}^{i+1} \pi_k Q_{ki} = 0 \quad \text{for } i = 0, 1, \ldots, N-1
$$

$$
\sum_{k=0}^{N} \pi_k Q_{kN} = 0 \quad \text{for } i = N.
$$

We consider first the case when π_0 is known. Then from the first block equation we can write π_1 in terms of the known quantity π_0. We have

$$
\pi_0 Q_{00} + \pi_1 Q_{10} = 0
$$

and hence

$$
\pi_1 = -\pi_0 Q_{00} Q_{10}^{-1}.
$$

Similarly, from the second block equation we have

$$
\pi_0 Q_{01} + \pi_1 Q_{11} + \pi_2 Q_{21} = 0,
$$

and since by this time both π_0 and π_1 are known, we can express π_2 in terms of known quantities. We get

$$
\pi_2 = -(\pi_0 Q_{01} + \pi_1 Q_{11}) Q_{21}^{-1} = -\pi_0 Q_{01} Q_{21}^{-1} + \pi_0 Q_{00} Q_{10}^{-1} Q_{11} Q_{21}^{-1}.
$$

Continuing in this fashion, we may determine all the subvectors π_i in terms of the known subvector π_0. The process is essentially a block forward substitution procedure.

Algorithmically, the operation we perform at step k for $k = 0, 1, \ldots, N - 1$ is

$$\pi_{k+1} = - \left(\sum_{i=0}^{k} \pi_i Q_{ik} \right) Q_{k+1,k}^{-1}, \tag{5.1}$$

which is only possible when $Q_{k+1,k}^{-1}$ is nonsingular. Therefore it follows that the recursive procedure can be implemented only when all the subdiagonal blocks are nonsingular.

When the infinitesimal generator is block *lower* Hessenberg and the superdiagonal blocks are nonsingular, a *reverse* recursion is possible. In this case, if we know π_N, we may recursively compute $\pi_{N-1}, \pi_{N-2}, \ldots, \pi_0$. If the infinitesimal generator is block tridiagonal, which is often the case, then it may be possible to perform either a forward recursion (if the $Q_{k+1,k}$ are nonsingular and we know π_0) or a backward recursion (if the $Q_{k,k+1}$ are nonsingular and we know π_N).

It is not always necessary to know π_0 to implement a forward recursion (or π_N to implement backward recursion). Notice that only the first $N - 1$ block equations are actually used in writing all π_i in terms of π_0. The final block equation

$$\sum_{k=0}^{N} \pi_k Q_{kN} = 0 \tag{5.2}$$

may be used to compute the subvector (π_0 in this case) that is used in the recursion. In other words, the procedure is to write all π_i, $i \neq 0$, in terms of the still unknown π_0 and then to use the final block equation (5.2) to determine π_0. Once π_0 has been computed, the previously mentioned recursions (5.1) can be carried out to find the other π_i.

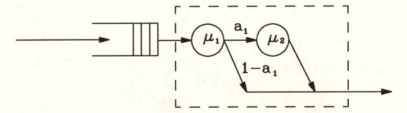

Figure 5.1: The $\lambda/C_2/1$ queue.

Example: The $\lambda/C_2/1/N$ Queue   We will illustrate these concepts by means of an example. Consider a single-server queueing system with Poisson arrivals at rate λ and service time distribution represented by a law of Cox of order 2. No more than N customers are permitted in the network. Customers that arrive when the network is full are lost. This queue is shown in Figure 5.1. The

Table 5.1: Infinitesimal Generator for $\lambda/C_2/1/3$ Queue

(0,1)	(0,2)	(1,1)	(1,2)	(2,1)	(2,2)	(3,1)	(3,2)
$-\lambda$	0	λ	0				
0	$-\lambda$	0	λ				
$(1-a_1)\mu_1$	0	$-(\mu_1+\lambda)$	$a_1\mu_1$	λ	0		
μ_2	0	0	$-(\mu_2+\lambda)$	0	λ		
		$(1-a_1)\mu_1$	0	$-(\mu_1+\lambda)$	$a_1\mu_1$	λ	0
		μ_2	0	0	$-(\mu_2+\lambda)$	0	λ
				$(1-a_1)\mu_1$	0	$-\mu_1$	$a_1\mu_1$
				μ_2	0	0	$-\mu_2$

states of the system may be represented by the ordered pair (η, k), in which η denotes the number of customers in the system and k $(= 1,2)$ denotes the current phase of service. When $\eta = 0$, the case in which there are no customers in the system, the value associated with the parameter k is irrelevant, and it is possible to combine states $(0, 1)$ and $(0, 2)$ into a single state. However, for clarity of exposition we will keep both for the present. We shall let $\pi(\eta, k)$ denote the stationary probability of being in state (η, k) and π_η the probability that the system contains η customers at equilibrium, i.e., $\pi_\eta = \sum_{k=1}^{2} \pi(\eta, k)$. The infinitesimal generator matrix for this system is block tridiagonal and is given in Table 5.1 for the case when $N = 3$.

Two features of this matrix are worthy of note. The first concerns those off-diagonal blocks that are nonsingular. The blocks below the diagonal, which represent departures from the queueing system, are singular, while the blocks above the diagonal, which represent arrivals to the system, are nonsingular. This means that only a *backward* recursion is possible, i.e., given π_3, we may determine π_2, π_1, and finally π_0. The second feature of interest is the manner in which the states $(0, 1)$ and $(0, 2)$ are handled. Both of these essentially represent the same state of the system. Notice, however, that it is possible for the system to enter state $(0, 1)$ only; the system can be in the state $(0, 2)$ only at system initialization, and once it leaves that state, it will never reenter it. The probability of finding an empty system must be obtained from state $(0, 1)$. The advantage of including state $(0, 2)$ in the matrix is that the off-diagonal blocks are all identical and, more importantly for the recursive method, all superdiagonal blocks are nonsingular. Since $(0, 2)$ is a transient state, its destination is unimportant and thus may be chosen to be any state or subset of states.

Let us now apply the recursive procedure to this example. We consider the specific case in which $\lambda = 1$, $\mu_1 = 2$, $\mu_2 = 3$, and $a_1 = 0.75$. From the last block column, we get

$$\pi_2 \begin{pmatrix} 1 & 0 \\ 0 & 1 \end{pmatrix} + \pi_3 \begin{pmatrix} -2.0 & 1.5 \\ 0.0 & -3.0 \end{pmatrix} = 0$$

which gives

$$\pi_2 = \pi_3 \begin{pmatrix} 2.0 & -1.5 \\ 0.0 & 3.0 \end{pmatrix}, \tag{5.3}$$

and if we knew the value of π_3, we could immediately calculate π_2. From the third block column, we have

$$\pi_1 \begin{pmatrix} 1 & 0 \\ 0 & 1 \end{pmatrix} + \pi_2 \begin{pmatrix} -3.0 & 1.5 \\ 0.0 & -4.0 \end{pmatrix} + \pi_3 \begin{pmatrix} 0.5 & 0.0 \\ 3.0 & 0.0 \end{pmatrix} = 0,$$

which gives

$$\pi_1 = \pi_3 \begin{pmatrix} 2.0 & -1.5 \\ 0.0 & 3.0 \end{pmatrix} \begin{pmatrix} 3.0 & -1.5 \\ 0.0 & 4.0 \end{pmatrix} + \pi_3 \begin{pmatrix} -0.5 & 0.0 \\ -3.0 & 0.0 \end{pmatrix} = \pi_3 \begin{pmatrix} 5.5 & -9.0 \\ -3.0 & 12.0 \end{pmatrix} \tag{5.4}$$

In a similar manner, from the second block column we may calculate π_0 as

$$\pi_0 = \pi_3 \begin{pmatrix} 20.00 & -44.25 \\ -18.00 & 52.50 \end{pmatrix}. \tag{5.5}$$

At this point we can write all the unknown π_i in terms of π_3, and now we proceed to use the first block column to find π_3. From

$$-\pi_0 \begin{pmatrix} 1 & 0 \\ 0 & 1 \end{pmatrix} + \pi_1 \begin{pmatrix} 0.5 & 0.0 \\ 3.0 & 0.0 \end{pmatrix} = 0,$$

we may write

$$\pi_3 \begin{pmatrix} -20.00 & 44.25 \\ 18.00 & -52.50 \end{pmatrix} + \pi_3 \begin{pmatrix} 5.5 & -9.0 \\ -3.0 & 12.0 \end{pmatrix} \begin{pmatrix} 0.5 & 0.0 \\ 3.0 & 0.0 \end{pmatrix} = 0$$

which simplifies to

$$\pi_3 \begin{pmatrix} -44.25 & 44.25 \\ 52.50 & -52.50 \end{pmatrix} = 0$$

To solve this equation, we need to fix one of the components of π_3 and solve for the other. Letting $\pi(3,2) = 1$, it follows that $\pi(3,1) = 52.5/44.25$. Having thus computed π_3 we can compute π_2, π_1, and finally π_0 by means of the matrix products (5.3), (5.4), and (5.5). Once completed, a final normalization gives the stationary probability vector. Notice that the elements of π_3 are precisely the values that cause the element $\pi(0,2)$ to be identically equal to zero.

The general procedure may be completely described in matrix terms. We consider the case of a block *lower* Hessenberg matrix, since this is the case that arises in the example of Section 5.2.2. Let us partition the infinitesimal generator matrix Q as

$$Q = \begin{pmatrix} V & W \\ X & Y \end{pmatrix}$$

$$= \begin{pmatrix} Q_{00} & Q_{01} & 0 & 0 & \cdots & 0 & 0 \\ Q_{10} & Q_{11} & Q_{12} & 0 & \cdots & 0 & 0 \\ Q_{20} & Q_{21} & Q_{22} & Q_{23} & \cdots & 0 & 0 \\ Q_{30} & Q_{31} & Q_{32} & Q_{33} & \cdots & 0 & 0 \\ \vdots & \vdots & \vdots & \vdots & \ddots & \vdots & \vdots \\ Q_{N-1,0} & Q_{N-1,1} & Q_{N-1,2} & Q_{N-1,3} & \cdots & Q_{N-1,N-1} & Q_{N-1,N} \\ Q_{N,0} & Q_{N,1} & Q_{N,2} & Q_{N,3} & \cdots & Q_{N,N-1} & Q_{N,N} \end{pmatrix}.$$

Notice that the following relation is true for all nonsingular W:

$$\begin{pmatrix} V & W \\ X & Y \end{pmatrix} \begin{pmatrix} I & 0 \\ -W^{-1}V & I \end{pmatrix} = \begin{pmatrix} 0 & W \\ X - YW^{-1}V & Y \end{pmatrix},$$

so that solving

$$(\pi_*, \pi_N) \begin{pmatrix} V & W \\ X & Y \end{pmatrix} = 0$$

is equivalent to solving

$$(\pi_*, \pi_N) \begin{pmatrix} 0 & W \\ S & Y \end{pmatrix} = 0,$$

where $S = X - YW^{-1}V$. This allows us to obtain π_N by solving $\pi_N(X - YW^{-1}V) = 0$. To do so, we first determine $YW^{-1}V$ by setting $W^{-1}V = Z$ and then solving $WZ = V$ for Z. Since W is block lower triangular, this is simply a block forward substitution. When Z has been computed, we may obtain the coefficient matrix S as $S = (X - YZ)$.

Once π_N, has been found, π_* may be obtained from the second set of equations

$$\pi_* W = -\pi_N Y,$$

this time using block back-substitution on W.

The numerical disadvantages of this recursive procedure stem from the fact that we need to find the inverse of matrices (or in practice, solve systems of equations with coefficient matrices) that are not necessarily diagonally dominant and that do not have the stable characteristics that the diagonal blocks possess. It is for this reason that the inversion of diagonal blocks in block Gaussian elimination is an acceptable and stable process, but the inversion of subdiagonal (or superdiagonal) blocks must be carefully monitored.

5.2.2 Example: A Telephone System with N Lines and K Operators

5.2.2.1 Model Description

Consider a telephone system that consists of N incoming lines and K operators ($N > K$). Calls arrive according to a Poisson process of rate λ; those arriving when all N lines are busy are lost. If a call is successful and acquires a line, it holds that line until one of the K operators becomes available and completes service for that caller. The service time required by the caller is assumed to be exponentially distributed with mean $1/\mu_1$. All callers are assumed to be identical. When service is completed, the caller departs and frees the line. However, the server does not immediately become free to serve another customer; rather, it is assumed that the server has some bookkeeping to perform first. It is further assumed that this bookkeeping operation is exponentially distributed with mean $1/\mu_2$. After completing the bookkeeping, the server becomes available to serve another caller. This model approximates fairly closely the case of a travel agency (or airline reservation company) whose personnel (the servers) assist prospective travelers (the callers) by providing them with information, making reservations, etc. Usually, after the customer hangs up, the travel agent needs to file some information and generally get things sorted out before beginning to serve the next customer.

We shall define a state of this system by the pair

$$(\eta, k) \qquad 0 \le \eta \le N; \qquad 0 \le k \le K,$$

where η is the number of customers in the system (equal to the number of busy lines) and k denotes the number of available servers (i.e., $k = K -$ (number of servers engaged in bookkeeping)). Notice that servers serving customers are considered to be available servers, so the number of available servers is not necessarily equal to the number of free servers. Let $\pi(\eta, k)$ denote the steady-state probability of being in state (η, k).

The events that trigger transitions and the rates at which these transitions occur from an arbitrary state (η, k) with $1 \le \eta \le N - 1$ and $1 \le k \le K - 1$ are

Arrivals: Increment the number of customers but leave the number
                                   of available servers unchanged.
                                   Transition is $(\eta, k) \to (\eta + 1, k)$ at rate λ.

Departures: Decrement the number of customers and the number of
                                   available servers.
                                   Transition is $(\eta, k) \to (\eta - 1, k - 1)$ at rate $\mu_1 \min(\eta, k)$.

Bookkeeping ends: Increments the number of available servers but leaves the
                                   number of customers unchanged.
                                   Transition is $(\eta, k) \to (\eta, k + 1)$ at rate $\mu_2(K - k)$.

The possible transitions to and from state (η, k) and their associated rates are
given in Figure 5.2.

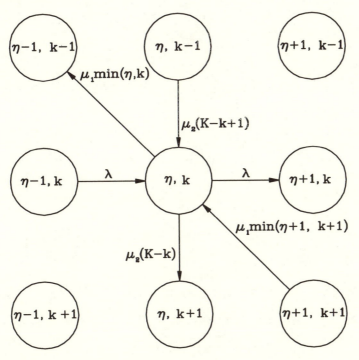

Figure 5.2: Transitions into and out of state $(\eta, \; k)$ when $1 \leq \eta \leq N - 1$ and
$1 \leq k \leq K - 1$.

The (nonboundary) global balance equations (the individual equations of $\pi Q = 0$)
for this system are

$$\{\lambda + \mu_1 \min(\eta, k) + \mu_2(K - k)\}\pi(\eta, k) =$$
$$\lambda\pi(\eta - 1, k) + \mu_2(K - k + 1)\pi(\eta, k - 1) + \mu_1\min(\eta + 1, k + 1)\pi(\eta + 1, k + 1),$$

for $\ 1 \leq \eta \leq N - 1 \ $ and $\ 1 \leq k \leq K - 1.$

Use of the boundary conditions yields the remaining equations.

In this example, it is convenient to order the states first by increasing order of the number of customers in the system, η, and secondly (for states with the same η value), by increasing order of the number of available servers, k. We shall let π_i denote the vector of length $K + 1$ given by

$$\pi_i = (\pi(i, 0), \ \pi(i, 1), \ \ldots, \ \pi(i, K)).$$

With this ordering, the infinitesimal generator matrix is block tridiagonal, and we have

$$(\pi_0, \pi_1, \ldots, \pi_N) \begin{pmatrix} A_0 & \Lambda & 0 & 0 & \ldots & 0 & 0 \\ B_1 & A_1 & \Lambda & 0 & \ldots & 0 & 0 \\ 0 & B_2 & A_2 & \Lambda & \ldots & 0 & 0 \\ 0 & 0 & B_3 & A_3 & \ldots & 0 & 0 \\ \vdots & \vdots & \vdots & \vdots & \ddots & \vdots & \vdots \\ 0 & 0 & 0 & 0 & \ldots & A_{N-1} & \Lambda \\ 0 & 0 & 0 & 0 & \ldots & B_N & A_N \end{pmatrix} = 0,$$

where Λ, A_i, and B_i are all $(K + 1) \times (K + 1)$ matrices defined as follows:

$$\Lambda = \lambda I;$$

$$A_i = \begin{pmatrix} a_{00}^{(i)} & a_{01}^{(i)} & 0 & \ldots & 0 & 0 \\ 0 & a_{11}^{(i)} & a_{12}^{(i)} & \ldots & 0 & 0 \\ 0 & 0 & a_{22}^{(i)} & \ddots & 0 & 0 \\ \vdots & \vdots & \vdots & \ddots & \vdots & \vdots \\ 0 & 0 & 0 & \ldots & a_{K-1,K-1}^{(i)} & a_{K-1,K}^{(i)} \\ 0 & 0 & 0 & \ldots & 0 & a_{K,K}^{(i)} \end{pmatrix},$$

where

$$a_{jj}^{(i)} = -[\lambda + \mu_1 \min(i, j) + \mu_2(K - j)]; \qquad j = 0, 1, \ldots, K,$$

$$a_{jj}^{(N)} = -[\mu_1 \min(N, j) + \mu_2(K - j)]; \qquad j = 0, 1, \ldots, K,$$

$$a_{j,j+1}^{(i)} = \mu_2(K - j); \qquad j = 0, 1, \ldots, K - 1$$

and

$$B_i = \begin{pmatrix} 0 & 0 & 0 & \ldots & 0 & 0 \\ b_{10}^{(i)} & 0 & 0 & \ldots & 0 & 0 \\ 0 & b_{21}^{(i)} & 0 & \ldots & 0 & 0 \\ \vdots & \vdots & \ddots & \ddots & \vdots & \vdots \\ 0 & 0 & 0 & \ldots & 0 & 0 \\ 0 & 0 & 0 & \ldots & b_{K,K-1}^{(i)} & 0 \end{pmatrix},$$

where $b_{jj-1}^{(i)} = \mu_1 \min(i,j)$ for $j = 1, 2, \ldots, K$.

Notice that the subblocks Λ correspond to arrivals. The subblocks B_i correspond to customer departures when there are i customers in the system. Diagonal blocks A_i contain diagonal elements, which equal the negated sum of all off-diagonal elements in their row, and superdiagonal elements, corresponding to additional servers becoming available (since this does not change the number of customers, it does not result in a transition out of the block).

5.2.2.2   Solution by Recursion

Given the structure of the global balance equations, it is possible to write all the subvectors π_i in terms of π_N and then to solve for π_N. We find

1. $\pi_{N-1}\Lambda + \pi_N A_N = 0$
    $\pi_{N-1} = -\pi_N A_N \Lambda^{-1}$

2. $\pi_{N-2}\Lambda + \pi_{N-1}A_{N-1} + \pi_N B_N = 0$
    $\pi_{N-2} = -\pi_{N-1}A_{N-1}\Lambda^{-1} - \pi_N B_N \Lambda^{-1}$
    $\pi_{N-2} = +\pi_N \{A_N \Lambda^{-1} A_{N-1} - B_N\}\Lambda^{-1}$

3. $\pi_{N-3} = \pi_N \{-A_N \Lambda^{-1} A_{N-1}\Lambda^{-1}A_{N-2} + B_N \Lambda^{-1}A_{N-2} + A_N \Lambda^{-1}B_{N-1}\}\Lambda^{-1},$

etc.

Continuing in this fashion, we obtain all π_i in terms of π_N. It only remains to determine π_N. Since both π_0 and π_1 are available in terms of π_N, the first equation,

$$\pi_0 A_0 + \pi_1 B_1 = 0,$$

may be written as

$$\pi_N S = 0,$$

where S is a singular matrix. Once this system of $(K+1)$ equations in $(K+1)$ unknowns is solved for π_N, the remaining π_i for $i = N-1, N-2, \ldots, 0$ may be computed from previous recurrence relations.

Consider as an example, the case when $N = 2$, $K = 2$, and the model parameters are assigned the numerical values $\lambda = 0.5$; $\mu_1 = 1.0$; and $\mu_2 = 10.0$. The infinitesimal generator matrix is

$$
\left(
\begin{array}{ccc|ccc|ccc}
-20.5 & 20.0 & 0.0 & 0.5 & 0.0 & 0.0 & 0.0 & 0.0 & 0.0 \\
0.0 & -10.5 & 10.0 & 0.0 & 0.5 & 0.0 & 0.0 & 0.0 & 0.0 \\
0.0 & 0.0 & -0.5 & 0.0 & 0.0 & 0.5 & 0.0 & 0.0 & 0.0 \\
\hline
0.0 & 0.0 & 0.0 & -20.5 & 20.0 & 0.0 & 0.5 & 0.0 & 0.0 \\
1.0 & 0.0 & 0.0 & 0.0 & -11.5 & 10.0 & 0.0 & 0.5 & 0.0 \\
0.0 & 1.0 & 0.0 & 0.0 & 0.0 & -1.5 & 0.0 & 0.0 & 0.5 \\
\hline
0.0 & 0.0 & 0.0 & 0.0 & 0.0 & 0.0 & -20.0 & 20.0 & 0.0 \\
0.0 & 0.0 & 0.0 & 1.0 & 0.0 & 0.0 & 0.0 & -11.0 & 10.0 \\
0.0 & 0.0 & 0.0 & 0.0 & 2.0 & 0.0 & 0.0 & 0.0 & -2.0
\end{array}
\right).
$$

Following the recursive procedure just outlined, we have, from $\pi_1 \Lambda + \pi_2 A_2 = 0$,

$$
\pi_1 = -\pi_2
\begin{pmatrix}
-20 & 20 & 0 \\
0 & -11 & 10 \\
0 & 0 & -2
\end{pmatrix}
\begin{pmatrix}
2 & 0 & 0 \\
0 & 2 & 0 \\
0 & 0 & 2
\end{pmatrix}
= \pi_2
\begin{pmatrix}
40 & -40 & 0 \\
0 & 22 & -20 \\
0 & 0 & 4
\end{pmatrix}
$$

giving us π_1 in terms of π_2. Now from $\pi_0 \Lambda + \pi_1 A_1 + \pi_2 B_2 = 0$, we have

$$
\pi_0 = -\pi_1 A_1 \Lambda^{-1} - \pi_2 B_2 \Lambda^{-1} =
$$

$$
-\pi_2
\begin{pmatrix}
40 & -40 & 0 \\
0 & 22 & -20 \\
0 & 0 & 4
\end{pmatrix}
\begin{pmatrix}
-20.5 & 20.0 & 0.0 \\
0.0 & -11.5 & 10.0 \\
0.0 & 0.0 & -1.5
\end{pmatrix}
\begin{pmatrix}
2 & 0 & 0 \\
0 & 2 & 0 \\
0 & 0 & 2
\end{pmatrix}
$$

$$
-\pi_2
\begin{pmatrix}
0 & 0 & 0 \\
1 & 0 & 0 \\
0 & 2 & 0
\end{pmatrix}
\begin{pmatrix}
2 & 0 & 0 \\
0 & 2 & 0 \\
0 & 0 & 2
\end{pmatrix},
$$

i.e.,

$$
\pi_0 = \pi_2
\begin{pmatrix}
1640 & -2520 & 800 \\
-2 & 506 & -500 \\
0 & -4 & 12
\end{pmatrix}
$$

and now we have π_0 in terms of π_2. Finally, from $\pi_0 A_0 + \pi_1 B_1 = 0$, by substituting for π_0 and π_1, we obtain

$$
0 = \pi_2
\begin{pmatrix}
1640 & -2520 & 800 \\
-2 & 506 & -500 \\
0 & -4 & 12
\end{pmatrix}
\begin{pmatrix}
-20.5 & 20.0 & 0.0 \\
0.0 & -10.5 & 10.0 \\
0.0 & 0.0 & -0.5
\end{pmatrix}
$$

$$
+\pi_2
\begin{pmatrix}
40 & -40 & 0 \\
0 & 22 & -20 \\
0 & 0 & 4
\end{pmatrix}
\begin{pmatrix}
0 & 0 & 0 \\
1 & 0 & 0 \\
0 & 1 & 0
\end{pmatrix},
$$

i.e.,

$$\pi_2 \begin{pmatrix} -33660 & 59260 & -25600 \\ 63 & -5373 & 5310 \\ 0 & 46 & -46 \end{pmatrix} = 0.$$

Setting $\pi(2,0) = 1$, we find that $\pi_2 = (1.0,\ 534.29,\ 61118.63)$, which leads to $\pi_1 = (40.0,\ 11714.38,\ 233788.73)$ and $\pi_0 = (571.42,\ 23356.22,\ 467078.56)$. These may now be normalized to obtain the final solution:

$$\pi = (.000716, .029261, .585162, .000050, .014676, .292894,$$
$$.000001, .000669, .076570).$$

If we refer to the matrix description, it may readily be verified that S may also be computed by a direct calculation of $X - YW^{-1}V$, where

$$X = \begin{pmatrix} 0 & 0 & 0 \\ 0 & 0 & 0 \\ 0 & 0 & 0 \end{pmatrix}, \qquad Y = \begin{pmatrix} 0 & 0 & 0 & -20 & 20 & 0 \\ 1 & 0 & 0 & 0 & -11 & 10 \\ 0 & 2 & 0 & 0 & 0 & -2 \end{pmatrix},$$

$$W^{-1} = \begin{pmatrix} 2.0 & 0.0 & 0.0 & 0.0 & 0.0 & 0.0 \\ 0.0 & 2.0 & 0.0 & 0.0 & 0.0 & 0.0 \\ 0.0 & 0.0 & 2.0 & 0.0 & 0.0 & 0.0 \\ 82.0 & -80.0 & 0.0 & 2.0 & 0.0 & 0.0 \\ 0.0 & 46.0 & -40.0 & 0.0 & 2.0 & 0.0 \\ 0.0 & 0.0 & 6.0 & 0.0 & 0.0 & 2.0 \end{pmatrix}, \quad \text{and}$$

$$V = \begin{pmatrix} -20.5 & 20.0 & 0.0 \\ 0.0 & -10.5 & 10.0 \\ 0.0 & 0.0 & -0.5 \\ 0.0 & 0.0 & 0.0 \\ 1.0 & 0.0 & 0.0 \\ 0.0 & 1.0 & 0.0 \end{pmatrix}.$$

Notice that the elements of the matrix S are relatively large. As a general rule of thumb, when elements become large, the user should beware, for this often indicates an undesirable growth in error, as we shall now see.

5.2.2.3 Stability Considerations

Let us observe what happens in the foregoing example as the value of N increases. The computed probability of being in state $(0,0)$ is given in the first column of Table 5.2. As before, $K = 2$, $\lambda = 0.5$, $\mu_1 = 1.0$, $\mu_2 = 10.0$, and all computations were performed in double precision. The negative probability obtained by the algorithm when $N = 20$ is obviously disturbing. The exact results are displayed

Table 5.2: Computed vs. Exact Probabilities

N	Computed $\hat{\pi}(0,0)$	Exact $\pi(0,0)$
2	.7158 93579e−3	.7158 93579e−3
3	.7006 39763e−3	.7006 39763e−3
4	.6968 25358e−3	.6968 25358e−3
5	.6958 52056e−3	.6958 52056e−3
8	.6955 21857e−3	.6955 21857e−3
10	.6955 16530e−3	.6955 16539e−3
12	.6955 15663e−3	.6955 16187e−3
14	.6954 74042e−3	.6954 16164e−3
16	.6808 16747e−3	.6955 16163e−3
20	−.1973 30147e+0	.6955 16163e−3

in the second column of Table 5.2. The first observed difference occurs when $N = 10$. It is instructive to observe the residuals. If π is computed exactly, then we should find that $\pi Q = 0$. However, when we multiply the computed value of π with Q, we find that the residual, $\|\hat{\pi}Q\|_2$, is as given in Table 5.3 and thus the error has increased even from the smallest values of N.

Table 5.3: Residuals

N:	2	3	4	5	8
Res:	$.240 \times 10^{-16}$	$.312 \times 10^{-15}$	$.185 \times 10^{-14}$	$.106 \times 10^{-13}$	$.338 \times 10^{-11}$

N:	10	12	14	16	20
Res:	$.465 \times 10^{-9}$	$.527 \times 10^{-7}$	$.523 \times 10^{-5}$	$.397 \times 10^{-3}$	$.509 \times 10^{+1}$

It would appear that the simple recursive scheme just outlined is rather unstable. Let us consider the jth block row. We have

$$\pi_{j-1}\Lambda + \pi_j A_j + \pi_{j+1} B_{j+1} = 0,$$

which gives

$$\pi_{j-1} = -(\pi_j A_j \Lambda^{-1} + \pi_{j+1} B_{j+1} \Lambda^{-1}), \qquad j = N-1,\ldots,1.$$

Note that the arrival rate λ is frequently small when compared to the service rates μ_1 and μ_2. In this case $A_j \Lambda^{-1}$ and $B_{j+1} \Lambda^{-1}$ will be large, since the elements of A_j and B_j are of order $\mu \times \min(j, k)$. The matrices $A_j\Lambda^{-1}$ and $B_{j+1}\Lambda^{-1}$ will therefore have large norms, and the substitution procedure is unstable, since small errors in π_{j+1} and π_j will be multiplied by large values in computing π_{j-1}.

In typical cases the values of $(\mu/\lambda)\min(\eta, k)$ are in the range $\{10, 10^2\}$, and the errors in this substitution procedure will satisfy relations such as

$$\epsilon_{j-1} \approx \{10, 10^2\} \times \epsilon_j \pm \{10, 10^2\}\epsilon_{j+1}.$$

Since ϵ_j are random variables, cancellation is almost impossible, so ϵ_{j-1} will be amplified by a factor of $\{10, 10^2\}$ of ϵ_j. From

$$\pi_{j-1} = -(\pi_j A_j\Lambda^{-1} + \pi_{j+1}B_{j+1}\Lambda^{-1}),$$

we may derive the following relationship among the errors induced during consecutive stages:

$$\epsilon_{j-1} = -(\epsilon_j A_j\Lambda^{-1} + \epsilon_{j+1}B_{j+1}\Lambda^{-1}).$$

This may be written as

$$(\epsilon_{j-1},\ \epsilon_j) = (\epsilon_j,\ \epsilon_{j+1}) \begin{pmatrix} -A_j\Lambda^{-1} & I \\ -B_{j+1}\Lambda^{-1} & 0 \end{pmatrix} \equiv (\epsilon_j,\ \epsilon_{j+1})\mathcal{E}_j,$$

and thus

$$(\epsilon_0,\ \epsilon_1) = (\epsilon_{N-1},\ \epsilon_N) \prod_{j=1}^{N-1} \mathcal{E}_j.$$

Hence, when the elements of $A_j\Lambda^{-1}$ and $B_j\Lambda^{-1}$ are large, we can only expect that

$$(\epsilon_0,\ \epsilon_1) \gg (\epsilon_{N-1},\ \epsilon_N).$$

In other words, the recursive procedure is unstable. Table 5.4 presents the computed residuals obtained with three different values of μ_2 and serve to illustrate the analysis. The values $K = 2$; $\lambda = 0.5$; and $\mu_1 = 1.0$ were used in these experiments.

The error growth is slow when μ_2 is small. When μ_2 is large, the growth of error is rapid. Similar tables may be produced by varying μ_1 or λ. The exact results presented here were obtained from the methods described later in Section 5.2.2.4.

To clarify the situation it is instructive to consider a well-known example of a recursive relation derived from the integral $\int_0^1 x^k e^{x-1}dx$. Let

$$E_k = \int_0^1 x^k e^{x-1}dx.$$

Table 5.4: Growth of Residuals under Recursive Procedure

N:	2	3	4	5	8
$\mu_2 = 2$	$.273 \times 10^{-16}$	$.280 \times 10^{-16}$	$.434 \times 10^{-16}$	$.292 \times 10^{-15}$	$.102 \times 10^{-13}$
$\mu_2 = 10$	$.240 \times 10^{-16}$	$.312 \times 10^{-15}$	$.185 \times 10^{-14}$	$.106 \times 10^{-13}$	$.338 \times 10^{-11}$
$\mu_2 = 50$	$.226 \times 10^{-16}$	$.116 \times 10^{-14}$	$.921 \times 10^{-13}$	$.357 \times 10^{-11}$	$.145 \times 10^{-6}$

N:	10	12	14	16	20
$\mu_2 = 2$	$.589 \times 10^{-13}$	$.302 \times 10^{-12}$	$.460 \times 10^{-11}$	$.110 \times 10^{-10}$	$.295 \times 10^{-8}$
$\mu_2 = 10$	$.465 \times 10^{-9}$	$.527 \times 10^{-7}$	$.523 \times 10^{-5}$	$.397 \times 10^{-3}$	$.509 \times 10^{+1}$
$\mu_2 = 50$	$.360 \times 10^{-3}$	$.552 \times 10^{-1}$	$.350 \times 10^{+2}$	Overflow	Overflow

Then

$$E_k = 1 - k \int_0^1 x^{k-1} e^{x-1} dx = 1 - k E_{k-1}.$$

Furthermore, $E_0 = 1 - 1/e$.

Let us examine the effect of evaluating this integral for $k = 0, 1, \ldots, K$ using the recurrence relation $E_k = 1 - k E_{k-1}$. This relation consists of only a single multiplication and a subtraction at each step and so is trivial to program. Using double-precision arithmetic on an IBM 3090 (in single-precision the same effect is observed more rapidly!), and taking $E_0 = 1 - 1/e$, we obtain the values given in Table 5.5. Note that

$$E_k \leq \frac{1}{e} \max_{0 \leq x \leq 1} (e^x) \int_0^1 x^k dx < \frac{1}{k+1}.$$

What is happening in this example is that E_0 is computed with a small error. This small error grows in each succeeding step of the recursion. When we compute $E_k = 1 - k E_{k-1}$, the error in E_{k-1} is multiplied by k. The small initial error in E_0 is magnified by $k!$ in the computation of E_k. The same growth of error (though not quite as dramatic) is what we observed in the previous Markov chain example.

From the unstable recursive formula for computing E_k, we can determine a *stable* recursion, one in which the error is *decreased* at each step rather than magnified. We may write

$$E_{k-1} = \frac{1 - E_k}{k},$$

so that if we know E_K, we may successfully compute $E_{K-1}, E_{K-2}, \ldots, E_1$, and finally E_0. This time the error at each step is diminished by a factor of k rather than magnified by k. In fact, we can use this situation to avoid computing an

Table 5.5: Growth of Error in Recursion

k	E_k
0	.6321 2055 8829
1	.3678 7944 1171
2	.2642 4111 7657
3	.2072 7664 7029
4	.1708 9341 1885
5	.1455 3294 0573
6	.1268 0235 6562
7	.1123 8350 4069
8	.1009 3196 7446
9	.0916 1229 2984
10	.0838 7707 0159
11	.0773 5222 8251
12	.0717 7326 0990
13	.0669 4760 7135
14	.0627 3350 0108
15	.0589 9749 8387
16	.0560 4002 5806
17	.0473 1956 1291
18	.1482 4789 6757
19	-1.8167 1003 8389
20	37.3342 0076 7778
21	-783.0182 1612 3346
22	17,227.4007 5471 3613
23	-396,229.2173 5841 3109
24	9,509,502.2166 0191 4490

initial E_K from the integral formula. By taking a starting point greater than K, say $K + l$, and setting $E_{K+l} = 0$ (or $E_{K+l} = 1/(K + l + 1)$), this initial error will be divided by $(K + l + 1)!/(K + 1)!$ by the time we compute E_K, so that E_K will be determined with high precision. To illustrate this effect, Table 5.6 shows the results obtained for various starting values. Observe how the errors decrease, even when the extremely incorrect value of 10^6 is chosen for E_{24}.

Having thus found a way in the integration example to convert an unstable recursive algorithm into a stable algorithm, we may well ask whether the same procedure could be adapted to the Markov chain example. Specifically, we would like to know whether it is possible to reverse the procedure and to determine all subvectors π_i, $i = 1, 2, \ldots, N$ in terms of π_0 rather than in terms of π_N, and if so, whether this reversed procedure is stable. Recall that the block Chapman–Kolmogorov equations have the form

$$\pi_{j-1}\Lambda + \pi_j A_j + \pi_{j+1} B_{j+1} = 0,$$

and our recursive procedure was based on the relationship

$$\pi_{j-1} = -(\pi_j A_j \Lambda^{-1} + \pi_{j+1} B_{j+1} \Lambda^{-1}).$$

We are now asking whether we can benefit by using a recursive relation based on

$$\pi_{j+1} = -(\pi_j A_j B_{j+1}^{-1} + \pi_{j-1} \Lambda B_{j+1}^{-1}).$$

Unfortunately, in this specific case the answer is no, for the matrices B_{j+1} are singular.

We wish to emphasize that great care must be taken in implementing recursive numerical procedures. In some instances they can be stable and efficient methods for determining the solution of Markov chains [131]. However, in many cases they turn out to be unstable. The reader interested in using these techniques should carefully investigate the matrix norms of the diagonal, subdiagonal, and superdiagonal blocks of the transition rate matrix before proceeding.

In the next section we discuss two stable methods for solving this particular Markov chain example that not only are stable but also require few is any more numerical operations than the recursive procedure.

Table 5.6: Reversed Recursive Formula Results

k	E_k	E_k	E_k
24	.000000000000	.040000000000	1000000.000000000000
23	.041666666667	.040000000000	-41666.625000000000
22	.041666666667	.041739130435	1811.635869565217
21	.043560606061	.043557312253	-82.301630434783
20	.045544733045	.045544889893	3.966744306418
19	.047722763348	.047722755505	-0.148337215321
18	.050119854561	.050119854973	.060438800806
17	.052771119191	.052771119168	.052197844400
16	.055719345930	.055719345931	.055753067976
15	.059017540879	$\Leftarrow$	.059015433251
14	.062732163941	$\Leftarrow$	.062732304450
13	.066947702576	$\Leftarrow$	.066947692539
12	.071773253648	$\Leftarrow$	.071773254420
11	.077352228863	$\Leftarrow$	.077352228798
10	.083877070103	$\Leftarrow$	.083877070109
9	.091612292990	$\Leftarrow$	.091612292989
8	.100931967446	$\Leftarrow$	$\Leftarrow$
7	.112383504069	$\Leftarrow$	$\Leftarrow$
6	.126802356562	$\Leftarrow$	$\Leftarrow$
5	.145532940573	$\Leftarrow$	$\Leftarrow$
4	.170893411885	$\Leftarrow$	$\Leftarrow$
3	.207276647029	$\Leftarrow$	$\Leftarrow$
2	.264241117657	$\Leftarrow$	$\Leftarrow$
1	.367879441171	$\Leftarrow$	$\Leftarrow$
0	.632120558829	$\Leftarrow$	$\Leftarrow$

5.2.2.4 Block Solution Methods That Work

For block-structured matrices such as those under consideration, block variants of the basic direct and iterative methods are often attractive. Here we consider block Gaussian elimination and block Gauss–Seidel. Block methods based on other point iterative procedures may also be applied.

Block Gaussian Elimination: We shall assume that the system to be solved is given by

$$H x = 0,$$

where H is block upper Hessenberg and $x = \pi^T$. Block Gaussian elimination, when applied to H, may be viewed as follows. Assume that just prior to the beginning of the jth block elimination step, we have the following:

$$\left(
\begin{array}{ccccc|ccccc}
L_0 U_0 & H_{01} & H_{02} & \ldots & H_{0,j-1} & H_{0j} & H_{0,j+1} & \ldots & H_{0,N-1} & H_{0,N} \\
0 & L_1 U_1 & \tilde{H}_{12} & \ldots & \tilde{H}_{1,j-1} & \tilde{H}_{1j} & \tilde{H}_{1,j+1} & \ldots & \tilde{H}_{1,N-1} & \tilde{H}_{1,N} \\
0 & 0 & L_2 U_2 & \ldots & \tilde{H}_{2,j-1} & \tilde{H}_{2j} & \tilde{H}_{2,j+1} & \ldots & \tilde{H}_{2,N-1} & \tilde{H}_{2,N} \\
\vdots & \vdots & \vdots & \ddots & \vdots & \vdots & \vdots & \ddots & \vdots & \vdots \\
0 & 0 & 0 & \ldots & L_{j-1} U_{j-1} & \tilde{H}_{j-1,j} & \tilde{H}_{j-1,j+1} & \ldots & \tilde{H}_{j-1,N-1} & \tilde{H}_{j-1,N} \\
\hline
0 & 0 & 0 & \ldots & 0 & \tilde{H}_{j,j} & \tilde{H}_{j,j+1} & \ldots & \tilde{H}_{j,N-1} & \tilde{H}_{j,N} \\
0 & 0 & 0 & \ldots & 0 & H_{j+1,j} & H_{j+1,j+1} & \ldots & H_{j+1,N-1} & H_{j+1,N} \\
\vdots & \vdots & \vdots & \ddots & \vdots & \vdots & \vdots & \ddots & \vdots & \vdots \\
0 & 0 & 0 & \ldots & 0 & 0 & 0 & \ldots & H_{N-1,N-1} & H_{N-1,N} \\
0 & 0 & 0 & \ldots & 0 & 0 & 0 & \ldots & H_{N,N-1} & H_{N,N}
\end{array}
\right)$$

where $L_0 U_0$ is an LU decomposition of H_{00}; $L_i U_i$ is an LU decomposition of $\tilde{H}_{ii}$ for $i = 1, 2, \ldots, j-1$; and the tildes indicate blocks that have been modified during previous block elimination steps.

The jth step involves obtaining an LU decomposition of $\tilde{H}_{jj}$ and eliminating the block $H_{j+1,j}$. This is achieved by the block operation

$$\tilde{H}_{j+1,i} \leftarrow H_{j+1,i} - H_{j+1,j}(\tilde{H}_{jj})^{-1}\tilde{H}_{ji} \quad \text{for } i = j, j+1, \ldots, N.$$

Notice that the term $H_{j+1,j}(\tilde{H}_{jj})^{-1}$ plays the role of the multiplier in the point version and that $\tilde{H}_{j+1,i}$ becomes zero at $i = j$. If we assume that the blocks are all of order K, then the operations involved at the jth block step are as follows:

		Multiplications
LU decomposition:	$L_j U_j = \tilde{H}_{jj}$	$K^3/3 + 0(K^2)$
Form multiplier:	$G_j = H_{j+1,j} U_j^{-1} L_j^{-1}$	K^3
For $i = j + 1, \ldots, N$ form:	$\tilde{H}_{j+1,i} = H_{j+1,i} - G_j \tilde{H}_{ji}$	$(N - j)K^3$

The total number of multiplications involved in the elimination phase is

$$\frac{(N+1)K^3}{3} + (N)K^3 + \sum_{j=1}^{N}(N-j)K^3 + O(NK^2) = \frac{N^2 K^3}{2} + \frac{5NK^3}{6} + O(NK^2 + K^3).$$

The back-substitution phase requires a further $(N^2 K^2 + 3NK^2 + 2)/2$ multiplications, since the jth step of this phase involves computing x_j from the relation

$$L_j U_j x_j + \tilde{H}_{j,j+1} x_{j+1} + \cdots + \tilde{H}_{jN} x_N = 0,$$

given that $x_N, x_{N-1}, \ldots, x_{j+1}$ have already been found. The total operation count for block Gaussian elimination applied to an upper Hessenberg matrix is then given as

$$\frac{N^2 K^3}{2} + \frac{N^2 K^2}{2} + \frac{5NK^3}{6} + O(NK^2 + K^3).$$

This number reduces substantially when the infinitesimal generator is block tridiagonal. In this case the superdiagonal blocks $H_{j-1,j}, j = 1, \ldots, N$ are left unaltered by the decomposition, since blocks $H_{ij} = 0$ for all $j \geq 2$ and $i \leq j - 2$. Each step simplifies to determining an LU decomposition of $\tilde{H}_{jj}$ and then updating the next diagonal block $\tilde{H}_{j+1,j+1}$ according to the formula

$$\tilde{H}_{j+1,j+1} = H_{j+1,j+1} - H_{j+1,j} \tilde{H}_{jj}^{-1} H_{j,j+1}. \tag{5.6}$$

It may readily be verified that the number of multiplications involved in the reduction phase is

$$\frac{(N+1)K^3}{3} + (N)K^3 + (N)K^3 + O(NK^2) = \frac{7NK^3}{3} + \frac{K^3}{3} + O(NK^2).$$

For the back-substitution phase, the number of multiplications needed is $(N + 1)K(K + 1)/2 + NK^2$, which is approximately equal to $1.5NK^2$.

In most cases the submatrices are not full, as the foregoing analysis assumes. Often they possess a well-defined structure such as diagonal, bidiagonal, or tridiagonal, and advantage can be taken of this structure to reduce the number of multiplications even further.

Let us now consider the application of block Gaussian elimination to the previous example of a telephone system with N lines and K operators. Once again, the infinitesimal generator matrix in this example is

$$
\begin{pmatrix}
A_0 & \Lambda & 0 & 0 & \cdots & 0 & 0 \\
B_1 & A_1 & \Lambda & 0 & \cdots & 0 & 0 \\
0 & B_2 & A_2 & \Lambda & \cdots & 0 & 0 \\
0 & 0 & B_3 & A_3 & \cdots & 0 & 0 \\
\vdots & \vdots & \vdots & \vdots & \ddots & \vdots & \vdots \\
0 & 0 & 0 & 0 & \cdots & A_{N-1} & \Lambda \\
0 & 0 & 0 & 0 & \cdots & B_N & A_N
\end{pmatrix}
$$

Since we wish to solve $\pi Q = 0$, we write $Q^T \pi^T = Q^T x = 0$. We have

$$
\begin{pmatrix}
A_0^T & B_1^T & 0 & 0 & \cdots & 0 & 0 \\
\Lambda & A_1^T & B_2^T & 0 & \cdots & 0 & 0 \\
0 & \Lambda & A_2^T & B_3^T & \cdots & 0 & 0 \\
0 & 0 & \Lambda & A_3^T & \cdots & 0 & 0 \\
\vdots & \vdots & \vdots & \vdots & \ddots & \vdots & \vdots \\
0 & 0 & 0 & 0 & \cdots & A_{N-1}^T & B_N^T \\
0 & 0 & 0 & 0 & \cdots & \Lambda & A_N^T
\end{pmatrix}
\begin{pmatrix}
x_0 \\
x_1 \\
x_2 \\
x_3 \\
\vdots \\
x_{N-1} \\
x_N
\end{pmatrix}
= 0.
$$

Note that when the Λ blocks are reduced to zero, none of the matrices B_i^T will be altered, because each block above B_i^T is zero. Referring back to equation (5.6) and using the notation of the example, we have

$$
A_{j+1}^T = A_{j+1}^T - \Lambda (\tilde{A}_j^T)^{-1} B_{j+1}^T.
$$

Thus, at block step j it suffices to determine an LU decomposition of $\tilde{A}_j^T$, to update the next diagonal block A_{j+1}^T, and to set the Λ block in block row $j+1$ to zero. Algorithmically, we have

1. For $j = 0$ to $N - 1$ do

   - Form LU decomposition of $\tilde{A}_j^T$ such that $L_j U_j = \tilde{A}_j^T$.

   - *Update next diagonal block:* $\tilde{A}_{j+1}^T = A_{j+1}^T - \Lambda U_j^{-1} L_j^{-1} B_{j+1}^T$.

2. Form LU decomposition of $\tilde{A}_N^T$.

After these steps of block Gaussian elimination, the resulting matrix has the form

$$
\begin{pmatrix}
L_0 U_0 & B_1^T & 0 & 0 & \dots & 0 & 0 \\
0 & L_1 U_1 & B_2^T & 0 & \dots & 0 & 0 \\
0 & 0 & L_2 U_2 & B_3^T & \dots & 0 & 0 \\
\vdots & \vdots & \vdots & \vdots & \ddots & \vdots & \vdots \\
0 & 0 & 0 & 0 & \dots & B_{N-1}^T & 0 \\
0 & 0 & 0 & 0 & \dots & L_{N-1} U_{N-1} & B_N^T \\
0 & 0 & 0 & 0 & \dots & 0 & L_N U_N
\end{pmatrix}.
$$

Assuming that the diagonal blocks become full, the number of multiplications required to form each LU decomposition is approximately $(K+1)^3/3$; to update each diagonal block requires $(K+1)^2 + K(K+1)$. The total number of multiplications is thus approximately equal to $NK^3/3$. The back-substitution that must follow requires an additional $NK^2/2$ multiplications. Unlike the recursive procedure of Section 5.2.2.2, the results obtained by this method are highly accurate.

Block Gauss–Seidel: Block iterative methods, including block Gauss–Seidel, were considered in Chapter 3. At this point we wish to consider using block Gauss–Seidel on the specific example of the telephone system with N lines and K operators. This example possesses qualities that make this method particularly appropriate. Recall that the transpose of the infinitesimal generator is split into $Q^T = D - L - U$, where L and U are respectively strictly lower and strictly upper *block* triangular matrices and D is a *block* diagonal matrix, and that the iterative procedure is

$$
Dx^{(l+1)} = Lx^{(l+1)} + Ux^{(l)}, \qquad l = 0, 1, \dots
$$

In this example, these matrices are as follows:

$$
-L = \begin{pmatrix}
0 & 0 & 0 & 0 & \dots & 0 & 0 \\
\Lambda & 0 & 0 & 0 & \dots & 0 & 0 \\
0 & \Lambda & 0 & 0 & \dots & 0 & 0 \\
\vdots & \vdots & \vdots & \vdots & \ddots & \vdots & \vdots \\
0 & 0 & 0 & 0 & \dots & 0 & 0 \\
0 & 0 & 0 & 0 & \dots & \Lambda & 0
\end{pmatrix}, \quad
D = \begin{pmatrix}
A_0^T & 0 & 0 & 0 & \dots & 0 & 0 \\
0 & A_1^T & 0 & 0 & \dots & 0 & 0 \\
0 & 0 & A_2^T & 0 & \dots & 0 & 0 \\
\vdots & \vdots & \vdots & \vdots & \ddots & \vdots & \vdots \\
0 & 0 & 0 & 0 & \dots & A_{N-1}^T & 0 \\
0 & 0 & 0 & 0 & \dots & 0 & A_N^T
\end{pmatrix},
$$

$$
-U = \begin{pmatrix}
0 & B_1^T & 0 & 0 & \dots & 0 & 0 \\
0 & 0 & B_2^T & 0 & \dots & 0 & 0 \\
0 & 0 & 0 & B_3^T & \dots & 0 & 0 \\
\vdots & \vdots & \vdots & \vdots & \ddots & \vdots & \vdots \\
0 & 0 & 0 & 0 & \dots & 0 & B_N^T \\
0 & 0 & 0 & 0 & \dots & 0 & 0
\end{pmatrix}.
$$

This means that at each iteration we need to perform the operations

$$
\begin{array}{rcl}
A_0^T x_0^{(l+1)} &=& -B_1^T x_1^{(l)} \\
A_1^T x_1^{(l+1)} &=& -B_2^T x_2^{(l)} - \Lambda x_0^{(l+1)} \\
A_2^T x_2^{(l+1)} &=& -B_3^T x_3^{(l)} - \Lambda x_1^{(l+1)} \\
\vdots &=& \vdots \\
A_N^T x_N^{(l+1)} &=& -\Lambda x_{N-1}^{(l+1)}
\end{array}
$$

This is a total of N equation solves per iteration. Observe that the right-hand side in any of these N systems may be formed before it becomes necessary to compute the solution. Also, in this example the diagonal blocks A_i^T are already lower triangular, so that it is not necessary to compute their LU decompositions. Given the structure of the submatrices B_i and Λ, the right-hand sides may be computed in a total of approximately $2KN$ multiplications. Furthermore, each solve requires only $2K$ multiplications, since only the diagonal and superdiagonal elements are nonzero, so the total number of multiplications needed per iteration is approximately $4KN$ — a very small number.

Let us turn now to the question of the rate of convergence. The *forward* block Gauss–Seidel iteration method is equivalent to

$$
x^{(l+1)} = (D - L)^{-1} U x^{(l)}.
$$

Since the elements of D and U are of the same order of magnitude (the elements of L being small), the nonzero elements of the iteration matrix are expected to be close to 1. However, if we alter the ordering of the block variables so that block i becomes block $N - i$ (essentially giving *backward* Gauss–Seidel), we get

$$
\begin{pmatrix}
A_N^T & \Lambda & 0 & 0 & \cdots & 0 & 0 \\
B_N^T & A_{N-1}^T & \Lambda & 0 & \cdots & 0 & 0 \\
0 & B_{N-1}^T & A_{N-2}^T & \Lambda & \cdots & 0 & 0 \\
0 & 0 & B_{N-2}^T & A_{N-3}^T & \cdots & 0 & 0 \\
\vdots & \vdots & \vdots & \vdots & \ddots & \vdots & \vdots \\
0 & 0 & 0 & 0 & \cdots & A_1^T & \Lambda \\
0 & 0 & 0 & 0 & \cdots & B_1^T & A_0^T
\end{pmatrix}
\begin{pmatrix}
x_N \\
x_{N-1} \\
x_{N-2} \\
x_{N-3} \\
\vdots \\
x_1 \\
x_0
\end{pmatrix}
= 0.
$$

This is the effect achieved if the ith block equation

$$
\Lambda x_{i-1} + A_i^T x_i + B_{i+1}^T x_{i+1} = 0
$$

is written as

$$
B_{i+1}^T z_{i+1} + A_i^T z_i + \Lambda z_{i-1} = 0,
$$

Table 5.7: Convergence of Block Gauss-Seidel Iteration

N	Number of Iterations	
	Backward	Forward
3	15	16
4	17	19
5	19	21
6	20	23
8	21	26
10	22	29
20	23	43
30	24	55
40	25	67
50	26	79

where $z_i = x_{N-i}$, $i = 0, 1, \ldots, N$. Applying block Gauss–Seidel to this equivalent form yields

$$z^{(l+1)} =$$

$$-\begin{pmatrix} A_N^T & 0 & 0 & 0 & \cdots & 0 & 0 \\ B_N^T & A_{N-1}^T & 0 & 0 & \cdots & 0 & 0 \\ 0 & B_{N-1}^T & A_{N-2}^T & 0 & \cdots & 0 & 0 \\ 0 & 0 & B_{N-2}^T & A_{N-3}^T & \cdots & 0 & 0 \\ \vdots & \vdots & \vdots & \vdots & \ddots & \vdots & \vdots \\ 0 & 0 & 0 & 0 & \cdots & A_1^T & 0 \\ 0 & 0 & 0 & 0 & \cdots & B_1^T & A_0^T \end{pmatrix}^{-1} \begin{pmatrix} 0 & \Lambda & 0 & 0 & \cdots & 0 & 0 \\ 0 & 0 & \Lambda & 0 & \cdots & 0 & 0 \\ 0 & 0 & 0 & \Lambda & \cdots & 0 & 0 \\ 0 & 0 & 0 & 0 & \cdots & 0 & 0 \\ \vdots & \vdots & \vdots & \vdots & \ddots & \vdots & \vdots \\ 0 & 0 & 0 & 0 & \cdots & 0 & \Lambda \\ 0 & 0 & 0 & 0 & \cdots & 0 & 0 \end{pmatrix} z^{(l)}$$

Now, since the elements of A_j and B_j are larger than Λ, the elements of the iteration matrix will likely be smaller than with forward Gauss–Seidel, and the iterative process may converge more rapidly.

Table 5.7 presents the number of iterations required by both the forward and backward block Gauss–Seidel iteration methods as the size of N increases. The values of the different parameters in each case are $K = 3$; $\lambda = 0.5$; $\mu_1 = 1.0$; and $\mu_2 = 10.0$. It clearly illustrates the advantage of using backward block Gauss–Seidel iteration for this example, particularly in light of the small number of numerical operations needed per iteration. Again, unlike the results of the recursive procedure, the solution obtained is highly accurate.

5.3   The Matrix-Geometric Approach

5.3.1   Introduction

The solution of the $M/M/1$ queue may be shown to be geometric. Its infinitesimal generator is

$$
Q = \begin{pmatrix}
-\lambda & \lambda & & & & \\
\mu & -(\lambda+\mu) & \lambda & & & \\
& \mu & -(\lambda+\mu) & \lambda & & \\
& & \mu & -(\lambda+\mu) & \lambda & \\
& & & \mu & -(\lambda+\mu) & \lambda \\
& & & & \ddots & \ddots & \ddots
\end{pmatrix}.
$$

From the global balance equations $\pi Q = 0$, it is easy to verify that $\pi_1 = (\lambda/\mu)\pi_0 = \rho\pi_0$ where $\rho = \lambda/\mu$ and that, from induction,

$$
\pi_i = \rho\pi_{i-1} = \rho^i\pi_0, \qquad i = 1, 2, \ldots
$$

Since $\sum_{i=0}^{\infty} \pi_i = 1$, we have $\pi_0 \sum_{i=0}^{\infty} \rho^i = 1$, and therefore

$$
\pi_0 = \frac{1}{1 + \sum_{i=1}^{\infty} \rho^i}. \tag{5.7}
$$

For stability of this queueing system, we must have $\rho < 1$, and consequently the summation in equation (5.7) will converge. We find

$$
\pi_0 = \frac{1}{1 + \rho/(1-\rho)}, \quad \text{i.e., } \pi_0 = 1 - \rho.
$$

Hence

$$
\pi_i = (1 - \rho)\rho^i, \qquad i = 0, 1, 2, \ldots
$$

which is clearly the geometric distribution.

Much work has been carried out by Neuts [110] on infinite stochastic matrices whose stationary probability vector may be written in a *matrix-geometric* form. Additionally, he has developed numerical techniques to compute these vectors. It is to these types of examples that we now turn.

5.3.2   Matrix Geometric Solutions: The Matrix R

Consider a two-dimensional irreducible Markov chain with state space $\{(\eta, k),\ \eta \geq 0,\ 1 \leq k \leq K\}$. Let the states be ordered so that the parameter k varies most rapidly. Assume further that the stochastic transition probability matrix has the structure

$$P = \begin{pmatrix} B_0 & A_0 & 0 & 0 & \ldots \\ B_1 & A_1 & A_0 & 0 & \ldots \\ B_2 & A_2 & A_1 & A_0 & \ldots \\ \vdots & \vdots & \vdots & \vdots & \ddots \\ B_j & A_j & A_{j-1} & A_{j-2} & \ldots \\ \vdots & \vdots & \vdots & \vdots & \ddots \end{pmatrix}$$

in which all submatrices A_j and B_j for $j = 0, 1, 2, \ldots$ are square and of order K. Since P is stochastic, $Pe = e$, and so

$$B_j e + \sum_{i=0}^{j} A_j e = e$$

for all j. For simplicity, we shall assume that the matrix A, defined as $A \equiv \sum_{i=0}^{\infty} A_i$, is stochastic and irreducible. This is certainly the case in most applications in which the matrix-geometric approach would be used. The interested reader will find additional results in [110] for situations in which these conditions do not hold.

Let the stationary probability vector π be partitioned conformally with P, i.e.,

$$\pi = (\pi_0, \ \pi_1, \ \pi_2, \ \ldots)$$

where the subvectors π_i are given by

$$\pi_i = (\pi(i, 1), \ \pi(i, 2), \ \ldots, \ \pi(i, K)), \qquad i = 0, 1, \ldots$$

and $\pi(i, k)$ is the steady-state probability of state (i, k). From $\pi P = \pi$, we have

$$\pi_j = \sum_{i=0}^{\infty} \pi_{j+i-1} A_i \quad \text{for } j \geq 1,$$

and $\pi_0 = \sum_{j=0}^{\infty} \pi_j B_j$. Also, $\sum_{j=0}^{\infty} \pi_j e = 1$, since $\pi e = 1$. Neuts defines a sequence of matrices $\{R(l)\}$ as

$$R(0) = 0; \qquad R(l+1) = \sum_{i=0}^{\infty} [R(l)]^i A_i \quad \text{for } l \geq 0,$$

and under the conditions previously imposed (specifically, the irreducibility of A) he shows that $R(l) \leq R(l+1)$ for $l \geq 0$. Additionally, he proves that the sequence $\{R(l)\}$ converges to a matrix $R \geq 0$, for which the following theorems hold.

Theorem 5.1 (Neuts) *If the Markov chain is positive-recurrent, the matrices* R^i, $i \geq 1$, *are finite.*

Theorem 5.2 (Neuts) *If the Markov chain is positive-recurrent, the matrix* R *satisfies the equation*

$$R = \sum_{i=0}^{\infty} R^i A_i,$$

and is the minimal nonnegative solution to the matrix equation $X = \sum_{i=0}^{\infty} X^i A_i$.

Theorem 5.3 (Neuts) *If the Markov chain is positive-recurrent, then*

1. *For* $i \geq 0$, $\pi_{i+1} = \pi_i R$.

2. *The eigenvalues of* R *lie inside the unit disk.*

3. *The matrix* $\sum_{i=0}^{\infty} R^i B_i$ *is stochastic.*

4. *The vector* π_0 *is the positive left-invariant eigenvector of* $\sum_{i=0}^{\infty} R^i B_i$ *normalized by* $\pi_0 (I - R)^{-1} e = 1$.

Given that our interests lie in positive-recurrent Markov chains, it follows that the stationary probability vector π satisfying $\pi P = \pi$ may be written as

$$\pi = (\pi_0, \ \pi_0 R, \ \pi_0 R^2, \ \dots)$$

and π_0 is found from Theorem 5.3, part 4.

5.3.3   Implementation: Computing R and π_0

We now turn our attention to the computation of the matrix R and the subvector π_0. In all cases considered by Neuts, R can be determined by an iterative procedure, which we shall describe. Additionally, in certain instances, R can be obtained explicitly as a function of the elements of the infinitesimal generator. We shall take up this approach in Section 5.4.

To show how R may be obtained iteratively, consider the matrix equation of Theorem 5.2:

$$R = \sum_{i=0}^{\infty} R^i A_i. \tag{5.8}$$

Writing this as

$$R = A_0 + R A_1 + \sum_{i=2}^{\infty} R^i A_i$$

leads to

$$R = A_0(I - A_1)^{-1} + \sum_{i=2}^{\infty} R^i A_i (I - A_1)^{-1},$$

and R may be computed by means of the iterative procedure

$$R_{l+1} = A_0(I - A_1)^{-1} + \sum_{i=2}^{\infty} R_l^i A_i (I - A_1)^{-1}, \qquad l = 0, 1, \ldots, \qquad (5.9)$$

using $R_0 = 0$ to initiate the process. As is shown by Neuts, the sequence $\{R_l\}$ is monotone increasing and converges to R.

Since A_1 is substochastic, $(I - A_1)$ is nonsingular. In the implementation of an algorithm based on equation (5.9), the inversion of $(I - A_1)$ and the matrix multiplications $A_i(I - A_1)^{-1}$ need to be computed only once, just prior to the beginning of the iterations. Furthermore, in many applications $A_i = 0$ for $i \geq L$, where L is a small integer constant. In quasi-birth-death (block tridiagonal) processes, for example, $A_i = 0$ for $i \geq 3$, and so equation (5.9) simplifies to

$$R_{l+1} = V + R_l^2 W, \qquad l = 0, 1, \ldots$$

where $V = A_0(I - A_1)^{-1}$ and $W = A_2(I - A_1)^{-1}$.

Once R has been computed, it only remains to determine π_0. This vector may be computed from the first set of Chapman–Kolmogorov equations with π_i written in terms of $\pi_0 R^i$. From Theorem 5.3, we have

$$\pi_0 = \pi_0(B_0 + RB_1 + R^2 B_2 + \cdots). \qquad (5.10)$$

In many applications, $B_i = 0$ for all $i \geq L$, where again L is a small integer constant. The $(K \times K)$ matrix $(B_0 + RB_1 + R^2 B_2 + \cdots)$ is then easy to form, and the vector π_0 may be subsequently determined by direct computation. In other applications this will not be the case, and the method of computing $(B_0 + RB_1 + R^2 B_2 + \cdots)$ will be dictated by the problem itself. Numerous examples have been provided by Neuts and his colleagues. The vector π_0, once computed, should be normalized according to Theorem 5.3, so that $\pi_0(I - R)^{-1} e = 1$.

The foregoing discussion revolved around the stochastic transition probability matrix P. As pointed out by Neuts, there are no important mathematical differences between the theorems presented and corresponding theorems relating to the infinitesimal generator matrix Q. If Q is partitioned as

$$Q = \begin{pmatrix} C_0 & F_0 & 0 & 0 & \cdots \\ C_1 & F_1 & F_0 & 0 & \cdots \\ C_2 & F_2 & F_1 & F_0 & \cdots \\ \vdots & \vdots & \vdots & \vdots & \ddots \\ C_j & F_j & F_{j-1} & F_{j-2} & \cdots \\ \vdots & \vdots & \vdots & \vdots & \ddots \end{pmatrix},$$

in which all submatrices F_j and C_j for $j = 0, 1, 2, \ldots$ are $K \times K$, then once again the stationary probability vector satisfying $\pi Q = 0$ is given by $\pi = (\pi_0, \pi_1, \pi_2, \ldots)$, where $\pi_{i+1} = \pi_i R$. In this case, the iterative formula for the computation of R is obtained from

$$\sum_{i=0}^{\infty} R^i F_i = 0.$$

The iterative formula corresponding to equation (5.9) is given by

$$R_{l+1} = F_0(-F_1)^{-1} + \sum_{i=2}^{\infty} R_l^i F_i (-F_1)^{-1}, \qquad l = 0, 1, \ldots \qquad (5.11)$$

and the initial vector π_0 computed from

$$0 = \pi_0(C_0 + RC_1 + R^2 C_2 + \cdots). \qquad (5.12)$$

5.3.4   Example: The $\lambda/C_2/1$ Queue

We will illustrate these concepts by means of the same example used in Section 5.2.1. In this case, however, rather than associating a finite queue length with the Coxian server, we shall use an infinite queue. As before, the states are represented by the ordered pair (η, k) in which η denotes the number of customers in the system and $k \ (= 1, 2)$ denotes the current phase of service. The infinitesimal generator for this system is block tridiagonal. Its leading principal submatrix is given by Table 5.8.

Because Q is block tridiagonal, the iterative procedure, equation (5.11), for the computation of R simplifies to

$$R_{l+1} = \begin{pmatrix} \lambda & 0 \\ 0 & \lambda \end{pmatrix} \begin{pmatrix} \mu_1 + \lambda & -a_1 \mu_1 \\ 0 & \mu_2 + \lambda \end{pmatrix}^{-1} +$$

$$R_l^2 \begin{pmatrix} (1-a_1)\mu_1 & 0 \\ \mu_2 & 0 \end{pmatrix} \begin{pmatrix} \mu_1 + \lambda & -a_1 \mu_1 \\ 0 & \mu_2 + \lambda \end{pmatrix}^{-1},$$

since $C_i = F_{i+1} = 0$ for $i \geq 2$. If we consider the specific case in which $\lambda = 1$, $\mu_1 = 2$, $\mu_2 = 3$, and $a_1 = 0.75$, then setting $R_0 = 0$ and using the iterative equation

$$R_{l+1} = \begin{pmatrix} \frac{1}{3} & \frac{1}{8} \\ 0 & \frac{1}{4} \end{pmatrix} + R_l^2 \begin{pmatrix} \frac{1}{6} & \frac{1}{16} \\ 1 & \frac{3}{8} \end{pmatrix}, \qquad l = 0, 1, \ldots$$

we find

$$R_{50} = \begin{pmatrix} .61536118 & .23076044 \\ .15383707 & .30768890 \end{pmatrix}.$$

Table 5.8: Leading Principal Submatrix of Infinitesimal Generator

$(0,1)$	$(0,2)$	$(1,1)$	$(1,2)$	$(2,1)$	$(2,2)$	$(3,1)$	$(3,2)$
$-\lambda$	0	λ	0				
0	$-\lambda$	0	λ				
$(1-a_1)\mu_1$	0	$-(\mu_1+\lambda)$	$a_1\mu_1$	λ	0		
μ_2	0	0	$-(\mu_2+\lambda)$	0	λ		
		$(1-a_1)\mu_1$	0	$-(\mu_1+\lambda)$	$a_1\mu_1$	λ	0
		μ_2	0	0	$-(\mu_2+\lambda)$	0	λ
				$(1-a_1)\mu_1$	0	$-(\mu_1+\lambda)$	$a_1\mu_1$
				μ_2	0	0	$-(\mu_2+\lambda)$

In this case the iterative process was terminated when $\max_{ij}|(R_l - R_{l-1})_{ij}| <$ .000005. This is an example for which the matrix R may be computed explicitly. It is shown in Section 5.4 that

$$R = \begin{pmatrix} 2 & -1 \\ -1.5 & 4 \end{pmatrix}^{-T}$$

which, to eight decimal places, is given by

$$R = \begin{pmatrix} .61538462 & .23076923 \\ .15384615 & .30769231 \end{pmatrix}.$$

Therefore we see that

$$\max_{ij}|(R - R_{50})_{ij}| < .000024.$$

Once a sufficiently accurate approximation to the matrix R has been computed, the vector π_0 must be found. We have already seen that this may be computed from

$$0 = \pi_0(C_0 + RC_1 + R^2C_2 + \cdots).$$

We then have

$$0 = \pi_0(C_0 + RC_1) = \pi_0\left[\begin{pmatrix} -\lambda & 0 \\ 0 & -\lambda \end{pmatrix} + R\begin{pmatrix} (1-a_1)\mu_1 & 0 \\ \mu_2 & 0 \end{pmatrix}\right],$$

and using the values given above (including the exact value of R), we find

$$0 = \pi_0 \left[\left(\begin{array}{cc} -1 & 0 \\ 0 & -1 \end{array} \right) + \left(\begin{array}{cc} .61538462 & .23076923 \\ .15384615 & .30769231 \end{array} \right) \left(\begin{array}{cc} 0.5 & 0 \\ 3 & 0 \end{array} \right) \right] = \pi_0 \left(\begin{array}{cc} 0 & 0 \\ 1 & -1 \end{array} \right).$$

(5.13)

The initial probability vector π_0 may now be computed. It is given by $\pi_0 = (x, 0)$ for any $x \neq 0$. The correct normalization is found from $\pi_0(I - R)^{-1}e = 1$, which in this particular example is

$$(x, 0) \left(\begin{array}{cc} 3 & 1 \\ \frac{2}{3} & \frac{5}{3} \end{array} \right) \left(\begin{array}{c} 1 \\ 1 \end{array} \right) = 1.$$

From this we obtain $x = .25$.

Note that the coefficient matrix in equation (5.13), from which the subvector π_0 is computed, is an infinitesimal generator matrix. Had we gone the stochastic matrix route using equation (5.10), we would have found the coefficient matrix,

$$B_0 + RB_1 = \left(\begin{array}{cc} 1 & 0 \\ .25 & .75 \end{array} \right),$$

to be stochastic.

In many examples (and this is one of them), it is possible to compute π_0 more efficiently. The probability that the system is empty may be found from a simple examination of the system. It is given by $\pi_0 = 1 - \lambda \bar{s}$, where $\bar{s}$ is the mean service time. For the two-phase Coxian service station, $\bar{s} = 1/\mu_1 + a_1/\mu_2$, and thus

$$\pi_0 = 1 - (0.5 + 0.75/3.0) = .25.$$

5.3.5   Alternative Methods for Finding R

Equation (5.9) is not the only iterative approach for computing the matrix R. In fact, another procedure, based directly on equation (5.8), is given by

$$R_0 = 0; \qquad R_{l+1} = \sum_{i=0}^{\infty} R_l^i A_i, \quad \text{for } l = 0, 1, \ldots,$$

and this also may be shown to converge monotonically to R as $l \to \infty$. However, numerical experiments reported by Ramaswami in [125] show that convergence of this scheme can be excruciatingly slow. Unfortunately, the first proposed iterative procedure, equation (5.9), is only marginally better. Ramaswami reports improvements of approximately 20%. This difficulty in computing the matrix R is the major disadvantage of the matrix-geometric approach. It often takes very many iterations to obtain R to an acceptable accuracy.

There is currently some research being conducted into applying Newton and Newton-related methods to find R. Let

$$F(X) = X - \sum_{i=0}^{\infty} X^i A_i.$$

Then, solving $X = \sum_{i=0}^{\infty} X^i A_i$ is the same as solving $F(X) = 0$. The Newton–Kantorovich iterative procedure to compute a solution of $F(X) = 0$ is given by

$$R_{l+1} = R_l - \left(F'(R_l)\right)^{-1} F(R_l); \qquad l = 0, 1, \ldots$$

for some initial approximation R_0. It is shown in [125] that if we set

$$Y_l = - \left(F'(R_l)\right)^{-1} F(R_l),$$

then Y_l is the unique solution to the linear system

$$Y_l = \left[\sum_{i=0}^{\infty} R_l^i A_i - R_l\right] + \sum_{i=1}^{\infty} \sum_{j=0}^{i-1} R_l^j Y_l R_l^{i-1-j} A_i. \tag{5.14}$$

Based on these equations, Ramaswami develops an iterative procedure as follows. Consider equation (5.14) and truncate the second term at $i = 2$. This gives

$$Y_l = \left[\sum_{i=0}^{\infty} R_l^i A_i - R_l\right] + Y_l A_1 + (R_l Y_l + Y_l R_l) A_2.$$

The implicit nature of this equation makes it still rather difficult to solve, so the Y_l terms in the right-hand side are replaced by the approximation Z_l, where

$$Z_l = -F(R_l)(I - A_1)^{-1}.$$

Notice that this approximation has previously been used in developing the iterative scheme of equation (5.9). Indeed, writing

$$
\begin{aligned}
R_{l+1} &= R_l + Z_l = R_l - F(R_l)(I - A_1)^{-1} \\
&= R_l - \left[R_l - \sum_{i=0}^{\infty} R_l^i A_i\right](I - A_1)^{-1} \\
&= R_l - R_l(I - A_1)^{-1} + R_l A_1 (I - A_1)^{-1} + \sum_{i=0,\ i\neq 1}^{\infty} R_l^i A_i (I - A_1)^{-1} \\
&= R_l - R_l(I - A_1)(I - A_1)^{-1} + \sum_{i=0,\ i\neq 1}^{\infty} R_l^i A_i (I - A_1)^{-1} \\
&= \sum_{i=0,\ i\neq 1}^{\infty} R_l^i A_i (I - A_1)^{-1}
\end{aligned}
$$

returns us to our first iterative procedure, equation (5.9), for R.

The complete procedure is therefore as follows:

$$R_0 = 0; \qquad R_{l+1} = R_l + Y_l, \qquad l = 0, 1, \ldots \qquad (5.15)$$

where

$$Y_l = \left[\sum_{i=0}^{\infty} R_l^i A_i - R_l \right] + Z_l A_1 + (R_l Z_l + Z_l R_l) A_2$$

and

$$Z_l = -F(R_l)(I - A_1)^{-1}.$$

Reported experiments ([125]) indicate that approximately a 50% to 70% improvement in the time needed to compute R is achieved with this approach. However, the approximations that have been incorporated into the algorithm mean that it is not a Newton scheme, and the convergence obtained is not quadratic. Some recent work reported by Latouche and Ramaswami [86] appears promising in the development of quadratically convergent sequences for quasi-birth-death processes. It is to these applications that we now turn.

5.3.6   The Quasi-Birth-Death (QBD) Case

Quasi-birth-death (QBD) processes are generalizations of the simple birth-death process exemplified by the $M/M/1$ queue. Their infinitesimal generators are block tridiagonal. QBD processes have been, and continue to be, widely used in many fields of application. Furthermore, they were among the first to be analyzed numerically [41, 173]. Their relatively simple block structure makes them attractive as models on which to test new experimental procedures, and so it is with recursive solution procedures. Attempts to derive quadratically converging methods for the computation of Neuts's R matrix have thus far concentrated on QBD processes. Two approaches are currently under consideration by different research groups. The first falls into the "spectral decomposition" approach of Daigle and Lucantoni [34] and Elwalid, Mitra, and Stern [40]. The second is a logarithmic reduction process recently developed by Latouche and Ramaswami [86]. We shall consider only the second.

Consider a two-dimensional Markov chain with state space $\{(\eta, k), \eta \geq 0, 1 \leq k \leq K\}$ and stochastic transition probability matrix given by

$$P = \begin{pmatrix} B_0 & A_0 & 0 & 0 & 0 & \cdots \\ B_1 & A_1 & A_0 & 0 & 0 & \cdots \\ 0 & A_2 & A_1 & A_0 & 0 & \cdots \\ 0 & 0 & A_2 & A_1 & A_0 & \\ \vdots & \vdots & & \ddots & \ddots & \ddots \end{pmatrix}.$$

Recall that the set of K states with parameter $\eta = l$ is said to constitute *level l*.

The exposition of the work of Latouche and Ramaswami revolves around three matrices, G, R, and U, all of order K. Their elements have the following probabilistic interpretations:

- G_{ij} is the probability that starting from state $(1, i)$ the Markov chain visits level 0 and does so by visiting the state $(0, j)$.

- R_{ij} is the expected number of visits into $(1, j)$, starting from $(0, i)$, until the first return to level 0. This matrix is "Neuts's R matrix."

- U_{ij} is the *taboo probability* that starting from state $(1, i)$ the chain eventually returns to level 1 and does so by visiting state $(1, j)$, under *taboo* of level 0 (i.e., without visiting any state in level 0).

The matrices are characterized as the minimal nonnegative solutions of the non-linear equations

$$
\begin{aligned}
G &= A_2 + A_1 G + A_0 G^2 \\
R &= A_0 + R A_1 + R^2 A_2 \\
U &= A_1 + A_0 (I - U)^{-1} A_2
\end{aligned}
$$

and are related through the equations

$$
\begin{aligned}
G &= (I - U)^{-1} A_2 \\
R &= A_0 (I - U)^{-1} \\
U &= A_1 + A_0 G = A_1 + R A_2
\end{aligned}
$$

It therefore follows that once we have found one of the three, the other two are readily computable. Many numerical algorithms for computing the first are based on successive substitution, some of which we have seen in previous sections. Before quadratically convergent algorithms were developed, the following was considered among the most efficient for QBD processes [86].

Algorithm: QBD — Successive Substitution

1. *Initialize:*

   - Set $i = 1$ and $U = A_1$.
   - Compute $G = (I - U)^{-1} A_2$.

2. While $\|1 - G1\|_\infty < \epsilon$ do

   - Compute $i = i + 1$.
   - Compute $U = A_1 + A_0 G$.
   - Compute $G = (I - U)^{-1} A_2$.

3. Compute $R = A_0 (I - U)^{-1}$.

The convergence criterion at Step 2 is based on the assumption that G is a stochastic matrix. This will be true if the QBD process is recurrent. If the QBD is not recurrent, G is substochastic. Also, it may be verified that the matrix U is always substochastic and that the matrix R has spectral radius less than 1, if the QBD process is positive-recurrent, and equal to 1 otherwise. If the QBD is ergodic and $G(k)$ denotes the matrix computed during the kth iteration of the *while* loop of this algorithm, the sequence $\{G(k),\ k \geq 1\}$ forms a nondecreasing sequence that converges to G.

Latouche and Ramaswami show that the matrices $G(k)$ have a special probabilistic interpretation. Specifically, they show that the ij^{th} element of this matrix is the probability that, starting from state $(1, i)$, the chain eventually visits level 0, and does so by visiting the state $(0, j)$, *under taboo* of levels $k + 1$ and above. Initially the chain is not permitted to move higher than level 1; at each new iteration, it is permitted to visit one higher level. Based on these observations, Latouche and Ramaswami develop a *logarithmic reduction* algorithm, whereby the chain is allowed to proceed up to a multiple of twice the level attained at the previous step. Since the matrix G consists of all the trajectories that the chain may visit before visiting level 0 for the first time, this new approach provides a logarithmic reduction in the number of iterations required to achieve convergence; hence its name. The new algorithm is based on the equation

$$G = \sum_{k=0}^{\infty} \left(\prod_{i=0}^{k-1} B_0^{(i)} \right) B_2^{(k)}, \qquad (5.16)$$

where

$$
\begin{aligned}
B_i^{(0)} &= (I - A_1)^{-1} A_i, \quad \text{for } i = 0, 2 \\
B_i^{(k+1)} &= (I - B_0^{(k)} B_2^{(k)} - B_2^{(k)} B_0^{(k)})^{-1} (B_i^{(k)})^2
\end{aligned}
$$

for $i = 0, 2$ and $k \geq 0$. The interested reader should consult the original paper [86] for the formal development of these equations. The complete algorithm is given by

Algorithm: QBD — Logarithmic Reduction [86]

1. *Initialize:*

    - Set $i = 0$.
    - Compute $B_0 = (I - A_1)^{-1}A_0$.
    - Compute $B_2 = (I - A_1)^{-1}A_2$.
    - Set $S = B_2$ and $T = B_0$.

2. While $\|1 - S1\|_\infty < \epsilon$ do

    - Compute $i = i + 1$.
    - Compute $A_1^* = B_0B_2 + B_2B_0$, $A_0^* = B_0^2$, and $A_2^* = B_2^2$.
    - Compute $B_0 = (I - A_1^*)^{-1}A_0^*$ and $B_2 = (I - A_1^*)^{-1}A_2^*$.
    - Compute $S = S + TB_2$ and $T = TB_0$.

3. *Termination:*

    - Set $G = S$.
    - Compute $U = A_1 + A_0G$.
    - Compute $R = A_0(I - U)^{-1}$.

It may be shown that if $L(\epsilon)$ is the number of iterations required to satisfy the convergence criterion (ϵ) for the successive substitution algorithm, then the number of iterations required to compute G to the same accuracy using the logarithmic reduction algorithm is $\lfloor \log_2 L(\epsilon) \rfloor$. Thus, we should expect this algorithm to converge rapidly. In fact, Latouche and Ramaswami observe that to need more than 40 iterations the QBD process must be allowed to move beyond level $l = 2^{40} \approx 10^{12}$ during a first passage from level 1 to level 0. For realistic models the value of ϵ would have to be unreasonably small if this probability were to contribute to G. Finally, numerous examples as well as some theoretical results indicate that the logarithmic reduction algorithm is numerically stable [86].

Since the $\lambda/C_2/1$ example of Section 5.3.4 is a QBD process, we may apply the two algorithms described in this section to it. In all, four different algorithms for computing R were programmed in MATLAB and applied to this example. They are denoted as follows:

Table 5.9: Errors, $\|R_{\text{exact}} - R_{\text{iter}}\|_2$, in the $\lambda/C_2/1$ Queue

Iter	MG-Original	MG-Improved	QBD-SSub	QBD-LR
1	0.34312839769451	0.34312839769451	0.18168327834409	0.06923287740729
2	0.22574664252443	0.18947301357165	0.10877581067908	0.01452596126591
3	0.16258700550171	0.12703279214776	0.06923287740729	0.00084529885493
4	0.12254126626620	0.08998901062126	0.04565416814524	0.00000312417043
5	0.09495430686098	0.06572451436462	0.03077579745051	0.00000000004292
6	0.07496252810333	0.04896069923748	0.02104375591178	0.00000000000000
7	0.05997280511964	0.03697318497994	0.01452596126591	0.00000000000000
8	0.04845704417766	0.02819450539239	0.01009122379682	0.00000000000000
9	0.03944901267606	0.02165472972177	0.00704115825305	0.00000000000000
10	0.03230483823269	0.01672104887702	0.00492784817409	0.00000000000000

1. **MG-Original.** This implements the procedure defined by equation (5.9) for matrix-geometric problems.

2. **MG-improved.** This refers to the algorithm of Ramaswami defined by equation (5.15), again for matrix-geometric problems.

3. **QBD-SSub.** This denotes the successive substitution algorithm for a quasi-birth-death process.

4. **QBD-LR.** This is the logarithmic reduction algorithm for quasi-birth-death processes.

Figure 5.3 shows the errors $\|R_{\text{exact}} - R_{\text{iter}}\|_2$ in the approximate solutions, R_{iter}, computed at each iteration for each of these four methods. It turns out that the logarithmic reduction algorithm requires only six iterations to compute R correct to full machine precision! Although it should be remembered that these results apply only to this particular queueing system, it is difficult not to be enthusiastic about the performance of this algorithm of Latouche and Ramaswami. The actual values of these residuals during the first 10 iterations are given in Table 5.9.

5.4   Explicit Solution of Matrix-Geometric Problems

In the previous section we indicated that in certain cases the matrix R could be formulated explicitly. It is to this topic that we now turn. Explicit solutions for a class of single-node queueing models having block tridiagonal infinitesimal generators were independently developed by several authors [19, 92, 110, 126].

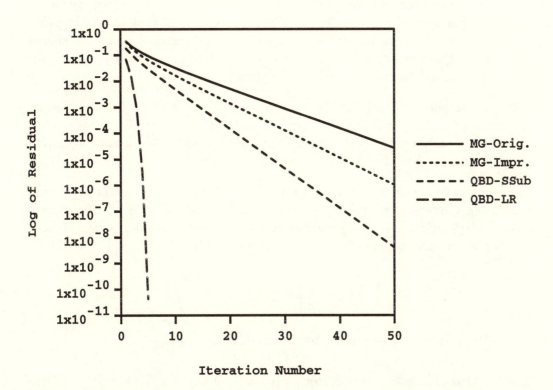

Figure 5.3: Residuals in all four algorithms in computing R for the $\lambda/C_2/1$ queue.

In these applications *aggregate balance equations* can be applied to reduce the global balance equations from a set of second-order difference equations to a set of first-order difference equations. This allows the stationary probability vector to be written explicitly as a function of the station parameters.

We shall see that while some queueing systems have block Hessenberg transition matrices that are amenable to solution either by the explicit technique or by the recursive technique, there are other, common queueing situations to which only one of the approaches is applicable. These situations may be detected by examining specific portions of the infinitesimal generator matrix [148].

The explicit solution approach requires that we develop a transformation procedure that allows a portion of the solution to be computed from a single, previously computed, adjacent portion. This is accomplished in two steps.

1. The first step is to generate the global balance equations, i.e., the Chapman–Kolmogorov equations which relate flow into and out of each state at equilibrium. Most often, when the explicit approach is applicable, these

are second-order difference equations. If the stationary probability vector π is partitioned conformally with the block structure of Q, stating that the global balance equations are second-order difference equations simply means that they specify π_η in terms of $\pi_{\eta-1}$ and $\pi_{\eta+1}$ (for $\eta \geq 1$), i.e., the infinitesimal generator is block tridiagonal.

2. The second step calls for the construction of another set of balance equations, which we shall call the *aggregate balance equations*. These are used to reduce each second-order difference equation (each global balance equation) to a first-order difference equation.

This aggregation permits us to define a $(K \times K)$ coefficient matrix $H(\eta)$ that relates the subvectors π_η and $\pi_{\eta-1}$ by $\pi_{\eta-1} = \pi_\eta H(\eta)$. More precisely, there is a family $H(\eta)$ of real $(K \times K)$ matrices whose elements are given explicitly as a function of the parameters of the station and for which

$$\pi_\eta = \left\{ \prod_{i=\eta+1}^{N} H(i) \right\} \pi_N \quad \text{for } 1 \leq \eta \leq N-1.$$

We shall illustrate this procedure by means of an example.

5.4.1   Queueing Systems for Which R May Be Explicitly Computed

5.4.1.1   Description

The models considered for explicit computation of R may be viewed as extensions of the well-known Coxian queues, which represent a service time distribution by a series of exponential stages arranged in tandem. Unlike the usual approach, whereby customers completing service at a given stage of service either terminate service altogether and leave the system or begin service at the next stage in a fixed sequence, these models allow the stages to be chosen in any order (i.e., both feed-forward and feed-backward among the stages is permitted).

The characteristics of the queueing system are as follows:

- The arrival rate $\lambda(\eta)$ depends on the number of customers in the system, η.

- The server is represented by K exponential phases with rates $\mu_1, \mu_2, \dots, \mu_K$.

- p_i is the probability that a customer initiates its service at phase i; ($\sum_{i=1}^{K} p_i = 1$).

- q_i is the probability that a customer, on completing service at phase i, will exit the queueing system.

- w_{ij} is the probability that a customer, on completing service at phase i, will proceed to phase j. Note that the matrix $W = [w_{ij}]$ is substochastic, since $\sum_{j=1}^{K} w_{ij} = 1 - q_i$ for $i = 1, 2, \ldots, K$, and there is at least one phase i for which $q_i \neq 0$.

We shall use the ordered pair (η, k) to represent the state of the system in which η customers are at the service facility (either waiting to be served or receiving service) and k denotes the current stage of service. Furthermore, we shall let $\pi(\eta, k)$ denote the stationary probability of the state (η, k).

5.4.1.2 The Global Balance Equations

The global balance equations (GBEs) are obtained by equating flow out of and into each state of the system. Let us consider a state (η, k) for $\eta > 1$ and $k = 1, 2, \ldots, K$. We first examine the ways we can get into this state.

1. We can enter state (η, k) if the system is in any of the states $(\eta + 1, j)$ for $j = 1, 2, \ldots, K$ (probability $\pi(\eta + 1, j)$), the customer in service departs from the system (at rate $\mu_j q_j$), and the next customer begins to be served in phase k (probability p_k). The flow from $(\eta + 1, j)$, for all j, to (η, k) is thus

$$p_k \sum_{j=1}^{K} \pi(\eta + 1, j) \mu_j q_j.$$

2. We can enter state (η, k) if the system is in state $(\eta - 1, k)$ and there is an arrival to the system. This is given by $\lambda(\eta - 1)\pi(\eta - 1, k)$.

3. Finally, we can enter state (η, k) if the system is in any state (η, j) for $j = 1, 2, \ldots, K$ and, on completion of the current service phase (j), the customer in service chooses to go next to service phase k (probability w_{jk}). This is given by

$$\sum_{j=1}^{K} \pi(\eta, j) \mu_j w_{jk}.$$

The total flow into state (η, k) is thus given by

$$p_k \sum_{j=1}^{K} \pi(\eta + 1, j) \mu_j q_j + \lambda(\eta - 1)\pi(\eta - 1, k) + \sum_{j=1}^{K} \pi(\eta, j) \mu_j w_{jk}.$$

Flow out of state (η, k) occurs if either a phase completion occurs or an arrival to the system takes place. This is given by

$$\text{Phase completion:} \quad \mu_k \pi(\eta, k)$$
$$\text{Arrival:} \quad \lambda(\eta) \pi(\eta, k)$$

The global balance equations are therefore:

$$(\lambda(\eta) + \mu_k)\pi(\eta, k) = p_k \sum_{j=1}^{K} \pi(\eta+1, j)\mu_j q_j + \lambda(\eta-1)\pi(\eta-1, k) + \sum_{j=1}^{K} \pi(\eta, j)\mu_j w_{jk}$$

$$\text{for} \quad \eta > 1 \text{ and } k = 1, 2, \ldots, K.$$

This equation will also hold for $\eta = 1$ if we define $\pi(0, k) = p_k \sum_{j=1}^{K} \pi(0, j)$.

Obviously the components $\pi(0, k)$ have no meaning by themselves. When there are zero customers in the system, it makes little sense to specify a service phase. The quantity $\sum_{j=1}^{K} \pi(0, j)$ defines the probability that there are zero customers in the system. In this case it may readily be verified that the initial balance equations are given by

$$\lambda(0) \sum_{j=1}^{K} \pi(0, j) = \sum_{j=1}^{K} \mu_j q_j \pi(1, j).$$

We wish to write the global balance equations in matrix form. For $\eta \geq 1$ we have

$$\lambda(\eta)\pi(\eta, 1) + \mu_1 \pi(\eta, 1) =$$
$$\left(\sum_{j=1}^{K} \pi(\eta+1, j)\mu_j q_j \right) p_1 + \lambda(\eta-1)\pi(\eta-1, 1) + \sum_{j=1}^{K} \pi(\eta, j)\mu_j w_{j1}$$
$$\lambda(\eta)\pi(\eta, 2) + \mu_2 \pi(\eta, 2) =$$
$$\left(\sum_{j=1}^{K} \pi(\eta+1, j)\mu_j q_j \right) p_2 + \lambda(\eta-1)\pi(\eta-1, 2) + \sum_{j=1}^{K} \pi(\eta, j)\mu_j w_{j2}$$

$$\vdots \qquad = \qquad \vdots$$

$$\lambda(\eta)\pi(\eta, K) + \mu_K \pi(\eta, K) =$$
$$\left(\sum_{j=1}^{K} \pi(\eta+1, j)\mu_j q_j \right) p_K + \lambda(\eta-1)\pi(\eta-1, K) + \sum_{j=1}^{K} \pi(\eta, j)\mu_j w_{jK}$$

Let

$$
\begin{aligned}
p &= (p_1, p_2, \ldots, p_K), \\
q^T &= (q_1, q_2, \ldots, q_K), \\
M &= \operatorname{diag}\{\mu_1, \mu_2, \ldots, \mu_K\}, \\
W &= [w_{ij}], \quad \text{and} \\
\pi(\eta) &= (\pi(\eta,1),\ \pi(\eta,2),\ldots,\pi(\eta,K)).
\end{aligned}
$$

The K global balance equations just given may then be written as

$$
\lambda(\eta)\pi(\eta) + \pi(\eta)M = (\pi(\eta+1)Mq)p + \lambda(\eta-1)\pi(\eta-1) + \pi(\eta)MW. \quad (5.17)
$$

This holds for $\eta \geq 1$. The initial equations are

$$
\lambda(0)\pi(0) = (\pi(1)Mq)p. \quad (5.18)
$$

5.4.1.3 The Aggregate Balance Equations

To determine the aggregate balance equations (ABEs), each term of equation (5.17) is multiplied on the right by e. We get

$$
\lambda(\eta)\pi(\eta)e + \pi(\eta)Me = (\pi(\eta+1))Mq)pe + \lambda(\eta-1)\pi(\eta-1)e + \pi(\eta)MWe.
$$

Rearranging, we have, since $pe = 1$,

$$
\lambda(\eta)\pi(\eta)e - \pi(\eta+1)Mq = \lambda(\eta-1)\pi(\eta-1)e - \pi(\eta)M(e - We).
$$

Using the fact that

$$
We = e - q
$$

gives

$$
\lambda(\eta)\pi(\eta)e - \pi(\eta+1)Mq = \lambda(\eta-1)\pi(\eta-1)e - \pi(\eta)Mq. \quad (5.19)
$$

Note carefully the relationship on both sides of equation (5.19). Replacing η with $\eta-1$ on the left-hand side makes the left-hand side identical to the right-hand side. Also, from equation (5.18), we have

$$
\lambda(0)\pi(0)e = (\pi(1)Mq)pe = \pi(1)Mq. \quad (5.20)
$$

Substituting $\eta = 1$ into (5.19) gives

$$
\lambda(1)\pi(1)e - \pi(2)Mq = \lambda(0)\pi(0)e - \pi(1)Mq,
$$

which reduces to

$$
\lambda(1)\pi(1)e = \pi(2)Mq
$$

on substituting from equation (5.20). Proceeding in this fashion, we next obtain

$$\lambda(2)\pi(2)e = \pi(3)Mq,$$

etc. The equations

$$\lambda(\eta)\pi(\eta)e = \pi(\eta + 1)Mq$$

are the aggregate balance equations. They may also be obtained by equating the probability of having a departure when there are $(\eta + 1)$ customers in the system to the probability of having an arrival when there are η customers in the system.

5.4.1.4   Reducing the Order of the GBEs

Notice that, whereas the global balance equations are second-order difference equations, i.e., when each equation involves terms in $(\eta - 1)$, η, and $(\eta + 1)$, the aggregate balance equations are first-order difference equations: they involve only terms in η and $(\eta + 1)$. For this example, it is possible to replace the terms in $(\eta + 1)$ in the global balance equations with terms in η, thereby reducing the global balance equations to first-order difference equations. Specifically, we have

GBE: $\lambda(\eta)\pi(\eta) + \pi(\eta)M = (\pi(\eta + 1)Mq)p + \lambda(\eta - 1)\pi(\eta - 1) + \pi(\eta)MW,$
ABE: $\lambda(\eta)\pi(\eta)e = \pi(\eta + 1)Mq,$

so that by multiplying both sides of the ABEs with p and substituting for the first term in the right-hand side of the GBEs we get

$$\lambda(\eta)\pi(\eta) + \pi(\eta)M = \lambda(\eta)\pi(\eta)ep + \lambda(\eta - 1)\pi(\eta - 1) + \pi(\eta)MW,$$

which includes terms in $(\eta - 1)$ and η only. Rearranging, we find

$$\lambda(\eta - 1)\pi(\eta - 1) = \pi(\eta)M - \pi(\eta)MW + \lambda(\eta)\pi(\eta) - \lambda(\eta)\pi(\eta)ep$$

$$= \pi(\eta)[M(I - W) + \lambda(\eta)(I - ep)],$$

or

$$\pi(\eta - 1) = \frac{1}{\lambda(\eta - 1)}\pi(\eta)[M(I - W) + \lambda(\eta)(I - ep)].$$

We may write this as

$$\pi(\eta - 1) = \pi(\eta)H(\eta),$$

where $H(\eta)$ is defined by

$$h_{ii}(\eta) = \frac{1}{\lambda(\eta - 1)}[\mu_i(1 - w_{ii}) + \lambda(\eta)(1 - p_i)];$$

$$h_{ij}(\eta) = -\frac{1}{\lambda(\eta - 1)}(\mu_i w_{ij} + \lambda(\eta)p_i) \quad \text{for } j \neq i.$$

It can be shown [19] that in all cases of practical interest the matrices $H(\eta)$ are nonsingular. We may thus write

$$\pi(\eta) = \pi(\eta - 1)H(\eta)^{-1}.$$

In the particular case when the station parameters are independent of η, $H(\eta) = H$ for all η, and

$$\begin{aligned}
\pi(\eta) &= \pi(\eta - 1)H^{-1}, \\
\pi(\eta) &= \pi(0)(H^{-1})^\eta,
\end{aligned}$$

which has a form similar to the scalar geometric distribution $\pi_i = \rho^i \pi_0$.

The problem of determining $\pi(0)$ remains. The techniques suggested for use in the iterative approach of Neuts can also be used here. Additionally, if the arrival rate is such that there exists an integer N for which $\lambda(\eta) = 0$ for all $\eta \geq N$, it is then possible to compute all the $\pi(i)$ as a function of some arbitrary value assigned to $\pi(0)$ and later to normalize the π's so that a probability vector is obtained.

5.4.1.5 Application to the $\lambda/C_2/1$ Queue

Consider now the example of the Coxian-2 server previously discussed. In this case we have $K = 2$, $\lambda(\eta) = \lambda$ for all η, $q_1 = (1 - a_1)$, $q_2 = 1$, $p_1 = 1$, $p_2 = 0$, $w_{12} = a_1$, and $w_{11} = w_{22} = w_{21} = 0$. We find that the matrix $H(\eta) = H$ is given by

$$H = \begin{pmatrix} \mu_1/\lambda & -1 \\ -a_1\mu_1/\lambda & 1 + \mu_2/\lambda \end{pmatrix}.$$

When we supply the numerical values that were used when examining the method of Neuts (i.e., $\lambda = 1$, $\mu_1 = 2$, $\mu_2 = 3$, and $a_1 = .75$), the matrix H is

$$H = \begin{pmatrix} 2 & -1 \\ -1.5 & 4 \end{pmatrix}.$$

Its inverse is given by

$$H^{-1} = \begin{pmatrix} 0.61538462 & 0.15384615 \\ 0.23076923 & 0.30769231 \end{pmatrix},$$

and it may be quickly verified that

$$\max_{ij}|(H^{-1} - R_{50}^T)_{ij}| < .000024$$

as indicated previously. Furthermore $R_l \to H^{-1}$ as $l \to \infty$. Neuts's R matrix has thus been computed explicitly.

5.4.2   Systems for Which R Must Be Computed Explicitly

We have shown that for a load-independent ($\lambda(\eta) = \lambda$, for all η) Coxian-2 server both the explicit approach and the iterative approach can be used to compute R. It is interesting to examine whether an iterative approach can be found when the arrival process is *load-dependent*.

For a CTMC whose infinitesimal generator Q has the form

$$
Q = \begin{pmatrix}
B_0 & A_0 & & & \\
B_1 & A_1 & A_0 & & \\
& A_2 & A_1 & A_0 & \\
& & A_2 & A_1 & A_0 \\
& & & \ddots & \ddots & \ddots
\end{pmatrix},
$$

the global balance equations are

$$
\pi_0 B_0 + \pi_1 B_1 = 0
$$

$$
\pi_{j-1} A_0 + \pi_j A_1 + \pi_{j+1} A_2 = 0 \qquad \text{for } j \geq 1.
$$

For example, for a Coxian-2 we have

$$
A_0 = \lambda I; \qquad A_1 = \begin{pmatrix} -\mu_1 - \lambda & a_1\mu_1 \\ 0 & -\mu_2 - \lambda \end{pmatrix}; \qquad A_2 = \begin{pmatrix} (1-a_1)\mu_1 & 0 \\ \mu_2 & 0 \end{pmatrix}.
$$

The steady-state probability distribution for this process is matrix-geometric in form, i.e.,

$$
\pi_j = \pi_0 R^j,
$$

so that the global balance equations may be written in terms of π_0 and R as

$$
\pi_0 B_0 + \pi_0 R B_1 = 0,
$$

$$
\pi_0 R^{j-1} A_0 + \pi_0 R^j A_1 + \pi_0 R^{j+1} A_2 = 0, \qquad j \geq 1,
$$

i.e.,

$$
\pi_0 R^{j-1}(A_0 + R A_1 + R^2 A_2) = 0, \qquad j \geq 1.
$$

Neuts has shown that the matrix R must satisfy $A_0 + R A_1 + R^2 A_2 = 0$ and that the matrices R_l, computed from

$$
R_{l+1} = A_0(-A_1)^{-1} + R_l^2 A_2(-A_1)^{-1},
$$

converge to R.

Now consider the case in which the blocks of the infinitesimal generator depend on η because of a load-dependent arrival process. The infinitesimal generator is now

$$
Q = \begin{pmatrix}
B_0 & A_{00} & & & \\
B_1 & A_{11} & A_{10} & & \\
& A_2 & A_{21} & A_{20} & \\
& & A_2 & A_{31} & A_{30} \\
& & & \ddots & \ddots & \ddots
\end{pmatrix}.
$$

For the Coxian-2 example, we have

$$
A_{j0} = \begin{pmatrix} \lambda_j & 0 \\ 0 & \lambda_j \end{pmatrix} \quad \text{and} \quad A_{j1} = \begin{pmatrix} -\mu_1 - \lambda_j & a_1\mu_1 \\ 0 & -\mu_2 - \lambda_j \end{pmatrix}.
$$

The blocks A_2 remain unchanged from the load-independent case. The global balance equations are

$$
\pi_{j-1}A_{j-1,0} + \pi_j A_{j1} + \pi_{j+1}A_2 = 0 \quad \text{for } j \geq 1.
$$

Since we assume that the process under consideration is irreducible and positive-recurrent, the existence of a unique stationary distribution is ensured. Assume that this distribution has the form

$$
\pi_j = \pi_0 \prod_{i=1}^{j} R(i).
$$

We may then write the global balance equations in terms of π_0 and $R(i)$ as

$$
\pi_0 \prod_{i=1}^{j-1} R(i)A_{j-1,0} + \pi_0 \prod_{i=1}^{j} R(i)A_{j1} + \pi_0 \prod_{i=1}^{j+1} R(i)A_2 = 0,
$$

i.e.,

$$
\pi_0 \prod_{i=1}^{j-1} R(i)\{A_{j-1,0} + R(j)A_{j1} + R(j)R(j+1)A_2\} = 0.
$$

These equations are satisfied if there are matrices $R(j)$ that satisfy

$$
A_{j-1,0} + R(j)A_{j1} + R(j)R(j+1)A_2 = 0.
$$

We know that for the generalized Coxian servers such matrices exist, because we can compute them explicitly. However, even in this simple case, it is not evident that these matrices can be computed iteratively. In fact, the singularity of the submatrices B_1 and A_2 would appear to preclude the use of such a solution method.

5.4.3   Systems for Which R Must Be Computed Iteratively

We now turn our attention to instances in which the explicit approach cannot be used. Such is the case for a simple exponential single-server queueing system on which a random process is imposed that affects either the arrival process or the service process. In general, systems of this type cannot be solved by computing R (or H) explicitly; they can be solved only by iteratively computing R. Consider a Markov chain with infinitesimal generator

$$
Q = \begin{pmatrix}
B_0 & A_0 & & & \\
C_1 & B_1 & A_1 & & \\
& C_2 & B_2 & A_2 & \\
& & C_3 & B_3 & A_3 \\
& & & \ddots & \ddots & \ddots
\end{pmatrix}.
$$

The global balance equations are

$$
\pi_{j-1}A_{j-1} + \pi_j B_j + \pi_{j+1}C_{j+1} = 0. \tag{5.21}
$$

The aggregate balance equations are

$$
\pi_j A_j e = \pi_{j+1} C_{j+1} e.
$$

Theorem 5.4 *If under the conditions given above, all of the matrices C_{j+1} are of rank 1, it is possible to find vector z_{j+1} satisfying $C_{j+1}ez_{j+1}^T = C_{j+1}$.*

Proof We use the fact that if A and B are any two matrices, the rank of AB is less than or equal to the rank of A and B. Let C be any matrix of rank 1. Since each row of a rank 1 matrix is some multiple of a specific row vector, we may write

$$
C = uv^T
$$

where v^T is the given row vector and the vector u contains the multipliers. Define $z^T = (v^T/v^T e)$. Then $Cez^T = uv^T e(v^T/v^T e) = uv^T = C$. The vector z^T will always exist unless $v^T e = 0$. Since the matrices C_{j+1} in the problem at hand are not identically zero, and since all the elements of the C_{j+1} are of the same sign, z_{j+1}^T must exist and be nonzero. $\square$

The existence of this vector z gives $\pi_{j+1}C_{j+1}ez_{j+1}^T = \pi_{j+1}C_{j+1}$ and allows us to replace $\pi_{j+1}C_{j+1}$ with $\pi_j A_j ez^T$ in the global balance equations (5.21), thereby reducing them to a set of first-order equations.

On the other hand, if some of the matrices C_{j+1} are of rank 2 or greater, such a substitution is impossible, since in this case a simple rank argument is sufficient to show that it is not possible to find a vector z^T satisfying

$$
C_{j+1}ez^T = C_{j+1}.
$$

The left-hand side has rank 1, the right-hand side has rank > 1.

A reciprocal analysis holds for the case when all of the A_j's have rank equal to 1. In this case it is possible to replace the term $\pi_{j-1}A_{j-1}$ in the global balance equations with terms in π_j.

5.4.4 Example: A Markovian Queue with N Servers Subject to Breakdowns and Repairs

As an example to illustrate the previous discussion, consider a queueing system with K servers that sometimes break down and that require repair by c repair crews. We define a state of the system by the pair (η, k), where η is the number of customers in the queue and k is the number of operating servers. The system parameters are

- $\lambda_k = $ Poisson arrival rate, which depends upon the number, k, of operating servers,

- $\mu = $ the service rate at each server,

- $\theta = $ the failure rate at each server,

- $\sigma = $ the repair rate at each server.

The parameters μ, θ, and σ are all exponentially distributed random variables. Let us first consider the case of one machine and one repaircrew. The global balance equations are

$$\pi_0 \begin{pmatrix} -\sigma - \lambda_0 & \sigma \\ \theta & -\theta - \lambda_1 \end{pmatrix} + \pi_1 \begin{pmatrix} 0 & 0 \\ 0 & \mu \end{pmatrix} = 0 \tag{5.22}$$

and

$$\pi_{j-1}A_0 + \pi_j A_1 + \pi_{j+1}A_2 = 0 \quad \text{for } j \geq 1, \tag{5.23}$$

where

$$A_0 = \begin{pmatrix} \lambda_0 & 0 \\ 0 & \lambda_1 \end{pmatrix}; \quad A_1 = \begin{pmatrix} -\sigma - \lambda_0 & \sigma \\ \theta & -\theta - \lambda_1 - \mu \end{pmatrix}; \quad A_2 = \begin{pmatrix} 0 & 0 \\ 0 & \mu \end{pmatrix}.$$

Notice that we may write equation (5.22) as $\pi_0(A_1 + A_2) + \pi_1 A_2 = 0$. Multiplying equations (5.22) and (5.23) on the right by e and using the fact that $(A_1 + A_2)e = -A_0 e$ gives

$$-\pi_0 A_0 e + \pi_1 A_2 e = 0,$$

and

$$\pi_{j-1}A_0 e + \pi_j A_1 e + \pi_{j+1}A_2 e = 0,$$

which together imply that

$$\pi_1 A_2 e + \pi_1 A_1 e + \pi_2 A_2 e = \pi_1 (A_2 + A_1) e + \pi_2 A_2 e = 0.$$

This results in $\pi_1 A_0 e = \pi_2 A_2 e$, and continuing in this fashion, we obtain the aggregate balance equations,

$$\pi_j A_0 e = \pi_{j+1} A_2 e.$$

Observe that there exists a $z^T = (0, 1)$ such that $A_2 e z^T = A_2$, since

$$\begin{pmatrix} 0 & 0 \\ 0 & \mu \end{pmatrix} \begin{pmatrix} 1 \\ 1 \end{pmatrix} (0, 1) = \begin{pmatrix} 0 & 0 \\ 0 & \mu \end{pmatrix} = A_2.$$

Consequently, by postmultiplying the aggregate balance equations by z^T, we obtain

$$\pi_j \begin{pmatrix} 0 & \lambda_0 \\ 0 & \lambda_1 \end{pmatrix} = \pi_{j+1} \begin{pmatrix} 0 & 0 \\ 0 & \mu \end{pmatrix},$$

which is just what is needed to reduce the global balance equations to an expression involving only π_j and π_{j-1}, i.e.,

$$\pi_{j-1} \begin{pmatrix} \lambda_0 & 0 \\ 0 & \lambda_1 \end{pmatrix} + \pi_j \begin{pmatrix} -\sigma - \lambda_0 & \sigma \\ \theta & -\theta - \lambda_1 - \mu \end{pmatrix} + \pi_j \begin{pmatrix} 0 & \lambda_0 \\ 0 & \lambda_1 \end{pmatrix} = 0.$$

We can now solve for π_j in terms of π_{j-1} as

$$\pi_j = \pi_{j-1} \begin{pmatrix} \lambda_0 & 0 \\ 0 & \lambda_1 \end{pmatrix} \begin{pmatrix} \sigma + \lambda_0 & -\sigma - \lambda_0 \\ -\theta & \theta + \mu \end{pmatrix}^{-1},$$

or, more compactly, $\pi_j = \pi_{j-1} R$, where

$$R = \begin{pmatrix} \lambda_0 (\theta + \mu) / \mu (\lambda_0 + \sigma) & \lambda_0 / \mu \\ \lambda_1 \theta / \mu (\lambda_0 + \sigma) & \lambda_1 / \mu \end{pmatrix}.$$

Now consider the situation in which there are two machines serviced by one repaircrew. The global balance equations in matrix form are

$$\pi_0 \begin{pmatrix} -\sigma - \lambda_0 & \sigma & 0 \\ \theta & -\theta - \sigma - \lambda_1 & \sigma \\ 0 & 2\theta & -2\theta - \lambda_2 \end{pmatrix} + \pi_1 \begin{pmatrix} 0 & 0 & 0 \\ 0 & \mu & 0 \\ 0 & 0 & 2\mu \end{pmatrix} = 0,$$

and for $j \geq 1$:

$$\pi_{j-1} \begin{pmatrix} \lambda_0 & 0 & 0 \\ 0 & \lambda_1 & 0 \\ 0 & 0 & \lambda_2 \end{pmatrix} + \pi_j \begin{pmatrix} \sigma - \lambda_0 & \sigma & 0 \\ \theta & -\theta - \sigma - \lambda_1 - \mu & \sigma \\ 0 & 2\theta & -2\theta - \lambda_2 - 2\mu \end{pmatrix}$$

$$+\pi_{j+1} \begin{pmatrix} 0 & 0 & 0 \\ 0 & \mu & 0 \\ 0 & 0 & 2\mu \end{pmatrix} = 0.$$

Multiplying on the right by e as before, we obtain

$$\pi_0 \begin{pmatrix} \lambda_0 \\ \lambda_1 \\ \lambda_2 \end{pmatrix} = \pi_1 \begin{pmatrix} 0 \\ \mu \\ 2\mu \end{pmatrix},$$

and

$$\pi_{j-1} \begin{pmatrix} \lambda_0 \\ \lambda_1 \\ \lambda_2 \end{pmatrix} + \pi_j \begin{pmatrix} -\lambda_0 \\ -\lambda_1 - \mu \\ -\lambda_2 - 2\mu \end{pmatrix} + \pi_{j+1} \begin{pmatrix} 0 \\ \mu \\ 2\mu \end{pmatrix} = 0,$$

from which we derive the aggregate balance equations,

$$\pi_j \begin{pmatrix} \lambda_0 \\ \lambda_1 \\ \lambda_2 \end{pmatrix} = \pi_{j+1} \begin{pmatrix} 0 \\ \mu \\ 2\mu \end{pmatrix}.$$

However, now we *cannot* find a z such that

$$\begin{pmatrix} 0 \\ \mu \\ 2\mu \end{pmatrix} z^T \stackrel{?}{=} \begin{pmatrix} 0 & 0 & 0 \\ 0 & \mu & 0 \\ 0 & 0 & 2\mu \end{pmatrix} = A_2$$

for A_2 is of rank 2. Therefore, R cannot be explicitly determined and instead must be computed iteratively, as in the approach adopted by Neuts and Lucantoni [111]. The structure of the aggregate balance equations makes it impossible to replace the terms in π_{j+1} in the global balance equation with an expression involving only π_j.

Chapter 6

Decompositional Methods

6.1 NCD Markov Chains

6.1.1 Introduction and Background

A decompositional approach to solving Markov chains is intuitively very attractive, since it appeals to the principle of divide and conquer — if the model is too large or complex to analyze in its entirety, it is divided into subsystems, each of which is analyzed separately, and a global solution is then constructed from the partial solutions. Ideally, the problem is broken into subproblems that can be solved independently, and the global solution is obtained by "pasting" the subproblem solutions together. Some of these methods have been applied to large economic models and to the analysis of computer systems. Additionally, domain decomposition methods, which are related to the methods discussed in this chapter, have long been used in the solution of systems of differential equations that arise in modelling various scientific and engineering processes.

Although it is rare to find Markov chains that can be divided into independent subchains, it is not unusual to have Markov chains in which this condition *almost* holds. In Markov modelling it is frequently the case that the state space of the model can be partitioned into disjoint subsets, with strong interactions among the states of a subset but with weak interactions among the subsets themselves. Such problems are sometimes referred to as *nearly completely decomposable (NCD)*, *nearly uncoupled*, or *nearly separable*. It is apparent that the assumption that the subsystems are independent and can therefore be solved separately does not hold. Consequently an error arises. We hope this error will be small if the assumption is approximately true.

The pioneering work on NCD systems was carried out by Simon and Ando [147] in investigating the dynamic behavior of linear systems as they apply to economic models. The concept was later extended to the performance analysis of computer systems by Courtois [28]. Error bounds and sensitivity issues have been considered by a number of authors [30, 32, 76, 152, 153].

The technique is founded on the idea that it is easy to analyze large systems in which all the states can be clustered into a small number of groups in which:

1. Interactions among the states of a group may be studied as if interactions among the groups do not exist, and

2. Interactions among groups may be studied without reference to the interactions that take place within the groups.

Simon and Ando showed that the dynamic behavior of an NCD system (in which the above conditions are approximated) may be divided into a *short-run* dynamics period and a *long-run* dynamics period. Specifically, they proved the following results:

1. In the short-run dynamics, the strong interactions within each subsystem are dominant and quickly force each subsystem to a local equilibrium almost independently of what is happening in the other subsystems.

2. In the long-run dynamics, the strong interactions within each subsystem maintain approximately the relative equilibrium attained during the short-run dynamics, but now the weak interactions among groups begin to become apparent and the whole system moves toward a global equilibrium. In this global equilibrium the relative values attained by the states at the end of the short-run dynamics period are maintained.

6.1.2 Definitions

Strong interactions among the states of a group and weak interactions among the groups themselves imply that the states of a nearly completely decomposable Markov chain can be ordered so that the stochastic matrix of transition probabilities has the form

$$P = \begin{pmatrix} P_{11} & P_{12} & \cdots & P_{1N} \\ P_{21} & P_{22} & \cdots & P_{2N} \\ \vdots & \vdots & \ddots & \vdots \\ P_{N1} & P_{N2} & \cdots & P_{NN} \end{pmatrix}, \tag{6.1}$$

in which the nonzero elements of the off-diagonal blocks are small compared to those of the diagonal blocks. The subblocks P_{ii} are square and of order n_i, for $i = 1, 2, \ldots, N$ and $n = \sum_{i=1}^{N} n_i$. We shall assume that

$$||P_{ii}|| = O(1), \qquad i = 1, 2, \ldots, N,$$
$$||P_{ij}|| = O(\epsilon), \qquad i \neq j,$$

where $||\cdot||$ denotes the spectral norm of a matrix and ϵ is a small positive number. The stationary probability vector π is given by

$$\pi P = \pi \quad \text{with } \pi e = ||\pi||_1 = 1.$$

As always, stationary probability vectors are taken to be row vectors. This is also the case for subvectors of stationary probability vectors and approximations to these vectors. All other vectors are column vectors.

6.1.3 Block Solutions

Let π be partitioned conformally with P, i.e., $\pi = (\pi_1, \pi_2, \ldots, \pi_N)$, and π_i is a (row) vector of length n_i. Notice that if the off-diagonal blocks are all zero, the matrix P is completely decomposable and we have

$$(\pi_1, \pi_2, \ldots, \pi_N) \begin{pmatrix} P_{11} & 0 & \cdots & 0 & 0 \\ 0 & P_{22} & \cdots & 0 & 0 \\ \vdots & \vdots & \ddots & \vdots & \vdots \\ 0 & 0 & \cdots & P_{N-1N-1} & 0 \\ 0 & 0 & \cdots & 0 & P_{NN} \end{pmatrix} = (\pi_1, \pi_2, \ldots, \pi_N).$$

Each π_i can be found directly from

$$\pi_i P_{ii} = \pi_i.$$

Following the reasoning of Simon and Ando, an initial attempt at determining the solution, in the more general case of nonzero off-diagonal blocks, is to assume that the system is completely decomposable and to compute the stationary probability distribution for each component.

A first problem that arises with this approach is that the P_{ii} are not stochastic but strictly substochastic. A possible solution to this problem is to make each P_{ii} stochastic by adding the probability mass that is to be found in the off-diagonal blocks P_{ij}, for $j = 1, \ldots, N$ and $j \neq i$, into the diagonal block P_{ii} on a row-by-row basis. This off-diagonal probability mass can be accumulated into the diagonal block in several ways. For example, it can be added into the diagonal elements of the diagonal block; it can be added into the reverse diagonal elements of the diagonal block; it can be distributed among the elements of a row of the diagonal block in a random fashion, etc. The way in which it is added into the diagonal block will have an effect on the accuracy of the results obtained. In particular, there exists a distribution of this probability mass that gives a diagonal block that is stochastic and whose stationary probability vector is, to a multiplicative constant, exactly equal to π_i. Unfortunately, it is not known how to determine this distribution without either a knowledge of the stationary probability vector π itself or extensive calculations — possibly in excess of those required to compute the exact solution. We shall return to this idea in the next section, when we discuss the concept of stochastic complements. In the meantime, a simpler solution to the problem of distributing the probability mass is to ignore it; i.e., to work directly with the substochastic matrices P_{ii} themselves. In other words, we may use the normalized eigenvector u_i, where $\|u_i\|_1 = 1$, corresponding to the Perron root (the eigenvalue of P_{ii} closest to 1) of block P_{ii} as the (conditional) probability vector of block i. Thus, the kth element of u_i approximates the probability of being in the kth state of block i, *conditioned* on the system occupying one of the states of block i.

For purposes of illustration, we shall use the 8×8 Courtois matrix already discussed in previous chapters:

$$P = \begin{pmatrix} .85 & .0 & .149 & .0009 & .0 & .00005 & .0 & .00005 \\ .1 & .65 & .249 & .0 & .0009 & .00005 & .0 & .00005 \\ .1 & .8 & .0996 & .0003 & .0 & .0 & .0001 & .0 \\ .0 & .0004 & .0 & .7 & .2995 & .0 & .0001 & .0 \\ .0005 & .0 & .0004 & .399 & .6 & .0001 & .0 & .0 \\ .0 & .00005 & .0 & .0 & .00005 & .6 & .2499 & .15 \\ .00003 & .0 & .00003 & .00004 & .0 & .1 & .8 & .0999 \\ .0 & .00005 & .0 & .0 & .00005 & .1999 & .25 & .55 \end{pmatrix}$$

An examination will reveal that it may be divided into three groups, with $n_1 = 3$, $n_2 = 2$, and $n_3 = 3$. The diagonal blocks, together with their Perron root and corresponding left-hand eigenvectors, are as follows:

$$P_{11} = \begin{pmatrix} .85 & .0 & .149 \\ .1 & .65 & .249 \\ .1 & .8 & .0996 \end{pmatrix}; \quad \lambda_{1_1} = .99911; \quad u_1 = (.40143, .41672, .18185),$$

$$P_{22} = \begin{pmatrix} .7 & .2995 \\ .399 & .6 \end{pmatrix}; \quad \lambda_{2_1} = .99929; \quad u_2 = (.57140, .42860),$$

$$P_{33} = \begin{pmatrix} .6 & .2499 & .15 \\ .1 & .8 & .0999 \\ .1999 & .25 & .55 \end{pmatrix}; \quad \lambda_{3_1} = .9999; \quad u_3 = (.24074, .55563, .20364).$$

To summarize this first part of the procedure, for each block i, $1 \leq i \leq N$, we determine the subvector u_i of length n_i from

$$u_i P_{ii} = \lambda_{i_1} u_i, \qquad u_i e = 1,$$

where λ_{i_1} is the Perron root of P_{ii} and u_i is its corresponding left-hand eigenvector. This provides us with approximate conditional probability vectors for each of the blocks.

6.1.4   The Coupling Matrix

A second problem that now arises is that once we have computed (an approximation to) the stationary probability vector for each block, simply concatenating them together will not give a probability vector. The elements of each subvector sum to 1. We still need to weight each of the probability subvectors by a quantity that is equal to the probability of being in that subblock of states. These weights are given by

$$(||\pi_1||_1, \ ||\pi_2||_1, \ \dots, \ ||\pi_N||_1),$$

since $||\pi_i||_1$ is exactly equal to the probability that the system is in one of the states of block i. The probability distributions u_i computed from the P_{ii} are conditional probabilities in the sense that they express the probability of being in a given state of the subset conditioned on the the system being in *one* of the states of that subset. By multiplying each u_i by $||\pi_i||_1$, we remove that condition. Of course, the vector π is not yet known, so it is not possible to compute the weights $||\pi_i||_1$. However, later we shall see how they may be approximated by using the probability vectors u_i computed from each of the individual P_{ii}.

It is possible to compute the probability of being in a given block of states if we have an $N \times N$ stochastic matrix whose ij^{th} element denotes the probability of a transition from block i to block j. In accordance with the Simon and Ando theory, this stochastic matrix characterizes the interactions among blocks. To construct this matrix we need to shrink each block P_{ij} of P down to a single element. This is accomplished by first replacing each row of each block by the sum of its elements. Mathematically, the operation performed for each block is $P_{ij}e$. The sum of the elements in row k of block P_{ij} gives the probability of leaving state k of block i and entering one of the states of block j. It no longer matters which particular state of block j is the destination state. In our example, this operation gives the 8×3 matrix

$$\begin{pmatrix} .999 & .0009 & .0001 \\ .999 & .0009 & .0001 \\ .9996 & .0003 & .0001 \\ .0004 & .9995 & .0001 \\ .0009 & .999 & .0001 \\ .00005 & .00005 & .9999 \\ .00006 & .00004 & .9999 \\ .00005 & .00005 & .9999 \end{pmatrix}.$$

Thus, for example, the element .9996 in position (3,1) gives the probability that on leaving state 3 we return to one of the states of the first block; the element .0003 in position (3,2) gives the probability that on leaving state 3 we move to one of the states of the second block, etc.

To complete the operation, we need to reduce each column subvector $P_{ij}e$ to a scalar. As we have just noted, the kth element of the vector $P_{ij}e$ is the probability of leaving state k of block i and entering block j. To determine the total probability of leaving (any state of) block i to enter (any state of) block j, we need to sum the elements of $P_{ij}e$ after each of these elements has been weighted by the probability that the system is in the state that element represents, given that the system is in one of the states of block i. These weighting factors may be obtained from the elements of the stationary probability vector. They are the components of $\phi_i \equiv \pi_i/\|\pi_i\|_1$. It therefore follows that in the construction of the $N \times N$ stochastic matrix the block P_{ij} of the original matrix is replaced by $\phi_i P_{ij}e$. For this particular matrix the stationary probability vector π, correct to five decimal places, is given by

$$\pi = (.08928, .09276, .04049, .15853, .11894, .12039, .27780, .10182).$$

This means that for the example under consideration we get

$$\begin{pmatrix} .40122 & .41683 & .18195 & & & & & \\ & & & .57135 & .42865 & & & \\ & & & & & .24077 & .55559 & .20364 \end{pmatrix} \times$$

$$\begin{pmatrix} .999 & .0009 & .0001 \\ .999 & .0009 & .0001 \\ .9996 & .0003 & .0001 \\ .0004 & .9995 & .0001 \\ .0009 & .999 & .0001 \\ .00005 & .00005 & .9999 \\ .00006 & .00004 & .9999 \\ .00005 & .00005 & .9999 \end{pmatrix} = \begin{pmatrix} .99911 & .00079 & .00010 \\ .00061 & .99929 & .00010 \\ .00006 & .00004 & .99990 \end{pmatrix}.$$

Thus, the element .00079 in position $(1,2)$ gives the probability that the system will enter one of the states of block 2 when it leaves one of the states of block 1.

The reduced $(N \times N)$ matrix A whose ij^{th} element is given by

$$a_{ij} = \frac{\pi_i}{\|\pi_i\|_1} P_{ij}e = \phi_i P_{ij}e$$

is often referred to as the *aggregation matrix* or the *coupling matrix*. Notice that it suffices to know each vector ϕ_i correct to a multiplicative constant only.

Theorem 6.1 *If P is an $n \times n$ irreducible stochastic matrix partitioned as*

$$P = \begin{pmatrix} P_{11} & P_{12} & \cdots & P_{1N} \\ P_{21} & P_{22} & \cdots & P_{2N} \\ \vdots & \vdots & \ddots & \vdots \\ P_{N1} & P_{N2} & \cdots & P_{NN} \end{pmatrix}$$

with square diagonal blocks, $(\pi_1, \pi_2, \ldots, \pi_N)P = (\pi_1, \pi_2, \ldots, \pi_N)$, and $\phi_i = \pi_i/ \|\pi_i\|_1$, then the $N \times N$ coupling matrix A whose entries are defined by

$$a_{ij} = \phi_i P_{ij} e$$

is stochastic and irreducible.

Proof Since the elements of P_{ij}, ϕ_i, and e are all real and nonnegative, it follows that the elements of A are all real and nonnegative. To show that A is stochastic, it now suffices to show that $Ae = e$. Summing across the elements of row i of A, we get

$$\sum_{j=1}^{N} a_{ij} = \sum_{j=1}^{N} \phi_i P_{ij} e = \phi_i \left(\sum_{j=1}^{N} P_{ij} e \right) = \phi_i e = 1.$$

That $\left(\sum_{j=1}^{N} P_{ij} e \right) = e$ follows from the fact that P is stochastic.

We now show that A is irreducible. Since $P_{ij} \geq 0, \phi_i > 0$, and $e > 0$, it follows that

$$a_{ij} = 0 \iff P_{ij} = 0.$$

Since P is irreducible, this implies that A must also be irreducible — otherwise, if A could be permuted to a block triangular form, so also could P. $\square$

Since A is a finite irreducible stochastic matrix, it possesses a unique stationary probability vector. Let ξ denote this vector, i.e., $\xi A = \xi$ and $\xi e = 1$. The ith component of ξ is the stationary probability of being in (one of the states of) block i. Notice, however, that

$$(\|\pi_1\|_1, \|\pi_2\|_1, \ldots, \|\pi_N\|_1)A = (\|\pi_1\|_1, \|\pi_2\|_1, \ldots, \|\pi_N\|_1)\times$$

$$\begin{pmatrix} \pi_1/\|\pi_1\|_1 & 0 & \cdots & 0 \\ 0 & \pi_2/\|\pi_2\|_1 & \cdots & 0 \\ \vdots & \vdots & \ddots & \vdots \\ 0 & 0 & \cdots & \pi_N/\|\pi_N\|_1 \end{pmatrix} P \begin{pmatrix} e & 0 & \cdots & 0 \\ 0 & e & \cdots & 0 \\ \vdots & \vdots & \ddots & \vdots \\ 0 & 0 & \cdots & e \end{pmatrix}$$

$$= \pi P \begin{pmatrix} e & 0 & \cdots & 0 \\ 0 & e & \cdots & 0 \\ \vdots & \vdots & \ddots & \vdots \\ 0 & 0 & \cdots & e \end{pmatrix} = \pi \begin{pmatrix} e & 0 & \cdots & 0 \\ 0 & e & \cdots & 0 \\ \vdots & \vdots & \ddots & \vdots \\ 0 & 0 & \cdots & e \end{pmatrix} = (\|\pi_1\|_1, \|\pi_2\|_1, \ldots, \|\pi_N\|_1).$$

Since ξ is the unique left-hand eigenvector of A, it follows that

$$\xi = (\|\pi_1\|_1, \|\pi_2\|_1, \ldots, \|\pi_N\|_1).$$

In an NCD system the coupling matrix will likely be a structured perturbation of the identity matrix. For example, in the Courtois example the (exact) coupling matrix is

$$\begin{pmatrix} .99911 & .00079 & .00010 \\ .00061 & .99929 & .00010 \\ .00006 & .00004 & .99990 \end{pmatrix}.$$

Its eigenvalues are 1.0; .9998; and .9985. The matrix is ill-conditioned, and any numerical solution procedure will experience difficulties. When N is not large, the direct GTH approach can be used effectively. However, when the coupling matrix is too large to permit solution by GTH and we are forced to adopt iterative procedures, the closeness of all the nonunit eigenvalues to 1 will likely result in extremely slow convergence. A possible approach to circumventing this problem is provided by Property 6.6 of stochastic matrices (Section 1.6.2 in Chapter 1). *For any irreducible Markov chain with stochastic transition probability matrix A, let*

$$A(\alpha) = I - \alpha(I - A)$$

where $\alpha \in \Re' \equiv (-\infty, \infty)\backslash\{0\}$. Then 1 is a simple eigenvalue of every $A(\alpha)$, and associated with this unit eigenvalue is a uniquely defined positive left-hand eigenvector of unit 1-norm, which is precisely the stationary probability vector π of A.

The example considered at that time was precisely the Courtois coupling matrix just given. Among the values of α considered was $\alpha_1 = 1/.00089$, which is the largest possible value of α for which $I - \alpha(I - A)$ is stochastic. We find

$$A(\alpha_1) = \begin{pmatrix} .0 & .88764 & .11236 \\ .68539 & .20225 & .11236 \\ .06742 & .04494 & .88764 \end{pmatrix},$$

with eigenvalues 1.0; .77528; and $-.68539$. These eigenvalues are certainly more conveniently distributed for iterative solution methods. As we mentioned in Chapter 1, the left-hand eigenvector corresponding to the unit eigenvalue for all nonzero values of α is

$$\pi = (.22333, .27667, .50000),$$

the stationary probability vector of the coupling matrix.

6.1.5 The NCD Approximation — A Rayleigh–Ritz Refinement Step

Now we see that if we can form the matrix A, we can determine its stationary probability vector and hence obtain the weights with which to multiply the approximate subvectors u_i. Although it looks as if we have come full circle, for to form A we need to know $\phi_i = \pi_i/||\pi_i||_1$ for $i = 1, 2, \ldots, N$, we can use the u_i as approximations, because we need to know these subvectors only to a multiplicative constant. We use

$$\phi_i = \pi_i/||\pi_i||_1 \approx u_i/||u_i||_1$$

and hence obtain an approximation A^* to the coupling matrix A as

$$(A^*)_{ij} = \frac{u_i}{||u_i||_1} P_{ij} e.$$

Consequently, the weights ξ_i can be estimated and an approximate solution to the stationary probability vector obtained. The complete procedure is as follows:

1. Analyze each diagonal block of P to find an approximation to the probability distribution of the states of block i (conditioned on being in block i); i.e., determine u_i from

$$u_i P_{ii} = \lambda_{i_1} u_i, \qquad u_i e = 1,$$

   where λ_{i_1} is the Perron root of P_{ii} and u_i is its corresponding left-hand eigenvector.

2. Form an approximation A^* to the aggregation matrix A, by evaluating

$$A^* = \begin{pmatrix} u_1 & 0 & \ldots & 0 \\ 0 & u_2 & \ldots & 0 \\ \vdots & \vdots & \ddots & \vdots \\ 0 & 0 & \ldots & u_N \end{pmatrix} \begin{pmatrix} P_{11} & P_{12} & \ldots & P_{1N} \\ P_{21} & P_{22} & \ldots & P_{2N} \\ \vdots & \vdots & \ddots & \vdots \\ P_{N1} & P_{N2} & \ldots & P_{NN} \end{pmatrix} \begin{pmatrix} e & 0 & \ldots & 0 \\ 0 & e & \ldots & 0 \\ \vdots & \vdots & \ddots & \vdots \\ 0 & 0 & \ldots & e \end{pmatrix}.$$

3. Determine the stationary probability vector of A^* and denote it ξ^*,

$$\text{i.e., } \xi^* A^* = \xi^*, \qquad \xi^* e = 1.$$

4. Form an approximate stationary probability vector π^* of P from

$$\pi^* = (\xi_1^* u_1, \, \xi_2^* u_2, \, \ldots, \, \xi_N^* u_N).$$

The reader will undoubtedly notice the similarity of this algorithm to the Rayleigh–Ritz procedure introduced in Section 4.1.1 of Chapter 4. Recall that the Rayleigh–Ritz procedure extracts an optimal approximation (in a certain Frobenius norm sense) to an eigenvector from a given subspace. Consequently, the NCD approximation given here is sometimes referred to as a *Rayleigh–Ritz refinement step* [85].

Returning to the example, the approximation A^* to the aggregation matrix A is

$$\begin{pmatrix} .40143 & .41672 & .18185 & & & & & \\ & & & .57140 & .42860 & & & \\ & & & & & .24074 & .55563 & .20364 \end{pmatrix} \times$$

$$\begin{pmatrix} .999 & .0009 & .0001 \\ .999 & .0009 & .0001 \\ .9996 & .0003 & .0001 \\ .0004 & .9995 & .0001 \\ .0009 & .999 & .0001 \\ .00005 & .00005 & .9999 \\ .00006 & .00004 & .9999 \\ .00005 & .00005 & .9999 \end{pmatrix} = \begin{pmatrix} .99911 & .00079 & .00010 \\ .00061 & .99929 & .00010 \\ .00006 & .00004 & .99990 \end{pmatrix}.$$

Note that in this example the *approximate* coupling matrix A^* is, to five decimal places, equal to the *exact* coupling matrix A. The stationary probability vector of A^*, i.e., ξ^*, is given by

$$\xi^* = (.22252, .27748, .50000),$$

and consequently the approximate stationary probability vector π^* of P is

$$\pi^* = (.08932, .09273, .04046, .15855, .11893, .12037, .27781, .10182).$$

Compare this to the exact solution:

$$\pi = (.08928, .09276, .04049, .15853, .11894, .12039, .27780, .10182).$$

6.2   Stochastic Complementation

6.2.1   Definition

In the previous section we mentioned that it is possible to distribute the off-diagonal probability mass of P across the diagonal blocks P_{ii} in such a way that the stationary vector of the resulting stochastic block is exactly the vector π_i up to a multiplicative constant. The block obtained by this exact distribution is called the *stochastic complement* of P_{ii}. Despite the fact that forming this stochastic complement often requires an excessive amount of computation, it is useful to examine the concept for the insight it provides into the dynamic behavior of nearly completely decomposable systems. The term *stochastic complement* was coined by Meyer [99], and we shall use his notation. More details on the results given in this section may be found in his original paper. Later we shall show that stochastic complementation is related to block Gaussian elimination.

Definition 6.1 *For an irreducible stochastic matrix*

$$P = \begin{pmatrix} P_{11} & P_{12} \\ P_{21} & P_{22} \end{pmatrix}$$

in which the diagonal blocks are square, the stochastic complement of P_{11} is defined to be the matrix

$$S_{11} = P_{11} + P_{12}(I - P_{22})^{-1} P_{21}.$$

Similarly, the stochastic complement of P_{22} is defined to be

$$S_{22} = P_{22} + P_{21}(I - P_{11})^{-1} P_{12}.$$

The definition of stochastic complement can be generalized to more than two blocks. For example, the stochastic complement of P_{22} in

$$P = \begin{pmatrix} P_{11} & P_{12} & P_{13} \\ P_{21} & P_{22} & P_{23} \\ P_{31} & P_{32} & P_{33} \end{pmatrix}$$

is given by

$$S_{22} = P_{22} + (P_{21}, \ P_{23}) \begin{pmatrix} I - P_{11} & -P_{13} \\ -P_{31} & I - P_{33} \end{pmatrix}^{-1} \begin{pmatrix} P_{12} \\ P_{32} \end{pmatrix},$$

where we implicitly assume that the diagonal blocks are square. In general, we adopt the following definition.

Definition 6.2 (Stochastic Complement) *Let P be an $n \times n$ irreducible stochastic matrix partitioned as*

$$P = \begin{pmatrix} P_{11} & P_{12} & \cdots & P_{1N} \\ P_{21} & P_{22} & \cdots & P_{2N} \\ \vdots & \vdots & \ddots & \vdots \\ P_{N1} & P_{N2} & \cdots & P_{NN} \end{pmatrix} \tag{6.2}$$

in which all diagonal blocks are square. For a given index i, let P_i denote the principal block submatrix of P obtained by deleting the ith row and i^{th} column of blocks from P, and let P_{i*} and P_{*i} designate

$$P_{i*} = (P_{i1}, \ P_{i2}, \ \ldots, \ P_{i,i-1}, \ P_{i,i+1}, \ \ldots, \ P_{iN})$$

and

$$P_{*i} = \begin{pmatrix} P_{1i} \\ \vdots \\ P_{i-1,i} \\ P_{i+1,i} \\ \vdots \\ P_{Ni} \end{pmatrix}$$

That is, P_{i*} is the ith row of blocks with P_{ii} removed, and P_{*i} is the ith column of blocks with P_{ii} removed. The stochastic complement of P_{ii} in P is defined to be the matrix

$$S_{ii} = P_{ii} + P_{i*}(I - P_i)^{-1}P_{*i}.$$

6.2.2   Properties of the Stochastic Complement

It has been shown by Meyer that every stochastic complement in P is also an irreducible stochastic matrix. We have the following theorems:

Theorem 6.2 (Meyer) *Let P be an irreducible stochastic matrix partitioned as in (6.2). Each stochastic complement*

$$S_{ii} = P_{ii} + P_{i*}(I - P_i)^{-1}P_{*i}$$

is also a stochastic matrix.

Proof   For clarity of notation, we shall consider only the case when $N = 2$ and show that S_{11} is a stochastic matrix. The complete proof (see [99]) is a straightforward extension based on permuting rows and columns, so that the block P_{ii} is moved to the upper left-hand corner and the resulting matrix repartitioned to yield $N = 2$.

We have

$$P = \begin{pmatrix} P_{11} & P_{12} \\ P_{21} & P_{22} \end{pmatrix}.$$

We first show that S_{11} contains no negative elements. Since $P_{22} \geq 0$ is strictly substochastic, it follows that $(I - P_{22})$ is a nonsingular M-matrix, and as such $(I - P_{22})^{-1} \geq 0$ (Berman and Plemmons [10]). Furthermore, since $P_{11} \geq 0$, $P_{21} \geq 0$, and $P_{12} \geq 0$, it follows that

$$S_{11} = P_{11} + P_{12}(I - P_{22})^{-1}P_{21} \geq 0.$$

It only remains to show that the row sums of S_{11} are equal to 1. From

$$\begin{pmatrix} P_{11} & P_{12} \\ P_{21} & P_{22} \end{pmatrix} \begin{pmatrix} e \\ e \end{pmatrix} = \begin{pmatrix} e \\ e \end{pmatrix},$$

we have

$$P_{11}e + P_{12}e = e \quad \text{and} \quad P_{21}e + P_{22}e = e.$$

From the latter of these equations,

$$e = (I - P_{22})^{-1}P_{21}e, \tag{6.3}$$

and substituting this into $P_{11}e + P_{12}e = e$ gives

$$P_{11}e + P_{12}(I - P_{22})^{-1}P_{21}e = e,$$

i.e.,

$$S_{11}e = e,$$

which completes the proof. □

Theorem 6.3 (Meyer) *Let P be an irreducible stochastic matrix partitioned as in (6.2), and let*

$$\pi = (\pi_1, \ \pi_2, \ \ldots, \ \pi_N)$$

be the partitioned stationary probability vector for P where the sizes of the π_i's correspond to the sizes of the P_{ii}'s, respectively. If each π_i is normalized in order to produce a probability vector ϕ_i, i.e.,

$$\phi_i = \frac{\pi_i}{\pi_i e},$$

then

$$\phi_i S_{ii} = \phi_i$$

for each $i = 1, 2, \ldots, N$. That is, ϕ_i is a stationary probability vector for the stochastic complement S_{ii}.

Proof Once again we shall only consider the case when $N = 2$ and prove the results for S_{11}, since the complete proof reduces to this case. Let

$$P = \begin{pmatrix} P_{11} & P_{12} \\ P_{21} & P_{22} \end{pmatrix}.$$

From $0 = \pi(I - P)$ we have

$$0 = (\pi_1, \ \pi_2)\begin{pmatrix} I - P_{11} & -P_{12} \\ -P_{21} & I - P_{22} \end{pmatrix}\begin{pmatrix} I & 0 \\ (I - P_{22})^{-1}P_{21} & I \end{pmatrix} \tag{6.4}$$

$$= (\pi_1, \ \pi_2)\begin{pmatrix} I - P_{11} - P_{12}(I - P_{22})^{-1}P_{21} & -P_{12} \\ -P_{21} + (I - P_{22})(I - P_{22})^{-1}P_{21} & I - P_{22} \end{pmatrix} \tag{6.5}$$

$$= (\pi_1, \ \pi_2)\begin{pmatrix} I - S_{11} & -P_{12} \\ 0 & I - P_{22} \end{pmatrix}, \tag{6.6}$$

which shows that $\pi_1(I - S_{11}) = 0$. Normalizing π_1 so that $||\pi_1||_1 = 1$ yields the desired result; that is ϕ_1 is the stationary probability vector of S_{11}. □

Theorem 6.4 (Meyer) *Let P be an irreducible stochastic matrix partitioned as in (6.2). Then each stochastic complement*

$$S_{ii} = P_{ii} + P_{i*}(I - P_i)^{-1}P_{*i}$$

is also an irreducible stochastic matrix.

Proof   As before, we shall prove this result for the case in which $N = 2$. We have already considered the stochasticity of S_{11}. We now show that it is irreducible. By noting that

$$\begin{pmatrix} I & P_{12}(I - P_{22})^{-1} \\ 0 & I \end{pmatrix} \begin{pmatrix} I - P_{11} & -P_{12} \\ -P_{21} & I - P_{22} \end{pmatrix} \begin{pmatrix} I & 0 \\ (I - P_{22})^{-1}P_{21} & I \end{pmatrix}$$

$$= \begin{pmatrix} I - S_{11} & 0 \\ 0 & I - P_{22} \end{pmatrix},$$

it follows that

$$\mathrm{rank}(I - P) = \mathrm{rank}(I - S_{11}) + \mathrm{rank}(I - P_{22}).$$

Let P, P_{11}, and P_{22} have dimensions $n \times n$, $r \times r$, and $q \times q$, respectively, with $r + q = n$. Since P is irreducible and stochastic, $\mathrm{rank}(I - P) = n - 1$ and $\mathrm{rank}(I - P_{22}) = q$. Thus

$$\mathrm{rank}(I - S_{11}) = n - 1 - q = r - 1,$$

and $I - S_{11}$ has a one-dimensional null space; the left-hand null space of $I - S_{11}$ is spanned by ϕ_1, which is a strictly positive row vector; the right-hand null space is spanned by e. Since $\phi_1 e = 1$, the spectral projector associated with the unit eigenvalue $\lambda = 1$ for S_{11} must be

$$R_{r \times r} \equiv e\phi_1 \gg 0$$

where, in the notation of Chapter 4, $A \gg B$ if $a_{ij} > b_{ij}$, $\forall\, i, j$. Since every stochastic matrix is Cesáro-summable to the spectral projector associated with the unit eigenvalue, we must have

$$\lim_{n \to \infty} \frac{I + S_{11} + S_{11}^2 + \cdots + S_{11}^{n-1}}{n} = R \gg 0.$$

It is now evident that S_{11} cannot be reducible, for otherwise S_{11} could be permuted to a form

$$\begin{pmatrix} A & B \\ 0 & C \end{pmatrix},$$

and the limit R would necessarily contain zero entries.   □

Note, finally, that if the graph corresponding to the block P_{ii} is acyclic, then the stochastic complement must be a *primitive* (i.e., acyclic as well as irreducible) stochastic matrix. This follows from the fact that all of the elements of $P_{i*}(I - P_i)^{-1}P_{*i}$ are greater than or equal to zero and S_{ii} is formed by adding this matrix to P_{ii}. Adding arcs to a graph cannot increase its periodicity.

6.2.3 Computing Stationary Distributions by Stochastic Complementation

The stochastic complement of P_{ii} is formed by adding the quantity $P_{i*}(I - P_i)^{-1}P_{*i}$ to P_{ii}. Since

$$(I - P_i)^{-1}P_{*i}e = e$$

(equation (6.3)), it follows that forming the stochastic complement corresponds to taking the probability mass in each row of P_{i*} and distributing it over the elements in the corresponding row of P_{ii} according to the weights $(I - P_i)^{-1}P_{*i}$. This procedure is not an approximation, and thus, it follows that this distribution is the one we alluded to previously when we claimed that it is possible to distribute the off-diagonal probability mass to obtain a block whose stationary probability vector is π_i. We hasten to point out that the analysis does not depend in any sense on the matrix being nearly completely decomposable.

When each stochastic complement S_{ii} has been formed, the procedure to be followed to compute the stationary probability vector π follows exactly that described in Section 6.1. We have the following theorem.

Theorem 6.5 *Let P be an $n \times n$ irreducible stochastic matrix partitioned as*

$$P = \begin{pmatrix} P_{11} & P_{12} & \dots & P_{1N} \\ P_{21} & P_{22} & \dots & P_{2N} \\ \vdots & \vdots & \ddots & \vdots \\ P_{N1} & P_{N2} & \dots & P_{NN} \end{pmatrix}$$

with square diagonal blocks and whose stationary probability vector π is written as

$$\pi = (\xi_1\phi_1, \ \xi_2\phi_2, \ \dots, \ \xi_N\phi_N)$$

with $\phi_i e = 1$ for $i = 1, 2, \dots, N$. Then ϕ_i is the unique stationary probability vector for the stochastic complement,

$$S_{ii} = P_{ii} + P_{i*}(I - P_i)^{-1}P_{*i},$$

and

$$\xi = (\xi_1, \ \xi_2, \ \dots, \ \xi_N)$$

is the unique stationary probability vector for the $N \times N$ irreducible stochastic matrix A whose entries are defined by

$$a_{ij} = \phi_i P_{ij} e.$$

Proof From Theorem 6.4, S_{ii} is an irreducible stochastic matrix. It then follows immediately from Theorem 6.3 that ϕ_i is the *unique* stationary probability vector of S_{ii}. That the coupling matrix A is stochastic and irreducible is shown in Theorem 6.1. It has a simple unit eigenvalue. Consider the jth component of the product ξA:

$$(\xi A)_j = \sum_{i=1}^{N} \xi_i a_{ij} = \sum_{i=1}^{N} \pi_i P_{ij} e = \left(\sum_{i=1}^{N} \pi_i P_{ij} \right) e = \pi_j e = \xi_j,$$

and thus,

$$\xi A = \xi.$$

It is clear that ξ must be a probability vector, since each element is nonnegative and

$$\sum_{j=1}^{N} \xi_j = \sum_{j=1}^{N} \pi_j e = 1.$$

Therefore, since A is irreducible, ξ must be the unique stationary probability vector for A. $\qquad\qquad\square$

6.2.4 Relationship with Block Gaussian Elimination

It is interesting to note that the stochastic complement approach is related to block Gaussian elimination and, furthermore, can be applied to more general systems of linear equations than those that arise in Markov modelling.

We wish to compute the stationary probability vector π from the system of equations

$$0 = \pi(I - P).$$

We begin by permuting blocks of rows and blocks of columns to move the ith diagonal block, $I - P_{ii}$, to the lower right-hand corner of the matrix. In other words, we exchange block row i with block row $i + 1$, then $i + 2, \ldots, N$, and then perform similar block exchanges with the columns and obtain

$$0 = (\pi_1, \ldots, \pi_{i-1}, \pi_{i+1}, \ldots, \pi_N, \pi_i) \begin{pmatrix} I - P_{11} & \cdots & \cdots & -P_{1N} & -P_{1i} \\ \vdots & \vdots & \vdots & \vdots & \vdots \\ -P_{i-1,1} & \cdots & \cdots & -P_{i-1,N} & -P_{i-1,i} \\ -P_{i+1,1} & \cdots & \cdots & -P_{i+1,N} & -P_{i+1,i} \\ \vdots & \vdots & \vdots & \vdots & \vdots \\ -P_{N1} & \cdots & \cdots & I - P_{NN} & -P_{Ni} \\ \hline -P_{i1} & \cdots & \cdots & -P_{iN} & I - P_{ii} \end{pmatrix}.$$

In the notation of stochastic complements this may be written as

$$0 = (\pi_*, \pi_i) \begin{pmatrix} I - P_i & -P_{*i} \\ -P_{i*} & I - P_{ii} \end{pmatrix}.$$

Postmultiplying by

$$\begin{pmatrix} I & (I - P_i)^{-1} P_{*i} \\ 0 & I \end{pmatrix}$$

gives

$$0 = (\pi_*, \pi_i) \begin{pmatrix} I - P_i & 0 \\ -P_{i*} & I - P_{ii} - P_{i*}(I - P_i)^{-1} P_{*i} \end{pmatrix}. \tag{6.7}$$

This postmultiplication operation corresponds to a reduction step of block Gaussian elimination applied to

$$\begin{pmatrix} (I - P_i)^T & -P_{i*}^T \\ -P_{*i}^T & (I - P_{ii})^T \end{pmatrix},$$

the transpose of the coefficient matrix. The block multiplier is

$$P_{*i}^T (I - P_i)^{-T},$$

and premultiplying each term in the first block row by the multiplier yields

$$\left(P_{*i}^T (I - P_i)^{-T} (I - P_i)^T, \ -P_{*i}^T (I - P_i)^{-T} P_{i*}^T \right).$$

When this is added into the second block row, the second row changes to

$$\left(0, \ (I - P_{ii})^T - P_{*i}^T (I - P_i)^{-T} P_{i*}^T \right).$$

The reduced matrix is thus given by

$$\begin{pmatrix} (I - P_i)^T & -P_{i*}^T \\ 0 & (I - P_{ii})^T - P_{*i}^T (I - P_i)^{-T} P_{i*}^T \end{pmatrix},$$

which is the transposed form of the matrix in equation (6.7).

From (6.7), we see that

$$0 = \pi_i \left(I - \left(P_{ii} + P_{i*}(I - P_i)^{-1} P_{*i} \right) \right).$$

In other words, π_i is the stationary probability vector of the irreducible stochastic matrix $P_{ii} + P_{i*}(I - P_i)^{-1} P_{*i} \ (= S_{ii})$.

It is evident that this elimination decoupling method can be used to solve any linear system of equations so long as the matrices that correspond to $(I - P_i)$ can be inverted. Consider the system

$$\begin{pmatrix} A_{11} & A_{12} & \dots & A_{1N} \\ A_{21} & A_{22} & \dots & A_{2N} \\ \vdots & \vdots & \ddots & \vdots \\ A_{N1} & A_{N2} & \dots & A_{NN} \end{pmatrix} \begin{pmatrix} x_1 \\ x_2 \\ \vdots \\ x_N \end{pmatrix} = \begin{pmatrix} b_1 \\ b_2 \\ \vdots \\ b_N \end{pmatrix}.$$

Performing permutation operations similar to those just described, we obtain

$$\begin{pmatrix} A_i & A_{*i} \\ A_{i*} & A_{ii} \end{pmatrix} \begin{pmatrix} x_* \\ x_i \end{pmatrix} = \begin{pmatrix} b_* \\ b_i \end{pmatrix},$$

where A_i, A_{i*}, and A_{*i} are analogous to the matrices $I - P_i$, P_{i*}, and P_{*i}. A block elimination gives

$$\begin{pmatrix} A_i & A_{*i} \\ 0 & A_{ii} - A_{i*}A_i^{-1}A_{*i} \end{pmatrix} \begin{pmatrix} x_* \\ x_i \end{pmatrix} = \begin{pmatrix} b_* \\ b_i - A_{i*}A_i^{-1}b_* \end{pmatrix}.$$

In other words, we can obtain x_i by solving

$$(A_{ii} - A_{i*}A_i^{-1}A_{*i})x_i = b_i - A_{i*}A_i^{-1}b_*,$$

which, of course, can be carried out concurrently for $i = 1, 2, \ldots, N$. Consequently, any linear system of equations with nonsingular A_i's can be solved concurrently by this method. Furthermore, if the solution is unique, there is no need of further calculations to combine the x_i's to form the final solution x. In the case of stochastic matrices, the system of linear equations is homogeneous and an infinite number of solutions exist, so a final recombination is necessary to form the unique stationary probability vector.

6.2.5   Computational Aspects of Stochastic Complementation

To use the stochastic complement approach, the following steps need to be performed.

1. Partition P into an $N \times N$ block matrix with square diagonal blocks.

2. Form the stochastic complement of each of the diagonal blocks:

$$S_{ii} = P_{ii} + P_{i*}(I - P_i)^{-1}P_{*i}, \qquad i = 1, 2, \ldots, N.$$

3. Compute the stationary probability vector of each stochastic complement:

$$\phi_i S_{ii} = \phi_i; \qquad \phi_i e = 1, \qquad i = 1, 2, \ldots, N.$$

4. Form the coupling matrix A whose ij^{th} element is given by $a_{ij} = \phi_i P_{ij} e$.

5. Compute the stationary probability vector of A: $\xi A = \xi$; $\xi e = 1$.

6. Construct the stationary probability vector of P as

$$\pi = (\xi_1 \phi_1, \ \xi_2 \phi_2, \ \ldots, \ \xi_N \phi_N).$$

The reader will immediately observe the similarity of this algorithm with that described earlier for the case of NCD systems. The stochastic complement S_{ii} plays the role of P_{ii} in the previous algorithm. In fact, S_{ii} is formed by adding a correction term to P_{ii}, and this correction term, $P_{i*}(I - P_i)^{-1}P_{*i}$, will be small when the stochastic matrix P is nearly completely decomposable. Only steps 1 and 2 are not present in the algorithm for the NCD case, and the reason is clear. To take advantage of the fact that the matrix is NCD means that there is little choice in the way in which it must be partitioned. This has to be accomplished in such a way that nonzero elements in the off-diagonal blocks are small compared to those in the diagonal blocks. Step 2 is not included because the correction term is time-consuming to compute and (we hope) small enough not to cause substantial deviation in the computed result.

When implementing the stochastic complement approach, we may choose a partition that is convenient for us. However, trade-offs must be made. As the number of partitions increases, the sizes of the stochastic complements S_{ii} become smaller, and our task in step 3, that of computing the stationary distribution of each S_{ii}, becomes easier. On the other hand, the work involved in forming the stochastic complements themselves, step 2, and that of computing the stationary distribution of the aggregation matrix A, step 5, becomes greater. The opposite effect occurs when the number of partitions is small. Notice that the two extreme cases are $S_{11} = P$, when $N = 1$, and $A = P$, when $N = n$.

In forming a stochastic complement we need to compute $(I - P_i)^{-1}P_{*i}$ for all i, and this is likely to be the most time-consuming portion of the entire algorithm. It is normally constructed by setting

$$X = (I - P_i)^{-1}P_{*i}$$

and then solving

$$(I - P_i)X = P_{*i}$$

for X, rather than by explicitly forming $(I - P_i)^{-1}$. However, a better strategy may be to combine steps 2 and 3 and to work with the stochastic complement in the form $P_{ii} + P_{i*}(I - P_i)^{-1}P_{*i}$ rather than to compute S_{ii} explicitly. Whereas the matrix P is likely to be sparse, the stochastic complements will probably be full and thus create storage difficulties for even moderate-sized Markov chains. An implementation of the power method to find ϕ_i might proceed as follows; any other iterative method could be similarly implemented.

1. Given a kth approximation $\phi_i^{(k)}$ to ϕ_i,

   - Set $b = \phi_i^{(k)} P_{i*}$.
   - Solve $z(I - P_i) = b$ for z.
   - Set $\phi_i^{(k+1)} = z P_{*i}$.
   - Update $\phi_i^{(k+1)} = \phi_i^{(k+1)} + \phi_i^{(k)} P_{ii}$.
   - Normalize and test for convergence.

When solving $z(I - P_i) = b$ for z, we may use either direct or iterative methods. Iterative methods will, of course, preserve the sparsity pattern of the coefficient matrix. This implies that the iterative power method now involves an iterative method within it, but this need not lead to any difficulties. If the nonzero structure of $I - P_i$ is such that some form of LU decomposition can be implemented without excessive fill-in, this may be attractive. The LU decomposition would need to be performed only once, not for each iteration of the power method.

If the stochastic complement is explicitly formed, then an alternative approach is to observe that each S_{ii} is itself a stochastic matrix whose stationary probability vector ϕ_i must be computed. It therefore becomes possible to use the procedure of stochastic complementation itself to compute this vector. This process can be carried on for as long as desirable. One possible sequence of operations is as follows. Starting with an irreducible stochastic matrix $P^{(0)}$ of size $n \times n$, partition $P^{(0)}$ roughly in half as

$$P^{(0)} = \left(\begin{array}{cc} P_{11} & P_{12} \\ P_{21} & P_{22} \end{array} \right)$$

and construct two stochastic complements — say $P_1^{(1)}$ and $P_2^{(1)}$ — that are each irreducible stochastic matrices of order approximately $n/2$. Each of these may in turn be partitioned (roughly in half) to produce four stochastic complements $P_1^{(2)}, P_2^{(2)}, P_3^{(2)}, P_4^{(2)}$, each of which is of order approximately $n/4$. This process can continue until all stochastic complements are so small that their stationary probability vectors are easy to compute. These small stationary probability vectors are then successively coupled until the stationary probability vector π for the original chain is produced. When the Markov chain is partitioned into two chains, steps 4 and 5 simplify considerably, and we find

$$\xi_1 = \frac{\phi_2 P_{21} e}{\phi_1 P_{12} e + \phi_2 P_{21} e} \quad \text{and} \quad \xi_2 = \frac{\phi_1 P_{12} e}{\phi_1 P_{12} e + \phi_2 P_{21} e} = 1 - \xi_1.$$

Using the fact that $Pe = e$, these coupling factors can be expressed using only the diagonal blocks of P by replacing $P_{12} e$ with $e - P_{11} e$ and $P_{21} e$ with $e - P_{22} e$. In this case we have, for example,

$$\xi_1 = \frac{1 - \phi_2 P_{22} e}{2 - \phi_1 P_{11} e - \phi_2 P_{22} e}.$$

An interesting variant of this latter implementation occurs when the coefficient matrix is block tridiagonal and the number of blocks is a power of 2 ([158, 181]). Consider the following case with eight blocks.

$$
Q = \left(
\begin{array}{cccc|cccc}
A_1 & U_1 & & & & & & \\
L_2 & A_2 & U_2 & & & & & \\
& L_3 & A_3 & U_3 & & & & \\
& & L_4 & A_4 & U_4 & & & \\
\hline
& & & L_5 & A_5 & U_5 & & \\
& & & & L_6 & A_6 & U_6 & \\
& & & & & L_7 & A_7 & U_7 \\
& & & & & & L_8 & A_8
\end{array}
\right).
$$

When the state space is partitioned into two groups and the stochastic complement approach applied, it is necessary to compute the inverse of matrices (step 2 of the algorithm) such as

$$
\left(
\begin{array}{cccc}
A_1 & U_1 & & \\
L_2 & A_2 & U_2 & \\
& L_3 & A_3 & U_3 \\
& & L_4 & A_4
\end{array}
\right). \tag{6.8}
$$

However, when the matrix is first permuted to the form

$$
\left(
\begin{array}{cccc|cccc}
A_1 & & & & U_1 & & & \\
& A_3 & & & L_3 & U_3 & & \\
& & A_5 & & & L_5 & U_5 & \\
& & & A_7 & & & L_7 & U_7 \\
\hline
L_2 & U_2 & & & A_2 & & & \\
& L_4 & U_4 & & & A_4 & & \\
& & L_6 & U_6 & & & A_6 & \\
& & & L_8 & & & & A_8
\end{array}
\right)
$$

and the stochastic complement approach applied, instead of finding the inverse of matrices such as (6.8), we require the inverse of matrices like

$$
\left(
\begin{array}{cccc}
A_1 & & & \\
& A_3 & & \\
& & A_5 & \\
& & & A_7
\end{array}
\right)
$$

which is obviously much less expensive to compute, requiring only the inverses of the diagonal blocks. The stochastic complement thus formed is also an irreducible stochastic matrix *and possesses the same block tridiagonal form* as the original matrix. This means that when the original matrix is a power of 2, this process may be carried out until the matrix contains only two blocks. The solutions may now be computed and propagated backwards to yield the global solution.

Other variations, as well as hybrid techniques, are clearly possible. Some variations may be possible without even explicitly forming the complements S_{ii}. However, it is unlikely that stochastic complementation will prove to be a computationally efficient method for computing stationary probability vectors of Markov chains. To explore this point a little further, we shall examine the operation count when the matrix P is full and is partitioned in equal-sized blocks of order s. Therefore, P_i is of order $(n-s)$, P_{i*} has dimensions $s \times (n-s)$, and P_{*i} has dimensions $(n-s) \times s$. The following are the major operations that must be performed, together with the order of magnitude of the number of floating-point multiplications involved.

	Operation	*Count*
Step 2a.	Solve matrix equation:	$(n-s)^3/3 + (n-s)^2 s/2$
Step 2b.	Form S_{ii}:	$(n-s)s^2$
Step 3	Solve for ϕ_i:	$s^3/3$

Each of these operations must be performed for each value of i, and since they are independent operations, they may be executed on different processors. The total number of operations per value of i is given by

$$\frac{1}{3}\left[(n-s)^3 + \frac{3}{2}(n-s)^2 s + 3(n-s)s^2 + s^3\right] = \frac{n^3}{3} - \frac{1}{2}(n-s)^2 s.$$

The first term, $n^3/3$, is the order of the number of operations required to solve the original system of equations on a single processor. We may therefore assume that the stochastic complement approach affords a saving of approximately $(n-s)^2 s/2$ operations if multiple processors are available. Notice that the maximum savings occur when $s = n/3$, which implies that when the matrix is full, it is best to partition the original system into three almost equal parts. In that case the number of operations involved in the remaining steps (4, 5, and 6), will be approximately equal to $n^2 + 3n$, the number taken by step 4, since steps 5 and 6 are negligible for a coupling matrix of order 3.

However, it should be pointed out that the efficiency of this method when implemented on a parallel processor is not high. Normally, with three processors we would like to be able to complete the task in something close to $\frac{1}{3}$ the time of a single processor. Using three processors, it may be shown that the total time is only 0.78 the time required on a single processor, or, rephrasing this, the three processors are as effective as 1.3 processors. These numbers are obtained by observing that when $s = n/3$,

$$\frac{n^3}{3} - \frac{1}{2}(n - \frac{n}{3})^2\frac{n}{3} = \frac{n^3}{3} - \frac{2}{27}n^3,$$

so that instead of taking $9n^3/27$ operations on a single processor, it requires $7n^3/27$ on each of three processors.

Stochastic complementation is therefore unlikely to be computationally efficient, even on parallel processors. Its contribution lies more in the insight it provides into theoretical aspects of nearly completely decomposable systems.

6.3   Iterative Aggregation/Disaggregation Methods

6.3.1   Introduction

We have seen that the stochastic complement approach may be used to compute the stationary probability vector of an arbitrary irreducible stochastic matrix (not necessarily NCD), but that the computational cost may be excessive. In this section we shall consider iterative algorithms, based on the approximate decompositional method described in Section 6.1.5, that rapidly converge onto the exact solution when the Markov chain is nearly completely decomposable.

Recall that after performing the sequence of operations described in Section 6.1.5, the result is an approximation $(\xi_1^* u_1, \xi_2^* u_2, \ldots, \xi_N^* u_N)$ to the stationary probability vector π (Step 4 of the procedure in Section 6.1.5). The question now arises as to whether incorporating this approximation back into a second Rayleigh–Ritz step will give an even better approximation. Notice, however, that since the subvectors u_i are used to compute the aggregation matrix, using $\xi_i^* u_i$ in their place will have no effect on the probability vector we compute. In other words, the Rayleigh–Ritz step is stationary. However, it was found that applying a power step to the approximation before inserting it back into the decomposition method had a very salutary effect. Later this power step was replaced by a block Gauss–Seidel step and became known as a disaggregation step; forming and solving the coupling matrix A was the aggregation step.

There are many algorithms that have been presented in the literature that are labeled under various combinations of the words *iteration, aggregation*, and *disaggregation* [18, 22, 29, 42, 71, 81, 85, 94, 142, 143, 144, 163], etc. They are all intimately related. We shall present the basic approach first and then the approach described by Takahashi [164], since these are the most frequently used. Later we shall discuss implementation considerations and convergence characteristics.

6.3.2   The Basic IAD Algorithm

We begin by describing the algorithm generally referred to as the *KMS (Koury, McAllister, and Stewart)* algorithm. This is perhaps the most straightforward of the iterative aggregation/disaggregation (IAD) algorithms.

6.3.2.1   The KMS Algorithm

In this and the other IAD algorithms described in this chapter, the iteration number is indicated by a superscript in parentheses on the appropriate variable names.

Algorithm: KMS — Iterative Aggregation/Disaggregation

1. Let $\pi^{(0)} = (\pi_1^{(0)}, \pi_2^{(0)}, \ldots, \pi_N^{(0)})$ be a given initial approximation to the solution π, and set $m = 1$.

2. Compute $\phi^{(m-1)}$: For $i = 1, 2, \ldots, N$ compute
$$\phi_i^{(m-1)} = \pi_i^{(m-1)} / \|\pi_i^{(m-1)}\|_1.$$

3. Construct the aggregation matrix $A^{(m-1)}$ whose elements are given by
$$(A^{(m-1)})_{ij} = \phi_i^{(m-1)} P_{ij} e.$$

4. Solve the eigenvector problem
$$\xi^{(m-1)} A^{(m-1)} = \xi^{(m-1)}; \qquad \xi^{(m-1)} e = 1.$$

5. (a) Compute the row vector
$$z^{(m)} = (\xi_1^{(m-1)} \phi_1^{(m-1)}, \; \xi_2^{(m-1)} \phi_2^{(m-1)}, \; \ldots, \; \xi_N^{(m-1)} \phi_N^{(m-1)}).$$

   (b) *Find $\pi^{(m)}$*: For $k = 1, 2, \ldots, N$ solve the system of equations
$$\pi_k^{(m)} = \pi_k^{(m)} P_{kk} + \sum_{j<k} \pi_j^{(m)} P_{jk} + \sum_{j>k} z_j^{(m)} P_{jk}.$$

6. Conduct a test for convergence. If the estimated accuracy is sufficient, then stop and take $\pi^{(m)}$ to be the required solution vector. Otherwise set $m = m + 1$ and go to step 2.

Note that steps 1 through 5(a) are exactly the steps that are performed in the approximation procedure described in Section 6.1.5. If the approximation computed at step 5(a) is now fed back into step 2, exactly the same coupling matrix will be formed and the same approximate solution obtained. The block Gauss-Seidel operation at step 5(b) improves the approximation, forces an improved coupling matrix, and provides a framework that allows the process to converge rapidly to the solution. Under certain circumstances, advantage may be gained from carrying out more than one iteration of step 5(b) before returning to step 2. This is, in some respects, similar to the polynomial preconditioning that is used in conjunction with projection methods. However, in many cases, IAD methods converge so rapidly that solving the blocks more than once is unnecessary. The results of some experiments in this area may be found in [154].

Table 6.1: KMS and BGS Residuals for the Courtois NCD Matrix

Iteration	KMS Residual $1.0e-05\times$	BGS Residual $1.0e-05\times$
1	0.93581293961421	0.94805408435419
2	0.00052482104506	0.01093707688215
3	0.00000000280606	0.00046904081241
4	0.00000000000498	0.00002012500900
5	0.00000000000412	0.00000086349742
6	0.00000000000351	0.00000003705098
7	0.00000000000397	0.00000000158929
8	0.00000000000529	0.00000000006641
9	0.00000000000408	0.00000000000596
10	0.00000000000379	0.00000000000395

6.3.2.2 Application to the Courtois Matrix

As an example of this rapid convergence, let us return to the 8×8 Courtois example. A MATLAB implementation of the KMS algorithm may be found in Section 6.3.7, where a more extensive set of experiments is discussed. For this particular test case, a direct method is used to solve both the coupling matrix (of size 3×3) and the individual blocks (of sizes 3×3, 2×2, and 3×3, respectively). Table 6.1 presents the residuals obtained by the KMS algorithm, as well as the residuals generated by the standard block Gauss–Seidel algorithm specified in Chapter 3. This latter is provided simply for comparative purposes. Note that by iteration 4 the KMS algorithm has achieved the maximum precision available with double-precision MATLAB.

6.3.2.3 A Fixed-Point Theorem

We now wish to show that the stationary probability vector of the irreducible stochastic matrix P is a fixed point of the KMS algorithm. Let $(I - P)$ be split as

$$(I - P) = D - L - U,$$

where D, L, and U are respectively block diagonal, strictly block lower triangular, and strictly block upper triangular matrices, i.e.,

$$D = \text{diag}\{I - P_{11}, \ I - P_{22}, \ \ldots, \ I - P_{NN}\},$$

$$L_{ij} \; = \; P_{ij} \quad \text{if } i > j;$$
$$\; = \; 0 \quad \text{otherwise;}$$

$$U_{ij} \; = \; P_{ij} \quad \text{if } i < j;$$
$$\; = \; 0 \quad \text{otherwise.}$$

Let

$$D^{(m-1)} = \text{diag}\left\{ \frac{\xi_1^{(m-1)}}{||\pi_1^{(m-1)}||_1} I, \; \ldots, \; \frac{\xi_N^{(m-1)}}{||\pi_N^{(m-1)}||_1} I \right\}.$$

Then

$$z^{(m)} = \pi^{(m-1)} D^{(m-1)}.$$

Furthermore, the block Gauss–Seidel iteration method applied in step 5(b) can be written as

$$\pi^{(m)} = z^{(m)} L(D - U)^{-1},$$

so the KMS iterative aggregation/disaggregation algorithm is equivalent to the iterative formula

$$\pi^{(m)} = \pi^{(m-1)} D^{(m-1)} L(D - U)^{-1}. \tag{6.9}$$

We now show that π, the exact stationary probability vector, is a fixed point of (6.9).

Lemma 6.1 *The unique fixed point of (6.9) is the left-hand eigenvector π, corresponding to the unit eigenvalue of the matrix P.*

  Proof   Let $\pi^{(m-1)} = \pi$. Then, from the definitions of A, the exact coupling matrix, and ξ, its stationary probability vector, given in Section 6.1.4, $A^{(m-1)} = A$ and $\xi^{(m-1)} = \xi$. It then follows that $D^{(m-1)} = I$, so

$$\pi^{(m)} = \pi L(D - U)^{-1}. \tag{6.10}$$

But $\pi P = \pi$, or $\pi(D - L - U) = 0$, and consequently

$$\pi L(D - U)^{-1} = \pi. \tag{6.11}$$

Substituting (6.11) into (6.10) yields $\pi^{(m)} = \pi$, and hence π is a fixed point of (6.9).

  To prove uniqueness, let the row vector $v = (v_1, \; v_2, \; \ldots, \; v_N)$ with $||v||_1 = 1$ be a fixed point of (6.9). Then

$$v = v D^{(v)} L(D - U)^{-1},$$

i.e.,

$$v(D - U) = vD^{(v)}L, \tag{6.12}$$

where

$$D^{(v)} = \text{diag} \left\{ \frac{\xi_1^{(v)}}{||v_1||_1} I, \quad \ldots, \quad \frac{\xi_N^{(v)}}{||v_N||_1} I \right\}.$$

From (6.12) we have

$$v_1(I - P_{11}) = \sum_{i=2}^{N} \xi_i^{(v)} \frac{v_i}{||v_i||_1} P_{i1}.$$

Multiplying both sides by e allows us to write

$$||v_1||_1(1 - a_{11}^{(v)}) = ||v_1||_1 \left(\frac{v_1}{||v_1||_1}(I - P_{11})e \right) = \sum_{i=2}^{N} \xi_i^{(v)} \frac{v_i}{||v_i||_1} P_{i1}e. \tag{6.13}$$

Since $\xi^{(v)}$ is the unique solution of $\xi^{(v)}A^{(v)} = \xi^{(v)}$, the term on the right-hand side of (6.13) must be equal to

$$\xi_1^{(v)} \left(1 - \frac{v_1}{||v_1||_1} P_{11}e \right) = \xi_1^{(v)}(1 - a_{11}^{(v)}),$$

and hence,

$$||v_1||_1 = \xi_1^{(v)}.$$

Similarly, it may be shown that $||v_j||_1 = \xi_j^{(v)}$ for $j = 2, 3, \ldots, N$, and hence $D^{(v)} = I$. By repeating the argument on (6.11), it follows that $v = \pi$, and the desired uniqueness property is proved. □

6.3.3 The Takahashi IAD Algorithm

A second aggregation/disaggregation algorithm that is often used is that of Takahashi [164]. This is most often introduced by conditional probability arguments. As has been pointed out previously, if the vector $\phi = (\phi_1, \phi_2, \ldots, \phi_N)$ is known, then the solution may be computed by forming the aggregation matrix A, solving for ξ from $\xi A = \xi$ with $\xi e = 1$, and setting $\pi = (\xi_1\phi_1, \xi_2\phi_2, \ldots, \xi_N\phi_N)$. The method of Takahashi revolves around an observation that enables a particular subvector of ϕ — say, ϕ_k — to be computed if the vector ξ and all the other subvectors ϕ_i, $i \neq k$, are available.

We now turn our attention to the manner in which the subvector ϕ_k is computed. All states that lie outside the kth block are combined into a single "exterior" state, and an augmented k^{th} diagonal block is formed. The block P_{kk} is augmented by a row and a column that capture the interactions of the states of block k with the exterior state. If we denote this augmented block by W_k, we have

$$W_k = \left(\begin{array}{cc} P_{kk} & s_k \\ r_k^T & q_k \end{array} \right), \tag{6.14}$$

where s_k is an $(n_k \times 1)$ vector; r_k^T is a $(1 \times n_k)$ vector, and q_k is a scalar quantity. Since we wish W_k to be a stochastic matrix representing transitions within the states of block k and with its exterior, we must have

$$s_k = e - P_{kk}e. \tag{6.15}$$

The components of the vector r_k denote the probabilities of transitions from the exterior state to the states of block k. Consequently, it follows that

$$r_k^T = \frac{1}{1 - \xi_k} \sum_{j \neq k} \pi_j P_{jk} = \frac{1}{1 - \xi_k} \sum_{j \neq k} \xi_j \phi_j P_{jk}. \tag{6.16}$$

Note that $(1 - \xi_k)$ is the probability that the system is *not* in a state of block k. Note also that the ith component of $\sum_{j \neq k} \pi_j P_{jk}$ is the probability of entering the ith state of block k from outside this block. Thus, the ith component of r_k is the conditional probability of entering the ith state of block k from the exterior state (conditioned on the system being in the exterior state), as required. Finally, and once again because W_k is a stochastic matrix, we have

$$q_k = 1 - r_k^T e. \tag{6.17}$$

The matrix W_k can therefore be computed without a knowledge of ϕ_k. The observation made by Takahashi is that

$$(\pi_k, \ 1 - \xi_k) \left(\begin{array}{cc} P_{kk} & s_k \\ r_k^T & q_k \end{array} \right) = (\pi_k, \ 1 - \xi_k). \tag{6.18}$$

Indeed, multiplying out, we get

$$\pi_k P_{kk} + (1 - \xi_k) r_k^T = \pi_k P_{kk} + (1 - \xi_k) \frac{1}{(1 - \xi_k)} \sum_{j \neq k} \pi_j P_{jk} = \sum_{j=1}^{N} \pi_j P_{jk} = \pi_k,$$

and

$$\begin{aligned} \pi_k s_k + (1 - \xi_k) q_k &= \pi_k (e - P_{kk}e) + (1 - \xi_k)(1 - r_k^T e) \\ &= \pi_k e - \pi_k P_{kk} e - \sum_{j \neq k} \pi_j P_{jk} e + (1 - \xi_k) \\ &= (\pi_k - \sum_{j=1}^{N} \pi_j P_{jk})e + (1 - \xi_k) = (1 - \xi_k). \end{aligned}$$

It now becomes possible to compute ϕ_k. This leads to an iterative algorithm in which, from a given approximation $\phi^{(m-1)}$, first the vector $\xi^{(m-1)}$ is computed, and then a new approximation $\phi^{(m)}$ is computed one subvector at a time. The subvector $\phi_k^{(m)}$ is computed from $\xi^{(m)}$, $\phi_i^{(m)}$ (for values of $i < k$), and $\phi_i^{(m-1)}$ (for values of $i > k$). The following three steps must be executed:

1. Form the aggregation matrix A (whose elements are $a_{ij}^{(m-1)} = \phi_i^{(m-1)} P_{ij} e$), and solve for $\xi^{(m-1)}$.

2. For $k = 1, 2, \ldots, N$ do

   - Form the augmented stochastic matrix W_k given by equation (6.14) using the most recently available components of ϕ and equations (6.15), (6.16), and (6.17).
   - Solve W_k for π_k and ξ_k as indicated by equation (6.18).
   - Set $\phi_k^{(m)} = \frac{1}{\xi}\pi_k$.

3. Test for convergence and return to step 1 if necessary.

In IAD methods, the components ξ_k, for $k = 1, 2, \ldots, N$, of the stationary probability vector of the coupling matrix give the probabilities of being in the different blocks. This was the primary reason for introducing the coupling matrix in the first place. An examination of the Takahashi algorithm, specifically equation (6.18), shows that these components are obtained during step 2 of the algorithm — the result of solving an augmented set of $n_k + 1$ equations at step k. This leads us to speculate that it may be unnecessary to explicitly construct and then solve the coupling matrix (step 1) during each global iteration. After all, we automatically compute the ξ_k's one by one as we solve each of the augmented blocks. Numerical experiments, however, indicate that if step 1 is omitted after the initial iteration, the rate of convergence reduces to approximately that of block Gauss–Seidel and the effect of the aggregation step is lost. We shall provide some mathematical justification to explain this observation when we analyze the convergence of IAD methods in Section 6.4. Briefly, anticipating these results, we show that the approximations may be written as

$$\pi^{(k)} = \pi + X_2 g_2 + X_3 g_3,$$

where X_2 is an $n \times (N-1)$ matrix and X_3 is an $n \times (n-N)$ matrix. In terms of the dynamic behavior of the system as developed by Simon and Ando, the matrix X_2 is related to the *long-run dynamics*, whereas X_3 is related to the *short-run dynamics*. Since we require $\pi^{(k)}$ to converge to π, the effect of IAD algorithms is to successively reduce the components in X_2 and X_3. It is shown in Section 6.4 that the block Gauss–Seidel steps reduce the X_3 component by a factor of

ϵ at each iteration, but effect a lesser reduction on the X_2 component. Several consecutive steps of this part may be beneficial in certain examples [154]. The Rayleigh–Ritz (or coupling step) is effective in reducing the X_2 component, again by a factor of ϵ, but this step is stationary and cannot be applied more than once. Thus, both parts are needed, and the overall improvement per iteration is of order ϵ.

Given, then, that we wish to form the coupling matrix explicitly, the Takahashi algorithm may be set out along lines similar to those of the KMS algorithm. This allows us to focus on the similarities and the differences in both algorithms.

Algorithm: Takahashi

1. Let $\pi^{(0)} = (\pi_1^{(0)}, \pi_2^{(0)}, \ldots, \pi_N^{(0)})$ be a given initial approximation to the solution π, and set $m = 1$.

2. *Compute $\phi^{(m-1)}$*: For $i = 1, 2, \ldots, N$ compute

$$\phi_i^{(m-1)} = \pi_i^{(m-1)}/||\pi_i^{(m-1)}||_1.$$

3. Construct the aggregation matrix $A^{(m-1)}$ whose elements are given by

$$(A^{(m-1)})_{ij} = \phi_i^{(m-1)} P_{ij} e.$$

4. Solve the eigenvector problem

$$\xi^{(m-1)} A^{(m-1)} = \xi^{(m-1)}, \qquad \xi^{(m-1)} e = 1.$$

5. *Find $\phi^{(m)}$*: For $k = 1, 2, \ldots, N$ solve the system of equations

$$z_k^{(m)} = z_k^{(m)} P_{kk} + \sum_{j<k} \xi_j^{(m-1)} \phi_j^{(m)} P_{jk} + \sum_{j>k} \xi_j^{(m-1)} \phi_j^{(m-1)} P_{jk}, \qquad (6.19)$$

$$\phi_k^{(m)} = z_k^{(m)}/||z_k^{(m)}||_1.$$

6. Conduct a test for convergence. If the estimated accuracy is sufficient, then stop and take

$$\pi^{(m)} = \left(\xi_1^{(m-1)} \phi_1^{(m)}, \; \xi_2^{(m-1)} \phi_2^{(m)}, \; \ldots, \; \xi_N^{(m-1)} \phi_N^{(m)} \right)$$

to be the required solution vector. Otherwise set $m = m + 1$ and go to step 3.

Essentially, the only difference between this algorithm and the KMS algorithm is in step 5. Writing the Takahashi algorithm in matrix form allows us to examine this difference more clearly. The equation in step 5 may be written as

$$0 = \phi_k^{(m)}\left(||z_k^{(m)}||_1 - \xi_k^{(m-1)}\right)(I - P_{kk}) + \phi_k^{(m)}\xi_k^{(m-1)}(I - P_{kk})$$

$$-\sum_{j<k} \phi_j^{(m)}\xi_j^{(m-1)}P_{jk} - \sum_{j>k} \phi_j^{(m-1)}\xi_j^{(m-1)}P_{jk}; \qquad k = 1, 2, \ldots, N. \qquad (6.20)$$

Letting

$$\Omega^{(m-1)} = \text{diag}\{\xi_1^{(m-1)}I, \ \xi_2^{(m-1)}I, \ldots, \xi_N^{(m-1)}I\}$$

and

$$\Lambda^{(m-1)} = \text{diag}\{(||z_1^{(m)}||_1 - \xi_1^{(m-1)})(I - P_{11}), \ldots, (||z_N^{(m)}||_1 - \xi_N^{(m-1)})(I - P_{NN})\},$$

the matrix form of (6.20) becomes

$$\phi^{(m)}\Lambda^{(m-1)} + \phi^{(m)}\Omega^{(m-1)}(D - U) - \phi^{(m-1)}\Omega^{(m-1)}L = 0 \qquad (6.21)$$

where $I - P = D - L - U$ as before. Consequently, the method of Takahashi is equivalent to the iterative formula

$$\phi^{(m)} = \phi^{(m-1)}\left[\Omega^{(m-1)}L(D + \Omega^{(m-1)^{-1}}\Lambda^{(m-1)} - U)^{-1}\Omega^{(m-1)^{-1}}\right]. \qquad (6.22)$$

From (6.21) and (6.22), it is evident that this method is equivalent to a "modified" block Gauss–Seidel iteration procedure plus an aggregation step. At each iteration the aggregation step is used to determine row-scaling factors ξ_i, which are incorporated along with a diagonal correction term to the coefficient matrix $(I - P)$. This is followed by a single step of the block Gauss–Seidel method. The analysis clearly shows the difference between the method of Takahashi and the KMS method (compare equations (6.9) and (6.22)), although the algorithmic descriptions are very similar.

For the method of Takahashi, the following fixed-point lemma holds. The method of proof is similar to that of the previous lemma and is omitted.

Lemma 6.2 *The unique fixed point of (6.22) is the vector*

$$\phi = \left(\frac{\pi_1}{||\pi_1||_1}, \frac{\pi_2}{||\pi_2||_1}, \ldots, \frac{\pi_N}{||\pi_N||_1}\right),$$

where $\pi = (\pi_1, \ \pi_2, \ldots, \pi_N)$ *is the left-hand eigenvector of the matrix P corresponding to the unit eigenvalue.*

6.3.4 Other IAD Variants

Takahashi's algorithm is based on the concept of isolating, say, the kth block of n_k states and lumping all the states that remain into a single state. By constructing a stochastic matrix of order $n_k + 1$ that defines the interactions among the n_k states of the block and the exterior, it becomes possible to develop an iterative aggregation/disaggregation algorithm. Using the notation developed in considering stochastic complements, block k is analyzed as the leading principal submatrix of order n_k of the following stochastic matrix of order $n_k + 1$,

$$\begin{pmatrix} P_{kk} & P_{k*}e \\ \frac{\pi_*}{\pi_*e}P_{*e} & \frac{\pi_*}{\pi_*e}P_{*e}e \end{pmatrix}.$$

Other IAD variants (e.g., that of Vantilborgh [168]), exist whereby the states exterior to those of block k are lumped, not into a single state but into as many states as there are nearly decomposable components. Each individual block, other than the one being analyzed, is lumped into a distinct state. Assuming that the transition probability matrix of the Markov chain is structured as in equation (6.1), block k is analyzed as the leading principal submatrix of order n_k of the stochastic matrix of order $n_k + N - 1$ given by

$$\begin{pmatrix} P_{kk} & P_{k*}E_k \\ \Pi_k P_{*k} & \Pi_k P_k E_k \end{pmatrix},$$

where

$$E_k = \mathrm{diag}\{\overbrace{e, e, \ldots, e}^{N-1 \text{ terms}}\},$$

i.e., a block diagonal matrix of dimension $(n - n_k) \times (N - 1)$, whose i^{th} diagonal block is a vector of n_i ones, and

$$\Pi_k = \mathrm{diag}\left\{ \frac{\pi_1}{\|\pi_1\|_1}, \ldots, \frac{\pi_{k-1}}{\|\pi_{k-1}\|_1}, \frac{\pi_{k+1}}{\|\pi_{k+1}\|_1}, \ldots, \frac{\pi_N}{\|\pi_N\|_1} \right\},$$

i.e., a block diagonal matrix with dimensions $(N - 1) \times (n - n_k)$ whose diagonal blocks are row vectors of length n_i.

The differences that we have observed in the IAD methods considered to this point all relate to the manner in which solutions are computed for the individual blocks. They are not the only IAD methods that have been developed, but they capture the essence of all of the methods and are currently the most widely used.

6.3.5 Restructuring an NCD Matrix

When an IAD method is applied to an NCD Markov chain, it is important that the states of the Markov chain be ordered so that the stochastic transition probability matrix has the block structure given by equation (6.1). The procedures that determine a suitable state space ordering and that perform the subsequent symmetric row and column permutations of the transition matrix, to put it into the "normal" form of an NCD matrix, should serve as a preprocessor to an IAD algorithm. Only after such a permutation can we be assured that the resulting transition matrix has the property that directly reflects the structural characteristics of the NCD system.

There is insufficient information currently available about the performance of IAD algorithms to allow us to precisely define the best way in which the Markov chain should be partitioned. One possibility is to provide a user with the information that we can automatically deduce about the system and leave the actual choice to the user. The manner in which this has been incorporated into the MARCA [157] modelling package has proven to be rather successful.

In MARCA a parameter γ, called the decomposability factor, is introduced. This parameter is varied from 10^{-10} to 10^{-1}. For each different value, all elements of the stochastic matrix P (or, more correctly, a copy of P) that are less than or equal to this value are replaced by zero. The resulting matrix is treated as the representation of a directed graph (see Section 7.2, in the next chapter), and a search is initiated for its strongly connected components. The *magnitude* of the nonzero elements that are left in the matrix plays no role in this search. As the value of γ increases, the search for strongly connected components is applied to each of the strongly connected components found at the previous level. Strongly connected components must necessarily be nested as γ increases. For each value of γ, MARCA lists the number of partitions found and some information concerning the number of states in each of the partitions. We shall describe this procedure on the following (artificially constructed) 10×10 irreducible stochastic matrix.

$$
\begin{pmatrix}
\cdot & .5 & \cdot & .5 & \cdot & \cdot & \cdot & \cdot & \cdot & \cdot \\
\cdot & \cdot & .999994 & \cdot & \cdot & \cdot & .000006 & \cdot & \cdot & \cdot \\
.999995 & \cdot & \cdot & \cdot & \cdot & \cdot & \cdot & \cdot & \cdot & .000005 \\
\cdot & \cdot & \cdot & .019999 & .44 & .44 & .000003 & .000002 & .000003 & .000002 \\
.0001 & \cdot & \cdot & .99 & .0099 & \cdot & \cdot & \cdot & \cdot & \cdot \\
\cdot & .0002 & .0002 & .99 & .00959 & \cdot & \cdot & .000005 & .000005 & \cdot \\
\cdot & .000001 & \cdot & \cdot & \cdot & \cdot & .49 & .49 & .001 & .00099 \\
\cdot & .000003 & \cdot & \cdot & \cdot & .000007 & .49999 & .49 & .006 & .004 \\
\cdot & \cdot & \cdot & \cdot & .000001 & \cdot & \cdot & .009999 & \cdot & .99 \\
\cdot & \cdot & .000005 & .000005 & \cdot & \cdot & .00999 & .01 & .98 & \cdot
\end{pmatrix}
$$

As none of the nonzero elements of P is less than 10^{-6}, there is no change in the number of strongly connected components as γ increases from 10^{-10} through 10^{-6}. Since P is irreducible, there is exactly one strongly connected component, and it contains all 10 states. As γ now increases to 10^{-5}, some nonzero elements of P are set to zero. The modified matrix now has the following structure.

$$
\left(
\begin{array}{cccccc|cccc}
\cdot & .5 & \cdot & .5 & \cdot & \cdot & \circ & \cdot & \cdot & \cdot \\
\cdot & \cdot & .999994 & \cdot & \cdot & \cdot & \cdot & \cdot & \cdot & \cdot \\
.999995 & \cdot & \cdot & \cdot & \cdot & \cdot & \cdot & \cdot & \cdot & \cdot \\
\cdot & \cdot & \cdot & .019999 & .44 & .44 & \cdot & \cdot & \cdot & \cdot \\
.0001 & \cdot & \cdot & .99 & .0099 & \cdot & \cdot & \cdot & \cdot & \cdot \\
\circ & .0002 & .0002 & .99 & .00959 & \cdot & \cdot & \cdot & \cdot & \cdot \\ \hline
\circ & \circ & \circ & \cdot & \cdot & \cdot & .49 & .49 & .001 & .00099 \\
\cdot & \cdot & \cdot & \cdot & \cdot & \circ & .49999 & .49 & .006 & .004 \\
\circ & \cdot & \cdot & \cdot & \cdot & \cdot & \cdot & .009999 & \cdot & .99 \\
\cdot & \cdot & \cdot & \cdot & \cdot & \cdot & .00999 & .01 & .98 & \cdot
\end{array}
\right)
$$

There are clearly two strongly connected components; the first consists of states 1 through 6, and the second of states 7 through 10. From this point on, we may consider each group separately.

If γ is increased to 10^{-4}, only the (5,1) element is set to zero, and this does not alter the number or size of the strongly connected components. If we now consider each group and eliminate all elements less than or equal to 10^{-3}, we find the following structure.

$$
\left(
\begin{array}{ccc|ccc}
\cdot & .5 & \cdot & .5 & \cdot & \cdot \\
\cdot & \cdot & .999994 & \cdot & \cdot & \cdot \\
.999995 & \cdot & \cdot & \cdot & \cdot & \cdot \\ \hline
\cdot & \cdot & \cdot & .019999 & .44 & .44 \\
\cdot & \cdot & \cdot & .99 & .0099 & \cdot \\
\cdot & \cdot & \cdot & .99 & .00959 & \cdot
\end{array}
\right)
\;\text{and}\;
\left(
\begin{array}{cccc}
.49 & .49 & \cdot & \cdot \\
.49999 & .49 & .006 & .004 \\
\cdot & .009999 & \cdot & .99 \\
.00999 & .01 & .98 & \cdot
\end{array}
\right)
$$

The changes made in the second group, consisting of states 7 through 10, have no effect on the number of strongly connected components: there is only one, and it contains all the states. If we consider the group consisting of states 1 through 6, we see that although it apparently breaks into two groups, there is only one strongly connected component: the component consisting of states 4, 5, and 6. States 1, 2, and 3 are transient. Therefore, when we apply the algorithm to find the strongly connected components, it will return only a single component, consisting of states 4, 5, and 6. However, we must remember that we are simply looking for a means to order the states of the Markov chain. It suffices to collect any states not included in a strongly connected component into a single group, and then to proceed with the algorithm.

Let us now further increase the size of γ to 10^{-2}. The only effect of eliminating elements less than or equal to 10^{-2} is on the last group of four states, which is separated into two strongly connected components.

$$
\begin{pmatrix}
\cdot & .5 & \cdot & .5 & \cdot & \cdot \\
\cdot & \cdot & .999994 & \cdot & \cdot & \cdot \\
.999995 & \cdot & \cdot & \cdot & \cdot & \cdot \\
\hline
\cdot & \cdot & \cdot & .019999 & .44 & .44 \\
\cdot & \cdot & \cdot & .99 & \cdot & \cdot \\
\cdot & \cdot & \cdot & .99 & \cdot & \cdot
\end{pmatrix}
\quad \text{and} \quad
\begin{pmatrix}
.49 & .49 & \cdot & \cdot \\
.49999 & .49 & \cdot & \cdot \\
\hline
\cdot & \cdot & \cdot & .99 \\
\cdot & \cdot & .98 & \cdot
\end{pmatrix}.
$$

Increasing γ further to 10^{-1} results in no additional changes to the number and size of the partitions. We may summarize the results of this procedure in Table 6.2, which will be useful in helping to decide on an appropriate value of γ to use in determining the NCD structure of the Markov chain. Of course, once a particular value has been chosen, it is then necessary to permute the rows and columns of the matrix so that the normal NCD structure is generated. In this example the states are already in a correct order, but this will not usually be the case. Additionally, the probability matrix normally will be stored in a sparse compact form, which makes the permuting task somewhat more arduous.

Table 6.2: MARCA Partitioning Results

Value of γ	Number of Partitions	Number of States in each Partition			
$\leq 10^{-6}$	1	10			
10^{-5}	2	6,	4		
10^{-4}	2	6,	4		
10^{-3}	3	3,	3,	4	
10^{-2}	4	3,	3,	2,	2
10^{-1}	4	3,	3,	2,	2

To determine the strongly connected components of a directed graph efficiently, all the vertices and edges of the graph must be visited in a systematic fashion. Perhaps the best algorithm to use is that of Tarjan [165]. This is a *depth-first search* (DFS) algorithm, which searches in the forward (deeper) direction as long as possible. Further details, including coding information, may be found in [74, 1]. The complexity of the algorithm is $O(|V| + |E|)$, where $|V|$ is the number of vertices and $|E|$ is the number of edges in the graph. Recall that for the power method the cost of one iteration is given by $|E|$ double-precision multiplications (the number of nonzero elements in the matrix) plus indexing overhead due to the compact storage scheme. Therefore, the cost of the proposed ordering scheme cannot be considered excessive.

One further point should be noted concerning the implementation of recursive algorithms such as the DFS algorithm. Many scientific programmers use Fortran, which, unfortunately, does not currently support recursive procedures. To program a recursive procedure in Fortran, it is necessary to "mechanically" remove the recursion. This can always be done by introducing a *stack*. When the recursive procedure is called, an image of the current status and the returning address must be saved ("pushed") onto a stack before control is returned to the beginning of the program again. After the called procedure has terminated, the information must be "popped" back from the stack and a branch initiated to the return address.

6.3.6   Implementation Considerations

We now turn our attention to some implementation details of IAD algorithms. The two primary concerns must be the choice of suitable data structures and the selection of appropriate computational procedures to implement the different segments of the algorithm. We shall give only brief consideration to the first and concentrate more on the second.

Since the problems we wish to solve will generally be large and the matrix structure sparse, some form of compact storage must be used. Additionally, if we examine the block structure of the matrix, it is likely that many of the blocks will contain only zero elements, so the actual block structure itself may be kept in a compact form: information that indicates the position of blocks that contain at least one nonzero element. It is likely that this will lead to some form of nesting, e.g., compact storage scheme within compact storage scheme. There is no unique way to implement these, and thus we shall leave this aspect with the simple caveat that great care must be exercised.

We now move on to the second aspect: the choice of appropriate computational procedures to embed in the algorithm. We shall consider the specific case of the KMS algorithm. Similar comments may be made concerning the Takahashi and other IAD algorithms.

The critical points of the KMS algorithm (see Section 6.3.2.1) are steps 3, 4, and 5(b).

- Step 3 is the computation of the coupling matrix whose elements are given by

$$(A^{(m-1)})_{ij} = \phi_i^{(m-1)} P_{ij} e.$$

It is computationally more efficient to compute $P_{ij}e$ only once for each block and to store it somewhere for use in all future iterations. However, this is only possible when sufficient memory is available. When this is not the case, it is necessary to compute $P_{ij}e$ each time as and when it is needed.

Since the ϕ_i change at each iteration, the inner product $\phi_i^{(m-1)}(P_{ij}e)$ must also be formed at each iteration.

- Step 4 is the solution of the $N \times N$ problem

$$\xi^{(m-1)}A^{(m-1)} = \xi^{(m-1)}; \qquad \xi^{(m-1)}e = 1$$

where $A^{(m-1)}$ is the coupling matrix. There are many possible ways to compute the stationary probability vector ξ. After all, the coupling matrix is a stochastic matrix, and all of the methods considered in this text may be used. Thus, we may use sparse direct methods; single-vector iterations such as preconditioned power iterations and SOR; projection methods; and so on. However, in NCD systems A will be close to the identity matrix and may well be ill-conditioned, so that perhaps GTH should be the method of choice. A possible alternative, as indicated in Section 6.1.4, is to apply an iterative method to $A(\alpha) = I - \alpha(I - A)$, where α is chosen to separate the subdominant eigenvalues as much as possible from 1.

- Step 5(b) requires the solution of the following N systems of equations:

$$\pi_k^{(m)} = \pi_k^{(m)}P_{kk} + \sum_{j>k} z_j^{(m)}P_{jk} + \sum_{j<k} \pi_j^{(m)}P_{jk}, \qquad k = 1, 2, \ldots, N.$$

These may be written as $Bx = r$, where $B = (I - P_{kk})^T$; $x^T = \pi_k^{(m)}$; and

$$r^T = \sum_{j>k} z_j^{(m)}P_{jk} + \sum_{j<k} \pi_j^{(m)}P_{jk}, \qquad k = 1, 2, \ldots, N. \tag{6.23}$$

In all cases, P_{kk} is strictly substochastic, so B is nonsingular. The vector r will be of small norm if the system is NCD. Again, there are many possibilities for solving these blocks. They include both direct and iterative methods, as cited for the solution of the coupling matrix. However, as the coefficient matrix is nonsingular and the right-hand side is nonzero, methods designated uniquely toward eigenvalue problems, such as the preconditioned power method, are not appropriate.

The reader might notice that the block equations that must be solved in step 5(b) are very similar to the block equations that arise and need to be solved during each global iteration of a regular block iterative method, such as the block Gauss–Seidel algorithm that was analyzed in Section 3.3.1 of Chapter 3. Indeed, it was in this context that step 5(b) was introduced. Naturally, it follows that many of the implementation details recommended at that time are also applicable here. We briefly review some of these.

- It is not necessary that the same method be used for all of the blocks. For example, it might be appropriate to use a direct method for small blocks and an iterative method for large blocks.

- Since the coefficient matrix for each of the N blocks does not change from one iteration to the next, it suffices, when applying a direct method, to perform the LU decomposition only once. The LU factors may be stored and used in all subsequent iterations. The same is true for any incomplete ILU factors that are computed to precondition iterative methods.

- When the blocks are solved using an iterative method, we end up with nested iterative methods: the global or "outer" iterations of the IAD algorithm will have embedded within them, "inner" iterative methods for solving the blocks. In this case it is beneficial to choose the initial starting vector for inner iterations to be the solution computed during the previous global iteration.

- With iterative methods it is possible to ask for only a small number of decimal digits of accuracy for the block solutions during the initial outer iterations and to gradually increase the accuracy demanded as the number of outer iterations increases. An alternative approach is to carry out a fixed and small number (say 30 to 40) of (inner) iterations during each solution. During the initial outer iterations, the computed solution, although not very accurate, will be sufficient. Given that each time an inner system is solved, an increasingly better approximation is used as the initial vector, the computed solution becomes progressively more accurate.

One final point concerning the solution of the blocks needs to be considered. When the matrix P is NCD, the blocks are almost stochastic, and consequently, $(I - P_{kk})$ for $k = 1, 2, \ldots, N$ will be close to singular. This could lead to numerical difficulties. However, it is possible to apply the GTH techniques described in Chapter 2. An additional column may be carried along during the decomposition — a column that is initialized so that the sum of elements in each row is equal to zero. The elementary row operations applied during the decomposition preserve this zero-row-sum property. The additional column is used only for the purposes of helping compute the diagonal elements in a stable manner and has no probabilistic significance. Furthermore, the right-hand side r, given by equation (6.23), may be appended to the block to give an $(n_k + 1) \times (n_k + 1)$ matrix on which the GTH algorithm may be applied exactly as described in Chapter 2. For example, for the kth block, we need to solve

$$x^T(I - P_{kk}) = r^T.$$

This may be written as

$$(x^T, 1) \begin{pmatrix} P_{kk} - I \\ r^T \end{pmatrix} = 0,$$

from which it follows that

$$(x^T, 1) \begin{pmatrix} P_{kk} - I & P_{kk}e \\ r^T & \alpha \end{pmatrix} = (0, 0)$$

where $\alpha = 1 - r^T e$. Transposing, we get

$$\begin{pmatrix} P_{kk}^T - I & r \\ e^T P_{kk}^T & \alpha \end{pmatrix} \begin{pmatrix} x \\ 1 \end{pmatrix} = \begin{pmatrix} 0 \\ 0 \end{pmatrix},$$

and an implementation of GTH as suggested in Chapter 2 may be used to determine an LU decomposition of the coefficient matrix and at the same time update the right-hand side. Once the reduction has been completed, the back-substitution may be initiated using the leading principal $n_k \times n_k$ submatrix and the first n_k elements of column $n_k + 1$ as the updated right-hand side.

6.3.7 MATLAB Experiments

For illustrative purposes, let us perform some numerical experiments in MAT-LAB. We shall choose the NCD Courtois matrix as our example. The following MATLAB program allows us to change between using Gaussian elimination and using Gauss–Seidel iterations to solve the block equations. The direct method is used to solve the coupling matrix. In the input parameter list, P is the *structured* NCD stochastic probability matrix; ni is a vector whose k^{th} element denotes the number of states in the kth block; $itmax1$ specifies the number of outer iterations to perform; and $itmax2$ specifies the number of iterations performed when solving each block iteratively.

The KMS Algorithm

```
function [soln,res] = kms(P,ni,itmax1,itmax2)
[n,n] = size(P); [na,nb] = size(ni);

%%%%%%%    ITERATIVE AGGREGATION/DISAGGREGATION FOR pi*P = pi    %%%%%%%%%%%%%%%%

bl(1) = 1;                              %  Get beginning and end
for k = 1:nb, bl(k+1) = bl(k)+ni(k); end    %  points of each block
```

```
E = zeros(n,nb);                         %  Form (n x nb) matrix E
next = 0;                                %  (This is needed in forming
for i = 1:nb,                            %   the coupling matrix: A )
    for k = 1:ni(i); next = next+1; E(next,i) = 1; end
end
Pije = P*E;                      %  Compute constant part of coupling matrix

Phi = zeros(nb,n);               %  Phi, used in forming the coupling matrix,
for m = 1:nb,                    %  keeps normalized parts of approximation
    for j = 1:ni(m), Phi(m,bl(m)+j-1) = 1/ni(m); end
end

A = Phi*Pije;                            %  Form the coupling matrix A
AA = (A-eye(nb))';    en = [zeros(nb-1,1);1];
xi = inv([AA(1:nb-1,1:nb);ones(1,nb)])*en;   %  Solve the coupling matrix
z = Phi'*xi;                             %  Initial approximation

%%%%%%%%%%%%%%%%%%%%%%%%%%  BEGIN OUTER LOOP  %%%%%%%%%%%%%%%%%%%%%%%%%%%%%%%%%

for iter = 1:itmax1,

    for m = 1:nb,                         %  Solve all diag. blocks; Pmm y = b
        Pmm = P(bl(m):bl(m+1)-1,bl(m):bl(m+1)-1); %  Get coefficient block, Pmm
        b = z(bl(m):bl(m+1)-1)'*Pmm-z'*P(1:n,bl(m):bl(m+1)-1);     %  RHS

        x0 = rows(z,bl(m),bl(m+1)-1);       %  Get new starting vector
        [y,resid] = gs(Pmm'-eye(ni(m)),x0,b',itmax2); %  y is soln for block Pmm
        if m==1, resid; end
%%%     y = inv(Pmm'-eye(ni(m)))*b';  %  Substitute this line for the 2 lines
                                      %  above it, to solve Pmm by direct method

        for j = 1:ni(m), z(bl(m)+j-1) = y(j); end %  Update solution vector
        y = y/norm(y,1);                     %  Normalize y



        for j = 1:ni(m),
            Phi(m,bl(m)+j-1) = y(j);             %  Update Phi
        end
    end

    pi = z;
    res(iter) = norm((P'-eye(n))*pi,2);      %  Compute residual
    A = Phi*Pije; AA = (A-eye(nb))';         %  Form the coupling matrix A
    xi = inv([AA(1:nb-1,1:nb);ones(1,nb)])*en;  %  Solve the coupling matrix
    z  = Phi'*xi;                            %  Compute new approximation

end

soln = pi; res = res';
```

Table 6.3: Residuals in NCD Courtois Problem

Iteration	Residual: $\hat{\pi}(I - P)$
	$1.0e - 03\times$
1	0.14117911369086
2	0.00016634452597
3	0.00000017031189
4	0.00000000015278
5	0.00000000000014
6	0.00000000000007
7	0.00000000000006
8	0.00000000000003
9	0.00000000000003
10	0.00000000000006

Table 6.3 shows the global residuals when Gauss–Seidel is used to solve the three blocks and a direct method is used for the coupling matrix. The convergence is comparable to that observed previously in Section 6.3.2.2, when Gaussian elimination was applied to the blocks.

However, an interesting phenomenon appears when we examine the computed residuals at each of the 20 iterations of Gauss–Seidel. We shall show what happens for block 1. Similar results are found when the residuals for the other two blocks are examined. Table 6.4 shows the Gauss–Seidel residuals obtained during each of the first four global iterations. Observe that in each of the columns of Table 6.4, after about six iterations the rate of convergence becomes extremely slow. There is no benefit in performing iterations beyond this point. What we are observing is that during the inner iterative cycles, convergence is rapid until the residual reduces to approximately γ^k, where k is the outer iteration number, after which point it effectively stagnates. This suggests that we should not request an accuracy of better than γ^k during global iteration k. Iterations performed after this point appear to be wasted.

6.3.8   A Large Experiment

In this section we describe the results obtained when the KMS method is used to compute the stationary probability distribution of a queueing network model. More information on this particular set of experiments may be found in [161].

Table 6.4: Table of Residuals for Block 1, of Size 3×3

Inner	Global Iteration			
Iteration	1	2	3	4
1	0.0131607445	0.000009106488	0.00000002197727	0.00000000002345
2	0.0032775892	0.000002280232	0.00000000554827	0.00000000000593
3	0.0008932908	0.000000605958	0.00000000142318	0.00000000000151
4	0.0002001278	0.000000136332	0.00000000034441	0.00000000000037
5	0.0001468896	0.000000077107	0.00000000010961	0.00000000000011
6	0.0001124823	0.000000051518	0.0000000003470	0.00000000000003
7	0.0001178683	0.000000055123	0.0000000003872	0.00000000000002
8	0.0001156634	0.000000053697	0.0000000003543	0.00000000000002
9	0.0001155802	0.000000053752	0.0000000003596	0.0000000000002
10	0.0001149744	0.000000053446	0.0000000003562	0.00000000000002
11	0.0001145044	0.000000053234	0.0000000003552	0.00000000000002
12	0.0001140028	0.000000052999	0.0000000003535	0.00000000000002
13	0.0001135119	0.000000052772	0.0000000003520	0.00000000000002
14	0.0001130210	0.000000052543	0.0000000003505	0.00000000000002
15	0.0001125327	0.000000052316	0.0000000003490	0.00000000000002
16	0.0001120464	0.000000052090	0.0000000003475	0.00000000000002

6.3.8.1   The Model: An Interactive Computer System

The model illustrated in Figure 6.1 represents the system architecture of a time-shared, paged, virtual memory computer. It is the same model discussed by Stewart [155] and is similar to that of Vantilborgh [168]. This model was previously introduced in Chapter 2 to illustrate the effects of fill-in in direct methods, so the actual numerical values assigned to the model parameters were unimportant. When the numerical values are considered, it turns out that the model is nearly completely decomposable, which explains our interest in it at this point.

The system consists of a set of N terminals from which N users generate commands, a central processing unit (CPU), a secondary memory device (SM), and a filing device (FD). A queue of requests is associated with each device, and the scheduling is assumed to be FCFS (first-come, first-served). When a command is generated, the user at the terminal remains inactive until the system responds. Symbolically, a user, having generated a command, enters the CPU queue. The behavior of the process in the system is characterized by a compute time followed by a page fault, after which the process enters the SM queue, or an input/output (file) request, in which case the process enters the FD queue. Processes that terminate their service at the SM or FD queue return to the CPU

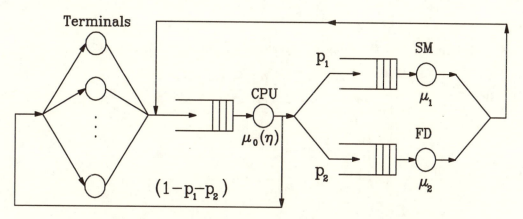

Figure 6.1: Illustration for interactive computer system.

queue. Symbolically, completion of a command is represented by a departure of the process from the CPU to the terminals.

The degree of multiprogramming at any time is given by $\eta = n_0 + n_1 + n_2$, where n_0, n_1 and n_2 are respectively the number of processes in the CPU, SM, and FD queues at that moment. If $(\mu_0(\eta))^{-1}$ is the mean service time at the CPU when the degree of multiprogramming is η, then the probabilities that a process leaving the CPU will direct itself to the SM device or to the FD device are respectively given by $p_1(\eta) = (\mu_0 q(\eta))^{-1}$ and $p_2(\eta) = (\mu_0 r(\eta))^{-1}$, where $q(\eta)$ is the mean compute time between page faults and $r(\eta)$ is the mean compute time between I/O requests. The probability that the process will depart from the CPU queue to the terminals is given by $p_0(\eta) = (\mu_0 c(\eta))^{-1} = 1 - p_1(\eta) - p_2(\eta)$, where $c(\eta)$ is the mean compute time of a process. For a process executing in memory space m, the parameter q is given by $q = \alpha(m)^k$, where α depends on the processing speed as well as on program characteristics, and k depends both on program locality and on the memory management strategy. We shall assume that the total primary memory available is of size M and that it is equally shared among processes currently executing in the system, in which case $q(\eta) = \alpha(M/\eta)^k$.

6.3.8.2 Numerical Results

The specific numerical values assigned to the parameters were as follows. The mean compute time between page faults, $q(\eta)$, was obtained by setting $\alpha = 0.01$, $M = 128$, and $k = 1.5$, so $p_1 \mu_0 = (q(\eta))^{-1} = 100(\eta/128)^{1.5}$. The mean compute time between two I/O requests, $r(\eta)$, was taken as 20 milliseconds, so $p_2 \mu_0 = 0.05$, and the mean compute time of a process, $c(\eta)$, was taken to be 500 milliseconds, giving $p_0 \mu_0 = 0.002$. The mean think time of a user at a terminal was estimated to be of the order of $\lambda^{-1} = 10$ seconds. The mean

Table 6.5: Partitioning in Interactive Computer System Model

	No. of Partitions, ngr	
γ	$n = 1,771$	$n = 23,426$
$\leq 10^{-6}$	1	1
10^{-5}	1	3
10^{-4}	7	51
10^{-3}	21	51
10^{-2}	231	23,426

Table 6.6: Sizes of the 51 Aggregates (50 Users, n=23,426)

1	3	6	10	15	21	28	36	45	55
66	78	91	105	120	136	153	171	190	210
231	253	276	300	325	351	378	406	435	465
496	528	561	595	630	666	703	741	780	820
861	903	946	990	1,035	1,081	1,128	1,176	1,225	1,275
1,326									

service time of the SM was taken as $\mu_1^{-1} = 5$ milliseconds and that of the FD as $\mu_2^{-1} = 30$ milliseconds. The model was solved for 20 users in the system, yielding a stochastic matrix of order 1,771 with 11,011 nonzero elements, and also for the case of 50 users, yielding a matrix of order 23,426 with 156,026 nonzero elements.

The number of groups obtained as a function of γ is shown in Table 6.5. As we have seen before, different values of γ lead to different decompositions, and hence to different numbers and sizes of block systems to be solved. In our examples we use a decomposition factor of $\gamma = 10^{-3}$. This decomposes the state space into 21 aggregates in the first case and into 51 aggregates in the latter. The sizes of the 51 aggregates for the larger case are listed in Table 6.6.

It turns out that the sizes of the 21 aggregates in the smaller example are just equal to the first 21 values of this table. This is not really surprising, for each group corresponds to a different number of busy terminals. For example, there is only one state in which all terminals are busy, three in which all but one are busy, six in which all but two are busy, and so on, and the decomposition is founded on the relatively fast speed of CPU, SM, and FD compared with the speed of interaction at a user terminal.

A direct method was used to solve all systems of size less than or equal to 100. This means that in both the large and the small model, Gaussian elimination was used to solve the coupling matrix. For blocks of size greater than 100, SOR with different values of the relaxation parameter was used. The initial approximation was taken to be e/n, and the stopping criterion required that the residual norm be less than $tol = 10^{-10}$. In other words, the algorithm terminated after the first iteration j for which the approximation $\pi^{(j)}$ satisfied

$$||\pi^{(j)}(I - P)||_2 \leq 10^{-10}.$$

The experiments were performed on a Sun SPARCstation 2 with 16 MB RAM. The results obtained are displayed in Tables 6.7 and 6.8. The times needed to search for and identify the partitions were only 0.08 and 1.19 seconds, respectively, for the small (20-terminal) and the large (50-terminal) examples. It took a little longer to perform the actual permutation of the matrices (stored in compact (a,ja,ia) format), namely 0.19 and 2.62 seconds, respectively. Clearly, this is not an excessive amount of time.

It was possible to use Gaussian elimination to solve *all* the blocks in the smaller example, for in this case the largest block is only of size 231. The KMS algorithm is seen to take 3 iterations, and the total time to compute the solution, 2.07 seconds, is the least of all the results presented. Insufficient memory made it impossible to use Gaussian elimination on all the blocks of the larger example.

Several comments may be made concerning the results obtained when SOR is used to solve blocks whose size is larger than 100. First, different values of the relaxation parameter cause a relatively large difference in number of iterations and total computation time. The relaxation parameter was fixed for all blocks during a particular run of the KMS algorithm; it is possible that this was a good value for some blocks but a poor choice for others. There was no attempt to optimize the choice of relaxation parameter.

Observe also the effect of increasing the value of $itmax2$, the number of iterations performed when solving the blocks by SOR. Results are presented for values of $itmax2 = 20$ and $itmax2 = 30$. Increasing the number of iterations decreases the total number of global iterations needed for convergence. However, the time spent per global iteration increases, so in some cases the overall time decreases and in others it increases.

Methods other than Gaussian elimination and SOR may be applied to the blocks. For example, if we use preconditioned GMRES with a subspace of size 10 and a threshold-based incomplete factorization using a threshold of .005, we find

n	$Iters$	$Time$	$Rnorm$
1,771	3	2.43	$.9840e-10$
23,426	2	28.12	$.7366e-10$

Table 6.7: Small Model: $n = 1,771$; $n_z = 11,011$; $ngr = 21$

	Parameters	itmax2 = 20			itmax2 = 30		
		Iters	Time	Rnorm	Iters	Time	Rnorm
GE		3	2.07	$0.122e-11$			
SOR	$\omega = 1.0$	17	9.31	$0.370e-10$	11	8.34	$0.785e-10$
	$\omega = 1.2$	11	6.14	$0.234e-10$	8	6.01	$0.764e-11$
	$\omega = 1.4$	6	3.49	$0.611e-11$	5	3.61	$0.587e-11$
	$\omega = 1.5$	5	2.80	$0.509e-11$	5	3.35	$0.359e-11$
	$\omega = 1.7$	6	3.16	$0.505e-11$	5	3.77	$0.229e-10$
	$\omega = 1.9$	11	6.01	$0.275e-10$	8	5.74	$0.101e-10$

Key to Tables	
n	Order of P
n_z	Number of nonzero elements in P
ngr	Number of partitions
Iter	Number of global iterations for residual norm $< 10^{-10}$
Time	Total time spent in the KMS algorithm (in CPU seconds)
Rnorm	The 2-norm of the product of the computed solution and $(I - P)$
itmax2	Number of SOR iterations performed when solving blocks
ω	Relaxation parameter for SOR

Table 6.8: Large Model: $n = 23,426$; $n_z = 156,026$; $ngr = 51$

	Parameters	itmax2 = 20			itmax2 = 30		
		Iters	Time	Rnorm	Iters	Time	Rnorm
SOR	$\omega = 1.0$	27	132.52	$0.926e-10$	18	111.36	$0.828e-10$
	$\omega = 1.2$	16	81.26	$0.223e-10$	10	69.01	$0.612e-10$
	$\omega = 1.4$	5	29.72	$0.474e-10$	4	31.72	$0.312e-10$
	$\omega = 1.5$	6	34.55	$0.216e-12$	5	35.65	$0.455e-11$
	$\omega = 1.7$	9	48.59	$0.402e-11$	6	44.30	$0.449e-10$
	$\omega = 1.9$	16	82.02	$0.188e-10$	10	68.06	$0.616e-10$

For key, see Table 6.7.

An extensive set of experiments is reported in [161].

6.3.8.3 Observations and Recommendations

Perhaps the most important point to stress is that for NCD problems, IAD methods are much more effective than methods that do not include an aggregation/disaggregation step. It was found that for NCD matrices an IAD method can be extremely efficient when an appropriate decomposition factor γ and suitable embedded numerical solvers and associated parameters are used. Unfortunately, we cannot give rules for finding the best methods or parameters, because the results are usually sensitive to the structure of the matrices and the nature of the problems. On the other hand, when the matrix is not NCD, there is little to recommend IAD methods. Numerical experiments indicate that the following rules of thumb seem to apply:

1. A large number of groups N (e.g., $N > 1,000$) should be avoided, since it often results in too many nonzero blocks and leads to excessive memory requirements. We should not allow γ to become too large, because we wish the off-diagonal blocks to have elements that are small compared to those of the diagonal blocks.

2. A small value of N (e.g., $N < 20$) should be avoided when it results in large-dimensional block systems that are themselves NCD. These may be difficult to solve, due to the poor separation of the dominant eigenvalues of the blocks.

3. It was found that the heuristic described for finding a decomposition factor γ for NCD systems gave satisfactory results. No experiments were carried out to determine the effect of a partitioning of the state space based on other criteria.

4. Gaussian elimination outperforms other methods when the sizes of the aggregation or disaggregation systems are small (e.g., less than 100). It is most effective when the nonzero elements of the blocks and coupling matrices lie close to the diagonal, since in these cases fill-in will be minimal.

5. For large-size aggregation or disaggregation systems, preconditioned GMRES and SOR (with an appropriately chosen overrelaxation parameter) always gave satisfactory results.

   - For preconditioned GMRES the value of m should not be chosen too large, since memory requirements and computation time per iteration increase rapidly. Neither should it be too small, for the number of iterations required at each step increases significantly. A value of $m = 10$ was often found to be satisfactory.

- For SOR the optimal ω is difficult to find before run time, particularly when each block system may possess its own optimal ω value. An adaptive technique to approximate this value might be useful in this regard.

6. Since the coupling matrix is itself a reduced-order stochastic matrix, it is possible to solve the system using the IAD method recursively. The investigation of a multilevel IAD method may be worthy of study when the queueing model is extremely large and hierarchical.

7. Based on the divide-and-conquer nature of the IAD method, this algorithm is well suited for parallel implementation. Nevertheless, the modular independence, in connection with its many synchronization points, needs careful investigation.

6.4 Convergence Properties and Behavior

6.4.1 Necessary Conditions for a "Regular" NCD Stochastic Matrix

To investigate the convergence properties of IAD methods and to get a better feeling for why they work, it is helpful to impose certain conditions or restrictions. In a certain sense, these conditions define a "regular" or "typical" nearly completely decomposable stochastic matrix. They allow us to investigate how the spectrum of a typical NCD stochastic matrix behaves as its off-diagonal blocks approach zero. Essentially we require that the smallest $(n - N)$ eigenvalues of P be bounded away from 1 as $\epsilon \to 0$ and that the $(N - 1)$ subdominant eigenvalues of P approach 1 no faster than $\epsilon \to 0$.

Condition 6.1 *P is an irreducible primitive stochastic matrix that has the form*

$$P = \begin{pmatrix} P_{11} & P_{12} & \dots & P_{1N} \\ P_{21} & P_{22} & \dots & P_{2N} \\ \vdots & \vdots & \ddots & \vdots \\ P_{N1} & P_{N2} & \dots & P_{NN} \end{pmatrix} \tag{6.24}$$

in which $\|P_{ii}\| = O(1)$, $i = 1, 2, \dots, N$, *and* $\|P_{ij}\| = O(\epsilon)$ *for* $i \neq j$.

As before, $\|.\|$ denotes the spectral norm of a matrix and ϵ is a small positive number.

Condition 6.2 *There is a constant $m_1 > 0$, independent of ϵ, such that $\|\pi_i\|_1 \geq m_1$, for $i = 1, 2, \ldots, N$.*

This condition is related to the asymptotic block irreducibility of P. As $\epsilon \to 0$, we do not want the coupling matrix $(a_{ij} = \phi_i P_{ij} e)$ to become reducible. If that were to happen some of the elements of the left-hand eigenvector, ξ, of A would approach zero. Condition 6.2 prevents this happening, for, as we have already seen,

$$\xi = (\|\pi_1\|_1, \ \|\pi_2\|_1, \ \ldots, \ \|\pi_N\|_1),$$

and thus all the components of ξ are $\geq m_1$. For example, consider the matrix

$$P = \begin{pmatrix} 1 - \epsilon^2 & \epsilon^2 & 0 & 0 \\ \epsilon^2 & 1 - \epsilon - \epsilon^2 & \epsilon & 0 \\ 0 & \epsilon^2 & 1 - 2\epsilon^2 & \epsilon^2 \\ 0 & 0 & \epsilon^2 & 1 - \epsilon^2 \end{pmatrix}.$$

Its stationary probability vector is given by

$$\pi = \frac{1}{2(1 + \epsilon)}(\epsilon, \ \epsilon, \ 1, \ 1),$$

so that when P is partitioned into (2×2) blocks it does not satisfy Condition 6.2.

The third condition relates to the eigenvalues of the diagonal blocks P_{ii}. We shall assume that, corresponding to each diagonal block P_{ii}, there exists only one dominant eigenvalue that is close to unity; the remaining eigenvalues of P_{ii} are assumed to be uniformly bounded away from 1. To describe this assumption more rigorously, let u_i be the left-hand eigenvector of P_{ii} corresponding to its dominant eigenvalue λ_{i_1}; then

$$\lambda_{i_1} = 1 - t_i \epsilon + o(\epsilon), \quad \text{for some } t_i > 0.$$

Furthermore, there exists an $(n_i \times n_i)$ nonsingular matrix

$$T_i = \begin{pmatrix} u_i \\ U_i \end{pmatrix}$$

such that

$$\begin{pmatrix} u_i \\ U_i \end{pmatrix} P_{ii} = \begin{pmatrix} 1 - t_i \epsilon + o(\epsilon) & 0 \\ 0 & H_i \end{pmatrix} \begin{pmatrix} u_i \\ U_i \end{pmatrix} \quad \text{for } i = 1, 2, \ldots, N \qquad (6.25)$$

where the norms of T_i and T_i^{-1} are bounded by some constant that is independent of ϵ. To ensure that the remaining eigenvalues of P_{ii} are uniformly bounded away from unity, we require the following condition:

Condition 6.3 *There is an M_2, independent of ϵ, for which*

$$\|(I - H_i)^{-1}\| < M_2 \quad for \ i = 1, 2, \ldots, N.$$

When each of the blocks P_{ii} of P satisfies the previous conditions, there exists a permutation matrix R and $T = \text{diag}\{T_1, T_2, \ldots, T_N\}$ such that

$$RTPT^{-1}R^{-1} = \left(\begin{array}{ccc|ccc} 1 - t_1\epsilon + o(\epsilon) & & O(\epsilon) & & & \\ & \ddots & & & 0 & \\ O(\epsilon) & & 1 - t_N\epsilon + o(\epsilon) & & & \\ \hline & & & H_1 & & O(\epsilon) \\ & 0 & & & \ddots & \\ & & & O(\epsilon) & & H_N \end{array}\right).$$

As a result of Gerschgorin's theorem and these conditions, the eigenvalues in the upper left-hand submatrix are of order $1 - O(\epsilon)$, while those of the lower right-hand submatrix are uniformly bounded away from 1.

One further condition is useful in proving the convergence of IAD methods. Since the matrix A is irreducible, it follows that there exist uniformly bounded matrices X and Y of dimension $N \times (N - 1)$ such that

$$W = (e, X) = \left(\begin{array}{c} \xi \\ Y^T \end{array}\right)^{-1},$$

where e and ξ are respectively right- and left-hand eigenvectors of A. It follows then that

$$W^{-1}AW = \left(\begin{array}{cc} 1 & 0 \\ 0 & J \end{array}\right). \tag{6.26}$$

In order to ensure that the subdominant eigenvalues of A remain bounded away from 1 by a quality proportional to ϵ, we shall assume

Condition 6.4 *There is a constant M_3 such that $\|(I - J)^{-1}\| < M_3\epsilon^{-1}$.*

To summarize, we have:

- **Condition 6.1:** P is an irreducible primitive stochastic matrix of block form (6.24) such that $\|P_{ii}\| = O(1)$ for $i = 1, 2, \ldots, N$ and $\|P_{ij}\| = O(\epsilon)$ for $i \neq j$.

- **Condition 6.2:** There is a constant $m_1 > 0$ such that $\|\pi_i\|_1 \geq m_1$ for $i = 1, 2, \ldots, N$.

- **Condition 6.3:** There is a constant M_2 such that $\|(I - H_i)^{-1}\| < M_2$ for $i = 1, 2, \ldots, N$.

- **Condition 6.4:** There is a constant M_3 such that $\|(I - J)^{-1}\| < M_3 \epsilon^{-1}$.

Our assumptions therefore imply that a nearly completely decomposable stochastic matrix of the form (6.24), satisfying Conditions 6.1 through 6.4, has exactly one unit eigenvalue; $N - 1$ eigenvalues that are close to unity; and $(n - N)$ eigenvalues that are uniformly bounded away from unity. We can therefore write the spectral decomposition of P as

$$P = yx^T + Y_2 P_2 X_2^T + Y_3 P_3 X_3^T, \tag{6.27}$$

where

$$
\begin{array}{ccc}
1 & N-1 & n-N \\
\text{column} & \text{columns} & \text{columns}
\end{array}
\qquad\qquad
\begin{array}{ccc}
1 & N-1 & n-N \\
\text{column} & \text{columns} & \text{columns}
\end{array}
$$

$$(\quad y \quad Y_2 \quad Y_3 \quad)^{-1} = (\quad x \quad X_2 \quad X_3 \quad)^T$$

and $y = e$; $x = \pi^T$. The columns of the matrices Y_i and X_i $(i = 2, 3)$ are respectively right and left invariant subspaces of P, and the P_i $(i = 2, 3)$ are the representations of P on the column spaces $\mathcal{R}(Y_i)$ with respect to the columns of Y_i.

Let z be any probability vector (as usual a row vector); i.e., $z_i > 0$ for $i = 1, 2, \ldots n$ and $ze = zy = 1$. The vector z may be written as a linear combination of x, X_2, and X_3 as

$$z^T = x + X_2 g_2 + X_3 g_3$$

where $g_2 = Y_2^T z^T$ and $g_3 = Y_3^T z^T$. Then

$$z P^k = x^T + g_2^T P_2^k X_2^T + g_3^T P_3^k X_3^T.$$

Since for small ϵ the eigenvalues of P_3 have modulus less than those of P_2, the matrix P_3^k approaches zero faster than P_2^k, at least asymptotically. Thus, the approach of zP^k to x^T as $k \to \infty$ proceeds in two stages:

- A fast transient, during which $g_3^T P_3^k X_3^T$ becomes negligible;

- A slow transient, during which $g_2^T P_2^k X_2^T$ becomes negligible.

The presence of the slow transient makes the computation of x ($= \pi^T$) difficult when basic iterative methods such as the power method or Gauss–Seidel are used. These methods reduce the component of z along X_3 satisfactorily, but they reduce the component along X_2 only slowly. Iteration and aggregation methods are effective in reducing the component along X_2.

6.4.2   Three Lemmas to Characterize Approximations

To establish the convergence of the iterative aggregation/disaggregation methods, we begin by proving three lemmas. The first concerns a comparison between the exact stationary probability vector π and the corresponding eigenvectors u_i of the matrices P_{ii}.

Lemma 6.3 *If $\pi = (\pi_1, \pi_2, \ldots, \pi_N)$ is the exact solution of $\pi P = \pi$; $\pi e = 1$, and u_i is the left-hand eigenvector of P_{ii} corresponding to the dominant eigenvalue $1 - t_i\epsilon + o(\epsilon)$, then there exists a constant $b_i = O(1)$ such that*

$$\|\pi_i - b_i u_i\|_1 = O(\epsilon).$$

Proof   Since $\pi_i(I - P_{ii}) = \sum_{j \neq i} \pi_j P_{ji}$ and $u_i(I - P_{ii}) = u_i(t_i\epsilon + o(\epsilon))$,

$$\pi_i - u_i = \left(\sum_{j \neq i} \pi_j P_{ji} + u_i(-t_i\epsilon + o(\epsilon)) \right) (I - P_{ii})^{-1}. \qquad (6.28)$$

Since $\begin{pmatrix} u_i \\ U_i \end{pmatrix}$ is an ($n_i \times n_i$) nonsingular matrix, and $\sum_{j \neq i} \pi_j P_{ji}$ is a row vector of length n_i, there exists a row vector c_i of length n_i such that

$$\sum_{j \neq i} \pi_j P_{ji} = c_i \begin{pmatrix} u_i \\ U_i \end{pmatrix}. \qquad (6.29)$$

Furthermore, since $\begin{pmatrix} u_i \\ U_i \end{pmatrix} = O(1)$ and $\sum_{j \neq i} \pi_j P_{ji} = O(\epsilon)$, it follows that $c_i = O(\epsilon)$. Substituting (6.29) into (6.28), we obtain

$$\pi_i - u_i = c_i \begin{pmatrix} u_i \\ U_i \end{pmatrix} (I - P_{ii})^{-1} + (-t_i\epsilon + o(\epsilon)) u_i (I - P_{ii})^{-1}. \qquad (6.30)$$

From equation (6.25), we have

$$\begin{pmatrix} u_i \\ U_i \end{pmatrix}(I - P_{ii}) = \begin{pmatrix} t_i\epsilon + o(\epsilon) & 0 \\ 0 & I - H_i \end{pmatrix}\begin{pmatrix} u_i \\ U_i \end{pmatrix} \quad \text{for } i = 1, 2, \ldots, N.$$

(6.31)

By taking the inverse of each side and then multiplying both on the right and on the left by $\begin{pmatrix} u_i \\ U_i \end{pmatrix}$, we obtain

$$\begin{pmatrix} u_i \\ U_i \end{pmatrix}(I - P_{ii})^{-1} = \begin{pmatrix} (t_i\epsilon + o(\epsilon))^{-1} & 0 \\ 0 & (I - H_i)^{-1} \end{pmatrix}\begin{pmatrix} u_i \\ U_i \end{pmatrix} \quad \text{for } i = 1, \ldots, N.$$

(6.32)

From (6.31), we have

$$u_i(I - P_{ii}) = (t_i\epsilon + o(\epsilon))u_i,$$

which leads to

$$u_i = (t_i\epsilon + o(\epsilon))\, u_i(I - P_{ii})^{-1}.$$

(6.33)

Substituting (6.32) and (6.33) into (6.30) yields

$$\pi_i - u_i = c_i\begin{pmatrix} (t_i\epsilon + o(\epsilon))^{-1} & 0 \\ 0 & (I - H_i)^{-1} \end{pmatrix}\begin{pmatrix} u_i \\ U_i \end{pmatrix} - u_i.$$

Therefore

$$\pi_i = c_i\begin{pmatrix} (t_i\epsilon + o(\epsilon))^{-1} & 0 \\ 0 & (I - H_i)^{-1} \end{pmatrix}\begin{pmatrix} u_i \\ U_i \end{pmatrix}.$$

It follows that

$$\pi_i = \frac{c_{i_1}}{t_i\epsilon + o(\epsilon)}u_i + (c_{i_2}, \ c_{i_3}, \ldots, c_{i_{n_i}})(I - H_i)^{-1}U_i,$$

and hence,

$$\pi_i - \frac{c_{i_1}}{t_i\epsilon + o(\epsilon)}u_i = (c_{i_2}, \ c_{i_3}, \ldots, c_{i_{n_i}})\,(I - H_i)^{-1}\,U_i.$$

Since $(c_{i_2}, \ c_{i_3}, \ldots, c_{i_{n_i}}) \approx O(\epsilon)$, $U_i \approx O(1)$, and $(I - H_i)^{-1} \approx O(1)$, it follows that

$$\left\|\pi_i - \left(\frac{c_{i_1}}{t_i\epsilon + o(\epsilon)}\right)u_i\right\| = O(\epsilon).$$

Finally, since $c_{i_1} \approx \tau_i\epsilon + o(\epsilon)$, $c_{i_1}(t_i\epsilon + o(\epsilon))^{-1} = \tau_i t_i^{-1} + O(1) = O(1)$, and consequently

$$\|\pi_i - b_i u_i\| = O(\epsilon),$$

where

$$b_i = \frac{c_{i_1}}{t_i\epsilon + o(\epsilon)} = O(1).$$

$\square$

The second lemma concerns the approximation $A^{(m-1)}$ to the matrix A.

Lemma 6.4 *If $\pi^{(m)}$ is an $O(\sigma)$ approximation to π, then the corresponding coupling matrix $A^{(m)}$ is an $O(\epsilon\sigma)$ approximation to A, where*

$$a_{ij}^{(m)} = \frac{\pi_i^{(m)}}{\|\pi_i^{(m)}\|_1} P_{ij} e.$$

Proof Since $\pi_i^{(m)} = \pi_i + O(\sigma)$, it follows that

$$\frac{\pi_i^{(m)}}{\|\pi_i^{(m)}\|_1} = \frac{\pi_i}{\|\pi_i\|_1} + O(\sigma).$$

For $i \neq j$ we have

$$a_{ij}^{(m)} = \frac{\pi_i^{(m)}}{\|\pi_i^{(m)}\|_1} P_{ij} e = \frac{\pi_i}{\|\pi_i\|_1} P_{ij} e + O(\sigma) P_{ij} e.$$

Therefore, from the definition of A we have

$$a_{ij}^{(m)} = a_{ij} + O(\epsilon\sigma), \quad \text{since } P_{ij} = O(\epsilon), \ i \neq j.$$

This, then, is the required result for $i \neq j$. When $i = j$, since both $A^{(m)}$ and A are stochastic matrices, we may write

$$a_{ii}^{(m)} = 1 - \sum_{i \neq j} a_{ij}^{(m)} = 1 - \sum_{i \neq j} a_{ij} + O(\sigma\epsilon) = a_{ii} + O(\sigma\epsilon),$$

which completes the proof of Lemma 6.4. □

The final lemma that we require concerns the accuracy of the approximate eigensolution $\xi^{(m-1)}$ to ξ.

Lemma 6.5 *If ξ and $\xi^{(m-1)}$ are, respectively, the left-hand eigenvectors of A and $A^{(m-1)}$ corresponding to the dominant eigenvalue, i.e., $\xi A = \xi$ and $\xi^{(m-1)} A^{(m-1)} = \xi^{(m-1)}$ with $\|\xi\|_1 = \|\xi^{(m-1)}\|_1 = 1$, and if $A^{(m-1)}$ is an $O(\sigma\epsilon)$ approximation to A, then*

$$\xi^{(m-1)} - \xi = O(\sigma).$$

Proof Using equation (6.26), we see that

$$(I - A) = W \begin{pmatrix} 0 & 0 \\ 0 & I - J \end{pmatrix} W^{-1}.$$

Since $\xi A = \xi$ and $\xi^{(m-1)} A^{(m-1)} = \xi^{(m-1)}$, we have

$$\xi^{(m-1)} A^{(m-1)} - \xi^{(m-1)} A + \xi^{(m-1)} A - \xi A = \xi^{(m-1)} - \xi,$$

i.e.,

$$(\xi^{(m-1)} - \xi)(I - A) = \xi^{(m-1)}(A^{(m-1)} - A).$$

Substituting for $(I - A)$ we find

$$(\xi^{(m-1)} - \xi)W \begin{pmatrix} 0 & 0 \\ 0 & I - J \end{pmatrix} = \xi^{(m-1)}(A^{(m-1)} - A)W.$$

Since $\|W\| = O(1)$, and $\|\xi^{(m-1)}\| = O(1)$,

$$(\xi^{(m-1)} - \xi)W \begin{pmatrix} 0 & 0 \\ 0 & I - J \end{pmatrix} = O(\sigma\epsilon). \tag{6.34}$$

The first element of the row vector $(\xi^{(m-1)} - \xi)W$ should be zero, since

$$\left((\xi^{(m-1)} - \xi)W\right)_1 = (\xi^{(m-1)} - \xi)e = \|\xi^{(m-1)}\|_1 - \|\xi\|_1 = 0.$$

If we denote the vector of length $N - 1$ consisting of elements 2 through N of $(\xi^{(m-1)} - \xi)W$ by ζ, then from (6.34)

$$\zeta(I - J) = O(\epsilon\sigma).$$

Since $(I - J)$ is nonsingular,

$$\|\zeta\| \leq O(\sigma\epsilon)\, \|(I - J)^{-1}\|,$$

and since $\|(I - J)^{-1}\| = O(\epsilon^{-1})$, we have

$$\zeta = O(\sigma),$$

and therefore

$$(\xi^{(m-1)} - \xi)W = O(\sigma).$$

Finally, since $\|W^{-1}\| = O(1)$, it follows that $(\xi^{(m-1)} - \xi) = O(\sigma)$, which completes the proof of the lemma. $\qquad\square$

6.4.3   A Convergence Theorem

We shall now use these lemmas to prove the convergence of the basic IAD algorithms.

Theorem 6.6 *Let P satisfy Conditions 6.1 through 6.4. Then the error in the approximate solution using the iterative aggregation/disaggregation algorithms is reduced by a factor of order ϵ at each iteration.*

Proof   We shall prove the theorem for the algorithm of Takahashi. The same procedure may be used to prove the convergence of the KMS algorithm. Let the error in the approximate solution vector $\pi^{(m-1)}$ be of order $O(\sigma)$, i.e.,

$$\pi^{(m-1)} = \pi + O(\sigma).$$

Then, from Lemma 6.4,

$$A^{(m-1)} = A + O(\epsilon\sigma),$$

and from Lemma 6.5,

$$\xi^{(m-1)} = \xi + O(\sigma).$$

Consider the formula for the method of Takahashi, equation (6.19), and let $k = 1$. We have

$$
\begin{aligned}
z_1^{(m)}(I - P_{11}) &= \sum_{j=2}^{N} \xi_j^{(m-1)} \phi_j^{(m-1)} P_{j1} \\
&= \sum_{j=2}^{N} (\xi_j + O(\sigma))(\phi_j + O(\sigma)) P_{j1} \\
&= \sum_{j=2}^{N} \xi_j \phi_j P_{j1} + \sum_{j=2}^{N} [\xi_j O(\sigma) + O(\sigma)\phi_j + O(\sigma^2)] P_{j1} \\
&= \pi_1(I - P_{11}) + \sum_{j=2}^{N} O(\sigma) P_{j1},
\end{aligned}
$$

since ξ_j and ϕ_j are both $O(1)$. Since $\begin{pmatrix} u_1 \\ U_1 \end{pmatrix}$ is nonsingular and $O(1)$, and $\sum_{j=2}^{N} O(\sigma) P_{j1}$ is a row vector whose components are of order $O(\sigma\epsilon)$, it follows that there is a row vector $\bar{c}$ whose components are of order $O(\sigma\epsilon)$, such that

$$\sum_{j=2}^{N} O(\sigma) P_{j1} = \bar{c} \begin{pmatrix} u_1 \\ U_1 \end{pmatrix},$$

and therefore

$$z_1^{(m)} - \pi_1 = \bar{c}\begin{pmatrix} u_1 \\ U_1 \end{pmatrix}(I - P_{11})^{-1} = \bar{c}\begin{pmatrix} (t_1\epsilon + o(\epsilon))^{-1}u_1 \\ (I - H_1)^{-1}U_1 \end{pmatrix}$$

$$= \frac{\bar{c}_1}{t_1\epsilon + o(\epsilon)}u_1 + (\bar{c}_2,\ \bar{c}_3,\ldots,\bar{c}_{n_1})(I - H_1)^{-1}U_1.$$

But, from Lemma 6.3,

$$u_1 = \frac{\pi_1}{b_1} + O(\epsilon),$$

and in view of $\bar{c} = O(\sigma\epsilon)$,

$$z_1^{(m)} - \alpha_1\pi_1 = \frac{\bar{c}_1 O(\epsilon)}{t_1\epsilon + O(\epsilon)} + (\bar{c}_2,\ \bar{c}_3,\ldots,\bar{c}_{n_1})(I - H_1)^{-1}U_1 = O(\sigma\epsilon),$$

where $\alpha_1 = 1 + O(\sigma)$. Therefore,

$$z_1^{(m)} = \alpha_1\pi_1 + O(\sigma\epsilon)$$

and

$$\phi_1^{(m)} = \frac{z_1^{(m)}}{\|z_1^{(m)}\|_1} = \frac{\alpha_1\pi_1 + O(\sigma\epsilon)}{\|\alpha_1\pi_1 + O(\sigma\epsilon)\|_1} = \frac{\pi_1}{\|\pi_1\|_1} + O(\sigma\epsilon).$$

In other words,

$$\phi_1^{(m)} = \phi_1 + O(\sigma\epsilon). \tag{6.35}$$

Suppose now that it has been shown that

$$\phi_j^{(m)} = \phi_j + O(\sigma\epsilon), \quad \text{for } j = 1, 2, \ldots, i - 1.$$

Then,

$$z_i^{(m)}(I - P_{ii}) = \sum_{j<i}\xi_j^{(m-1)}\phi_j^{(m)}P_{ji} + \sum_{j>i}\xi_j^{(m-1)}\phi_j^{(m-1)}P_{ji}$$

$$= \sum_{j<i}(\xi_j + O(\sigma))(\phi_j + O(\sigma\epsilon))P_{ji} + \sum_{j>i}(\xi_j + O(\sigma))(\phi_j + O(\sigma))P_{ji}$$

$$= \sum_{j<i}\xi_j\phi_j P_{ji} + \sum_{j>i}\xi_j\phi_j P_{ji} + \sum_{j<i}\left(O(\sigma)\phi_j + \xi_j O(\sigma\epsilon) + O(\sigma^2\epsilon)\right)P_{ji}$$

$$+ \sum_{j>i}\left(\xi_j O(\sigma) + O(\sigma)\phi_j + O(\sigma^2)\right)P_{ji}$$

$$= \pi_i(I - P_{ii}) + \sum_{j\neq i}O(\sigma)P_{ji}.$$

Repeating the argument used to derive (6.35) yields directly

$$\pi_i^{(m)} = \pi_i + O(\sigma\epsilon).$$

Hence we may conclude that

$$\pi^{(m)} = \pi + O(\sigma\epsilon),$$

which completes the proof of the theorem.   □

Thus, if the maximum degree of coupling (ϵ) between aggregates is sufficiently small, and the initial approximation is sufficiently good (σ), these methods are convergent under Conditions 6.1 through 6.4. Furthermore, the theorem indicates that for nearly uncoupled stochastic matrices the convergence will be rapid. The error in the approximate solution is reduced by a factor of order ϵ at each iteration.

Chapter 7

P-Cyclic Markov Chains

7.1 Introduction

Markov chains sometimes possess the property that the number of single-step
transitions that must be made on leaving any state to return to that state, by
any path, is a multiple of some integer $p > 1$. These models are said to be *periodic of period p* or *cyclic of index p*. We shall use both terms interchangeably.
Bonhoure [12] has shown that Markov chains that arise from queueing network
models frequently possess this property. In this chapter, we shall examine such
Markov chains and show how advantage may be taken of their periodicity to reduce the amount of computer memory and computation time needed to compute
stationary probability vectors.

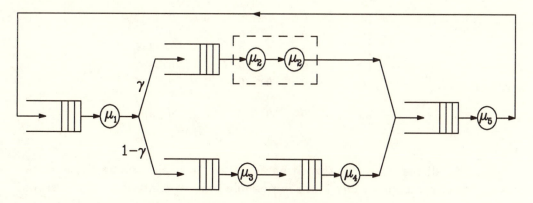

Figure 7.1: A cyclic queueing network.

Consider, as an example, the queueing network model shown in Figure 7.1. All
the stations contain a single exponential server except the second, which consists
of a two-phase Erlang server. The service rates are given by μ_i, $i = 1, 2, \ldots, 5$,
and are as marked on the figure. A state of the system is completely described

by the vector

$$(\eta_1, \ \eta_2, \ k, \ \eta_3, \ \eta_4, \ \eta_5),$$

in which η_i denotes the number of customers at station i, and k denotes the service phase at the Erlang server. For example, when the number of customers in the network is given by $N = 2$, then the system contains 20 states, shown in Table 7.1.

Table 7.1: States of Cyclic Queueing Network

	η_1	η_2	k	η_3	η_4	η_5
1.	2	0	0	0	0	0
2.	1	1	1	0	0	0
3.	1	1	2	0	0	0
4.	1	0	0	1	0	0
5.	1	0	0	0	1	0
6.	1	0	0	0	0	1
7.	0	2	1	0	0	0
8.	0	2	2	0	0	0
9.	0	1	1	1	0	0
10.	0	1	1	0	1	0
11.	0	1	1	0	0	1
12.	0	1	2	1	0	0
13.	0	1	2	0	1	0
14.	0	1	2	0	0	1
15.	0	0	0	2	0	0
16.	0	0	0	1	1	0
17.	0	0	0	1	0	1
18.	0	0	0	0	2	0
19.	0	0	0	0	1	1
20.	0	0	0	0	0	2

Figure 7.2 displays all the transitions that are possible among these states. Observe that a return to any state is possible only in a number of transitions that is some positive (nonzero) multiple of 4. For example, the following paths from state 5 back to state 5 are possible:

```
5 --> 6 --> 1 --> 4 --> 5  of length 4,
5 -->10 -->13 -->19 -->20 --> 6 --> 1 --> 4 --> 5  of length 8,
etc.
```

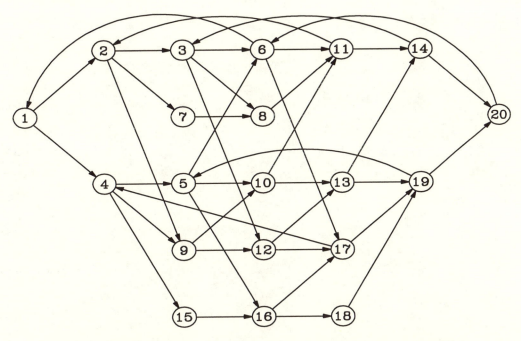

Figure 7.2: Possible state transitions of queueing network model.

Let us now group the states according to their generated distance from the initial state, 1. This is shown in Figure 7.3, where the groups have been labeled $A, B, \ldots, G$. We have

$$
\begin{aligned}
A &= \{1\} \\
B &= \{2, 4\} \\
C &= \{3, 7, 5, 9, 15\} \\
D &= \{6, 8, 10, 12, 16\} \\
E &= \{11, 13, 17, 18\} \\
F &= \{14, 19\} \\
G &= \{20\}
\end{aligned}
$$

We shall use the term *preclass* to designate a set of states $A, B, \ldots, G$. If we display the transitions among these preclasses, we find the interactions among them to be as indicated in Figure 7.4. If we now rearrange the preclasses as indicated in Figure 7.4, we obtain the sets C_1, C_2, C_3, and C_4, where

$$
\begin{aligned}
C_1 &= A \cup E \\
C_2 &= B \cup F \\
C_3 &= C \cup G
\end{aligned}
$$

$$C_4 \;=\; D$$

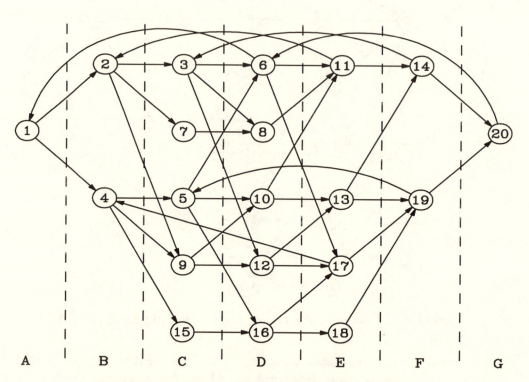

Figure 7.3: The preclasses of the Markov chain.

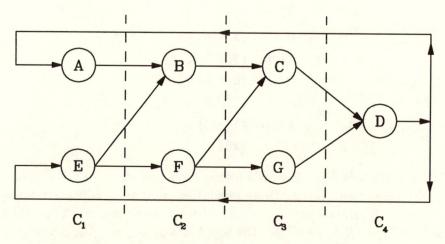

Figure 7.4: The periodic classes of the Markov chain.

The sets C_1, C_2, C_3, C_4 are called the *periodic classes* of the Markov chain; transitions among them are shown in Figure 7.5.

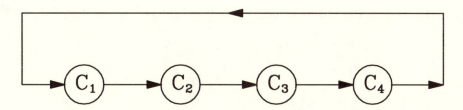

Figure 7.5: Transitions among the periodic classes.

Figure 7.6 shows the structure of the infinitesimal generator when the states are ordered according to their periodic classes. The order of the states within any particular periodic class is unimportant.

New ordering: 1, 11, 13, 17, 18; 2, 4, 14, 19; 3, 7, 5, 9, 15, 20; 6, 8, 10, 12, 16.
Periodic class: C_1 C_2 C_3 C_4

We indicate a nonzero element of the infinitesimal generator by the symbol $\times$, a zero element by $\cdot$, and diagonal elements, which are equal to the negated sum of the off-diagonal elements, by an asterisk.

```
⎛ *  ·  ·  ·  ·  │ ×  ×  ·  ·  │ ·  ·  ·  ·  ·  ·  │ ·  ·  ·  ·  · ⎞
│ ·  *  ·  ·  ·  │ ×  ·  ×  ·  │ ·  ·  ·  ·  ·  ·  │ ·  ·  ·  ·  · │
│ ·  ·  *  ·  ·  │ ·  ·  ×  ×  │ ·  ·  ·  ·  ·  ·  │ ·  ·  ·  ·  · │
│ ·  ·  ·  *  ·  │ ·  ×  ·  ×  │ ·  ·  ·  ·  ·  ·  │ ·  ·  ·  ·  · │
│ ·  ·  ·  ·  *  │ ·  ·  ·  ×  │ ·  ·  ·  ·  ·  ·  │ ·  ·  ·  ·  · │
│ ·  ·  ·  ·  ·  │ *  ·  ·  ·  │ ×  ×  ·  ×  ·  ·  │ ·  ·  ·  ·  · │
│ ·  ·  ·  ·  ·  │ ·  *  ·  ·  │ ·  ·  ×  ×  ×  ·  │ ·  ·  ·  ·  · │
│ ·  ·  ·  ·  ·  │ ·  ·  *  ·  │ ×  ·  ·  ·  ·  ×  │ ·  ·  ·  ·  · │
│ ·  ·  ·  ·  ·  │ ·  ·  ·  *  │ ·  ·  ×  ·  ·  ×  │ ·  ·  ·  ·  · │
│ ·  ·  ·  ·  ·  │ ·  ·  ·  ·  │ *  ·  ·  ·  ·  ·  │ ×  ×  ·  ×  · │
│ ·  ·  ·  ·  ·  │ ·  ·  ·  ·  │ ·  *  ·  ·  ·  ·  │ ·  ×  ·  ·  · │
│ ·  ·  ·  ·  ·  │ ·  ·  ·  ·  │ ·  ·  *  ·  ·  ·  │ ×  ·  ×  ·  × │
│ ·  ·  ·  ·  ·  │ ·  ·  ·  ·  │ ·  ·  ·  *  ·  ·  │ ·  ·  ×  ×  · │
│ ·  ·  ·  ·  ·  │ ·  ·  ·  ·  │ ·  ·  ·  ·  *  ·  │ ·  ·  ·  ·  × │
│ ·  ·  ·  ·  ·  │ ·  ·  ·  ·  │ ·  ·  ·  ·  ·  *  │ ×  ·  ·  ·  · │
│ ×  ×  ·  ×  ·  │ ·  ·  ·  ·  │ ·  ·  ·  ·  ·  ·  │ *  ·  ·  ·  · │
│ ·  ×  ·  ·  ·  │ ·  ·  ·  ·  │ ·  ·  ·  ·  ·  ·  │ ·  *  ·  ·  · │
│ ·  ×  ×  ·  ·  │ ·  ·  ·  ·  │ ·  ·  ·  ·  ·  ·  │ ·  ·  *  ·  · │
│ ·  ·  ×  ×  ·  │ ·  ·  ·  ·  │ ·  ·  ·  ·  ·  ·  │ ·  ·  ·  *  · │
⎝ ·  ·  ·  ×  ×  │ ·  ·  ·  ·  │ ·  ·  ·  ·  ·  ·  │ ·  ·  ·  ·  * ⎠
```

Figure 7.6: Infinitesimal generator of cyclic Markov chain.

This structure shows that the matrix may be written in the form

$$\begin{pmatrix} D_1 & Q_1 & \cdot & \cdot \\ \cdot & D_2 & Q_2 & \cdot \\ \cdot & \cdot & D_3 & Q_3 \\ Q_4 & \circ & \circ & D_4 \end{pmatrix}$$

in which the D_i are square diagonal submatrices. In this chapter we shall show how to make use of this structure to compute stationary probability vectors efficiently. We begin by introducing some elementary concepts from the theory of graphs.

7.2   Directed Graphs and P-Cyclic Matrices

7.2.1   Graph Terminology and Definitions

A *directed graph* or *digraph* $G(V,E)$ is an ordered pair of sets V and E. V is a nonempty set of vertices (or nodes), and E is a set of edges (or arcs). Each element of E consists of an ordered pair of vertices of V. As the name indicates, graphs may be conveniently represented graphically; each vertex is represented by a point and each edge by a line joining two vertices. As an example, the directed graph $G(V,E)$, whose sets V and E are as follows:

$$V = \{v_1, v_2, v_3, v_4\}; \qquad E = \{e_1, e_2, e_3, e_4, e_5, e_6\}$$

with

$$e_1 = \{v_1, v_2\}, \qquad e_2 = \{v_3, v_2\}, \qquad e_3 = \{v_2, v_4\},$$

$$e_4 = \{v_1, v_3\}, \qquad e_5 = \{v_4, v_1\}, \qquad e_6 = \{v_3, v_3\},$$

may be represented graphically as shown in Figure 7.7

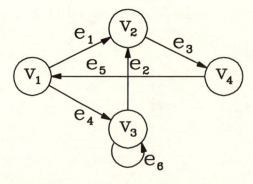

Figure 7.7: A sample directed graph.

The nomenclature of directed graphs is not always consistent from one text to another. We adopt the following. A directed graph is *finite* if it contains only a finite number of vertices. An edge that joins a vertex to itself is called a *loop* (e.g., e_6 in Figure 7.7), whereas an edge with different beginning and ending vertices is called a *link*. A *path* of G is an alternating sequence of vertices and edges:

$$\Gamma = (v_0,\ e_1,\ v_1,\ e_2,\ v_2,\ e_3,\ v_3,\ \ldots,\ e_k,\ v_k).$$

The vertices v_0 and v_k are the *origin* and *terminus* of the path Γ, and k is its *length*. When $k > 0$ and $v_0 = v_k$, the path is said to be *closed*, and in that case the ordering of the terms of Γ is defined to a circular permutation.

If all the terms in the sequence of edges $(e_1, e_2, \ldots, e_k)$ are distinct, the path Γ is said to be *simple*. In other words, in following a simple path, the same edge is never taken more than once. If all the terms in the sequences of vertices $(v_0, v_1, \ldots, v_k)$ are distinct, the path is said to be an *elementary* path. In other words, in following an elementary path, no vertex is traversed more than once. It follows that an elementary path must be simple, and its origin and terminus are distinct. Consider the path

$$\Gamma = (v_1,\ e_1,\ v_2,\ e_3,\ v_4,\ e_5,\ v_1,\ e_4,\ v_3)$$

in the example of Figure 7.7. In this path the sequence of edges (e_1, e_3, e_5, e_4) is distinct, and thus the path is a simple path of length 4. It is not an elementary path, because the vertex v_1 appears twice.

A closed simple path is called a *cycle*. A cycle is said to be an *elementary cycle* if all the terms in the sequence $(v_0, v_1, \ldots, v_{k-1})$ are distinct. Thus, the length of an elementary cycle is equal to the number of vertices it contains. A loop is an elementary cycle of length 1. The path

$$\Gamma = (v_2,\ e_3,\ v_4,\ e_5,\ v_1,\ e_4,\ v_3,\ e_2,\ v_2)$$

is an elementary cycle of length 4. A simple cycle (or path) is often specified by writing only the sequence of vertices in the cycle (or path). Thus, the cycle just mentioned would be written as $\Gamma = (v_2,\ v_4,\ v_1,\ v_3,\ v_2)$ or even as $\Gamma = (2,\ 4,\ 1,\ 3,\ 2)$. In this notation, the first vertex is often not repeated at the end of the sequence, so we simply write $\Gamma = (2,\ 4,\ 1,\ 3)$.

Two vertices v_i and v_j are said to be *connected* if there is a path from v_i to v_j. The vertices v_i and v_j are *strongly connected* if there is a path from v_i to v_j and a path from v_j to v_i. A set of vertices of G that are strongly connected to each other forms a *strongly connected component* of G. If all the vertices of G belong to a single strongly connected component, then G is said to be *strongly connected*. When G is strongly connected, there is an elementary path from every vertex of

G to every other vertex. By inspection, the directed graph G given in Figure 7.7 is strongly connected.

With every directed graph G, we may associate a square matrix A, of order equal to the number of vertices, whose ij^{th} element is nonzero iff there is an edge from vertex i to vertex j. A matrix representation of the graph given in Figure 7.7 is

$$A = \begin{pmatrix} 0 & 1 & 1 & 0 \\ 0 & 0 & 0 & 1 \\ 0 & 1 & 1 & 0 \\ 1 & 0 & 0 & 0 \end{pmatrix}.$$

Conversely, with every square matrix A we may associate a directed graph $G(A)$. The value of each nonzero element of A is sometimes assigned to the corresponding edge of the graph. The concept of a strongly connected directed graph and that of an irreducible matrix are related. We have the following theorem, the proof of which is left to the reader.

Theorem 7.1 *An $(n \times n)$ matrix A is irreducible if and only if its directed graph, $G(A)$, is strongly connected.*

Definition 7.1 *The period p of a finite, strongly connected directed graph G is defined as*

$$p = \gcd(l_1, \ldots, l_c, \ldots, l_C),$$

where C is the number of elementary cycles in G; l_c represents the length, in terms of the number of edges, of cycle c; and gcd is the greatest common divisor. G is said to be periodic iff $p > 1$. If $p = 1$, G is said to be aperiodic or primitive.

The elementary cycles of the graph in Figure 7.7 are (3); (1, 2, 4); (1, 3, 2, 4); and circular permutations thereof. These have lengths 1, 3, and 4, respectively, implying that this graph is aperiodic. On the other hand, the graph whose matrix is given as

$$A = \begin{pmatrix} 0 & 1 & 1 & 0 \\ 0 & 0 & 0 & 1 \\ 0 & 0 & 0 & 1 \\ 1 & 0 & 0 & 0 \end{pmatrix},$$

has circular permutations of (1, 2, 4) and (1, 3, 4) as elementary cycles and is thus periodic with period $p = 3$.

Lemma 7.1 *In a finite, strongly connected directed graph G of period $p > 1$, the vertices can be partitioned into p periodic classes, C_1, C_2, ..., C_p, in which, for some i, each edge in G connects a vertex of C_i to a vertex of the next periodic class, $C_{(i \bmod p)+1}$* [1].

---

[1] Recall that $p \bmod p = 0$.

This implies that any vertex in a given periodic class can be reached only from another vertex of the same class by an elementary path whose length is a nonzero multiple of p.

Proof The proof is by construction. The periodic classes are formed in a piecemeal fashion. Each class is initially set equal to the empty set. An initial vertex v_1 is chosen and assigned to periodic class C_1. We now consider all edges that originate at v_1 and assign their terminal vertices to periodic class C_2. (The edges considered so far lead from C_1 to C_2.) Each edge that originates at one of the vertices just assigned to periodic class C_2 is now examined and its terminus assigned to periodic class $C_{(2 \bmod p)+1}$. (Again, all edges lead from periodic class C_2 to its successor.) The procedure continues in this fashion. At each step, the edges that originate at vertices just assigned to periodic class C_i (and not those that may have previously been there) are examined and their terminal vertices assigned to periodic class $C_{(i \bmod p)+1}$ (assuming that they have not already been assigned to that class). Since G is strongly connected, there is a path from v_1 to all other vertices, and thus all vertices will eventually be considered and assigned to a periodic class. However, depending on the structure of G, it is likely that several passes will have to be made over the p periodic classes before all are included.

This process will eventually terminate, since G is finite and each vertex is considered only once as an originator of edges. On the other hand, a given vertex may be the terminus of several different edges, and it may be feared that such a vertex could be assigned to a number of different periodic classes depending on the class of the originating vertex. However, the periodicity of G prevents this from happening. Since G is periodic of period p, a vertex that is the terminus of an elementary path of length i from any given vertex v_1 can only be reached from v_1 by paths of length $k \times p + i$, $k \in \mathcal{N}$. The construction procedure traces out elementary cycles in which a given vertex is always a distance of $k \times p + i$, $k \in \mathcal{N}$ from v_1, no matter which path is taken, and therefore will always be assigned to periodic class C_i. It follows that once all the vertices have been assigned a periodic class, the edges that originate in class C_i, for $i = 1, 2, \ldots, p$, must necessarily terminate in class $C_{(i \bmod p)+1}$ as required. $\square$

It follows immediately from this lemma that, corresponding to any finite, strongly connected directed graph G of period p, there exists a permutation matrix P such that

$$PAP^T = \begin{pmatrix} 0 & A_{12} & 0 & \cdots & 0 & 0 \\ 0 & 0 & A_{23} & \cdots & 0 & 0 \\ 0 & 0 & 0 & \cdots & 0 & 0 \\ \vdots & \vdots & \vdots & \ddots & \vdots & \vdots \\ 0 & 0 & 0 & \cdots & 0 & A_{p-1,p} \\ A_{p1} & 0 & 0 & \cdots & 0 & 0 \end{pmatrix}, \tag{7.1}$$

in which the diagonal submatrices are square and zero.

7.2.2 Primitive and Cyclic Matrices

7.2.2.1 Definitions

The existence of the permutation matrix specified in equation (7.1) is sometimes used to define a p-cyclic matrix, thereby eliminating any need to refer to an underlying graphical representation. We have

Definition 7.2 *Let $A \geq 0$ be an irreducible $(n \times n)$ matrix and let P be a permutation matrix such that PAP^T has the form (7.1). If $p = 1$, then A is said to be* primitive; *otherwise A is said to be* cyclic of index p.

An alternative and equivalent definition (see [170]) is the following:

Definition 7.3 *Let $A \geq 0$ be an irreducible $(n \times n)$ matrix, and let p be the number of eigenvalues of A of modulus $\rho(A)$. If $p = 1$, then A is* primitive. *If $p > 1$, then A is* cyclic of index p.

The representation of a nonnegative irreducible cyclic matrix of index p can be permuted into several different but equivalent forms. The following is called the *normal form* of an irreducible, cyclic matrix $A \geq 0$ of index p (> 1):

$$PAP^T = \begin{pmatrix} 0 & 0 & 0 & \cdots & 0 & A_{1p} \\ A_{21} & 0 & 0 & \cdots & 0 & 0 \\ 0 & A_{32} & 0 & \cdots & 0 & 0 \\ \vdots & \vdots & \vdots & \ddots & \vdots & \vdots \\ 0 & 0 & 0 & \cdots & 0 & 0 \\ 0 & 0 & 0 & \cdots & A_{p,p-1} & 0 \end{pmatrix}. \tag{7.2}$$

The diagonal submatrices are square and zero. When A is in form (7.2), it is said to be *consistently ordered*.

Definition 7.4 *An $(n \times n)$ complex matrix A (not necessarily nonnegative or irreducible) is* weakly cyclic of index p (> 1) *if there exists an $(n \times n)$ permutation matrix P such that PAP^T is of the form (7.2).*

Note that a matrix can be simultaneously weakly cyclic of different indices. We now consider some properties of primitive and cyclic matrices.

7.2.2.2 Some Properties of Primitive Matrices

Proofs of the following lemmas and the theorem by Frobenius may be found in [170]. Bear in mind that a primitive matrix is necessarily nonnegative and irreducible.

Lemma 7.2 *If* $A > 0$ *is an* $(n \times n)$ *matrix, then* A *is primitive.*

Lemma 7.3 *If* A *is primitive, then* $\rho(A)$ *is a simple eigenvalue of* A *and is the only eigenvalue of* A *of modulus* $\rho(A)$.

Lemma 7.4 *If* A *is primitive, then* A^m *is also primitive for all positive integers* m.

Lemma 7.5 *If* $A \geq 0$ *is an irreducible* $(n \times n)$ *matrix with* $a_{ii} > 0$ *for all* $1 \leq i \leq n$, *then* $A^{n-1} > 0$.

Theorem 7.2 (Frobenius) *Let* $A \geq 0$ *be an* $(n \times n)$ *matrix. Then* $A^m > 0$ *for some positive integer* m *if and only if* A *is primitive.*

If A is an $(n \times n)$ primitive matrix, then by Theorem 7.2 some positive integer power of A is positive, and it is clear from the irreducibility of A that all subsequent powers of A are also positive. Thus, there exists a least positive integer $\gamma(A)$, called the *index of primitivity* of A, for which $A^{\gamma(A)} > 0$. As an example, let $A \geq 0$ be an irreducible $(n \times n)$ matrix with $a_{ii} > 0$ for all $1 \leq i \leq n$. It follows then that A is primitive (since it is irreducible and acyclic), and $\gamma(A) \leq n - 1$. Indeed, if A has exactly $d \geq 1$ nonzero diagonal entries, then A is primitive, and it is known (e.g., see [170]) that

$$\gamma(A) \leq 2n - d - 1.$$

7.2.2.3 Some Properties of *p*-Cyclic Matrices

If $A \geq 0$ is an irreducible $(n \times n)$ cyclic matrix of index p, then from part (c) of the Perron–Frobenius theorem (Chapter 1, Section 5.3) the spectrum of A includes the p eigenvalues

$$\alpha_k = \rho(A)\beta_k, \qquad k = 0, 1, \ldots, p - 1$$

where

$$\beta_k = e^{2\pi k i/p} \tag{7.3}$$

and $i = \sqrt{-1}$. Thus, a cyclic matrix A of index p has precisely p eigenvalues of modulus $\rho(A)$, and these are given by the roots of the equation

$$\lambda^p - \rho^p(A) = 0.$$

Furthermore, *all* nonzero eigenvalues of A appear in p-tuples, each member of a tuple having the same multiplicity. In other words, if $\mu \in \sigma(A) \setminus \{0\}$, then $\mu\beta_k \in \sigma(A)$ for $k = 1, 2, \ldots, p-1$, with the same multiplicity as μ (see [170], Theorem 2.3). The eigenvalues $\mu\beta_k$ for $k = 0, 1, \ldots, p-1$, are said to be in the same *cyclic class* of A.

An interesting relationship among the eigenvectors of an irreducible, cyclic stochastic matrix corresponding to eigenvalues of the same cyclic class was observed in a theorem of Courtois and Semal [31], an immediate generalization of which we now provide.

Theorem 7.3 *Let A be a consistently ordered cyclic matrix of index p. Let $\psi = (\psi_1^T, \psi_2^T, \ldots, \psi_p^T)^T$ be a right-hand eigenvector associated with an eigenvalue $\mu \neq 0$ of A, where the partitioning of ψ is conformal with the partitioning of A. Then the right-hand eigenvector of A corresponding to the eigenvalue $\mu^{(k)} = \mu\beta_k$, $k = 1, 2, \ldots, p-1$, where β_k is given in (7.3), is (up to a single multiplicative constant)*

$$\psi^{(k)} = (\psi_1^T, \ \mu\beta_k^{-1}\psi_2^T, \ \mu\beta_k^{-2}\psi_3^T, \ \ldots, \ \mu\beta_k^{-(p-1)}\psi_p^T)^T.$$

For a proof, see [31]. The fact that the subvectors of eigenvectors in the same cyclic class are essentially identical plays an important role in the convergence of certain iterative methods, (e.g., see Section 7.7).

Theorem 7.4 *Let A be an $(n \times n)$ weakly cyclic matrix of index $p > 1$. Then A^{jp} is completely reducible for every $j \geq 1$; i.e., there exists an $(n \times n)$ permutation matrix P such that*

$$PA^{jp}P^T = \begin{pmatrix} C_1^j & 0 & 0 & \cdots & 0 & 0 \\ 0 & C_2^j & 0 & \cdots & 0 & 0 \\ 0 & 0 & C_3^j & \cdots & 0 & 0 \\ \vdots & \vdots & \vdots & \ddots & \vdots & \vdots \\ 0 & 0 & 0 & \cdots & C_{p-1}^j & 0 \\ 0 & 0 & 0 & \cdots & 0 & C_p^j \end{pmatrix} \tag{7.4}$$

where each diagonal submatrix C_i^j is square and

$$\rho(C_1) = \rho(C_2) = \cdots = \rho(C_p) = \rho^p(A).$$

Moreover, if A is nonnegative, irreducible, and cyclic of index p, then each submatrix C_i, for $i = 1, 2, \ldots, p$, is primitive.

The proof may be found in [170].

7.3   *p*-Cyclic Markov Chains

7.3.1   The Embedded Markov Chain

We saw in Chapter 1, Section 4.3, that with every continuous-time Markov chain with infinitesimal generator Q, it is possible to define a discrete-time chain at state departure instants, the so-called *embedded* Markov chain. Defining S as

$$s_{ij} = \frac{q_{ij}}{-q_{ii}} \quad \text{for} \;\; i \neq j$$
$$s_{ij} = 0 \qquad \text{for} \;\; i = j,$$

then S is stochastic and is the transition probability matrix of the embedded Markov chain. All of its diagonal elements are zero. Furthermore, if Q is irreducible, so also is S. If x is the stationary probability vector of the embedded Markov chain, then π, the stationary probability vector of the original CTMC, may be found from

$$\pi = \frac{-x D_Q^{-1}}{\|x D_Q^{-1}\|_1} \quad \text{where } D_Q = \mathrm{diag}\{Q\}. \tag{7.5}$$

7.3.2   Markov Chains with Periodic Graphs

Every Markov chain, whether continuous-time or discrete-time, may be associated with a directed graph. The vertices of the graph correspond to the states of the Markov chain, and the edges correspond to transitions among states. Notice that with the exception of loops, the graph of a CTMC is identical to that of its embedded chain. With this in mind, and in view of the fact that the diagonal elements of an infinitesimal generator are negative and equal to the negated sum of its off-diagonal elements, we shall say that a CTMC is periodic of period p if its embedded chain is periodic of period p.

Consider a CTMC with infinitesimal generator Q, and let S be the transition probability matrix of its embedded Markov chain. Assume also that S is periodic with period p and hence can be permuted to the following periodic form:

$$S = \begin{pmatrix} 0 & S_1 & 0 & \cdots & 0 & 0 \\ 0 & 0 & S_2 & \cdots & 0 & 0 \\ 0 & 0 & 0 & \cdots & 0 & 0 \\ \vdots & \vdots & \vdots & \ddots & \vdots & \vdots \\ 0 & 0 & 0 & \cdots & 0 & S_{p-1} \\ S_p & 0 & 0 & \cdots & 0 & 0 \end{pmatrix}.$$

The permutation operation consists uniquely of reordering the states, and since S is periodic, this reordering defines a partition of the set of states into p *classes* in such a way that the only transitions possible are from a state of class i to a state of class $(i \bmod p) + 1$. This partition is also defined when $p = 1$; in that case there is a single class, which contains all the states. With this reordering of the states, the matrix Q has the structure

$$
Q = \begin{pmatrix}
D_1 & Q_1 & 0 & \cdots & 0 & 0 \\
0 & D_2 & Q_2 & \cdots & 0 & 0 \\
0 & 0 & D_3 & \cdots & 0 & 0 \\
\vdots & \vdots & \vdots & \ddots & \vdots & \vdots \\
0 & 0 & 0 & \cdots & D_{p-1} & Q_{p-1} \\
Q_p & 0 & 0 & \cdots & 0 & D_p
\end{pmatrix},
$$

in which the diagonal blocks D_i are themselves diagonal matrices. Partitioning π and x according to this periodic structure,

$$
\pi = (\pi_1, \pi_2, \ldots, \pi_i, \ldots, \pi_p) \quad \text{and} \quad x = (x_1, x_2, \ldots, x_i, \ldots, x_p),
$$

we find that

$$
S_i = -D_i^{-1} Q_i \tag{7.6}
$$

and

$$
\pi_i = \frac{-x_i D_i^{-1}}{\|x D_Q^{-1}\|_1}. \tag{7.7}
$$

7.3.3   Computation of the Periodicity

When working with Markov chains that are periodic, it is necessary to have an efficient means of determining their periodicity and the associated periodic classes so that the infinitesimal generator may be symmetrically permuted into normal cyclic form. Advantage can be taken of the periodic structure only when Q is permuted to this form.

Let $G_M = (V, E)$ be the graph of an irreducible CTMC. From this graph we shall construct a second graph $G_C = (W, F)$, whose period is the same as that of G_M and which usually contains substantially fewer vertices than G_M. The algorithm is as follows.

Algorithm for the Construction of G_C

1. *Initialize:*

   - Set $W = \emptyset$, $F = \emptyset$, and $k = 1$.
   - Choose an arbitrary initial state $v_0 \in V$.
   - Set $W_1 = \{v_0\}$, and $W = W_1$.

2. *Loop:*

- Set $W_{k+1} = \emptyset$.
- For each state $e \in W_k$ do
    - For each state $e' \in V$ such that $(e, e') \in E$ do
        * If $\exists\ k' \leq k$ such that $e' \in W_{k'}$, then
          Set $F = F \cup \{(W_k, W_{k'})\}$.
        else
          Set $W_{k+1} = W_{k+1} \cup \{e'\}$.
- If $W_{k+1} \neq \emptyset$, then
    - Set $W = W \cup W_{k+1}$.
    - Set $F = F \cup \{(W_k, W_{k+1})\}$.
    - Set $k = k + 1$.
    - Go to step 2.

  else

    - Stop.

This algorithm constructs a graph G_C in which each vertex is a set of states (a *preclass*) of V. Its construction necessitates one pass across all the vertices of G_M; during this pass, identifiers can be associated with each vertex to facilitate the grouping of vertices into periodic classes. Additionally, the periodicity may be obtained by performing the greatest common divisor (gcd) operation on pairs of integers. Relating the definitions in this algorithm to the example of Section 1, if G_M is the 20-vertex graph displayed in Figure 7.2, then the nine vertices A, B, C,..., G of Figures 7.3 and 7.4 are the vertices W_1, W_2,..., W_9 of G_C. The reader may wish to keep this example close by while reading through the following lemmas and theorem. The proofs, which for the most part are fairly obvious, may be found in [14]. Recall that with each path c of a graph, we associate its length as a number of edges, l_c. Since G_M is strongly connected, we have the following properties.

Lemma 7.6 *With each elementary path between states e and e' of G_M, there corresponds a path (not necessarily elementary) of G_C of the same length connecting the two preclasses that contain the states e and e'.*

Lemma 7.7 *G_C is strongly connected.*

Lemma 7.8 *With every elementary cycle c of G_M, there corresponds a set of elementary cycles of G_C, $\{c_i\}$, such that*

$$\sum_i l_{c_i} = l_c.$$

Lemma 7.9 *If two states v_1 and v_2 belong to the same preclass in G_C, then v_1 and v_2 belong to the same periodic class of G_M.*

Lemma 7.10 *Let p be the periodicity of G_M and c be an elementary cycle of G_C. Then c is of length $\eta \times p$, $\eta \in \mathcal{N}^+$.*

We are now in a position to state the principal result.

Theorem 7.5 *If G_M is strongly connected, then G_C has the same periodicity as G_M.*

As mentioned previously, the construction of G_C corresponds to a pass through G_M from some initial state. In many modelling experiments, this can be incorporated into the generation procedure of the states and transition rate matrix of the Markov chain, for it is not unusual to generate states and transition matrix from some given initial state (see [157]); a separate, independent pass to compute G_C is unnecessary. Furthermore, it is not actually necessary to compute G_C explicitly to compute its period. Indeed, only the lengths of the elementary cycles of G_C are needed, and all of the elementary cycles of G_C appear during the construction of the Markov chain. Thus, if at stage k of the algorithm, state e' originating from state $e \in W_k$, $(e, e') \in E$, already belongs to an existing preclass $W_{k'}$, we form an elementary cycle of G_C of length $k - k' + 1$. Since the gcd is an associative operation, it suffices to compute the gcd of these lengths to compute the period of G_C.

In addition to the computation of the period of the Markov chain, each state is assigned an integer value that denotes its preclass. When the generation of the Markov chain is complete and its periodicity ascertained, a simple folding of the preclasses suffices to yield the *periodic classes* of the chain. This gives a permutation vector (an ordering of the states) that can be used to transform the infinitesimal generator into normal cyclic form.

Numerical results presented later show that periodic structure in CTMCs offers considerable advantage in both computation time and memory requirements. Since the cost of finding the period of a Markov chain and permuting the infinitesimal generator to normal periodic form is low, it is always beneficial to use this periodicity property when it is present.

7.4 Numerical Methods Applied to *p*-Cyclic Matrices

As always, we wish to compute the stationary probability vector π of a finite-state, continuous-time Markov chain whose infinitesimal generator is Q. Our aim in this section is to compare the efficiency of some numerical methods for finding π when the matrix Q is permuted to normal cyclic form and when the same methods are applied to Q in its original, nonpermuted form [13].

Let us first note that there will be no difference when using methods for which the ordering of the state space has no importance, for example, the power method, the method of Arnoldi, and GMRES. As a result, we consider only methods in which ordering is important. In particular, we shall consider direct LU decompositions and Gauss–Seidel iteration.

7.4.1 Direct Methods

Consider a direct LU decomposition of Q. The principle of the decomposition is to express Q as the product of two matrices L and U, i.e., $LU = Q$, L being a lower triangular matrix and U an upper triangular matrix. When Q is periodic, we have

$$
Q = \begin{pmatrix}
D_1 & Q_1 & 0 & \cdots & 0 & 0 \\
0 & D_2 & Q_2 & \cdots & 0 & 0 \\
0 & 0 & D_3 & \cdots & 0 & 0 \\
\vdots & \vdots & \vdots & \ddots & \vdots & \vdots \\
0 & 0 & 0 & \cdots & D_{p-1} & Q_{p-1} \\
Q_p & 0 & 0 & \cdots & 0 & D_p
\end{pmatrix},
\tag{7.8}
$$

where the diagonal matrices are themselves diagonal, and thus, with the exception of the last block row, Q is already upper triangular. An LU decomposition must therefore have the following form:

$$
\begin{pmatrix}
I & 0 & \cdots & 0 & 0 \\
0 & I & \cdots & 0 & 0 \\
\vdots & \vdots & \ddots & \vdots & \vdots \\
0 & 0 & \cdots & I & 0 \\
L_1 & L_2 & \cdots & L_{p-1} & L_p
\end{pmatrix}
\begin{pmatrix}
D_1 & Q_1 & \cdots & 0 & 0 \\
0 & D_2 & \cdots & 0 & 0 \\
\vdots & \vdots & \ddots & \vdots & \vdots \\
0 & 0 & \cdots & D_{p-1} & Q_{p-1} \\
0 & 0 & \cdots & 0 & U_p
\end{pmatrix}
$$

$$
= \begin{pmatrix}
D_1 & Q_1 & \cdots & 0 & 0 \\
0 & D_2 & \cdots & 0 & 0 \\
\vdots & \vdots & \ddots & \vdots & \vdots \\
0 & 0 & \cdots & D_{p-1} & Q_{p-1} \\
Q_p & 0 & \cdots & 0 & D_p
\end{pmatrix}
$$

This decomposition is known once the blocks $L_1, L_2, \ldots, L_p$ and U_p are known, and these are easily determined. We have

$$L_1 = Q_p D_1^{-1}$$

and, for $i = 2, 3, \ldots, p-1$,

$$L_i = (-L_{i-1} Q_{i-1}) D_i^{-1}$$

and the only inverses needed so far are those of diagonal matrices. The blocks L_p and U_p may be obtained by performing a decomposition of the right-hand side of

$$L_p U_p = (D_p - L_{p-1} Q_{p-1}) = D_p + (-1)^{p-1} Q_p (D_1^{-1} Q_1)(D_2^{-1} Q_2) \ldots (D_{p-1}^{-1} Q_{p-1}). \tag{7.9}$$

In fact, the only block that is actually needed is U_p, since it is the only one that is used in the back-substitution. In certain cases, it may be advantageous to compute the right-hand side of (7.9) explicitly. This depends on the dimensions of the individual Q_i and on the order in which the multiplications are performed.

The amount of fill-in generated in producing matrices L and U is likely to be much lower when Q is in periodic form, for it is necessarily restricted to the final block row. Therefore, savings in time and memory will generally result from using the periodic structure of Q, as opposed to a structure resulting from an arbitrary ordering of the states. The numerical experiments described later support this claim.

7.4.2   The Gauss–Seidel Iterative Method

We now apply the method of Gauss–Seidel to the infinitesimal generator Q of the CTMC when Q is in normal cyclic form. Gauss–Seidel consists of applying the iterative scheme

$$\pi^{(k+1)}(D_Q - U_Q) = \pi^{(k)} L_Q,$$

where $\pi^{(k)} = (\pi_1^{(k)}, \pi_2^{(k)}, \ldots, \pi_p^{(k)})$ is the approximation to $\pi = (\pi_1, \pi_2, \ldots, \pi_p)$ at step k. Due to the special periodic structure of Q, this yields

$$\begin{aligned}
\pi_1^{(k+1)} D_1 &= -\pi_p^{(k)} Q_p \\
\pi_2^{(k+1)} D_2 &= -\pi_1^{(k+1)} Q_1 \\
&\vdots \quad \vdots \quad \vdots \\
\pi_p^{(k+1)} D_p &= -\pi_{p-1}^{(k+1)} Q_{p-1}
\end{aligned} \tag{7.10}$$

from which we may deduce

$$\pi_p^{(k+1)} D_p = -\pi_p^{(k)} Q_p (-D_1^{-1} Q_1) \cdots (-D_{p-1}^{-1} Q_{p-1}). \tag{7.11}$$

Using equations (7.6) and (7.7), equation (7.11) may be written as

$$\pi_p^{(k+1)} = \pi_p^{(k)} S_p S_1 S_2 \cdots S_{p-1} = \pi_p^{(k)} R_p \tag{7.12}$$

where $R_p \equiv S_p S_1 \cdots S_{p-1}$. In Lemma 7.11 we show that R_p is stochastic, irreducible, and acyclic. Therefore, convergence of the iterative scheme (7.12), and as a result, convergence of Gauss–Seidel, is guaranteed. The results of some experiments are presented later.

Before closing on the method of Gauss–Seidel applied to Q, it is interesting to recall from Chapter 3 the results of Mitra and Tsoucas [100] relating to the concept of high-stepping transitions. Given a particular ordering of the states, high-stepping transitions are those transitions from any state m to another state m' for which order$(m) <$ order(m'), where order(m) denotes the position of state m in the list of states. Mitra and Tsoucas show that Gauss–Seidel iterations have the property of "leaping" over the individual Markov chain steps so as to cover an entire high-stepping random walk in one iteration. So, orderings that make high-stepping transitions likely, and hence make high-stepping random walks longer, will make the Gauss–Seidel method visit more states in one iteration and are expected to converge faster to the stationary distribution. In cyclic Markov chains whose states are arranged in order of their periodic class, the only states that lead to low-stepping transitions are those in the last periodic class. When this number is small and the period large, we should therefore expect rapid convergence.

7.4.3   Numerical Experiments

7.4.3.1   The Model

We now describe the results obtained when Gaussian elimination and Gauss–Seidel iteration are applied to the CTMC derived from the queueing model presented in Figure 7.8. This is a closed queueing network consisting of six exponential service stations and N customers. The embedded Markov chain is periodic with $p = 4$. The routing of customers among the stations is probabilistic. A customer leaving station 2 goes to station 4 with probability γ and to station 3 with probability $1 - \gamma$. A customer leaving station 5 goes to station 1 with probability 0.5 and to station 6 with the same probability, 0.5. The service rates at the various stations are as follows:

μ_1	μ_2	μ_3	μ_4	μ_5	μ_6
1.0	2.0	μ_3	4.0	5.0	6.0

We consider two very different cases for γ and μ_3. In the first case, we take $\gamma = 0.5$ and $\mu_3 = 3.0$. In the second, we take $\gamma = 0.9995$ and $\mu_3 = 0.003$. The Markov chain corresponding to the second case is nearly completely decomposable, whereas the first is not. We choose these two distinct cases, since it

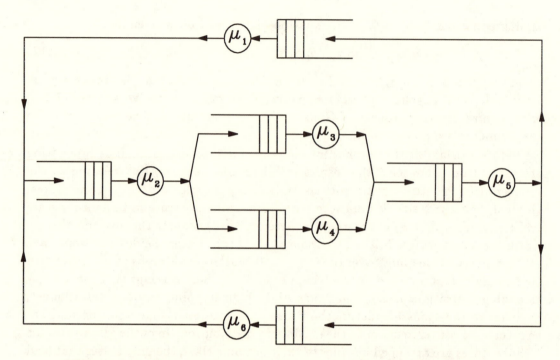

Figure 7.8: The queueing network used for numerical experiments.

is well known that numerical iterative methods are sensitive to the degree of decomposability of the Markov chain.

The results were obtained by means of the software package MARCA (MARkov Chain Analyzer), running on a Sun SPARCstation 2 with 16 MB of memory. We performed comparative tests by applying numerical methods to the matrix in the original ordering generated by MARCA and when it was permuted to normal cyclic form.

7.4.3.2   Gaussian Elimination

With Gaussian elimination, the computation time and memory requirements are independent of the actual values assigned to the parameters of the queueing network model. The results presented in Table 7.2 are therefore applicable to both the NCD and non-NCD cases. In this table and in the tables that follow, n denotes the order of the matrix Q and n_z the number of nonzero elements in Q. The number in the "Fill-in" columns gives the number of elements in the upper triangular matrix that results from the LU decomposition. It may be taken as a measure of memory requirements.

The improvement factors, as measured by the ratio of results obtained in experiments on the nonperiodic ordering to the periodic ordering, range from 3.0

Table 7.2: Results for Gaussian Elimination

N	n	n_z	Nonperiodic		Periodic	
			Fill-in	Time	Fill-in	Time
5	252	1,260	11,523	0.58	3,799	0.20
6	462	2,478	35,621	2.72	10,160	0.73
8	1,287	7,623	237,214	40.22	57,838	9.56

Table 7.3: Results for Gauss–Seidel Iteration

N	n	n_z	Nonperiodic		Periodic	
			Iterations	Time	Iterations	Time
5	252	1,260	73	0.38	67	0.35
6	462	2,478	103	0.96	95	0.89
8	1,287	7,623	178	4.82	165	4.44

to 4.1 for memory requirements and from 2.9 to 4.2 for computation time. Over many examples, it was observed that the improvement factor was often roughly equal to the periodicity of the Markov chain.

7.4.3.3   Gauss–Seidel

The results that are presented in Table 7.3 correspond to the non-NCD case, simply because in the NCD case the number of iterations needed by Gauss–Seidel to achieve the requested precision of 10^{-10} was much too large (in excess of 1,000 iterations).

It appears that the ordering imposed by MARCA is practically as efficient as the periodic ordering. This may be explained by the fact that the ordering assigned by MARCA is well suited to Gauss–Seidel iterations. Indeed, this ordering also takes the flow of the Markov chain into account and thus leads to long high-stepping transitions [100]. These results should therefore be interpreted, not as a poor performance by the periodic ordering, but rather as a desirable characteristic of the ordering provided by MARCA. Nevertheless, the periodic ordering guarantees the convergence of Gauss–Seidel and is by itself a good reason to choose this ordering.

We do not make comparisons between the results obtained with Gaussian elimination and those obtained with Gauss–Seidel. This is in keeping with our objective of studying only the effects of structuring the matrices according to their periodicity on different numerical solution methods.

7.5   Reduced Schemes

7.5.1   Reduced Schemes Associated with a Stochastic Matrix

We now show that, due to the special periodic structure of matrices Q and S, it is possible to derive a system of equations involving only one subvector of the stationary probability vector. Solving this *reduced* system involves fewer numerical operations than solving the original system, and once one subvector has been found, the others can be obtained by means of a single (reduced) matrix-vector multiply. Let us first consider the matrix S of the embedded Markov chain. S is a stochastic, irreducible, and periodic matrix of period p with a unique left-hand eigenvector x associated with its unit eigenvalue. We have

$$x = xS, \qquad x > 0, \qquad ||x||_1 = 1.$$

From

$$(x_1, x_2, \ldots, x_p) = (x_1, x_2, \ldots, x_p) \begin{pmatrix} 0 & S_1 & 0 & \cdots & 0 & 0 \\ 0 & 0 & S_2 & \cdots & 0 & 0 \\ 0 & 0 & 0 & \cdots & 0 & 0 \\ \vdots & \vdots & \vdots & \ddots & \vdots & \vdots \\ 0 & 0 & 0 & \cdots & 0 & S_{p-1} \\ S_p & 0 & 0 & \cdots & 0 & 0 \end{pmatrix}$$

we get

$$x_i = x_{i-1} S_{i-1} \quad \text{for } i = 2, 3, \ldots, p \quad \text{and} \quad x_1 = x_p S_p. \tag{7.13}$$

Since S is stochastic, $||S_i||_1 = 1$ for all $i = 1, 2, \ldots, p$, so

$$||x_1||_1 \le ||x_p||_1 \le ||x_{p-1}||_1 \le \cdots \le ||x_2||_1 \le ||x_1||_1.$$

Therefore, since $||x||_1 = 1$, it follows that

$$||x_i||_1 = \frac{1}{p} \quad \text{for } i = 1, 2, \ldots, p. \tag{7.14}$$

The matrix $R \equiv S^p$ is a completely decomposable (block diagonal) stochastic matrix given by

$$R = \begin{pmatrix} R_1 & 0 & \cdots & 0 & 0 \\ 0 & R_2 & \cdots & 0 & 0 \\ \vdots & \vdots & \ddots & \vdots & \vdots \\ 0 & 0 & \cdots & R_{p-1} & 0 \\ 0 & 0 & \cdots & 0 & R_p \end{pmatrix}$$

where

$$
\begin{aligned}
R_1 &= S_1 S_2 \cdots S_{p-1} S_p \\
R_2 &= S_2 S_3 \cdots S_p S_1 \\
&\vdots \quad \vdots \quad \vdots \\
R_p &= S_p S_1 \cdots S_{p-2} S_{p-1}
\end{aligned}
$$

Notice that each R_i, $i = 1, 2, \ldots, p$, is a matrix of p-step transition probabilities of the embedded Markov chain. Since S is irreducible and p-cyclic, the only eigenvalues of S having modulus 1 are the pth roots of unity,

$$
\beta_k = e^{2k\pi i/p}, \qquad k = 0, 1, \ldots, p - 1,
$$

and therefore, because $(\beta_k)^p = 1$ for $k = 0, 1, \ldots, p-1$, R has exactly p eigenvalues equal to 1 and no others of modulus 1. We have the following lemma:

Lemma 7.11 *Each R_i for $i = 1, 2, \ldots, p$ is a stochastic, irreducible, and acyclic matrix.*

Proof

1. R_i is stochastic, since R is stochastic and completely decomposable.

2. Since S is irreducible, there is, in particular, a path between any two states of a given periodic class. Since S is cyclic of period p, the length of this path is a multiple of p. Each transition in R corresponds to p consecutive steps in S. Therefore there is a path in R between the two states. This implies that each submatrix of R is irreducible.

3. The matrix R possesses only p eigenvalues of modulus 1 (in fact, all equal to 1); one for each submatrix. Thus, it is not possible for any submatrix to possess more than one eigenvalue of unit modulus. Since R_i is stochastic (with dominant eigenvalue equal to 1), this implies that R_i cannot be cyclic.

$\square$

Writing out equation (7.13) we have

$$
\begin{aligned}
x_1 &= x_p S_p \\
x_2 &= x_1 S_1 \\
&\vdots \quad \vdots \quad \vdots \\
x_{p-1} &= x_{p-2} S_{p-2} \\
x_p &= x_{p-1} S_{p-1},
\end{aligned}
$$

from which we may deduce, for example, that

$$x_p = x_p S_p S_1 S_2 \cdots S_{p-1}$$

which leads to an equation of the form

$$x_p = x_p R_p \tag{7.15}$$

with

$$R_p = S_p S_1 S_2 \cdots S_{p-1}. \tag{7.16}$$

Equation (7.15) is called the *reduced scheme* associated with the stochastic matrix S, because it involves only a part of the original probability vector x. It is evident from equation (7.15) that x_p is an eigenvector corresponding to the unit eigenvalue of the stochastic matrix R_p. Notice that it is sufficient to work with the reduced scheme to obtain the complete solution x, since once x_p is obtained, the other subvectors x_i may be determined from x_p by the simple matrix-vector multiplications indicated in equation (7.13).

7.5.2   Reduced Schemes Associated with an Infinitesimal Generator

Whereas equation (7.15) relates to a stochastic matrix, it is also possible to derive a similar equation using the infinitesimal generator. Given that Q is in normal periodic form, equation (7.8), $\pi Q = 0$ leads to

$$
\begin{aligned}
\pi_1 D_1 &= -\pi_p Q_p \\
\pi_2 D_2 &= -\pi_1 Q_1 \\
&\vdots \\
\pi_p D_p &= -\pi_{p-1} Q_{p-1},
\end{aligned}
\tag{7.17}
$$

from which we may derive

$$\pi_p D_p = -\pi_p Q_p (-D_1^{-1}) Q_1 \cdots (-D_{p-1}^{-1}) Q_{p-1}.$$

This results in an equation of the form

$$\pi_p T_p = 0, \tag{7.18}$$

where

$$T_p = D_p + Q_p(-D_1^{-1}Q_1) \cdots (-D_{p-1}^{-1}Q_{p-1}). \tag{7.19}$$

Equation (7.18) is called the *reduced scheme* associated with the infinitesimal generator Q. It involves only a part of the original probability vector π. Again, it is sufficient to work with the reduced scheme, since once π_p is obtained, the other subvectors π_i may be determined from π_p by matrix-vector multiplications.

It is worthwhile noticing that both reduced schemes are related. Indeed, matrices T_p and R_p are related by

$$T_p = D_p(I - R_p), \qquad (7.20)$$

since equation (7.19) may be written as

$$T_p = D_p(I - (-D_p^{-1}Q_p)(-D_1^{-1}Q_1)\cdots(-D_{p-1}^{-1}Q_{p-1}) \qquad (7.21)$$

which, using equations (7.6) and (7.16), leads to (7.20). Notice that since R_p is stochastic and irreducible, it follows from (7.20) that T_p is an irreducible infinitesimal generator.

From the point of view of numerical computation, it is of interest to use either of the reduced schemes (7.15) and (7.18) rather than the corresponding original equations, $x = xS$ and $\pi Q = 0$. All of the blocks R_i have the same spectrum [84]. Since the eigenvalues of R are those of S raised to the power p, the spectrum of R_i is often much more suitable for iterative methods than that of S. Clearly, the same is true for T_p.

7.5.3 Numerical Methods Based on the Reduced Scheme

We now consider the application of some numerical methods to the reduced scheme and show how they may be used to reduce computational complexity. We shall not consider methods that require the explicit construction of T_p (or R_p). Depending on the nonzero structure of the individual blocks, it may or may not be beneficial to form T_p explicitly. We shall leave this for the reader to explore. Our choice therefore eliminates LU decompositions and Gauss–Seidel iterations. On the other hand, methods that need only perform multiplication with the periodic matrix, such as the power and preconditioned power methods or the methods of Arnoldi and GMRES, may be used.

7.5.3.1 Iterative Methods

The power method on S consists of applying the iterative scheme

$$x^{(k+1)} = x^{(k)}S.$$

For an arbitrary initial vector $x^{(0)}$, this will not converge, because S possesses p eigenvalues of unit modulus. There will, however, be convergence of the $x_i^{(k)}$ toward the individual x_i, correct to a multiplicative constant. Since we know from equation (7.14) that $||x_i||_1 = 1/p$ for each i, each subvector may be correctly normalized, and therefore it becomes possible to determine x with the power method applied to the matrix S. Note also that if $x^{(0)}$ is such that $||x_i^{(0)}||_1 = 1/p$ for all $i = 1, 2, \ldots, p$, then the power method will converge directly toward x.

However, it is possible to go p times faster. It is evident from equation (7.15) that x_p is the eigenvector corresponding to the unit eigenvalue of the stochastic matrix R_p. It is therefore possible to apply the power method to this matrix to determine the vector x_p:

$$x_p^{(k+1)} = x_p^{(k)} S_p S_1 S_2 \cdots S_{p-1} = x_p^{(k)} R_p. \qquad (7.22)$$

This will converge p times faster, since the spectrum of R_p is that of S raised to the power p. We call this the *accelerated* power method. Its convergence is guaranteed, because the stochastic matrix R_p is irreducible and acyclic. It is not necessary to compute the matrix R_p explicitly. Instead, the multiplications of (7.22) may be performed succcessively. In this way, the total cost of a step is equal to the number of nonzero elements in the matrix S.

Let us now compare the regular power method on S and the accelerated power method. The amount of computation per iteration is identical for both methods; it is equal to the number of nonzero elements in S. On the other hand, the application of one step of the accelerated power method is identical to p steps of the regular power method. Therefore, the accelerated power method will converge p times faster with the same amount of computation per iteration. This result may be interpreted as a consequence of the better spectral properties of R_p as compared to those of S.

Other iterative methods such as Arnoldi or GMRES, whose only involvement with the coefficient matrix is in multiplying it with a vector, can also be applied to the reduced scheme (7.18). Again, these methods will make use of the better spectral properties of T_p as compared to those of Q. Moreover, the construction of the orthonormal basis of the Krylov subspace requires less computation time, since the dimension of the reduced scheme is of the order n/p, where n is the number of states of the Markov chain. Notice that since all the matrices T_i have the same nonzero eigenvalues, it is advantageous to work with the smallest of them (which cannot exceed n/p).

7.5.3.2 Preconditioning

We may also apply preconditioned methods to the reduced schemes. Consider $\pi_p T_p = 0$. If we were to implement an ILU decomposition directly on T_p, we would need T_p explicitly, but this we have ruled out. There is, however, an alternative way of obtaining an ILU decomposition of T_p. Indeed, it is apparent from equations (7.9) and (7.21) that $L_p U_p$ is an LU decomposition of T_p. Now, it may easily be verified that performing an ILU decomposition of Q (in periodic form)

Table 7.4: Results for Power Iteration with *ILUTH* Preconditioning: the Non-NCD Case

N	η	n_z	T'hold	Nonperiodic			Periodic		
				Fill-in	Iters	Time	Fill-in	Iters	Time
5	252	1,260	0.02	2,564	26	14.3	531	23	6.7
			0.01	3,754	16	12.5	1,075	13	6.0
6	462	2,478	0.01	6,926	26	36.0	2,069	20	16.2
			0.005	10,570	17	35.9	3,601	10	15.2
8	1,287	7,623	0.01	20,389	50	200.0	6,337	45	102.0
			0.005	31,099	33	192.0	11,889	22	88.5

provides approximations $\tilde{L}_p$ and $\tilde{U}_p$ to L_p and U_p, respectively. As a result, $\tilde{L}_p \tilde{U}_p$ provides an *ILU* decomposition of T_p. Thus, performing an *ILU* decomposition of Q implicitly provides an *ILU* decomposition of T_p. Preconditioned methods can thus be used on the reduced schemes without explicitly forming T_p or R_p.

7.5.4 Numerical Experiments

We shall use the same cyclic queueing network example used previously in Section 7.4.3.1. This time we consider preconditioned power iterations and Arnoldi's method, both with threshold-based *ILU* preconditioning.

We make comparisons between the preconditioned methods applied to the reduced scheme and the preconditioned methods applied to the original matrix. We do not compare these results with those obtained when preconditioned methods are applied to the matrix in normal periodic form; the results obtained on the reduced scheme were always superior. The (unpreconditioned) power method is not considered either, because we have already seen that the gain in computing time is exactly p. Also, as far as the unpreconditioned Arnoldi method is concerned, the improvement in computation time was found to be greater than p and the improvement in memory equal to p. These results are not presented here, because the preconditioned version of this method is much more efficient.

The experiments of this section were carried out on a Sun 3/80 with an arithmetic coprocessor; this is a slower machine than the SPARCstation 2, so the computation times should not be compared against those of the previous section.

7.5.4.1 *ILUTH*-Preconditioned Power Iterations

Let us consider the non-NCD case first. The results obtained with this method are provided in Table 7.4. In these examples it is abundantly clear that the

Table 7.5: Results for Power Iteration with *ILUTH* Preconditioning: the NCD Case

				Nonperiodic			Periodic		
N	η	n_z	T'hold	Fill-in	Iters	Time	Fill-in	Iters	Time
5	252	1,260	0.002	5,621	22	23.7	1,064	22	8.6
			0.00002	9,308	6	18.0	2,013	6	6.2
6	462	2,478	0.002	13,975	76	172.7	2,854	27	24.1
			0.00002	24,813	7	61.4	5,471	8	20.7
8	1,287	7,623	0.0001	68,008	98	1,073.7	15,977	40	166.2
			0.0005	77,837	35	543.2	16,512	26	128.4

periodic version of the preconditioned power method is superior to the nonperiodic version. The savings in memory, as measured by the ratio of the nonperiodic to the periodic versions, is from 2.6 to 4.8. Computation time is reduced by a factor between 2.0 and 2.4. It is known that the choice of a numerical value for the threshold has an important influence on methods that use *ILUTH* preconditioning. However, no matter what threshold was chosen, the periodic version outperformed the nonperiodic version.

We now consider the NCD case. The results are presented in Table 7.5. The benefits in terms of decreased computation time and memory requirements are seen to be better than in the non-NCD case. This may be explained, to a certain extent, by the fact that smaller thresholds have been chosen in these NCD cases. The memory improvement factors are now between 4.2 and 5.3. The computation time improvement factor is between 2.8 and 7.2. The difference in fill-in between the two versions is also substantially larger. Notice, finally, that the savings in computation time in this method results more from the decrease in the number of nonzero elements in the preconditioner than from any decrease in the number of iterations.

7.5.4.2 Arnoldi's Method with *ILUTH* Preconditioning

The results obtained by the method of Arnoldi are presented in Tables 7.6 and 7.7 for the non-NCD and NCD cases, respectively. This method requires an additional parameter, the size of the Krylov subspace. We chose a value of 10 for this parameter. The conclusions are similar to those of the preconditioned power method. Notice, however, that here the improvements are even better. For the non-NCD case the improvement factors for memory lie between 3.2 and 10.4, and the improvement factors for computation time lie between 2.0 and 2.4.

Table 7.6: Results for Preconditioned (ILUTH) Arnoldi: the Non-NCD Case

				Nonperiodic			Periodic		
N	η	n_z	T'hold	Fill-in	Iters	Time	Fill-in	Iters	Time
5	252	1,260	0.05	1,399	20	17.5	134	20	7.2
			0.02	2,564	10	11.6	531	10	5.5
6	462	2,478	0.05	2,627	20	33.7	216	20	14.1
			0.01	6,926	10	27.3	2,069	10	14.0
8	1,287	7,623	0.02	13,758	20	125.3	2,871	20	59.5
			0.01	20,389	20	148.7	6,337	20	74.5

Table 7.7: Results for Preconditioned (*ILUTH*) Arnoldi: the NCD Case

				Nonperiodic			Periodic		
N	η	n_z	T'hold	Fill-in	Iters	Time	Fill-in	Iters	Time
5	252	1,260	0.01	3,404	20	24.5	820	20	9.9
			0.005	4,721	10	17.0	1,023	10	6.5
6	462	2,478	0.005	9,696	20	59.5	2523	20	23.9
			0.002	13,975	10	48.2	2,854	10	16.3
8	1,287	7,623	0.002	49,720	30	385.2	14,602	20	117.9
			0.001	68,009	20	396.7	15,977	10	89.5

For the NCD case the improvement factors for memory are from 3.4 to 4.9, and for computation time from 2.5 to 4.4.

7.6   IAD Methods for NCD, *p*-Cyclic Markov Chains

Iterative aggregation/disaggregation (IAD) methods were discussed in Chapter 6. We now investigate how advantage may be derived from the periodicity property when IAD methods are used for periodic Markov chains. To handle the added complexity of periodicity, the notation of this section differs slightly from the notation used previously.

7.6.1 Iterative Aggregation/Disaggregation (IAD) Methods

In Chapter 6 we saw that in IAD methods the state space is partitioned into a set of N blocks. Each iteration of an IAD procedure consists of an aggregation or coupling step followed by a disaggregation step. The coupling step involves generating an $N \times N$ stochastic matrix of block transition probabilities and then determining its stationary probability vector. When the number of blocks is small with respect to the number of states — and this is often the case — the computational complexity of the coupling step is negligible compared to that of the disaggregation step.

The disaggregation step, which in many methods is related to a single iteration of block Gauss–Seidel, computes an approximation to the probability of each state of block j, for $1 \leq j \leq N$. The part of the disaggregation step corresponding to block j is obtained by solving the system of equations

$$\pi^j Q^j = b^j, \tag{7.23}$$

where π^j is an approximate vector of probabilities of block j; Q^j is a diagonal submatrix of the infinitesimal generator Q corresponding to the states of block j; and b^j is a vector obtained from the previous aggregation and disaggregation step. IAD methods have proven to be especially efficient when applied to nearly completely decomposable Markov chains.

7.6.2 Ordering for Periodicity and Decomposability

Consider a CTMC whose associated graph is periodic of period p. We assume that the set of states is partitioned into N blocks according to the degree of decomposability of the Markov chain. Notice that this partitioning of the state space into blocks is not at all related to a partitioning of the state space based on the periodic structure of the Markov chain.

Consider the submatrix Q^j corresponding to block j. The set of vertices of G^j, the graph associated with this submatrix, is a subset of the set of vertices of the graph of the CTMC. As a result, the length of all elementary cycles in G^j is a multiple of p. Therefore, it is possible to order the states corresponding to block j so that the submatrix Q^j appears in the following form.

$$Q^j = \begin{pmatrix} D_1^j & Q_1^j & 0 & \cdots & 0 & 0 \\ 0 & D_2^j & Q_2^j & \cdots & 0 & 0 \\ 0 & 0 & D_3^j & \cdots & 0 & 0 \\ \vdots & \vdots & \vdots & \ddots & \vdots & \vdots \\ 0 & 0 & 0 & \cdots & D_{p-1}^j & Q_{p-1}^j \\ Q_p^j & 0 & 0 & \cdots & 0 & D_p^j \end{pmatrix}. \tag{7.24}$$

Notice that the diagonal blocks D_i^j are themselves diagonal matrices and that the structure of Q^j is identical to that of Q when Q is given in periodic form.

Thus, given a decomposability-based partitioning of the state space into blocks, it is possible to permute the elements of each block so that it has the form (7.24). The periodic class of each state is determined as usual during the generation of the matrix, and it suffices to permute the states of each block accordingly. The net result is that the states are ordered, first with respect to the degree of decomposability of the Markov chain, and secondly within each block according to their periodic class.

7.6.3 Application to *p*-Cyclic Matrices

In solving the systems of equations $\pi^j Q^j = b^j$ for $j = 1, 2, \ldots, N$, during each disaggregation step it is possible to take advantage of the periodic structure of the matrices Q^j. First, when the methods described in Section 7.4 are applied to equation (7.23), they can take advantage of the periodic form of Q^j. When the *LU* decomposition method is used, the savings in computation time and memory requirements can be especially significant.

Second, the reduced schemes described in Section 7.5 for the original system $\pi Q = 0$ may also be applied to $\pi^j Q^j = b^j$. Indeed, let us partition the vector π^j of conditional probabilities of block j according to the periodic structure of Q^j, i.e., $\pi^j = (\pi_1^j, \pi_2^j, \ldots, \pi_p^j)$. Equation (7.23) may then be decomposed into

$$\begin{aligned}
\pi_1^j D_1^j &= -\pi_p^j Q_p^j + b_1^j \\
\pi_2^j D_2^j &= -\pi_1^j Q_1^j + b_2^j \\
&\vdots \quad \vdots \quad \vdots \\
\pi_p^j D_p^j &= -\pi_{p-1}^j Q_{p-1}^j + b_p^j.
\end{aligned}$$

Consider, for instance, the subvector π_p^j. From the above set of equations, it follows that

$$\begin{aligned}
\pi_p^j D_p^j = \quad &-\pi_p^j Q_p^j (-D_1^j)^{-1} Q_1^j \cdots (-D_{p-1}^j)^{-1} Q_{p-1}^j + \\
&b_1^j (-D_1^j)^{-1} Q_1^j \cdots (-D_{p-1}^j)^{-1} Q_{p-1}^j + \\
&b_2^j (-D_2^j)^{-1} Q_2^j \cdots (-D_{p-1}^j)^{-1} Q_{p-1}^j + \\
&\cdots + \\
&b_{p-1}^j (-D_{p-1}^j)^{-1} Q_{p-1}^j + \\
&b_p^j,
\end{aligned}$$

which leads to

$$\pi_p^j \left(D_p^j + Q_p^j(-D_1^j)^{-1}Q_1^j \cdots (-D_{p-1}^j)^{-1}Q_{p-1}^j \right) =$$
$$b_1^j(-D_1^j)^{-1}Q_1^j \cdots (-D_{p-1}^j)^{-1}Q_{p-1}^j +$$
$$b_2^j(-D_2^j)^{-1}Q_2^j \cdots (-D_{p-1}^j)^{-1}Q_{p-1}^j +$$
$$\cdots +$$
$$b_{p-1}^j(-D_{p-1}^j)^{-1}Q_{p-1}^j +$$
$$b_p^j,$$

i.e., an equation of the type

$$\pi_p^j T_p^j = c_p^j. \tag{7.25}$$

This equation is the reduced scheme associated with equation (7.23). It involves only a part of π^j, namely π_p^j. It is similar to equation (7.18) except that the right-hand term is nonzero. By using a form of nesting, the right-hand side c_p^j may be formed in a number of operations that is equal to the number of nonzero elements in the matrix Q^j. Despite this nonzero right-hand side, the methods discussed in Section 7.4 can still be applied, and the conclusions reached there concerning the saving in memory requirements and computation time are also applicable. Notice in particular that it is possible to apply preconditioned methods to this reduced scheme. An *ILU* decomposition of T_p^j is now obtained by performing an *ILU* decomposition of Q^j in periodic form.

In summary, it is still possible to take advantage of the periodicity property when using aggregation/disaggregation methods. Since in many instances most of the complexity of IAD methods is due to the disaggregation steps, it should be expected that the savings achieved will be of the same order of magnitude as that obtained by a periodic method over a nonperiodic method applied to the entire matrix Q.

7.7 Block SOR and Optimum Relaxation

7.7.1 Introduction

Given the system of equations

$$Ax = b; \qquad A \in \Re^{n \times n}; \qquad x, b \in \Re^n, \tag{7.26}$$

and the usual block decomposition

$$A = D - L - U \tag{7.27}$$

where D, L, and U are block diagonal, strictly lower block triangular, and strictly upper block triangular matrices, respectively, and D is nonsingular, the block SOR method with relaxation parameter $\omega \neq 0$ is defined as

$$Dx^{(m)} = Dx^{(m-1)} + \omega(Lx^{(m)} - Dx^{(m-1)} + Ux^{(m-1)} + b), \qquad m = 1, 2, \dots \quad (7.28)$$

The method can be described equivalently as

$$x^{(m)} = \mathcal{L}_\omega x^{(m-1)} + c, \qquad m = 1, 2, \dots \quad (7.29)$$

where

$$\mathcal{L}_\omega \equiv (D - \omega L)^{-1}[(1 - \omega)D + \omega U)] \quad (7.30)$$

and

$$c \equiv \omega(D - \omega L)^{-1}b. \quad (7.31)$$

For general nonsingular systems, SOR converges if and only if the spectral radius of the iteration matrix is strictly less than 1, i.e., if and only if $\rho(\mathcal{L}_\omega) < 1$. The associated convergence factor is then $\rho(\mathcal{L}_\omega)$. Few results are available on how the relaxation parameter affects convergence. For the special but important case of "matrices with property A" Young [183] derived an optimum value, ω_{opt}. This was generalized by Varga [169, 170] to systems having an associated block Jacobi matrix, $J \equiv D^{-1}(L + U)$, that is weakly cyclic of index p. For such matrices Varga established the important relationship

$$(\lambda + \omega - 1)^p = \lambda^{p-1}\omega^p\mu^p \quad (7.32)$$

between the eigenvalues μ of J and the eigenvalues λ of $\mathcal{L}_\omega$. Furthermore, when all the eigenvalues of J^p satisfy $0 \leq \mu^p \leq \rho(J^p) < 1$, he shows that ω_{opt} is the unique positive solution of the equation

$$[\rho(J)\omega]^p = p^p(p-1)^{1-p}(\omega - 1) \quad (7.33)$$

in the interval $(1, p/(p-1))$. This yields a convergence factor equal to

$$\rho(\mathcal{L}_\omega) = (p-1)(\omega_{\text{opt}} - 1). \quad (7.34)$$

Similar results have been obtained when the eigenvalues of J^p are nonpositive, i.e., $-1 < -\rho(J)^p < \mu^p \leq 0$ [23, 53, 112, 113, 123, 176], as well as when they are both positive and negative [38].

All of these results pertain only to *nonsingular* systems of equations. Hadjidimos [64] examined the singular case ($\det(A) = 0$ and $b \in \mathcal{R}(A)$). Under the assumptions that J is weakly cyclic of index p, that the eigenvalues of J^p are nonnegative, and that J has a simple unit eigenvalue, he showed that ω_{opt} is the unique root of (7.33) in the same interval as in the nonsingular case, but with $\rho(J)$ replaced by $\tilde{\rho}(J)$, the maximum of the moduli of the eigenvalues of J, excluding those of modulus 1; i.e., $\tilde{\rho}(J) \equiv \max\{|\lambda|, \ \lambda \in \sigma(J), \ |\lambda| \neq 1\}$.

We are interested in using block SOR to compute the stationary probability distribution of an irreducible Markov chain for which the infinitesimal generator has the form

$$
Q = \begin{pmatrix}
Q_{11} & Q_{12} & 0 & \cdots & 0 & 0 \\
0 & Q_{22} & Q_{23} & \cdots & 0 & 0 \\
\vdots & \vdots & \vdots & \ddots & \vdots & \vdots \\
0 & 0 & 0 & \cdots & Q_{p-1,p-1} & Q_{p-1,p} \\
Q_{p1} & 0 & 0 & \cdots & 0 & Q_{pp}
\end{pmatrix}. \tag{7.35}
$$

The subblocks Q_{ii}, $i = 1, 2, \ldots, p$ are square. Notice that, in this case, the diagonal blocks of the infinitesimal generator are not necessarily diagonal submatrices, and thus the Markov chain may be *aperiodic*. Let

$$
Q = \hat{D} - \hat{L} - \hat{U}, \tag{7.36}
$$

where $\hat{D}$, $\hat{L}$ and $\hat{U}$ are block diagonal, strictly lower block triangular, and strictly upper block triangular matrices, respectively.

Theorem 7.6 *The matrix*

$$
\hat{J} = \hat{D}^{-1}\left(\hat{L} + \hat{U}\right) \tag{7.37}
$$

is an irreducible cyclic stochastic matrix of period p.

Proof   Indeed, since $-Q$ is a singular M-matrix, $-\hat{D}_{ii}$, $i = 1, 2, \ldots, p$ are nonsingular M-matrices, and therefore $-\hat{D}_{ii}^{-1}$ exist and are strictly positive. Since $-(\hat{L} + \hat{U})$ is nonnegative (it is the off-diagonal part of an infinitesimal generator), we conclude that $\hat{J} = \hat{D}^{-1}\left(\hat{L} + \hat{U}\right)$ is nonnegative. Furthermore, using the fact that $Qe = 0$, we have

$$
Qe = \hat{D}(I - \hat{D}^{-1}(\hat{L} + \hat{U}))e = \hat{D}(I - \hat{J})e = 0
$$

and so $\hat{J}e = e$. Along with nonnegativity, this proves that $\hat{J}$ is a stochastic matrix. The irreducibility results from the fact that $-Q$ is an irreducible M-matrix. Lastly, observe that, by construction, $\hat{J}$ is cyclic. The period is p, for otherwise some of the blocks in $\hat{J}$ could be partitioned further, or equivalently, Q would be k-cyclic for some $k > p$.   □

The proofs of the properties of M-matrices used in proving Theorem 7.6 may be found in [10]. Notice that if the Q_{ii} are diagonal, then $\hat{J}$ is the embedded Markov chain and corresponds to the matrix S defined in Section 7.4.1. We now put the system of equations into the form (7.26) by writing

$$Q^T x = 0, \qquad \|x\|_1 = 1, \tag{7.38}$$

with $x = \pi^T$. Let Q^T be partitioned as in (7.27). Obviously, referring to (7.36), $D = \hat{D}^T$, $L = \hat{U}^T$ and $U = \hat{L}^T$. The block Jacobi matrix J associated with the system (7.38) is given by

$$J = D^{-1}(L + U) \tag{7.39}$$

and is consistently ordered. We may write

$$J = \begin{pmatrix} 0 & 0 & 0 & \cdots & 0 & B_1 \\ B_2 & 0 & 0 & \cdots & 0 & 0 \\ 0 & B_3 & 0 & \cdots & 0 & 0 \\ \vdots & \vdots & \vdots & \ddots & \vdots & \vdots \\ 0 & 0 & 0 & \cdots & 0 & 0 \\ 0 & 0 & 0 & \cdots & B_p & 0 \end{pmatrix}.$$

It follows immediately that

$$J = D^{-1}\hat{J}^T D.$$

In the following sections we show that when the value of the relaxation parameter ω is chosen in such a way that $\mathcal{L}_\omega$ violates the conditions for semiconvergency, SOR "converges" to a vector that has subvectors parallel to the corresponding subvectors of the stationary probability vector π. We refer to this as *extended convergence*. It is straightforward to compute the actual weights that should multiply the subvectors produced by SOR. Optimal values for ω may be found outside the usual range. Furthermore, numerical experiments suggest that these values are much more robust than the "usual" optimum in $(1, p/(p-1))$.

7.7.2 A 3-Cyclic Queueing Network Example

Throughout this section, we shall make use of the cyclic queueing network shown in Figure 7.9, with three service centers in series. Each center contains a single exponential server and an infinite queue. A state of this system may be represented by a three-component vector (η_1, η_2, η_3), where η_i denotes the number of customers at the ith service station. With three customers in the network ($N = 3$), the total number of states generated is 10. These 10 states, and the transitions that are possible among them, are illustrated in Figure 7.10. Notice that the first four states on the left are repeated on the right, so all transition arrows head in the same direction.

N=3

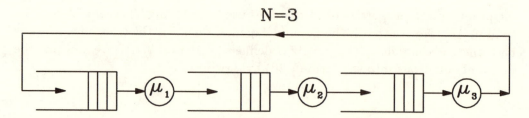

Figure 7.9: Queueing model example.

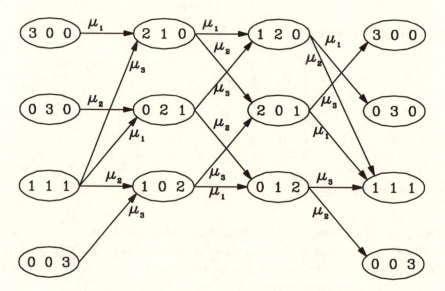

Figure 7.10: State space and transitions.

The infinitesimal generator for this system is given by

$$
Q = \begin{pmatrix}
-\sigma_1 & \cdot & \cdot & \cdot & \mu_1 & \cdot & \cdot & \cdot & \cdot & \cdot \\
\cdot & -\sigma_2 & \cdot & \cdot & \cdot & \mu_2 & \cdot & \cdot & \cdot & \cdot \\
\cdot & \cdot & -\sigma_3 & \cdot & \mu_3 & \mu_1 & \mu_2 & \cdot & \cdot & \cdot \\
\cdot & \cdot & \cdot & -\sigma_4 & \cdot & \cdot & \mu_3 & \cdot & \cdot & \cdot \\
\cdot & \cdot & \cdot & \cdot & -\sigma_5 & \cdot & \cdot & \mu_1 & \mu_2 & \cdot \\
\cdot & \cdot & \cdot & \cdot & \cdot & -\sigma_6 & \cdot & \mu_3 & \cdot & \mu_2 \\
\cdot & \cdot & \cdot & \cdot & \cdot & \cdot & -\sigma_7 & \cdot & \mu_3 & \mu_1 \\
\cdot & \mu_1 & \mu_2 & \cdot & \cdot & \cdot & \cdot & -\sigma_8 & \cdot & \cdot \\
\mu_3 & \cdot & \mu_1 & \cdot & \cdot & \cdot & \cdot & \cdot & -\sigma_9 & \cdot \\
\cdot & \cdot & \mu_3 & \mu_2 & \cdot & \cdot & \cdot & \cdot & \cdot & -\sigma_{10}
\end{pmatrix}.
$$

Assigning specific numerical values to the parameters of this model, $\mu_1 = .2$, $\mu_2 = .3$, and $\mu_3 = .5$, this becomes

$$
Q = \left(
\begin{array}{cccc|ccc|ccc}
-.2 & \cdot & \cdot & \cdot & .2 & \cdot & \cdot & \cdot & \cdot & \cdot \\
\cdot & -.3 & \cdot & \cdot & \cdot & .3 & \cdot & \cdot & \cdot & \cdot \\
\cdot & \cdot & -1.0 & \cdot & .5 & .2 & .3 & \cdot & \cdot & \cdot \\
\cdot & \cdot & \cdot & -.5 & \cdot & \cdot & .5 & \cdot & \cdot & \cdot \\
\cdot & \cdot & \cdot & \cdot & -.5 & \cdot & \cdot & .2 & .3 & \cdot \\
\cdot & \cdot & \cdot & \cdot & \cdot & -.8 & \cdot & .5 & \cdot & .3 \\
\cdot & \cdot & \cdot & \cdot & \cdot & \cdot & -.7 & \cdot & .5 & .2 \\
\cdot & .2 & .3 & \cdot & \cdot & \cdot & \cdot & -.5 & \cdot & \cdot \\
.5 & \cdot & .2 & \cdot & \cdot & \cdot & \cdot & \cdot & -.7 & \cdot \\
\cdot & \cdot & .5 & .3 & \cdot & \cdot & \cdot & \cdot & \cdot & -.8
\end{array}
\right).
$$

The embedded Markov chain is

$$
\hat{J} = \left(
\begin{array}{cccc|ccc|ccc}
\cdot & \cdot & \cdot & \cdot & 1.0 & \cdot & \cdot & \cdot & \cdot & \cdot \\
\cdot & \cdot & \cdot & \cdot & \cdot & 1.0 & \cdot & \cdot & \cdot & \cdot \\
\cdot & \cdot & \cdot & \cdot & 5/10 & 2/10 & 3/10 & \cdot & \cdot & \cdot \\
\cdot & \cdot & \cdot & \cdot & \cdot & \cdot & 1.0 & \cdot & \cdot & \cdot \\
\cdot & \cdot & \cdot & \cdot & \cdot & \cdot & \cdot & 2/5 & 3/5 & \cdot \\
\cdot & \cdot & \cdot & \cdot & \cdot & \cdot & \cdot & 5/8 & \cdot & 3/8 \\
\cdot & \cdot & \cdot & \cdot & \cdot & \cdot & \cdot & \cdot & 5/7 & 2/7 \\
\cdot & 2/5 & 3/5 & \cdot & \cdot & \cdot & \cdot & \cdot & \cdot & \cdot \\
5/7 & \cdot & 2/7 & \cdot & \cdot & \cdot & \cdot & \cdot & \cdot & \cdot \\
\cdot & \cdot & 5/8 & 3/8 & \cdot & \cdot & \cdot & \cdot & \cdot & \cdot
\end{array}
\right),
$$

and the Jacobi iteration matrix J is

$$
J = \left(
\begin{array}{cccc|ccc|ccc}
\cdot & \cdot & \cdot & \cdot & \cdot & \cdot & \cdot & \cdot & 5/2 & \cdot \\
\cdot & \cdot & \cdot & \cdot & \cdot & \cdot & \cdot & 2/3 & \cdot & \cdot \\
\cdot & \cdot & \cdot & \cdot & \cdot & \cdot & \cdot & 3/10 & 2/10 & 5/10 \\
\cdot & \cdot & \cdot & \cdot & \cdot & \cdot & \cdot & \cdot & \cdot & 3/5 \\
2/5 & \cdot & 1.0 & \cdot & \cdot & \cdot & \cdot & \cdot & \cdot & \cdot \\
\cdot & 3/8 & 2/8 & \cdot & \cdot & \cdot & \cdot & \cdot & \cdot & \cdot \\
\cdot & \cdot & 3/7 & 5/7 & \cdot & \cdot & \cdot & \cdot & \cdot & \cdot \\
\cdot & \cdot & \cdot & \cdot & 2/5 & 1.0 & \cdot & \cdot & \cdot & \cdot \\
\cdot & \cdot & \cdot & \cdot & 3/7 & \cdot & 5/7 & \cdot & \cdot & \cdot \\
\cdot & \cdot & \cdot & \cdot & \cdot & 3/8 & 2/8 & \cdot & \cdot & \cdot
\end{array}
\right).
$$

Notice that the matrix J is *not* a stochastic matrix. The eigenvalues of J (which must be the same as those of $\hat{J}$), and those of J^p, as computed by MATLAB, are given in Table 7.8. Throughout this section we shall refer to this example to illustrate various properties of cyclic matrices in relation to the SOR method. Notice that both J and J^p possess complex eigenvalues.

20

80 **Chapter 7**

Table 7.8: Eigenvalues of Jacobi Iteration Matrix

Eigenvalues of J	Eigenvalues of J^3
$-0.5000 + 0.8660i$	1.0000
$-0.5000 + 0.8660i$	1.0000
1.0000	1.0000
$-0.4535 + 0.3904i$	$0.1140 + 0.1814i$
$-0.4535 - 0.3904i$	$0.1140 - 0.1814i$
$-0.1113 + 0.5880i$	$0.1140 + 0.1814i$
$-0.1113 - 0.5880i$	$0.1140 - 0.1814i$
$0.5648 + 0.1976i$	$0.1140 + 0.1814i$
$0.5648 - 0.1976i$	$0.1140 - 0.1814i$
0.0000	0.0000

7.7.3 A p-step Iteration Procedure

We now consider a p-step iterative method involving J for solving $Ax = b$. Its convergence is closely related to the convergence of SOR [63]. Consider the *p-step relaxation*

$$x^{(m)} = \omega J x^{(m-1)} + (1-\omega)x^{(m-p)} + \omega c', \qquad m \geq p \qquad (7.40)$$

where $J = D^{-1}(L+U)$ is the Jacobi matrix associated with (7.26) and $c' = D^{-1}b$. Let the iterates $x^{(m)} = \left(x_1^{(m)^T}, x_2^{(m)^T}, \ldots, x_p^{(m)^T}\right)^T$ be partitioned conformally with J. By considering successive block equations arising from consecutive iterations of the p-step method, it is immediately apparent that

$$
\begin{aligned}
x_1^{(m)} &= \omega B_1 x_p^{(m-1)} + (1-\omega)x_1^{(m-p)} + \omega c'_1 \\
x_2^{(m+1)} &= \omega B_2 x_1^{(m)} + (1-\omega)x_2^{(m-p+1)} + \omega c'_2 \\
&\vdots \qquad \vdots \qquad \vdots \\
x_p^{(m+p-1)} &= \omega B_p x_{p-1}^{(m+p-2)} + (1-\omega)x_p^{(m-1)} + \omega c'_p
\end{aligned}
\qquad (7.41)
$$

If, on the other hand, we apply one SOR iteration to the vector

$$y^{(m-1)} = \left(x_1^{(m-p)^T}, \ x_2^{(m-p+1)^T}, \ \ldots, \ x_p^{(m-1)^T}\right)^T,$$

we get

$$y^{(m+p-1)} = \left(x_1^{(m)^T}, \ x_2^{(m+1)^T}, \ \ldots, \ x_p^{(m+p-1)^T}\right)^T.$$

Table 7.9: p-Step Iterations $(x^{(i)})$ with $\omega = 1.1$

	$i = 1$	$i = 2$	$i = 3$	$i = 4$	$i = 5$	$i = 6$	$i = 7$	$i = 8$
$x_1^{(i)}$	**0.2650**	0.3082	0.3992	**0.4021**	0.3604	0.3505	**0.3328**	0.3371
$x_2^{(i)}$	**0.0633**	0.0956	0.0765	**0.0914**	0.0991	0.1066	**0.1071**	0.1026
$x_3^{(i)}$	**0.1000**	0.0953	0.0870	**0.0849**	0.0888	0.0896	**0.0914**	0.0911
$x_4^{(i)}$	**0.0560**	0.0288	0.0204	**0.0144**	0.0188	0.0167	**0.0198**	0.0214
$x_5^{(i)}$	0.1440	**0.2166**	0.2304	0.2570	**0.2487**	0.2332	0.2271	**0.2221**
$x_6^{(i)}$	0.0588	**0.0436**	0.0556	0.0496	**0.0567**	0.0597	0.0637	**0.0636**
$x_7^{(i)}$	0.1157	**0.0811**	0.0575	0.0455	**0.0432**	0.0509	0.0508	**0.0544**
$x_8^{(i)}$	0.1440	0.1180	**0.1333**	0.1482	0.1558	**0.1585**	0.1535	0.1543
$x_9^{(i)}$	0.1157	0.1488	**0.1559**	0.1423	0.1420	**0.1356**	0.1357	0.1327
$x_{10}^{(i)}$	0.0588	0.0461	**0.0303**	0.0329	0.0284	**0.0323**	0.0353	0.0374

In other words, after one SOR step is applied to $y^{(m-1)}$, the block components are exactly the same as if a whole sweep (equation (7.41)), involving p iterations of the p-step relaxation (7.40), were applied. We therefore see that when solving a nonhomogeneous system with nonsingular coefficient matrix, SOR converges exactly when the p-step iteration converges, and SOR is p times faster. This is not necessarily true in our case. The system that we wish to solve, equation (7.38), is a homogeneous system with a coefficient matrix of order n and rank $n - 1$. The solution of (7.38) is specified only up to a multiplicative constant, and different iterates of the p-step iteration may carry different multiplicative constants. Although SOR may converge in the standard sense to the true solution, a method based on concatenating parts of successive iterates of the p-step iteration may "converge" to a vector composed of subvectors parallel to their counterparts in the solution but multiplied by different factors, so that the computed "solution" will not be parallel to the solution of (7.38).

To illustrate this, we have printed the first eight iterations of the p-step iterative procedure (Table 7.9) and the same number of iterations of SOR (Table 7.10) when these methods are applied to the 10-state example of Section 7.7.2. In both cases the initial vector was chosen as $x_i^{(0)} = y_i^{(0)} = 0.1$, for $i = 1, 2, \ldots, 10$. The value of the relaxation parameter was chosen as $\omega = 1.1$. Notice how successive blocks of the SOR iterates come from different iterates of the p-step iterations. Thus, for example, the first block of $y^{(1)}$ is the same as the first block of $x^{(1)}$, the second block of $y^{(1)}$ is the same as the second block of $x^{(2)}$, the third block of $y^{(1)}$ is the same as the third block of $x^{(3)}$, etc.

Although we chose a value of ω such that $0 < \omega < 2$, any nonzero value would

Table 7.10: SOR Iterations $(y^{(i)})$ with $\omega = 1.1$

	$i = 1$	$i = 2$	$i = 3$	$i = 4$	$i = 5$	$i = 6$	$i = 7$	$i = 8$
$y_1^{(i)}$	0.2650	0.4021	0.3328	0.3348	0.3425	0.3406	0.3401	0.3404
$y_2^{(i)}$	0.0633	0.0914	0.1071	0.1007	0.1002	0.1010	0.1009	0.1008
$y_3^{(i)}$	0.1000	0.0849	0.0914	0.0913	0.0906	0.0908	0.0908	0.0908
$y_4^{(i)}$	0.0560	0.0144	0.0198	0.0231	0.0217	0.0217	0.0218	0.0218
$y_5^{(i)}$	0.2166	0.2487	0.2221	0.2255	0.2278	0.2269	0.2268	0.2270
$y_6^{(i)}$	0.0436	0.0567	0.0636	0.0603	0.0602	0.0606	0.0605	0.0605
$y_7^{(i)}$	0.0811	0.0432	0.0544	0.0557	0.0542	0.0544	0.0545	0.0545
$y_8^{(i)}$	0.1333	0.1585	0.1519	0.1503	0.1514	0.1514	0.1513	0.1513
$y_9^{(i)}$	0.1559	0.1356	0.1338	0.1367	0.1363	0.1361	0.1362	0.1362
$y_{10}^{(i)}$	0.0303	0.0323	0.0380	0.0364	0.0361	0.0363	0.0363	0.0363

give similar results. For example, with $\omega = 3.1$, and starting with the initial vector, $x^{(0)} = y^{(0)} = e_1$, we find the results shown in Tables 7.11 and 7.12.

We now proceed to make this discussion more concrete by examining the eigensystems of the iteration matrices associated with the p-step iteration and with SOR. The p-step relaxation can be transformed to the equivalent first-order relaxation

$$ r^{(m)} = T r^{(m-1)} + \omega c'', \qquad m = p, p+1, \ldots \qquad (7.42) $$

where

$$ \Re^{np} \ni r^{(m)} = \left(x^{(m)^T}, \; x^{(m-1)^T}, \; \ldots, \; x^{(m-p+1)^T} \right)^T, \qquad (7.43) $$

$$ \Re^{np} \ni c'' = \left(c'^T, \; 0^T, \; \ldots, \; 0^T \right)^T, $$

$$ \Re^{np,np} \ni T = \begin{pmatrix} \omega J & 0 & \ldots & 0 & (1-\omega)I \\ I & 0 & \ldots & 0 & 0 \\ 0 & I & \ldots & 0 & 0 \\ \vdots & \vdots & \ddots & \vdots & \vdots \\ 0 & 0 & \ldots & I & 0 \end{pmatrix}, \qquad (7.44) $$

and I is the identity matrix of order n. Since we are interested in solving the homogeneous problem (7.38), both methods (7.29) and (7.42) reduce to the power method applied to the appropriate iteration matrix. In view of Theorem 7.3, we seek to relate the eigensystems of $\mathcal{L}_\omega$ and T with the eigensystem of J. The next two theorems perform this task.

Table 7.11: p-Step Iterations $(x^{(i)})$ with $\omega = 3.1$

	$i=1$	$i=2$	$i=3$	$i=4$	$i=5$	$i=6$
$x_1^{(i)}$	**−2.1000**	−2.1000	10.6676	**−22.4019**	−22.4019	194.3079
$x_2^{(i)}$	**0**	0	3.1777	**−6.6732**	−6.6732	69.1915
$x_3^{(i)}$	**0**	0	2.4514	**−5.1479**	−5.1479	60.3198
$x_4^{(i)}$	**0**	0	0	**0**	0	16.7901
$x_5^{(i)}$	1.2400	**−2.6040**	−2.6040	18.2230	**−38.2684**	−38.2684
$x_6^{(i)}$	0	**0**	0	5.5939	**−11.7472**	−11.7472
$x_7^{(i)}$	0	**0**	0	3.2568	**−6.8393**	−6.8393
$x_8^{(i)}$	0	1.5376	**−3.2290**	−3.2290	36.7087	**−77.0883**
$x_9^{(i)}$	0	1.6474	**−3.4596**	−3.4596	27.9626	**−58.7214**
$x_{10}^{(i)}$	0	0	**0**	0	9.0269	**−18.9566**

Table 7.12: SOR Iterations $(y^{(i)})$ with $\omega = 3.1$

	$i=1$	$i=2$	$i=3$ 1.0e+03 ×
$y_1^{(i)}$	−2.1000	−22.4019	−0.4080
$y_2^{(i)}$	0	−6.6732	−0.1453
$y_3^{(i)}$	0	−5.1479	−0.1267
$y_4^{(i)}$	0	0	−0.0353
$y_5^{(i)}$	−2.6040	−38.2684	−0.8183
$y_6^{(i)}$	0	−11.7472	−0.2424
$y_7^{(i)}$	0	−6.8393	−0.2320
$y_8^{(i)}$	−3.2290	−77.0883	−1.6043
$y_9^{(i)}$	−3.4596	−58.7214	−1.4776
$y_{10}^{(i)}$	0	−18.9566	−0.4218

Theorem 7.7 *Let μ_j, for $j = 1, 2, \ldots, n$, be the n eigenvalues of J, and let ψ_j be a right-hand eigenvector of J associated with the eigenvalue μ_j, i.e., $J\psi_j = \mu_j\psi_j$. Then each of the np eigenvalues of T is one of the np zeros of the n polynomials of degree p*

$$f_{\omega,\mu_j}(\lambda) \equiv \lambda^p - \omega\mu_j\lambda^{p-1} - (1-\omega), \qquad j = 1, 2, \ldots, n.$$

Furthermore, if $\lambda_{j,k}$ is the kth root of $f_{\omega,\mu_j}(\lambda)$, then the corresponding right-hand eigenvector of T is (up to a single multiplicative constant)

$$\left(\lambda_{j,k}^{p-1}\psi_j^T, \ \lambda_{j,k}^{p-2}\psi_j^T, \ \ldots, \ \lambda_{j,k}\psi_j^T, \psi_j^T\right)^T.$$

Proof Let λ be an eigenvalue of T and $z = (z_1, z_2, \ldots, z_p)^T$ be an associated eigenvector. Then $Tz = \lambda z$, or, by considering (7.44), $\omega J z_1 + (1-\omega)z_p = \lambda z_1$ and $z_i = \lambda z_{i+1}$, $\quad i = 1, 2, \ldots, p-1$. Therefore

$$z_i = \lambda^{p-i}z_p, \qquad i = 1, 2, \ldots, p \tag{7.45}$$

and hence

$$\lambda^{p-1}\omega J z_p + (1-\omega)z_p = \lambda^p z_p,$$

or, in other words,

$$J z_p = \frac{\lambda^p - (1-\omega)}{\omega\lambda^{p-1}}z_p.$$

This last relation shows that z_p is a right-hand eigenvector of J corresponding to some eigenvalue μ_j. Therefore,

$$\mu_j = \frac{\lambda^p - (1-\omega)}{\omega\lambda^{p-1}},$$

which immediately leads to $f_{\omega,\mu_j}(\lambda) = 0$. The proof follows from relation (7.45). □

Returning to the 10-state example, and using a relaxation parameter of $\omega = 3.1$, we find the roots of $f_{\omega,\mu_j}(\lambda) \equiv \lambda^p - \omega\mu_j\lambda^{p-1} - (1-\omega)$, for each of the 10 different eigenvalues μ_j of J, are as given in Table 7.13. Tables 7.14 and 7.15 show the eigenvectors corresponding to two of the eigenvalues of J and some of the related eigenvectors of T. Notice that the eigenvector of T corresponding to $\lambda_{j,k}$ satisfies (up to a single multiplicative constant)

$$\left(\lambda_{j,k}^2\psi_j^T, \ \lambda_{j,k}\psi_j^T, \ \psi_j^T\right)^T,$$

where ψ_j is a right-hand eigenvector of J associated with the eigenvalue μ_j.

The connection between the SOR iterations (7.29) and the p-step relaxation (7.42) is further illuminated by the next theorem.

Table 7.13: Roots of $f_{\omega,\mu_j}(\lambda)$ for Eigenvalues μ_j of J

Eigenvalues μ_j of J	Roots of $f_{\omega,\mu_j}(\lambda)$	Modulus
$-0.5000 + 0.8660i$	$-1.4198 + 2.4591i$	2.8396
	$-0.5000 + 0.8660i$	1.0000
	$0.3698 - 0.6405i$	0.7396
$-0.5000 - 0.8660i$	$-1.4198 - 2.4591i$	2.8396
	$-0.5000 - 0.8660i$	1.0000
	$0.3698 + 0.6405i$	0.7396
1.0000	2.8396	2.8396
	1.0000	1.0000
	-0.7396	0.7396
$-0.4535 + 0.3904i$	$-1.7751 + 0.7959i$	1.9454
	$0.3648 - 0.8059i$	0.8847
	$0.0044 + 1.2202i$	1.2202
$-0.4535 - 0.3904i$	$-1.7751 - 0.7959i$	1.9454
	$0.3648 + 0.8059i$	0.8847
	$0.0044 - 1.2202i$	1.2202
$-0.1113 + 0.5880i$	$0.1983 + 1.9353i$	1.9454
	$0.5156 - 0.7189i$	0.8847
	$-1.0589 + 0.6063i$	1.2202
$-0.1113 - 0.5880i$	$0.1983 - 1.9353i$	1.9454
	$0.5156 + 0.7189i$	0.8847
	$-1.0589 - 0.6063i$	1.2202
$0.5648 + 0.1976i$	$1.5768 + 1.1394i$	1.9454
	$1.0546 - 0.6139i$	1.2202
	$-0.8804 + 0.0870i$	0.8847
$0.5648 - 0.1976i$	$1.5768 + 1.1394i$	1.9454
	$1.0546 - 0.6139i$	1.2202
	$-0.8804 + 0.0870i$	0.8847
0.0000	-1.2806	1.2806
	$0.6403 + 1.1090i$	1.2806
	$0.6403 - 1.1090i$	1.2806

Table 7.14: Some Eigenvalues and Eigenvectors of J

Eigenvalues	
1.0000	0.0000

Corresponding Eigenvectors	
0.7012	0.8809
0.2078	0.2349
0.1870	-0.3524
0.0449	0.2114
0.4675	0.0000
0.1247	0.0000
0.1122	0.0000
0.3117	0.0000
0.2805	0.0000
0.0748	0.0000

Theorem 7.8 *If λ is an eigenvalue of T, then λ^p is an eigenvalue of $\mathcal{L}_\omega$. Furthermore, if μ_j is the eigenvalue of J to which λ is related by Theorem 7.7, then the right-hand eigenvector of $\mathcal{L}_\omega$ corresponding to the eigenvalue λ^p is (up to a multiplicative constant) $\psi' = \left(\psi_1^T,\ \lambda\psi_2^T,\ \ldots,\ \lambda^{p-1}\psi_p^T\right)^T$, where again $\psi = \left(\psi_1^T,\ \psi_2^T,\ \ldots,\ \psi_p^T\right)^T$ is a right-hand eigenvector of J corresponding to the eigenvalue μ_j.*

Proof   We only need to prove that

$$\mathcal{L}_\omega \psi' = \lambda^p \psi'.$$

Let

$$y = \mathcal{L}_\omega \psi'.$$

From (7.30), we have

$$(D - \omega L)\, y = [(1 - \omega)D + \omega U]\, \psi'$$

or

$$\left(I - \omega D^{-1}L\right) y = \left[(1 - \omega)I + \omega D^{-1}U\right] \psi',$$

Table 7.15: Some Eigenvalues and Eigenvectors of T

Eigenvalues			
2.8396	1.0000	−0.7396	−1.2806
Corresponding Eigenvectors			
−0.6570	0.4049	−0.2823	−0.6258
−0.1947	0.1200	−0.0836	−0.1669
−0.1752	0.1080	−0.0753	0.2503
−0.0420	0.0259	−0.0181	−0.1502
−0.4380	0.2699	−0.1882	0.0000
−0.1168	0.0720	−0.0502	0.0000
−0.1051	0.0648	−0.0452	0.0000
−0.2920	0.1799	−0.1255	0.0000
−0.2628	0.1619	−0.1129	0.0000
−0.0701	0.0432	−0.0301	0.0000
−0.2314	0.4049	0.3817	0.4887
−0.0686	0.1200	0.1131	0.1303
−0.0617	0.1080	0.1018	−0.1955
−0.0148	0.0259	0.0244	0.1173
−0.1542	0.2699	0.2545	0.0000
−0.0411	0.0720	0.0679	0.0000
−0.0370	0.0648	0.0611	0.0000
−0.1028	0.1799	0.1696	0.0000
−0.0925	0.1619	0.1527	0.0000
−0.0247	0.0432	0.0407	0.0000
−0.0815	0.4049	−0.5161	−0.3816
−0.0241	0.1200	−0.1529	−0.1018
−0.0217	0.1080	−0.1376	0.1526
−0.0052	0.0259	−0.0330	−0.0916
−0.0543	0.2699	−0.3441	0.0000
−0.0145	0.0720	−0.0918	0.0000
−0.0130	0.0648	−0.0826	0.0000
−0.0362	0.1799	−0.2294	0.0000
−0.0326	0.1619	−0.2064	0.0000
−0.0087	0.0432	−0.0551	0.0000

and by (7.39) and (7.2), it follows that

$$y_1 = (1 - \omega)\psi_1 + \omega\lambda^{p-1}B_1\psi_p \tag{7.46}$$

and

$$y_i = \omega B_i y_{i-1} + (1 - \omega)\lambda^{i-1}\psi_i, \qquad i = 2, 3, \ldots, p. \tag{7.47}$$

But since $J\psi = \mu_j\psi$, we have

$$B_1\psi_p = \mu_j\psi_1$$

and

$$B_i\psi_{i-1} = \mu_j\psi_i, \qquad i = 2, 3, \ldots, p,$$

so that (7.46) becomes

$$y_1 = \left[(1 - \omega) + \omega\lambda^{p-1}\mu_j\right]\psi_1.$$

In view of $f_{\omega,\mu_j}(\lambda) = 0$, this gives

$$y_1 = \lambda^p\psi_1 = \lambda^p\psi'_1.$$

By using (7.47) inductively, we get

$$y_i = \lambda^{p+i-1}\psi_i = \lambda^p\psi'_i, \qquad i = 2, 3, \ldots, p,$$

which gives the desired result.   □

Theorem 7.8 seems to contradict the fact that there are np eigenvalues of T, while $\mathcal{L}_\omega$ is only of order n. Notice, however, that if Λ_{μ_j} is the set containing the p eigenvalues of T corresponding to the eigenvalue μ_j of J, viz.,

$$\Lambda_{\mu_j} \equiv \left\{\lambda \mid f_{\omega,\mu_j}(\lambda) = 0\right\},$$

then the set of the p eigenvalues of T corresponding to some other eigenvalue of J in the same cyclic class with μ_j (i.e., the set of eigenvalues $\mu_j\beta_k$ with $\beta_k = e^{2\pi ki/p}$; $k = 1, 2, \ldots, p-1$) is

$$\Lambda_{\mu_j\beta_k} = \left\{\lambda \mid f_{\omega,\mu_j\beta_k}(\lambda) = 0\right\} = \left\{\lambda\beta_k \mid \lambda \in \Lambda_{\mu_j}\right\}; \qquad k = 1, 2, \ldots, p-1.$$

It follows immediately that if $\lambda' \in \Lambda_{\mu_j}$ and $\lambda'' \in \Lambda_{\mu_j\beta_k}$ then $(\lambda')^p = (\lambda'')^p$. In other words, the p eigenvalues of J in the same cyclic class produce (via the roots of the polynomials $f_{\omega,\mu_j\beta_k}(\lambda)$, $k = 0, 1, \ldots, p-1$) only p, not p^p, eigenvalues of $\mathcal{L}_\omega$. By using Theorems 7.3 and 7.8, one also sees that the corresponding eigenvectors coincide as they should.

Table 7.16: Eigenvalues of T^3 and $\mathcal{L}_\omega$ in 10-State Example

Eigenvalues of T^3	Eigenvalues of $\mathcal{L}_\omega$
$-2.2206 + 7.0195i$	22.8955
$-2.2206 - 7.0195i$	$-2.2206 + 7.0195i$
22.8955	$-2.2206 - 7.0195i$
22.8955	$-0.0195 + 1.8168i$
22.8955	$-0.0195 - 1.8168i$
$-2.2206 + 7.0195i$	-2.1000
$-2.2206 + 7.0195i$	1.0000
$-2.2206 + 7.0195i$	$-0.6623 + 0.2017i$
$-2.2206 - 7.0195i$	$-0.6623 - 0.2017i$
$-0.0195 + 1.8168i$	-0.4045
$-0.0195 - 1.8168i$	
$-0.0195 + 1.8168i$	
$-0.0195 - 1.8168i$	
$-0.0195 + 1.8168i$	
$-0.0195 - 1.8168i$	
$1.0000 + 0.0000i$	
$1.0000 - 0.0000i$	
1.0000	
-2.1000	
-2.1000	
-2.1000	
$-0.6623 + 0.2017i$	
$-0.6623 - 0.2017i$	
$-0.6623 + 0.2017i$	
$-0.6623 - 0.2017i$	
$-0.6623 + 0.2017i$	
$-0.6623 - 0.2017i$	
-0.4045	
-0.4045	
-0.4045	

Table 7.17: Eigenvalues and Eigenvectors of SOR Iteration Matrix

Eigenvalues			
22.8955	−2.1000	1.0000	−0.4045
Corresponding Eigenvectors			
0.1851	−0.8809	0.7012	−0.8038
0.0548	−0.2349	0.2078	−0.2382
0.0494	0.3524	0.1870	−0.2143
0.0118	−0.2114	0.0449	−0.0514
0.3504	0.0000	0.4675	0.3963
0.0934	0.0000	0.1247	0.1057
0.0841	0.0000	0.1122	0.0951
0.6633	0.0000	0.3117	−0.1954
0.5969	0.0000	0.2805	−0.1759
0.1592	0.0000	0.0748	−0.0469

Table 7.18: Eigenvalues and Eigenvectors of Jacobi Iteration Matrix

Eigenvalues	
1.0000	0.0000
Corresponding Eigenvectors	
0.7012	0.8809
0.2078	0.2349
0.1870	−0.3524
0.0449	0.2144
0.4675	0.0000
0.1247	0.0000
0.1122	0.0000
0.3117	0.0000
0.2805	0.0000
0.0748	0.0000

Returning to the 10-state example with the relaxation parameter given by $\omega = 3.1$, the eigenvalues of T^3 and of the SOR iteration matrix $\mathcal{L}_\omega$ are given in Table 7.16. Some selected eigenvectors of $\mathcal{L}_\omega$ and J are given in Tables 7.17 and 7.18. Note that the eigenvectors satisfy the properties specified in the theorem. For example, the eigenvector of $\mathcal{L}_\omega$ corresponding to the eigenvalue 22.8955 is equal to a scalar multiple, $(.1851/.7012)$, of the vector

$$(.7012, .2078, .1870, 0449, \quad 2.8396 \times (.4675, .1247, .1122),$$

$$2.8396^2 \times (.3117, .2805, .0746) \,)^T.$$

Notice also that the eigenvectors corresponding to the unit eigenvalues of both J and $\mathcal{L}_\omega$ are identical.

7.7.4 Convergence Conditions for Block SOR

The observations of the previous section reveal that by examining the roots of $f_{\omega,\mu_j}(\lambda)$, we can determine the eigenvalues of $\mathcal{L}_\omega$ associated with *all* $\mu_j\beta_k$. The roots of $f_{\omega,1}(\lambda)$ are of particular importance, since, as Theorems 7.3 and 7.8 dictate, the eigenvectors of $\mathcal{L}_\omega$ that correspond to the pth power of these roots have subvectors parallel to the corresponding subvectors of the Perron eigenvector of J. We therefore proceed to develop some notation that allows us to describe the properties of these roots.

Let us define $\alpha(\omega)$ as

$$\alpha(\omega) = \max\left\{|\lambda| \mid f_{\omega,1}(\lambda) = 0\right\}, \tag{7.48}$$

that is, the maximum modulus of any eigenvalue of the p-step iteration matrix that is associated with the unit eigenvalue of J. Notice that $\alpha(\omega) \geq 1$ for all ω, since, trivially, $f_{\omega,1}(1) = 0$ for all ω.

Also, let $\theta(\omega)$ be the maximum among the moduli of eigenvalues of T that correspond to all eigenvalues of J, other than those that belong to the same cyclic class as the unit eigenvalue, i.e.,

$$\theta(\omega) = \max\left\{|\lambda| \mid f_{\omega,\mu}(\lambda) = 0, \ \mu \in \sigma(J), \ |\mu| < 1\right\}. \tag{7.49}$$

Finally, let $\mathcal{A}(\omega)$ be

$$\mathcal{A}(\omega) = \left\{\lambda \mid f_{\omega,1}(\lambda) = 0, \ |\lambda| = \alpha(\omega)\right\}, \tag{7.50}$$

i.e., the set of all eigenvalues of T that correspond to the unit eigenvalue of J and have the maximum modulus $\alpha(\omega)$. From the definition of $\alpha(\omega)$ in (7.48), it follows immediately that $|\mathcal{A}(\omega)|$, the cardinality of $\mathcal{A}(\omega)$, is greater than or equal to 1.

In our example,

- $\{|\lambda| \mid f_{\omega,1}(\lambda) = 0\}$ is the set $\{2.8396,\ 1.0000,\ 0.7396\}$, and hence, $\alpha(\omega) = 2.8396$.

- The set $\{|\lambda| \mid f_{\omega,\mu}(\lambda) = 0,\ \mu \in \sigma(J),\ |\mu| < 1\} = \{1.9454,\ 0.8847,\ 1.2202,\ 1.2806\}$, and so $\theta(\omega) = 1.9454$.

- $\mathcal{A}(\omega) = \{\lambda \mid f_{\omega,1}(\lambda) = 0,\ |\lambda| = \alpha(\omega)\} = \{2.8396\}$.

Since, for all ω, $1 \in \sigma(\mathcal{L}_\omega)$, the usual conditions for semiconvergence (Chapter 3, Theorem 6.2) are

A1: $\rho(\mathcal{L}_\omega) = 1$

A2: All elementary divisors associated with the unit eigenvalue are linear, i.e., $\mathrm{rank}(I - \mathcal{L}_\omega) = \mathrm{rank}(I - \mathcal{L}_\omega)^2$.

A3: If $\lambda \in \sigma(\mathcal{L}_\omega)$ and $|\lambda| = 1$, then $\lambda = 1$.

In the light of our previous definitions these conditions may be restated as

B1: $\alpha(\omega) = 1$ and $\theta(\omega) < \alpha(\omega)$.

B2: 1 is a simple root of $f_{\omega,1}(\lambda) = 0$.

B3: $|\mathcal{A}(\omega)| = 1$.

It is evident now that the SOR method will converge to the solution of the system (7.38) for exactly those ω values that satisfy conditions B1 through B3. Consider, however, the following set of more relaxed constraints:

C1: $\theta(\omega) < \alpha(\omega)$.

C2: If $\lambda \in \mathcal{A}(\omega)$, then λ is a simple root of $f_{\omega,1}(\lambda) = 0$.

Obviously, the set of values for ω that satisfies conditions B1 through B3 is a subset of the set that satisfies C1 and C2. Assume now that for some specific ω, conditions C1 and C2 are satisfied, while B1 through B3 are not. Then there are some (perhaps more than one) eigenvalues of $\mathcal{L}_\omega$ that have modulus $\alpha(\omega)^p \geq 1$. Since condition C2 is satisfied, all such eigenvalues have multiplicity 1, and hence SOR will converge to a linear combination of the eigenvectors associated with these eigenvalues. Theorem 7.8, in conjunction with Theorem 7.3, guarantees that this linear combination will maintain subvectors that are parallel to the corresponding subvectors of the stationary probability vector.

It is easy to see that conditions C1 and C2 are *necessary* for the subvectors to be parallel to the subvectors of the solution of (7.38). It is shown in Section 7.7.5 that besides the exceptional value $\omega = 0$ (in which case the SOR iteration degenerates to a null iteration) there are at most two distinct values of ω that fail to satisfy condition C2. One of these values is $\omega = p/(p-1)$; the second is a negative value and occurs only when p is odd. However, even in these two special cases it may be shown that SOR converges to a vector whose subvectors are parallel to the corresponding subvectors of the stationary probability vector. This allows us to enunciate the following theorem.

Theorem 7.9 *The SOR method converges to a vector that has subvectors parallel to the corresponding subvectors of the solution of (7.38), if and only if the relaxation parameter ω is chosen so that condition C1 is satisfied. The associated convergence factor is $(\theta(\omega)/\alpha(\omega))^p$. The SOR method converges in the standard sense, if and only if conditions B1 through B3 are satisfied.*

We wish to stress that for an ω value that satisfies conditions C1 and C2 but does not satisfy B1 through B3, SOR does *not* converge in the standard sense; the SOR relaxations converge to a vector that is *different* from the solution of (7.38). It happens, though, that, given the vector to which SOR converges, the desired solution can be trivially obtained (see Section 7.7.8). In this sense, the method converges to the desired solution. Any reference to convergence in the sequel, unless explicitly stated otherwise, is to be understood as convergence in this extended sense.

Note, also, that since we relaxed condition B1 to condition C1, we do not need to insist on $\rho(\mathcal{L}_\omega) = 1$. In particular, we do not have to impose the restriction $0 < \omega < 2$. Any value of $\omega \in \Re \setminus \{0\}$ that satisfies condition C1 will guarantee convergence. In the special case of block Gauss–Seidel, we have $\omega = 1$ and, as is easy to see, the roots of $f_{1,\mu}(\lambda)$ are μ, with multiplicity 1, and 0, with multiplicity $p-1$. It follows immediately that block Gauss–Seidel converges (in the standard sense) and the associated convergence factor is $\tilde{\rho}(J^p)$, where $\tilde{\rho}(J)$ is the maximum among the moduli of the eigenvalues of J, other than those belonging to the cyclic class of the unit eigenvalue. Notice that it follows from Theorem 7.8 that the Gauss–Seidel iteration matrix is not cyclic, unlike that of Jacobi.

Continuing with the example, iterations 2 through 8 of the block SOR method with $\omega = 3.1$ (and normalized so that $||y^{(i)}||_1 = 1$) yield the vectors shown in Table 7.19. The exact solution (the stationary probability vector of the 3-cyclic Markov chain) is given by

$$\pi = (.2791, .0827, .0744, .0178, .1861, .0496, .0447, .1241, .1117, .0298)^T.$$

This exact solution is obtained when the eighth approximation obtained by block SOR is modified so that the first four components are multiplied by 3.3913; the

Table 7.19: Iterations 2 through 8 of Block SOR

2	3	4	5	6	7	8
0.0911	0.0740	0.0829	0.0830	0.0821	0.0823	0.0823
0.0271	0.0264	0.0236	0.0244	0.0245	0.0244	0.0244
0.0209	0.0230	0.0219	0.0219	0.0220	0.0220	0.0219
0.0000	0.0064	0.0056	0.0051	0.0053	0.0053	0.0053
0.1557	0.1485	0.1573	0.1563	0.1556	0.1558	0.1559
0.0478	0.0440	0.0404	0.0415	0.0417	0.0415	0.0415
0.0278	0.0421	0.0376	0.0369	0.0375	0.0374	0.0374
0.3136	0.2911	0.2938	0.2956	0.2950	0.2949	0.2950
0.2389	0.2681	0.2679	0.2648	0.2654	0.2656	0.2655
0.0771	0.0765	0.0691	0.0705	0.0710	0.0708	0.0708

next three by 1.1937; and the final three by 0.4207. The determination of these three scalar multiples is considered in Section 7.7.8.

7.7.5   Applicable Values for the Relaxation Parameter

We have seen that the SOR relaxations converge exactly when conditions C1 *and* C2 are satisfied. We now turn our attention to determining the values of ω for which these conditions hold. We begin by considering condition C2 first. The following two theorems determine the quantity $\alpha(\omega)$ for different ω ranges. The proofs may be found in [84].

Theorem 7.10 *For $\omega > 0$:*

a) *If $0 < \omega \leq p/(p-1)$, then $\alpha(\omega) = 1$ and $\mathcal{A}(\omega) = \{1\}$. If $0 < \omega < p/(p-1)$, then $f_{\omega,1}(\lambda)$ has 1 as a simple root. If $\omega = p/(p-1)$, then 1 is a double root.*

b) *If $\omega > p/(p-1)$, then $\alpha(\omega) = \gamma(\omega) > 1$ and $\mathcal{A}(\omega) = \{\gamma(\omega)\}$, where $\gamma(\omega)$ is the unique, simple, positive root of $f_{\omega,1}(\lambda)$ in the interval $(\omega(p-1)/p, \omega)$. Furthermore, the relation*

$$\left(\frac{p-1}{p}\omega\right)^p > (p-1)(\omega-1) \tag{7.51}$$

holds.

Notice that when $\omega = p/(p-1)$, the dominant root, 1, is a double root and therefore does not satisfy condition C2. However, since there is a single eigenvector associated with the unit eigenvalue of $\mathcal{L}_\omega$, it has an associated Jordan block of size 2. It may be shown that the *principal vector* corresponding to this eigenvalue also has subvectors parallel to the solution vector. Since SOR will converge onto a linear combination of eigenvector and principal vector, the solution will have subvectors parallel to the subvectors of the stationary distribution vector.

Notice also that if we restrict ourselves to convergence in the usual sense, only values of ω in the range $(0, p/(p-1))$ are candidates, consistent with the already known results. The next theorem investigates negative values of ω.

Theorem 7.11 *For $\omega < 0$,*

a) If p is even, then $\alpha(\omega) = -\gamma(\omega) > 1$ and $\mathcal{A}(\omega) = \{\gamma(\omega)\}$, where $\gamma(\omega)$ is the unique, simple, negative root of $f_{\omega,1}(\lambda)$ in $\left[\omega - 1, -(1-\omega)^{1/p}\right)$.

b) If p is odd, then let $\omega^\dagger$ be the unique, negative root of

$$g(x) = \left(x \frac{p-1}{p}\right)^p - (p-1)(x-1) \tag{7.52}$$

in $[-p, -p/(p-1))$. Then

i) If $\omega < \omega^\dagger < -p/(p-1)$, then $\alpha(\omega) = -\gamma(\omega) > 1$ and $\mathcal{A}(\omega) = \{\gamma(\omega)\}$, where $\gamma(\omega)$ is the unique, simple, negative root of $f_{\omega,1}(\lambda)$ in the interval $(\omega, \omega(p-1)/p)$. The relation

$$\left(\frac{p-1}{p}\omega\right)^p < (p-1)(\omega - 1) \tag{7.53}$$

holds.

ii) If $\omega = \omega^\dagger$, then $\gamma(\omega^\dagger) = \omega^\dagger(p-1)/p$ and the root $\gamma(\omega^\dagger)$ is a double root.

iii) If $\omega^\dagger < \omega < 0$, then let $t^\dagger$ be the unique solution of

$$U^p_{p-1}(t) = -\frac{\omega^p}{1-\omega} U^{p-1}_{p-2}(t) \tag{7.54}$$

in $(\cos(\pi/p), 1)$, where $U_{p-1}(t)$ and $U_{p-2}(t)$ are the Chebyshev polynomials of the second kind, of degrees $p-1$ and $p-2$, respectively. Also, let

$$\gamma^p(\omega) = -(1-\omega)U_{p-2}(t^\dagger) < -(1-\omega). \tag{7.55}$$

Then $\alpha(\omega) = |\gamma(\omega)|$; $\mathcal{A}(\omega) = \left\{\gamma(\omega)e^{i\cos^{-1}t^\dagger}, -\gamma(\omega)e^{i\cos^{-1}(-t^\dagger)}\right\}$, and the roots in $\mathcal{A}(\omega)$ are simple. Furthermore, the relation

$$\left(\frac{p-1}{p}\omega\right)^p > (p-1)(\omega - 1) \tag{7.56}$$

holds.

Notice that, since for $\omega < 0$ we always have $\alpha(\omega) > 1$, no negative values for the ω parameter are candidates for convergence in the standard sense. From this second theorem we see that condition C2 is satisfied by all real, negative values of the ω parameter, except, when p is odd, for $\omega = \omega^\dagger$. With this value, $f_{\omega,1}(\lambda)$ has a double, as opposed to a simple, dominant root. Again, there is a single eigenvector associated with the eigenvalue $\gamma^p(\omega^\dagger)$ of $\mathcal{L}_\omega$ and a Jordan block of size 2 associated with this eigenvalue. Again it may be shown that the principal vector associated with this eigenvalue also has subvectors parallel to the solution vector, and SOR will converge in the extended sense, as previously.

Notice that when ω is chosen in the ranges,

$$\omega \in \left(-\infty, \omega^\dagger\right) \cup (0, p/(p-1)) \cup (p/(p-1), +\infty),$$

the set $\mathcal{A}(\omega)$ contains a single simple eigenvalue. For an ω value that lies in the range $(\omega^\dagger, 0)$, $\mathcal{A}(\omega)$ contains two simple eigenvalues.

Table 7.20 presents the eigenvalues that belong to the set $\mathcal{A}^3(\omega)$ as ω varies from -4.0 to 3.0 for the 10×10, 3-cyclic example. We present the cube of the elements of this set, since this gives the eigenvalues of the SOR iteration matrix. For comparison purposes, the table also presents the cube of the largest eigenvalue of T that does not correspond to a unit eigenvalue of J. The modulus of this eigenvalue defines the quantity $\theta(\omega)$. In this example the polynomial $g(x)$ defined in Theorem 7.11 is given by

$$g(x) = \frac{8x^3}{27} - 2x + 2 = 0,$$

which has a root at $x = -3.0$ and a double root at $x = 1.5$. It follows that $\omega^\dagger = -3.0$ in this example.

We have thus addressed condition C2 and shown that it holds for all real, nonzero values of ω, with the exception of the two points $\omega = \omega^\dagger$ and $\omega = p/(p-1)$, and even these two values do not create difficulties. We now turn our attention to condition C1. For condition C1 to be valid, for some given ω, we require that all eigenvalues of T, excluding those related to the eigenvalues of J with modulus 1, have modulus strictly less than $\alpha(\omega)$. Let us define $\tilde{\sigma}(J)$ as

$$\tilde{\sigma}(J) = \sigma(J) \setminus \{\beta_k \mid k = 0, 1, \ldots, p-1\}$$

with β_k as in (7.3). Also, let $\tilde{\rho}(J)$ be

$$\tilde{\rho}(J) = \max\{|\lambda| \mid \lambda \in \tilde{\sigma}(J)\} = \max\{|\lambda| \mid \lambda \in \sigma(J), \ |\lambda| < 1\}. \tag{7.57}$$

We shall consider the case when all the eigenvalues of J^p are real and nonnegative. This assumption implies that

$$\tilde{\sigma}(J) \subset \left\{\tilde{\rho}(J)e^{2k\pi i/p} \mid 0 < \tilde{\rho}(J) < 1, \ k = 0, 1, \ldots, p-1\right\}$$

Table 7.20: Eigenvalues in $\mathcal{A}^3$ and Cube of Largest Eigenvalue outside Unit Periodic Class, as Function of ω

ω	$\mathcal{A}^3$	θ^3	ω	$\mathcal{A}^3$	θ^3
−4.0	−47.3607	$5.7817 \pm 15.5510i$	0.0	1.0000	1.0000
−3.9	−42.8750	$5.8948 \pm 14.6640i$	0.1	1.0000	$0.9545 \pm 0.0199i$
−3.8	−38.6075	$5.9828 \pm 13.8122i$	0.2	1.0000	$0.9046 \pm 0.0401i$
−3.7	−34.5478	$6.0468 \pm 12.9946i$	0.3	1.0000	$0.8496 \pm 0.0604i$
−3.6	−30.6838	$6.0881 \pm 12.2102i$	0.4	1.0000	$0.7885 \pm 0.0807i$
−3.5	−27.0000	$6.1079 \pm 11.4581i$	0.5	1.0000	$0.7200 \pm 0.1006i$
−3.4	−23.4753	$6.1074 \pm 10.7374i$	0.6	1.0000	$0.6423 \pm 0.1199i$
−3.3	−20.0769	$6.0878 \pm 10.0471i$	0.7	1.0000	$0.5527 \pm 0.1379i$
−3.2	−16.7430	$6.0502 \pm 9.3866i$	0.8	1.0000	$0.4463 \pm 0.1535i$
−3.1	−13.3147	$5.9957 \pm 8.7549i$	0.9	1.0000	$0.3127 \pm 0.1653i$
−3.0	$-8.0000 \pm 0.0000i$	$5.9256 \pm 8.1514i$	1.0	1.0000	$0.1140 \pm 0.1814i$
−2.9	$-6.8445 \pm 3.5315i$	$5.8408 \pm 7.5752i$	1.1	1.0000	$-0.1121 \pm 0.3196i$
−2.8	$-5.7760 \pm 4.6379i$	$5.7425 \pm 7.0257i$	1.2	1.0000	$-0.2805 \pm 0.4775i$
−2.7	$-4.7915 \pm 5.2626i$	$5.6316 \pm 6.5022i$	1.3	1.0000	$-0.4380 \pm 0.6438i$
−2.6	$-3.8880 \pm 5.6160i$	$5.5093 \pm 6.0040i$	1.4	1.0000	$-0.5910 \pm 0.8235i$
−2.5	$-3.0625 \pm 5.7876i$	$5.3765 \pm 5.5305i$	1.5	1.0000	$-0.7407 \pm 1.0189i$
−2.4	$-2.3120 \pm 5.8274i$	$5.2341 \pm 5.0810i$	1.6	1.4454	$-0.8872 \pm 1.2315i$
−2.3	$-1.6335 \pm 5.7679i$	$5.0831 \pm 4.6549i$	1.7	1.9857	$-1.0302 \pm 1.4622i$
−2.2	$-1.0240 \pm 5.6320i$	$4.9244 \pm 4.2516i$	1.8	2.6269	$-1.1692 \pm 1.7119i$
−2.1	$-0.4805 \pm 5.4369i$	$4.7589 \pm 3.8705i$	1.9	3.3750	$-1.3033 \pm 1.9813i$
−2.0	$0.0000 \pm 5.1962i$	$4.5874 \pm 3.5110i$	2.0	4.2361	$-1.4319 \pm 2.2713i$
−1.9	$0.4205 \pm 4.9206i$	$4.4107 \pm 3.1726i$	2.1	5.2162	$-1.5542 \pm 2.5825i$
−1.8	$0.7840 \pm 4.6192i$	$4.2298 \pm 2.8546i$	2.2	6.3214	$-1.6692 \pm 2.9156i$
−1.7	$1.0935 \pm 4.2997i$	$4.0452 \pm 2.5565i$	2.3	7.5577	$-1.7761 \pm 3.2713i$
−1.6	$1.3520 \pm 3.9684i$	$3.8579 \pm 2.2778i$	2.4	8.9312	$-1.8739 \pm 3.6503i$
−1.5	$1.5625 \pm 3.6309i$	$3.6684 \pm 2.0178i$	2.5	10.4480	$-1.9617 \pm 4.0533i$
−1.4	$1.7280 \pm 3.2921i$	$3.4775 \pm 1.7761i$	2.6	12.1141	$-2.0385 \pm 4.4811i$
−1.3	$1.8515 \pm 2.9562i$	$3.2860 \pm 1.5521i$	2.7	13.9356	$-2.1032 \pm 4.9345i$
−1.2	$1.9360 \pm 2.6268i$	$3.0944 \pm 1.3453i$	2.8	15.9184	$-2.1548 \pm 5.4142i$
−1.1	$1.9845 \pm 2.3071i$	$2.9033 \pm 1.1550i$	2.9	18.0686	$-2.1923 \pm 5.9210i$
−1.0	$2.0000 \pm 2.0000i$	$2.7134 \pm 0.9808i$	3.0	20.3923	$-2.2146 \pm 6.4558i$
−0.9	$1.9855 \pm 1.7079i$	$2.5253 \pm 0.8220i$			
−0.8	$1.9440 \pm 1.4328i$	$2.3396 \pm 0.6782i$			
−0.7	$1.8785 \pm 1.1765i$	$2.1567 \pm 0.5489i$			
−0.6	$1.7920 \pm 0.9406i$	$1.9772 \pm 0.4333i$			
−0.5	$1.6875 \pm 0.7262i$	$1.8017 \pm 0.3311i$			
−0.4	$1.5680 \pm 0.5342i$	$1.6306 \pm 0.2416i$			
−0.3	$1.4365 \pm 0.3653i$	$1.4644 \pm 0.1643i$			
−0.2	$1.2960 \pm 0.2200i$	$1.3036 \pm 0.0987i$			
−0.1	$1.1495 \pm 0.0982i$	$1.1487 \pm 0.0441i$			
0.0	1.0000	1.0000			

with $\tilde{\sigma}(J)$ and $\tilde{\rho}(J)$ as defined above. For some given ω we may write condition C1 as

$$\text{If } |\zeta'| \geq \alpha(\omega) \quad \text{then } f_{\omega,\mu}(\zeta') \neq 0, \ \mu \in \tilde{\sigma}(J).$$

This becomes

$$1 - (1-\omega)\frac{1}{\zeta'^p} \neq \omega\mu\frac{1}{\zeta'}, \qquad |\zeta'| \geq \alpha(\omega), \qquad \mu \in \tilde{\sigma}(J)$$

or by setting $\zeta = 1/\zeta'$,

$$\mu \neq \frac{1 - (1-\omega)\zeta^p}{\omega\zeta}, \qquad |\zeta| \leq \frac{1}{\alpha(\omega)}, \qquad \mu \in \tilde{\sigma}(J). \qquad (7.58)$$

It is shown in [84] that equation (7.58) holds for all values of ω, so we may state the following theorem:

Theorem 7.12 *If the eigenvalues of J^p are real and nonnegative, then block SOR converges for all $\omega \in \Re \setminus \{0\}$. Convergence in the standard sense is obtained for all $\omega \in (0, p/(p-1))$.*

The ω range that gives standard semiconvergence is consistent with the results in [64]. The techniques used to prove this theorem are extendible to the case when the eigenvalues of J^p are real but not necessarily nonnegative. They do not apply when J^p has complex eigenvalues, as in the case of our 10×10 example. Table 7.21 contains the values of $\alpha(\omega)^3$ and $\theta(\omega)^3$ for different values of ω. Note in particular that $\alpha(\omega)^3$ is less than $\theta(\omega)^3$ for $-3 \leq \omega < 0$.

7.7.6   Convergence Testing in the Extended Sense

A procedure often employed for convergence testing is to normalize and compare successive vector approximations until the relative difference, as measured by some vector norm, satisfies a prespecified tolerance criterion. This will likely be satisfactory when block SOR converges to a single eigenvector; for example, when ω is such that

$$\omega \in (-\infty, \omega^\dagger) \cup (0, \frac{p}{p-1}) \cup (\frac{p}{p-1}, +\infty)$$

and $\alpha(\omega) > \theta(\omega)$. It is unlikely to be satisfactory for other values of ω, because in these cases SOR converges to a linear combination of eigenvectors or of eigenvectors and principal vectors, and convergence is achieved only in a subvector sense. In this case it is necessary to normalize each subvector separately and to apply a relative convergence test to each subvector (of the current and previous iterate). The largest difference may be taken as a measure of the precision obtained and compared against the requested tolerance to determine whether convergence (in the subvector sense) has been achieved. Before the method proceeds to the

Table 7.21: Convergence Factors as a Function of ω

ω	$\alpha(\omega)^3$	$\theta(\omega)^3$	Ratio
-100	9.9970×10^5	2.1412×10^5	.2142
-90	7.2873×10^5	1.5607×10^5	.2142
-80	5.1176×10^5	1.0959×10^5	.2141
-70	3.4279×10^5	0.7339×10^5	.2141
-60	2.1582×10^5	0.4619×10^5	.2140
-50	1.2485×10^5	0.2670×10^5	.2139
-40	6.3877×10^4	1.3649×10^4	.2137
-30	2.6907×10^4	0.5737×10^4	.2132
-20	7.9368×10^3	1.6820×10^3	.2119
-10	966.6230	199.9523	.2069
-9	698.5685	144.0959	.2063
-8	484.4953	100.0658	.2065
-7	318.3919	66.6990	.2095
-6	194.2341	42.7865	.2203
-5	105.9615	26.7849	.2528
-4	47.3607	16.5910	.3503
-3	8.0000	10.0776	1.2597
-2	5.1962	5.7768	1.1117
-1	2.8284	2.8852	1.0201
0	1.0000	1.0000	1.0000
1	1.0000	0.2143	.2143
2	4.2361	2.6850	.6338
3	20.3923	6.8251	.3347
4	54.4955	13.6219	.2500
5	112.5685	24.7258	.2197
6	200.6231	42.0006	.2094
7	324.6653	67.0485	.2065
8	490.6990	101.2203	.2063
9	704.7265	145.7832	.2069
10	972.7494	201.9945	.2077
20	7.9429×10^3	1.6850×10^3	.2121
30	2.6913×10^4	0.5740×10^4	.2133
40	6.3883×10^4	1.3653×10^4	.2137
50	1.2485×10^5	0.2671×10^5	.2139
60	2.1582×10^5	0.4619×10^5	.2140
70	3.4279×10^5	0.7339×10^5	.2141
80	5.1176×10^5	1.0959×10^5	.2141
90	7.2873×10^5	1.5607×10^5	.2142
100	9.9970×10^5	2.1413×10^5	.2142

next iteration of SOR, the most recent approximation should be normalized in its entirety.

The fact that the block SOR method converges for values of the relaxation parameter outside the normal range, $0 < \omega < 2$, may at first seem to be rather strange. However, it should be remembered that when applied to a homogeneous system of equations, SOR is just the power method applied to a certain iteration matrix, and it will converge to the eigenvector corresponding to the dominant eigenvalue. For *all* values of $\omega \neq 0$, the SOR iteration matrix possesses a unit eigenvalue whose corresponding eigenvector is the stationary probability vector we are trying to compute. For values of ω outside the normal range, SOR does not converge onto this vector but onto a different vector. The curious fact is not that it converges, but that the vector to which it converges may be used to compute the eigenvector corresponding to the unit eigenvalue.

7.7.7 Optimal Convergence Factors and Associated ω values

In this section we specify ω values that yield optimal convergence for block SOR when all the eigenvalues of J^p are nonnegative reals. We leave out the case where $\tilde{\rho}(J) = 0$, since it must be evident that the optimal choice in this case is $\omega = 1$ (Gauss–Seidel), which yields a zero convergence factor (infinite convergence rate). The following theorem in proved in [84].

Theorem 7.13 *Let J, the block Jacobi iteration matrix derived from the system of equations (7.38), be of the form (7.2). Let J^p have only nonnegative eigenvalues, and let $\tilde{\rho}(J) > 0$. Define the polynomial $h(\omega)$ of degree p to be*

$$h(\omega) \equiv \left(\frac{p-1}{p} \tilde{\rho}(J)\omega \right)^p - (p-1)(\omega - 1) = 0$$

and let ω_0 and ω_+ be the only positive roots of $h(\omega)$ in $(1, \ p/(p-1))$ and $(p\gamma, \ p^{p/(p-1)}\gamma)$ respectively, where $\gamma = [1/(p-1)]\tilde{\rho}(J)^{-p/(p-1)}$. If the SOR relaxations are employed with ω equal to any of these two values, then the optimal SOR convergence factor $r_*^p = (p-1)(\omega_0 - 1)$ is obtained. In addition, if p is odd, let ω_- be the unique negative root of $h(\omega)$ in $(-p^2\gamma, -p\gamma)$. Then ω_-, as well as ω_0 and ω_+, yields the optimal convergence factor r_*^p. SOR converges in the standard sense only for $\omega = \omega_0$.

This theorem reveals that, judging just from the point of view of asymptotic convergence, the newly introduced ω ranges that yield convergence in the extended sense cannot improve the convergence rate. However, numerical experiments reveal that the convergence factor of SOR appears to be much less sensitive to small perturbations of ω around the values ω_+ and ω_- than around ω_0. This facilitates the choice of a "good" value for the relaxation parameter.

The 3-cyclic example with which we have been working does *not* satisfy the conditions of this theorem, since J^3 possesses complex eigenvalues. However, an empirical study of the eigenvalues of the SOR iteration matrix does reveal that it apparently possesses three values of ω at which the optimum convergence rate is attained. Table 7.22 attempts to pinpoint the location of these values of ω. It can be seen that $\omega_0 = 1.0155$, $\omega_- = -8.6$, and $\omega_+ = 7.6$ all yield a convergence factor of .2062.

One feature of Table 7.22 that is worthy of comment is the manner in which the convergence factor varies as a function of ω around the optimal ω values. It is evident that the convergence factor is much more sensitive around the "usual" optimum ω_0 than it is around the points ω_+ and ω_-. This has been observed in many experiments, and very recent results reported in [65] lend theoretical support to this observation. Furthermore, as the absolute value of ω increases, the convergence factor remains practically invariant, increasing only very slowly. This behavior is of particular value when the quantity $\bar{\rho}(J)$ is not known in advance and the optimal ω parameters cannot be computed. In this case, a crude approximation of the parameter ω_+ or ω_- can be used, and as SOR proceeds, an adaptive procedure can be used to fine-tune it. During this adaptive procedure, the convergence factor will not fall excessively far from the optimal. On the other hand, the optimum ω in the range $0 < \omega < p/(p-1)$ is much less robust, for small fluctuations may lead to much slower convergence rates.

Care must be exercised when large numerical values of ω are used. Theorem 7.8 shows that SOR converges to a vector whose subvectors are multiplied by factors that are powers of $\alpha(\omega)$. This quantity increases as the magnitude of ω increases. If ω is large, normalization of the succesive iterates may cause the initial subvectors (which are multiplied by smaller powers of $\alpha(\omega)$) to underflow. This phenomenon becomes particularly severe when p is large.

Fortunately, it sometimes happens that when p is large, the dimensions of the subblocks of Q are small, and this permits us to adopt an alternative strategy. The matrix J^p is a block diagonal matrix whose diagonal blocks are permuted products of the blocks of J. Thus it follows that the spectra of the diagonal blocks of J^p, excluding the zero eigenvalues, are identical. When the dimensions of the blocks of Q are small, we may explicitly form the smallest diagonal block of J^p and compute its entire spectrum. This in turn allows us to compute $\bar{\rho}(J)$ and hence obtain an accurate estimation of ω_0. This value may be used for the relaxation parameter, thereby reducing the possibility of underflow.

Table 7.22: Location of Optimum Relaxation Parameters

ω	$\alpha(\omega)^3$	$\theta(\omega)^3$	Ratio	ω	$\alpha(\omega)^3$	$\theta(\omega)^3$	Ratio
0.5	1.0000	0.7270	.7270	−8.0	484.4953	100.0658	.2065
0.6	1.0000	0.6534	.6534	−8.1	503.6448	103.9699	.2064
0.7	1.0000	0.5696	.5696	−8.2	523.2799	107.9818	.2064
0.8	1.0000	0.4719	.4719	−8.3	543.4068	112.1027	.2063
0.9	1.0000	0.3537	.3537	−8.4	564.0314	116.3336	.2063
1.0	1.0000	0.2143	.2143	−8.5	585.1598	120.6758	.2062
1.01	1.0000	0.2073	.2073	−8.6	606.7980	125.1304	.2062
1.011	1.0000	0.2070	.2070	−8.7	628.9519	129.6985	.2062
1.012	1.0000	0.2067	.2067	−8.8	651.6276	134.3814	.2062
1.013	1.0000	0.2065	.2065	−8.9	674.8312	139.1801	.2062
1.014	1.0000	0.2063	.2063	−9.0	698.5685	144.0959	.2063
1.015	1.0000	0.2062	.2062				
1.016	1.0000	0.2062	.2062				
1.017	1.0000	0.2063	.2063				
1.018	1.0000	0.2064	.2064				
1.019	1.0000	0.2066	.2066				
1.02	1.0000	0.2069	.2069				
1.03	1.0000	0.2135	.2135				
1.04	1.0000	0.2256	.2256				
1.05	1.0000	0.2412	.2412				
1.06	1.0000	0.2589	.2589				
1.07	1.0000	0.2779	.2779	ω	$\alpha(\omega)^3$	$\theta(\omega)^3$	Ratio
1.08	1.0000	0.2977	.2977				
1.09	1.0000	0.3180	.3180	7.0	324.6653	67.0485	.2065
1.1	1.0000	0.3387	.3387	7.1	339.2800	70.0339	.2064
1.2	1.0000	0.5538	.5538	7.2	354.3206	73.1119	.2063
1.3	1.0000	0.7787	.7787	7.3	369.7932	76.2838	.2063
1.4	1.0000	1.0136	1.0136	7.4	385.7037	79.5507	.2062
1.5	1.0000	1.2597	1.2597	7.5	402.0580	82.9141	.2062
1.6	1.4454	1.5178	1.0501	7.6	418.8624	86.3750	.2062
1.7	1.9857	1.7887	.9008	7.7	436.1226	89.9349	.2062
1.8	2.6269	2.0730	.7891	7.8	453.8448	93.5949	.2062
1.9	3.3750	2.3716	.7027	7.9	472.0349	97.3563	.2062
2.0	4.2361	2.6850	.6338	8.0	490.6990	101.2203	.2063

7.7.8   Computing the Subvector Multipliers

As commented upon previously, if block SOR is employed with an ω parameter value for which $\alpha(\omega) > \theta(\omega)$ but outside the interval $(0, p/(p-1))$, then convergence is to a vector that is *not* the stationary probability vector. The subvectors of the vector to which it converges are known (from Theorem 7.8) to be parallel to the corresponding subvectors of the stationary probability vector. Let

$$\nu = (\nu_1^T, \nu_2^T, \ldots, \nu_p^T)^T$$

be the vector obtained when each subvector of the SOR computed solution is normalized so that its elements sum to 1. Each ν_i, $i = 1, 2, \ldots, p$ is the probability distribution of the Markov chain being in a particular state of the subset of states defined by block Q_{ii}, *conditioned upon the fact that the Markov chain is in that subset*. The stationary probability vector may now be written as

$$\pi = (\tau_1 \nu_1^T, \tau_2 \nu_2^T, \ldots, \tau_p \nu_p^T), \tag{7.59}$$

where τ_i, $i = 1, 2, \ldots, p$ are appropriate positive constants[2]. Notice that τ_i must be the probability that the Markov chain is in one of the states defined by block Q_{ii}. The vector

$$\tau = (\tau_1, \tau_2, \ldots, \tau_p)$$

is easily seen, from the observations above, to be a probability vector. Moreover, it is the stationary probability vector of yet another Markov chain whose infinitesimal generator we call G, i.e.,

$$\tau G = 0 \tag{7.60}$$

and $g_{ij} = \nu_i^T (I + Q)_{ij} e$. The elements of G define transitions from a whole block of states of the original Markov chain to another whole block of states, treating these blocks as if they were single states. This approach was used in the previous chapter to construct the "coupling" matrix for nearly completely decomposable Markov chains. The matrix G is irreducible if the original chain is irreducible. Since we have assumed irreducibility of Q, equation (7.60) and the normalizing condition $\tau e = 1$ uniquely determine the vector τ.

System (7.60) is of dimension p and is trivial to solve, since p is typically small. The fact that Q is in p-cyclic form (7.35) makes this computation even easier. It is possible, for example, to assign an arbitrary value to τ_1, implement the recursion

$$\tau_i = -\tau_{i-1} \frac{\nu_{i-1}^T Q_{i-1,i} e}{\nu_i^T Q_{ii} e} \qquad i = 2, 3, \ldots, p, \tag{7.61}$$

---

[2]We transpose ν_i in this vector because we assume that the ν_i are column vectors derived from SOR, while π is a row vector.

and then normalize τ so that its elements sum to 1. If the original chain is NCD, the quantities $\nu_{i-1}^T Q_{i-1,i} e$ may be very small, and numerical difficulties may arise in the application of (7.61). If this is the case, however, it may be more appropriate to use iterative aggregation/disaggregation techniques to solve the original problem (7.38), rather than block SOR. The forward recursive scheme (7.61) (or a respective backward scheme) appears to be a stable and efficient way (requiring fewer operations than a single block SOR iteration step) to find the τ_i.

Returning one last time to the 10×10, 3-cyclic example, it was seen that the approximation obtained after eight iterations of block SOR with $\omega = 3.1$ (last column of Table 7.19) is given by

$$\begin{pmatrix} .0823 & .0244 & .0219 & .0053 & | & .1559 & .0415 & .0374 & | & .2950 & .2655 & .0708 \end{pmatrix}.$$

The vector $(\tau_1 \nu_1^T, \tau_2 \nu_2^T, \tau_3 \nu_3^T)$, obtained by normalizing each of the subvectors so that they sum to 1, is given by

$$\begin{pmatrix} .6146 & .1821 & .1639 & .0393 & | & .6637 & .1770 & .1593 & | & .4673 & .4206 & .1122 \end{pmatrix}.$$

The probability vector τ may now be computed from the infinitesimal generator:

$$\begin{pmatrix} .6146 & .1821 & .1639 & .0393 & 0 & 0 & 0 & 0 & 0 & 0 \\ 0 & 0 & 0 & 0 & .6637 & .1770 & .1593 & 0 & 0 & 0 \\ 0 & 0 & 0 & 0 & 0 & 0 & 0 & .4673 & .4206 & .1122 \end{pmatrix} \times$$

$$\begin{pmatrix} -.2 & \cdot & \cdot & \cdot & .2 & \cdot & \cdot & \cdot & \cdot & \cdot \\ \cdot & -.3 & \cdot & \cdot & \cdot & .3 & \cdot & \cdot & \cdot & \cdot \\ \cdot & \cdot & -1.0 & \cdot & .5 & .2 & .3 & \cdot & \cdot & \cdot \\ \cdot & \cdot & \cdot & -.5 & \cdot & \cdot & .5 & \cdot & \cdot & \cdot \\ \cdot & \cdot & \cdot & \cdot & -.5 & \cdot & \cdot & .2 & .3 & \cdot \\ \cdot & \cdot & \cdot & \cdot & \cdot & -.8 & \cdot & .5 & \cdot & .3 \\ \cdot & \cdot & \cdot & \cdot & \cdot & \cdot & -.7 & \cdot & .5 & .2 \\ \cdot & .2 & .3 & \cdot & \cdot & \cdot & \cdot & -.5 & \cdot & \cdot \\ .5 & \cdot & .2 & \cdot & \cdot & \cdot & \cdot & \cdot & -.7 & \cdot \\ \cdot & \cdot & .5 & .3 & \cdot & \cdot & \cdot & \cdot & \cdot & -.8 \end{pmatrix} \begin{pmatrix} 1 & 0 & 0 \\ 1 & 0 & 0 \\ 1 & 0 & 0 \\ 1 & 0 & 0 \\ 0 & 1 & 0 \\ 0 & 1 & 0 \\ 0 & 1 & 0 \\ 0 & 0 & 1 \\ 0 & 0 & 1 \\ 0 & 0 & 1 \end{pmatrix}$$

$$= \begin{pmatrix} -.3611 & .3611 & 0 \\ 0 & -.5850 & .5850 \\ .6178 & 0 & -.6178 \end{pmatrix}.$$

It may readily be verified that the stationary probability vector is given by

$$\tau = (.4541, .2804, .2655).$$

(This is also the result obtained if the recursive approach specified by equation (7.61) is used.) Hence, the computed approximation to the stationary probability vector is

$$
\begin{aligned}
[\,.4541 &\times \;(.6146, .1821, .1639, .0393), \\
.2804 &\times \;(.6637, .1770, .1593), \\
.2655 &\times \;(.4673, .4206, .1122)\,]
\end{aligned}
$$

$$= (.2791, .0827, .0744, .0178, .1861, .0496, .0447, .1241, .1117, .0298)$$

which is the same as the solution, π, provided earlier.

Chapter 8

Transient Solutions

8.1   Introduction

If $\pi_i(t)$ denotes the probability that a Markov chain with infinitesimal generator Q is in state i at time t, and $\pi(t)$ is the vector of all such probabilities, then it may be shown directly from the Chapman–Kolmogorov differential equations (see Chapter 1) that

$$\pi(t) = \pi(0)e^{Qt}.$$

Here, e^{Qt} is the matrix exponential defined by

$$e^{Qt} = \sum_{k=0}^{\infty}(Qt)^k/k!$$

Depending on the numerical approach adopted, $\pi(t)$ may be computed by first forming e^{Qt} and then premultiplying this with the initial probability vector $\pi(0)$. Such is the case with the decompositional and the matrix-scaling and -powering methods described in Section 8.3. In other cases, $\pi(t)$ may be computed directly without explicitly forming e^{Qt}. This is the approach taken by the ordinary differential equation (ODE) solvers of Section 8.4. Some methods allow for both possibilities, e.g., the uniformization method of the next section and the Krylov subspace methods presented in Section 8.5.

Moler and Van Loan [101] discuss nineteen *dubious* ways to compute the exponential of a relatively small-order matrix. A major problem in all these methods is that the accuracy of the approximations depends heavily on the norm of the matrix [88]. Thus, when the norm of Q is large or t is large, attempting to compute e^{Qt} directly is likely to yield unsatisfactory results. It becomes necessary to divide the interval $[t_0, t]$ into subintervals $[t_0, t_1, t_2, \ldots, t]$ and to compute the transient solution at each intermediate time t_j using the solution at time t_{j-1} as the initial vector. It often happens that this is exactly what is required by a user who can, as a result, see the evolution of certain system performance measures with time.

For Markov chain problems, methods currently used to obtain transient solutions are based either on readily available differential equation solvers such as the Adams formulae and backward differentiation formulae (BDF) or on the method of uniformization (also called Jensen's method or the method of randomization). Most methods experience difficulty when both $\max_j |q_{jj}|$ (the largest exit rate from any state) and t (the time at which the solution is required) are large, and there appears to be little to recommend a single method for all situations. Only the relatively novel approach of using Krylov subspaces is promising over a large range of parameters. It has the possibility of becoming the method of choice.

We shall begin by considering the *uniformization method*. This method has attracted much attention, is extremely simple to program and often outperforms other methods, particularly when the solution is needed at a single time point close to the origin. Papers of general interest in this area include [11], [93], [127], [128], and [167].

8.2 Uniformization

8.2.1 The Basic Concepts

Consider any continuous-time Markov chain $\{X(t), t \geq 0\}$ on a countable state space S and assume that it is defined by the pair Q^X (the infinitesimal generator matrix) and $\pi(0)$ (the initial state probability distribution vector). The uniformization procedure ([57], [58], [61], [79], [96]) may be applied to this process so long as the diagonal elements of Q^X are bounded; i.e., so long as there exists a Γ such that

$$|q_{ii}^X| \leq \Gamma < \infty, \quad \text{for all } i \in S.$$

For computational purposes it is usual to take $\Gamma = \max_i |q_{ii}^X|$.

Let $\pi(0) = (\pi_0(0), \pi_1(0), \ldots)$ be an initial probability vector and $\pi(t)$ the probability vector of the continuous-time process $X(t)$ at time t. Let

$$p_{ij}^X(t) = \text{Prob}\{X(t) = j | X(0) = i\}$$

and let P^X be the matrix of all such probabilities. Then

$$\pi(t) = \pi(0)P^X(t). \tag{8.1}$$

Recall from Section 1.4.4 of Chapter 1 that the forward Kolmogorov differential equations may be written as

$$\frac{dP^X(t)}{dt} = P^X(t)Q^X,$$

from which we may deduce that

$$P^X(t) = e^{tQ^X}.$$

Also, notice that for a given $\pi(0)$ we have

$$\frac{d[\pi(0)P^X(t)]}{dt} = \pi(0)P^X(t)Q^X,$$

and substituting from equation (8.1) gives

$$\frac{d\pi(t)}{dt} = \pi(t)Q^X$$

with solution

$$\pi(t) = \pi(0)e^{tQ^X}.$$

Now consider the (discretized) stochastic transition probability matrix P, defined as

$$P = I + \frac{1}{\Gamma}Q^X.$$

Then $Q^X = \Gamma(P - I)$, and hence,

$$e^{tQ^X} = e^{t\Gamma P - t\Gamma I} = e^{-t\Gamma}e^{(t\Gamma)P},$$

since $e^{-(t\Gamma)I} = e^{-t\Gamma}I$ and P and I commute. Therefore, we obtain

$$P^X(t) = e^{tQ^X} = \sum_{k=0}^{\infty} \frac{(\Gamma t)^k}{k!} e^{-\Gamma t} P^k.$$

The formula

$$P^X(t) = \sum_{k=0}^{\infty} P^k e^{-\Gamma t} \frac{(\Gamma t)^k}{k!} \tag{8.2}$$

is called the *uniformization equation*. The transient distribution at time t is obtained by computing an approximation to the infinite summation

$$\pi(t) = \pi(0) \sum_{k=0}^{\infty} P^k e^{-\Gamma t} \frac{(\Gamma t)^k}{k!} = \sum_{k=0}^{\infty} \pi(0) P^k e^{-\Gamma t} \frac{(\Gamma t)^k}{k!} \tag{8.3}$$

Observe that the uniformization formula incorporates the distribution function of the Poisson process,

$$\text{Prob}\{N(t) = k\} = \frac{(\Gamma t)^k}{k!} e^{-\Gamma t}, \qquad k \geq 0, \ t \geq 0.$$

It is in reference to this fact that the uniformization formula is most often developed, rather than by the purely algebraic approach that we adopted. The continuous-time process $X(t)$ is interpreted as a discrete-time process (with transition probability matrix P), which is embedded in a Poisson process of rate Γ, i.e., the probability of having k transitions of the discrete process during a time interval t follows a Poisson process with expectation Γt.

8.2.2   The Truncation Error

Among the numerical advantages of the uniformization technique are the ease with which it can be translated into computer code and the control it gives over the truncation error. We shall first discuss the truncation error. In implementing the uniformization method, we need to truncate the series in (8.3); i.e., we approximate $\pi(t)$ as

$$\pi(t) \approx \sum_{k=0}^{K} \pi(0)P^k e^{-\Gamma t}\frac{(\Gamma t)^k}{k!}$$

Let

$$\pi^*(t) = \sum_{k=0}^{K} \pi(0)P^k e^{-\Gamma t}\frac{(\Gamma t)^k}{k!} \qquad (8.4)$$

and let $\delta(t) = \pi(t) - \pi^*(t)$. For any consistent vector norm $||.||$, $||\delta(t)||$ is the truncation error. It is not difficult to bound this error numerically. We have

$$
\begin{aligned}
||\pi(t) - \pi^*(t)||_\infty &= ||\sum_{k=0}^{\infty} \pi(0)P^k e^{-\Gamma t}\frac{(\Gamma t)^k}{k!} - \sum_{k=0}^{K} \pi(0)P^k e^{-\Gamma t}\frac{(\Gamma t)^k}{k!}||_\infty \\
&= ||\sum_{k=K+1}^{\infty} \pi(0)P^k e^{-\Gamma t}\frac{(\Gamma t)^k}{k!}||_\infty \\
&\leq \sum_{k=K+1}^{\infty} e^{-\Gamma t}\frac{(\Gamma t)^k}{k!} = \sum_{k=0}^{\infty} e^{-\Gamma t}\frac{(\Gamma t)^k}{k!} - \sum_{k=0}^{K} e^{-\Gamma t}\frac{(\Gamma t)^k}{k!} \\
&= 1 - \sum_{k=0}^{K} e^{-\Gamma t}\frac{(\Gamma t)^k}{k!} \qquad (8.5)
\end{aligned}
$$

If we choose K sufficiently large that $1 - \sum_{k=0}^{K} e^{-\Gamma t}(\Gamma t)^k/k! \leq \epsilon$, or equivalently, that

$$\sum_{k=0}^{K} \frac{(\Gamma t)^k}{k!} \geq \frac{1-\epsilon}{e^{-\Gamma t}}, \qquad (8.6)$$

where ϵ is some prespecified truncation criterion, then it follows that

$$||\pi(t) - \pi^*(t)||_\infty \leq \epsilon.$$

This does not necessarily imply that the *computed* solution will satisfy this criterion, for the analysis does not take roundoff error into account. In the next section we shall discuss implementation details and see that roundoff error is not generally a problem.

8.2.3   Implementation

In implementing the uniformization method, we may code equation (8.4) exactly as it appears (taking advantage, of course, of the relationships that exist between consecutive terms in the summation), or we may partition it into time steps $t_0, t_1, t_2, \ldots, t_m = t$ and write code to implement

$$\pi(t_{i+1}) = \sum_{k=0}^{K_i} \pi(t_i) P^k e^{-\Gamma(t_{i+1}-t_i)} \Gamma^k (t_{i+1} - t_i)^k / k! \qquad (8.7)$$

successively for $i = 0, 1, \ldots, m - 1$. This second approach is the obvious way to perform the computation if the transient solution is *required* at various points $t_1, t_2, \ldots$ between the initial time t_0 and the final time t. It may be computationally more expensive if the transient solution is required only at a single terminal point. However, even in this case it may become necessary to adopt the second approach when the numerical values of Γ and t are such that the computer underflows when computing $e^{-\Gamma t}$. Such instances can be detected a priori and appropriate action taken. For example, one may decide not to allow values of Γt to exceed 100. When such a situation is detected, the time t may be divided into $l = 1 + \lfloor \Gamma t / 100 \rfloor$ equal intervals and the transient solution computed at times $t/l, 2t/l, 3t/l, \ldots, t$.

To examine this more carefully, let the interval $[t_0, t]$ be divided into l equal panels of length h, i.e., $h = (t - t_0)/l$ and $t_{i+1} = t_i + h$ for $i = 0, 1, \ldots, l - 1$, with $t_l = t$. Then, from equation (8.7),

$$\pi(t_{i+1}) = \pi(t_i) T_i, \qquad i = 0, 1, \ldots, l - 1 \qquad (8.8)$$

where

$$T_i = \sum_{k=0}^{K_i} P^k e^{-\Gamma h} \frac{(\Gamma h)^k}{k!}.$$

It is evident that the matrices T_i have a common matrix factor, $\sum_{k=0}^{K} P^k e^{-\Gamma h} \times (\Gamma h)^k / k!$ where $K = \min_{0 \le i \le l-1} K_i$. Indeed, the K_i differ only when the step sizes differ. If the number of states in the Markov chain is small, it is possible to compute and store this matrix factor initially and then use it in all the steps, thereby obtaining a considerable savings in computation time. Furthermore, an efficient Horner-like scheme may be devised for its computation. This scheme is discussed in Section 8.3, which is devoted exclusively to the case in which the number of states in the Markov chain is small.

When the interval $[t_0, t]$ is divided into l subintervals, care must be taken to ensure that the criterion for determining the value of K_i in equation (8.7) reflects the fact that ϵ may no longer be an appropriate value to use. If ϵ has been designated as the truncation error over the entire interval, then smaller values must be used to ensure that the sum of all l truncation errors does not exceed ϵ.

It is more appropriate in this case to substitute ϵ/l for ϵ in equation (8.6). Let ϵ_i be the truncation error and r_i the roundoff error that occurs in moving from time t_i to t_{i+1}. Also, at time t_i, let $\pi(t_i)$ denote the exact solution and π_i the computed solution. On successive steps, the newly computed solution is obtained from equation (8.8), and we have

$$\pi_1 = \pi(t_0)T_0 + \epsilon_0 + r_0 = \pi(t_1) + \epsilon_0 + r_0,$$

$$\pi_2 = \pi_1 T_1 = (\pi(t_0)T_0 + \epsilon_0 + r_0)T_1 + \epsilon_1 + r_1 = \pi(t_0)T_0 T_1 + (\epsilon_0 + r_0)T_1 + \epsilon_1 + r_1$$

until finally,

$$
\begin{aligned}
\pi_l = \quad & \pi(t_0)T_0 T_1 \cdots T_{l-1} \\
+ \; & (\epsilon_0 + r_0)T_1 T_2 \cdots T_{l-1} \\
+ \; & (\epsilon_1 + r_1)T_2 T_3 \cdots T_{l-1} \\
+ \; & \vdots \\
+ \; & (\epsilon_{l-2} + r_{l-2})T_{l-1} + (\epsilon_{l-1} + r_{l-1}).
\end{aligned}
\tag{8.9}
$$

If the value of K_i is chosen at each step so that the truncation error at that step is less than ϵ, it then follows that

$$\|T_i\| \leq \sum_{k=0}^{K_i} \|P\|^k e^{-\Gamma h}\frac{(\Gamma h)^k}{k!} \leq 1 - \epsilon.$$

Therefore, from (8.9), we have

$$
\begin{aligned}
\|\pi(t) - \pi_l\| \;\; & < \;\; \max_i \|\epsilon_i + r_i\| \sum_{k=1}^{l-1}(1 - \epsilon)^k \\
& < \;\; \max_i (\|\epsilon_i\| + \|r_i\|)l.
\end{aligned}
$$

In other words, the truncation error and the roundoff error may be magnified l times in the worst case. Hence, it is necessary to use ϵ/l to determine K_i so that the truncation error over the entire error is bounded by ϵ.

The following algorithm computes the transient solution $\pi(t)$ at time t given the probability distribution $\pi(0)$ at time $t = 0$; P, the stochastic transition probability matrix of the discrete-time Markov chain; Γ, the parameter of the Poisson process; and ϵ, a tolerance criterion. In the algorithm, it suffices to replace t with $t_{k+1} - t_k$ and $\pi(0)$ with $\pi(t_k)$ to compute $\pi(t_{k+1})$, the solution at time t_{k+1}. This implementation is designed for the case in which the number of states in the Markov chain is large. The only operation involving the matrix P is its multiplication with a vector.

Algorithm: Uniformization

1. *Use equation (8.6) to compute K, the number of terms in the summation:*

   - Set $K = 0$; $\xi = 1$; $\sigma = 1$; $\eta = (1 - \epsilon)/e^{-\Gamma t}$.
   - While $\sigma < \eta$ do
     - Compute $K = K + 1$; $\xi = \xi \times (\Gamma t)/K$; $\sigma = \sigma + \xi$.

2. *Approximate $\pi(t)$ from equation (8.4):*

   - Set $\pi = \pi(0)$; $y = \pi(0)$.
   - For $k = 1$ to K do
     - Compute $y = yP \times (\Gamma t)/k$; $\pi = \pi + y$.
   - Compute $\pi(t) = e^{-\Gamma t}\pi$.

This algorithm requires approximately $K(n+n_z)$ multiplications, where n is the number of states in the Markov chain; n_z is the number of nonzero elements in P; and K is the smallest integer that satisfies equation (8.6). Memory requirements are modest; in addition to storage for the initial vector $\pi(0)$ and the matrix P (which may conveniently be kept in a compact form), only two work arrays, each of length n, are needed. A number of factors contribute to the fact that rounding error is generally not a problem with the uniformization algorithm. Observe that the numerical operations involve no negative numbers, nor are there any subtractions. Notice also that the calculation of the matrix power series itself is stable, because $\|P\| \leq 1$ and the coefficient series converges to 1. Additionally, the transition matrix from t_i to t_{i+1}, $\sum_{k=0}^{K_i} P^k e^{-\Gamma(t_{i+1}-t_i)}\Gamma^k(t_{i+1} - t_i)^k/k!$, has a norm that is less than or equal to 1. Therefore, any roundoff error that may occur in initial calculations is not likely to grow.

8.3 Methods Applicable to Small State Spaces

The uniformization method that we have just examined is likely to prove highly satisfactory when Γt is not large. However, in certain applications, such as reliability modelling, it may become necessary to examine the transient behavior at a time t that is very large indeed. Values of $t = 10^8$ have been quoted in the literature [91]. Additionally, the transition rates applicable to certain problems may be such that Γ also is large. In such cases, the quantity Γt is too large to permit the efficient computation of $\pi(t) = \pi(0)e^{Qt}$ by an approximation around zero.

In this section, we shall consider only the case when the number of states n is small, which we take to mean that the transition matrices are sufficiently small to be stored and manipulated as full two-dimensional arrays. Since Q can be stored in full, we shall seek to compute the matrix e^{Qt} *before* applying it to the vector $\pi(0)$. Our concern shall be with algorithms whose computational complexity is proportional to n^3 at most and that require on the order of n^2 (double-precision) memory locations. Several algorithms fall into this category; indeed, in what is arguably the most widely quoted paper in the area, Moler and Van Loan [101] examine 19 "dubious ways to compute the exponential of a matrix" under the assumption that the matrices are of order less than a few hundred and hence can be stored as two-dimensional arrays. We shall consider two classes of method, namely matrix decompositional and matrix-powering methods, since these are likely to be the most effective for Markov chain problems.

8.3.1   Matrix Decompositional Methods

The matrix decompositional methods we consider are based on similarity transformations applied to the infinitesimal generator Q. For nonsingular S, we obtain a matrix B such that

$$B = SQS^{-1}.$$

It then follows from the power series definition of the matrix exponential that

$$e^{Qt} = S^{-1}e^{Bt}S.$$

The objective is to find a matrix S such that e^{Bt} is easy to compute. For example, a diagonal matrix B would seem to be ideal, since for any diagonal matrix $D = \text{diag}\{d_{11}, d_{22}, \ldots, d_{nn}\}$,

$$e^{tD} = \text{diag}\{e^{td_{11}}, e^{td_{22}}, \ldots, e^{td_{nn}}\}.$$

A number of possibilities may be considered, including eigenvalue/eigenvector decompositions and Schur decompositions.

8.3.1.1   Eigenvalue/Eigenvector Decompositions

A first approach is to take S to be the matrix whose rows are the left-hand eigenvectors of Q, i.e., $S = [s_1^T, s_2^T, \ldots, s_n^T]^T$, where $s_iQ = \lambda_i s_i$, and λ_i, $i = 1, 2, \ldots, n$ are the eigenvalues of Q. We shall assume for the moment that Q is *not defective* — that it contains a set of n linearly independent (left-hand) eigenvectors. This is the case, for example, when all the eigenvalues of Q are distinct. We may then write

$$SQ = \Lambda S,$$

where $\Lambda = \text{diag}\{\lambda_1, \lambda_2, \ldots, \lambda_n\}$, and hence

$$e^{Qt} = S^{-1}e^{\Lambda t}S.$$

The right-hand side is now easy to compute, since $e^{\Lambda t} = \mathrm{diag}\{e^{\lambda_1 t}, e^{\lambda_2 t}, \ldots, e^{\lambda_n t}\}$. When we seek the probability distribution at time t given an initial distribution $\pi(0)$, we obtain

$$\pi(t) = \pi(0) S^{-1} e^{\Lambda t} S = \sum_{i=1}^{n} \alpha_i e^{\lambda_i t} s_i. \tag{8.10}$$

The α_i are obtained by solving the system of equations $\alpha S = \pi(0)$, which simply states that the initial vector is a linear combination of the eigenvectors, S. It will be immediately recognized that the great advantage of equation (8.10) is that it allows us to compute the probability distribution at any time t very efficiently. Once the eigensolution has been computed, and the coefficients α_i of the initial distribution in the base S determined, the solution at time t is obtained in one vector-matrix multiplication. The bad news, of course, is that the matrix Q may not have a complete set of n linearly independent eigenvectors. In this case, it is not possible to decompose Q exactly as before, since there will be no invertible matrix of eigenvectors. Nor is it possible to avoid this problem by using the *Jordan canonical form*, since rounding errors make the numerical computation of this impossible [101]. Problems also arise when the matrix is nearly defective, for then the condition number of S will be large, and initially small rounding errors will be greatly magnified in the final computed result. For these reasons, we cannot recommend an eigenvalue/eigenvector decomposition approach for computing transient solutions of Markov chain problems.

8.3.1.2 The Schur Decomposition

A possible means of overcoming the problem that arises in the eigenvector approach when the matrix Q is defective or nearly defective is to use similarity transformations that are orthogonal. Thus, we seek an orthogonal matrix O and a triangular matrix W such that

$$OQO^T = W.$$

Since $O^T = O^{-1}$ for any matrix O that is orthogonal, it follows that the eigenvalues of Q are to be found on the diagonal of W. This is called the *Schur decomposition* of Q. If Q contains complex eigenvalues, then both O and W must be taken to be complex, and the transpose replaced with the complex conjugate. However, it is also possible to implement the algorithm using only real arithmetic, and in that case W becomes quasi-triangular — it may contain 2×2 blocks representing a complex pair of eigenvalues on the diagonal.

Given that the Schur decomposition is available, we have

$$e^{Qt} = O^T e^{Wt} O$$

and hence the only remaining problem is the computation of e^{Wt}. A number of possibilities present themselves. One is to try to compute the exponential of the triangular matrix W directly. Some work of Parlett on computing functions of triangular matrices [116] may be useful in this regard. Alternatively, methods that compute the exponential of a matrix by using rational approximations (such as Chebyshev) to the exponential function may prove to be useful [135]. If the QR algorithm of Francis [44] is used to compute the Schur decomposition, then one possibility is to follow the route normally taken by the QR algorithm in computing eigenvectors; that is, to construct a triangular matrix R and a diagonal matrix Λ (the diagonal elements of W) such that

$$RW = \Lambda R.$$

The left-hand eigenvectors are then computed as $V = RO$. This means we can now write

$$Q = O^T R^{-1} \Lambda R O$$

and thus

$$\pi(0)e^{Qt} = \pi(0)O^T R^{-1} e^{\Lambda t} R O.$$

Solving $\alpha R = \pi(0)O^T$ for α shows that $\pi(t)$ may be obtained for any t by computing

$$\pi(t) = \alpha e^{\Lambda t} V,$$

where $V = RO$. An estimate of the condition number of the triangular matrix R (and hence of V) may be readily obtained. This provides a convenient mechanism for detecting conditions under which the computed results may not be accurate.

8.3.2   Matrix Scaling and Powering

Matrix-scaling and -powering methods arise from a property that is unique to the exponential function, viz.:

$$e^{Qt} = \left(e^{Qt/2}\right)^2. \tag{8.11}$$

The basic idea is to compute e^{Qt_0} for some small value t_0 such that $t = 2^m t_0$ and subsequently to form e^{Qt} by repeated application of the relation (8.11). It is in the computation of the initial e^{Qt_0} that methods differ, and we shall return to this point momentarily.

Let Q be the infinitesimal generator of an ergodic, continuous-time Markov chain, and let $\pi(0)$ be the probability distribution at time $t = 0$. We seek $\pi(t)$, the transient solution at time t. Let us introduce an integer m and a time $t_0 \neq 0$ for which $t = 2^m t_0$. Then

$$\pi(t) = \pi(2^m t_0).$$

Writing $t_j = 2t_{j-1}$, we shall compute the matrices e^{Qt_j} for $j = 0, 1, \ldots, m$ and consequently, by multiplication with $\pi(0)$, the transient solution at times

$t_0, 2t_0, 2^2 t_0, \ldots, 2^m t_0 = t$. Note that $P(t_j) \equiv e^{Qt_j}$ is a stochastic matrix and that, from the Chapman–Kolmogorov equations,

$$P(t_j) = P(t_{j-1})P(t_{j-1}). \tag{8.12}$$

Thus, once $P(t_0)$ has been computed, each of the remaining $P(t_j)$ may be computed from equation (8.12) by squaring the previous $P(t_{j-1})$. Unlike the uniformization method, which computes the transient solution at a single point in time only, matrix-powering methods, in the course of their computation, provide the transient solution at the intermediate times $t_0, 2t_0, 2^2 t_0, \ldots, 2^{m-1} t_0$. However, a disadvantage of matrix-powering methods, besides computational costs proportional to n^3 and memory requirements of n^2, is that repeated squaring may induce rounding error buildup, particularly in instances in which $m \gg 1$.

We now turn our attention to the computation of e^{Qt_0}. Since t_0 is small, methods based on approximations around zero become possible candidates. Thus we may use the uniformization method itself. This yields the uniformized power method, which we shall consider next. Another possibility that we shall examine is that of rational Padé approximations around the origin.

8.3.2.1 Uniformized Powering

To compute $P(t_0)$ by means of the uniformization method, we use (see equation (8.2))

$$P(t_0) = \sum_{k=0}^{K} P^k e^{-\Gamma t_0} \frac{(\Gamma t_0)^k}{k!}. \tag{8.13}$$

For this to converge quickly, we shall choose t_0 sufficiently small that $\Gamma t_0 \le 0.1$. Since t_0 must also satisfy $t_0 = t/2^m$, setting $m = \lceil \log_2(10\Gamma t) \rceil$ satifies both these criteria. In [91], the recommended value for m is

$$m = \lfloor \log_2[4(\zeta + 3)\Gamma t] \rfloor$$

where ζ is the maximum number of nonzero elements in any row of P. Since $\zeta \ge 2$, this will always give a larger value for m.

Once a value of m (and hence t_0) has been chosen, equation (8.13) may be used to find $P(t_0)$. As before, K, the number of terms in the summation, must first be computed from equation (8.6). Notice that during the computation of $P(t_0)$, advantage may be taken of a Horner-like procedure. (This was not incorporated into the algorithm presented in Section 8.2.3, since the assumption made at that time was that the number of states in the Markov chain was large. The Horner-like scheme is appropriate only when the coefficient matrix is small enough to be stored in a two-dimensional array.) Writing

$$Z_1 = \frac{\Gamma t_0}{K} I$$

and using the recurrence relation

$$Z_{k+1} = \frac{\Gamma t_0}{K - k}(Z_k P + I), \qquad k = 1, 2, \ldots, K - 1, \qquad (8.14)$$

$P(t_0)$ may be found as

$$P(t_0) = e^{-\Gamma t_0}(Z_K P + I).$$

With $P(t_0)$ computed, the remaining $P(t_k)$ are computed directly from equation (8.12).

The following algorithm is a possible implementation of the uniformized powering procedure. The variables $\pi(0)$, P, Γ, and ϵ are all as defined in the uniformization algorithm of Section 8.2.

Algorithm: Uniformized Powering

1. *Find appropriate values of m and t_0:*

   - Compute $m = \lceil \log_2(10\Gamma t) \rceil$; $t_0 = t/2^m$.

2. *Use equation (8.6) to compute K, the number of terms in the summation:*

   - Set $K = 0$; $\xi = 1$; $\sigma = 1$.
   - Compute $\eta = (1 - \epsilon)/e^{-\Gamma t_0}$.
   - While $\sigma < \eta$ do
     - $K = K + 1$; $\xi = \xi \times (\Gamma t_0)/K$; $\sigma = \sigma + \xi$.

3. *Approximate $P(t_0)$ using the uniformization equation and a Horner-type recursion, equation (8.14):*

   - Compute $Z = (\Gamma t_0/K)I$.
   - For $k = 1$ to $K - 1$ compute
     - $Z = \Gamma t_0(ZP + I)/(K - k)$.
   - Compute $P(t_0) = e^{-\Gamma t_0}(ZP + I)$.

4. *Raise matrix to power 2^m by squaring m times and then compute $\pi(t)$:*

   - For $j = 1$ to m compute
     - $P(t_j) = P(t_{j-1})P(t_{j-1})$.
   - Compute $\pi(t) = \pi(0)P(t_m)$.

In this algorithm the major computational burden is in steps 3 and 4. Step 3 requires approximately $K(nn_z + n)$ multiplications, where, as before, n is the number of states in the Markov chain; n_z is the number of nonzero elements in P; and K is the smallest integer that satisfies equation (8.6) (with t replaced by t_0). Step 4 requires $mn^3 + n^2$ multiplications, where $m = \lceil log_2(10\Gamma t) \rceil$. In addition to storage for the initial probability vector $\pi(0)$ and the stochastic transition probability matrix P (a compact scheme may be used), step 3 requires a double-precision array with n^2 elements. In step 4, two such arrays are needed.

8.3.2.2 Rational Padé Approximations

The (p, q) Padé approximant to the matrix exponential e^X is, by definition, the unique (p, q) rational function $R_{pq}(X)$,

$$R_{pq}(X) \equiv \frac{N_{pq}(X)}{D_{pq}(X)}$$

which matches the Taylor series expansion of e^X through terms to the power $p + q$. Its coefficients are determined by solving the algebraic equations

$$\sum_{j=0}^{\infty} \frac{X^j}{j!} - \frac{N_{pq}(X)}{D_{pq}(X)} = O\left(X^{p+q+1}\right),$$

which yields

$$N_{pq}(X) = \sum_{j=0}^{p} \frac{(p+q-j)!p!}{(p+q)!j!(p-j)!} X^j$$

and

$$D_{pq}(X) = \sum_{j=0}^{q} \frac{(p+q-j)!q!}{(p+q)!j!(q-j)!} (-X)^j.$$

For more detailed information on Padé approximants, the interested reader should consult the text by Baker [8].

A major disadvantage of Padé approximants is that they are accurate only near the origin and so should not be used when $\|X\|_2$ is large. However, since we shall be using them in the context of a matrix-scaling and -powering procedure, we may (and shall) choose t_0 so that $\|Qt_0\|_2$ is sufficiently small that the Padé approximant to e^{Qt_0} may be obtained with acceptable accuracy, even for relatively low-degree approximants.

When $p = q$, we obtain the *diagonal Padé approximants*, and there are two main reasons why this choice is to be recommended. First, they are more stable [101]. In Markov chain problems all the eigenvalues of $X = Qt$ are to be found in the left half-plane. In this case the computed approximants $R_{pq}(X)$ for $p \neq q$ have larger rounding errors, because either $p > q$ and cancellation problems may arise, or $p < q$ and $D_{pq}(X)$ may be badly conditioned. Second, we obtain a higher-order method with the same amount of computation. To compute $R_{pq}(X)$ with $p < q$ requires about qn^3 flops and yields an approximant that has order $p + q$. To compute $R_{qq}(X)$ requir es essentially the same number of flops but produces an approximant of order $2q > p + q$. Similar statements may be made when $p > q$.

For diagonal Padé approximants we find

$$R_{pp}(X) = \frac{N_{pp}(X)}{N_{pp}(-X)} \tag{8.15}$$

where

$$N_{pp}(X) = \sum_{j=0}^{p} \frac{(2p-j)!p!}{(2p)!j!(p-j)!} X^j \equiv \sum_{j=0}^{p} c_j X^j.$$

The coefficients c_j can be conveniently constructed by means of the recursion

$$c_0 = 1; \quad c_j = c_{j-1} \frac{p+1-j}{j(2p+1-j)}.$$

For actual implementation purposes, the following irreducible form offers a considerable savings in computation time at the expense of additional memory locations.

$$R_{pp}(X) = \begin{cases} I + 2 \dfrac{X \sum_{k=0}^{p/2-1} c_{2k+1} X^{2k}}{\sum_{k=0}^{p/2} c_{2k} X^{2k} - X \sum_{k=0}^{p/2-1} c_{2k+1} X^{2k}} & \text{if } p \text{ is even} \\[4mm] -I - 2 \dfrac{\sum_{k=0}^{(p-1)/2} c_{2k} X^{2k}}{X \sum_{k=0}^{(p-1)/2} c_{2k+1} X^{2k} - \sum_{k=0}^{(p-1)/2} c_{2k} X^{2k}} & \text{if } p \text{ is odd} \end{cases}$$

$$\tag{8.16}$$

Thus, for even values of p,

$$R_{pp}(X) = I + 2 \frac{S_e}{T_e - S_e},$$

where

$$S_e = c_1 X + c_3 X^3 + \cdots + c_{p-1} X^{p-1} \quad \text{and} \quad T_e = c_0 + c_2 X^2 + c_4 X^4 + \cdots + c_p X^p,$$

while for odd values of p,

$$R_{pp}(X) = -\left(I + 2 \frac{S_o}{T_o - S_o} \right),$$

where now

$$S_o = c_0 + c_2 X^2 + c_4 X^4 + \cdots + c_{p-1} X^{p-1} \quad \text{and} \quad T_o = c_1 X + c_3 X^3 + \cdots + c_p X^p.$$

These computations may be conveniently combined, and they cry out for a Horner-type evaluation procedure. Indeed, Horner evaluations of the numerator and the denominator in equation (8.16) need only one-half the operations of a straightforward implementation of equation (8.15).

The following algorithm, adapted from that presented in [118], implements a Padé variant of the matrix-powering and -scaling approach for the computation of e^X. To compute the transient solution at time t of a Markov chain with generator Q and initial state $\pi(t_0)$, it suffices to apply this algorithm with $X = Qt$ and then to form $\pi(t_0)R$, where R is the approximation to e^X computed by the algorithm.

Algorithm: Padé Approximation for e^X

1. *Find appropriate scaling factor:*

    - Compute $m = \max(0, \lfloor \log \|X\|_\infty / \log 2 \rfloor + 1)$.

2. *Compute coefficients and initialize:*

    - Set $c_0 = 1$.
    - For $j = 1, 2, \ldots, p$ do
        - Compute $c_j = c_{j-1} \times (p + 1 - j)/(j(2p + 1 - j))$.
    - Compute $X1 = 2^{-m}X$; $X2 = X1^2$; $T = c_p I$; $S = c_{p-1} I$.

3. *Application of Horner scheme:*

    - Set $odd = 1$.
    - For $j = p - 1, \ldots, 2, 1$ do
        - if $odd = 1$, then
            * Compute $T = T \times X2 + c_{j-1} I$.
          else
            * Compute $S = S \times X2 + c_{j-1} I$.
        - Set $odd = 1 - odd$.
    - If $odd = 0$, then
        - Compute $S = S \times X1$; $R = I + 2 \times (T - S)^{-1} \times S$.
      else
        - Compute $T = T \times X1$; $R = -(I + 2 \times (T - S)^{-1} \times S$.

4. *Raise matrix to power 2^m by repeated squaring:*

    - For $j = 1$ to m do
        - Compute $R = R \times R$.

This algorithm requires a total of approximately $(p + m + \frac{4}{3})n^3$ multiplications. It may be implemented with three double-precision arrays, each of size n^2, in addition to the storage required for the matrix itself.

This leaves us with the choice of p. In the appendix of [101], a backward error analysis of the Padé approximation is presented, in which it is shown that if $\|X\|_2/2^m \leq \frac{1}{2}$, then

$$\left[R_{pp}(2^{-m}X)\right]^{2^m} = e^{X+E},$$

where

$$\frac{\|E\|_2}{\|X\|_2} \;\leq\; \left(\frac{1}{2}\right)^{2p-3} \frac{(p!)^2}{(2p)!(2p+1)!} \;\approx\; \begin{cases} 0.77 \times 10^{-12} & (p = 5) \\ 0.34 \times 10^{-15} & (p = 6) \\ 0.11 \times 10^{-18} & (p = 7) \\ 0.27 \times 10^{-22} & (p = 8) \end{cases} \qquad (8.17)$$

This suggests that relatively low-degree Padé approximants are adequate. However, the above analysis does not take rounding error into account. This aspect has been examined by Ward [175], who proposes certain criteria for selecting appropriate values for some computers. Additionally, a discussion on "the degree of best rational approximation to the exponential function" is provided by Saff [138]. Finally, numerical experiments on Markov chains by Philippe and Sidje [118] find that even values of p are better than odd values and that the value $p = 6$ is generally satisfactory.

8.4   Ordinary Differential Equation (ODE) Solvers

In a continuous-time Markov chain with infinitesimal generator Q and initial probability distribution $\pi(0)$, the probability distribution at time t is given by

$$\pi(t) = \pi(0)e^{Qt}.$$

The vector $\pi(t)$ is the solution of the Chapman–Kolmogorov differential equations

$$\frac{d\pi(t)}{dt} = \pi(t)Q \qquad (8.18)$$

with initial conditions $\pi(t = 0) = \pi(0)$. The solution of ordinary differential equations (ODEs) has been a subject of extensive research, and it is therefore appropriate for us to examine the possibilities of applying ODE techniques to the determination of transient solutions of Markov chains. An immediate advantage of such an approach is that, unlike uniformization and the other methods considered in this chapter, numerical methods for the solution of ODEs are applicable to *nonhomogeneous* Markov chains, i.e., Markov chains whose infinitesimal generators are a function of time, $Q(t)$.

The available literature on the numerical solution of differential equations is vast, and it would be impossible for us to give anything other than a brief introduction in a single section. The reader is referred to any of the many excellent texts in the area, and in particular to the two volumes by Hairer, Norsett, and Wanner [67] and Hairer and Wanner [68]. Anyone seriously interested in applying ODE methods to the transient solution of Markov chains should keep these two volumes within close reach. Therefore, instead of vainly attempting to provide an extensive treatment of the topic in this section, we shall be content to describe the most important concepts and to provide an understanding of just what it means to solve an ODE numerically. We shall describe the factors that distinguish one class of numerical solution method from another and generally give examples in each class. In particular, we shall indicate numerical methods that have already been used in Markov chain problems, together with their advantages and disadvantages. We will also provide references to implementations (and Fortran source code) for methods that are currently considered by researchers in the ODE arena to be among the best available.

8.4.1 Ordinary Differential Equations

We begin our analysis with a discussion of ordinary differential equations and what it means to solve them from a numerical point of view. An ODE is an equation that involves some unknown function together with one or more of its derivatives. Its *order* is the order of the highest derivative, and its *degree* is the power to which the highest derivative is raised. An ODE is said to be *homogeneous* if every term contains the dependent variable; otherwise it is said to be *nonhomogeneous*. Thus, from equation (8.18), it is apparent that our concern is with *first-order, first-degree, homogeneous* ordinary differential equations. We shall restrict our attention to first-order differential equations and write the general form as

$$y' \equiv \frac{dy}{dt} = f(t, y).$$

A function $y(t)$ is called a solution of this equation if for all t, $y'(t) = f(t, y(t))$. One of the most notable features that distinguishes a first-order differential equation from a nondifferential equation (such as $y = f(t)$ for example) is that instead of being able to compute the *value* of the function at any point t, we have access to a *direction vector* or *slope* at any point in the (t, y) plane. In a sense we may think of the plane as representing a *force field* into which a particle is placed. The movement of the particle depends on the force acting on it, and this varies according to its position in the field. The particle moves in the direction of the force and traces out a path. In general, different paths are traced according to the position in the plane at which the particle is initially placed. Each possible path is a *solution curve*.

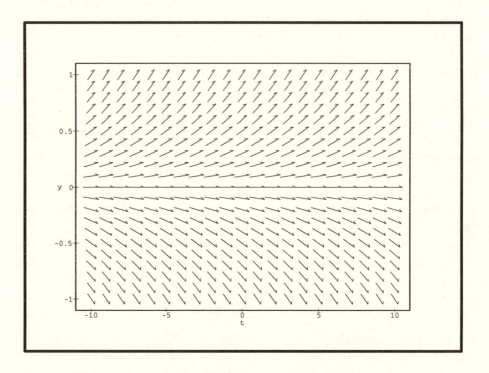

Figure 8.1: Direction vectors for $y' = .5y$.

For example, Figure 8.1 shows the direction vectors for the equation

$$y' = \lambda y \tag{8.19}$$

in the particular case when $\lambda = .5$. Notice that since $f(t, y)$ is a function of y only, the direction of the slope at any point y is the same for all values of t. In this case, it is known that the analytic solution is given by

$$y(t) = ce^{\lambda t}, \qquad \lambda \neq 0, \tag{8.20}$$

where c is a *constant of integration*. In the absence of constraints, this constant of integration may be chosen arbitrarily, and hence equation (8.20) represents an infinity of solutions to the differential equation (8.19). In other words, the constant of integration generates a *family of solution curves*; each member of the family essentially corresponds to an initial placement of a particle in the plane. Figures 8.2 and 8.3 show some solution curves for the two cases $\lambda > 0$ and $\lambda < 0$, respectively. If along with the differential equation $y' = \lambda y$, we are also given an *initial condition* such as $y(t = 0) = y_0$, then we may uniquely determine the constant of integration and hence a unique solution: $y(t) = y_0 e^{\lambda t}$.

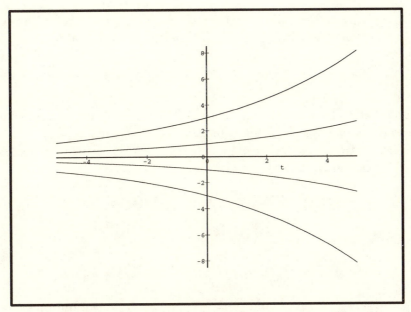

Figure 8.2: Solution curves for $y' = \lambda y$; $\lambda > 0$.

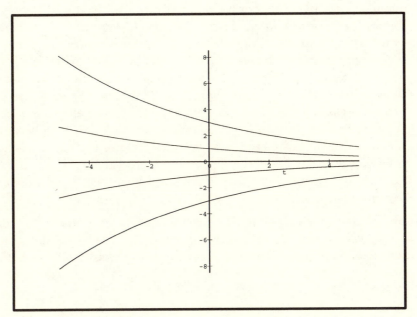

Figure 8.3: Solution curves for $y' = \lambda y$; $\lambda < 0$.

8.4.2 Numerical Solutions and Stability

Given a first-order differential equation $y' = f(t, y)$ and an initial condition $y(t_0) = y_0$, a solution is a *differentiable* function $y(t)$ such that

$$y(t_0) = y_0; \qquad \frac{d}{dt}y(t) = f(t, y(t)). \tag{8.21}$$

When f satisfies certain conditions relating to continuity and boundedness, the existence and uniqueness of the solution may be established.

The extension to higher dimensions is immediate. A first-order *system* of ordinary differential equations of dimension n is given as

$$
\begin{aligned}
y_1' &= f_1(t, y_1, \ldots, y_n); & y_1(t_0) &= y_{10} \\
y_2' &= f_2(t, y_1, \ldots, y_n); & y_2(t_0) &= y_{20} \\
&\;\;\vdots & &\;\;\vdots \\
y_n' &= f_n(t, y_1, \ldots, y_n); & y_n(t_0) &= y_{n0}
\end{aligned}
$$

Writing $y = (y_1, y_2, \ldots, y_n)^T$ and $f = (f_1, f_2, \ldots, f_n)^T$, we obtain $y' = f(t, y)$ as before. In what follows, all results are applicable to the vector case unless otherwise noted.

Numerical procedures to compute the solution of equation (8.21) attempt to follow a unique solution curve from its value at an initially specified point to its value at some other prescribed point τ. This usually involves a discretization procedure on the interval $[0, \tau]$ and the computation of approximations to the solution at the intermediate points. Given a discrete set of points (or *mesh*) $\{0 = t_0, t_1, t_2, \ldots, t_\eta = \tau\}$ in $[0, \tau]$, we shall denote the exact solution of the differential equation (8.21) at time t_i by $y(t_i)$. The *step size* or *panel width* at step i is defined as $h_i = t_i - t_{i-1}$. A numerical method generates a sequence $\{y_1, y_2, \ldots, y_\eta\}$ such that y_i is an approximation[1] to $y(t_i)$. In some instances the solution at time τ is all that is required, and in that case the computed solution is taken to be y_η. In other cases the value of $y(t)$ for all $t_0 \leq t \leq \tau$ is required, often in the form of a graph, and this is obtained by fitting a suitable curve to the values $y_0, y_1, y_2, \ldots, y_\eta$.

If the solution is continuous and differentiable, then in a small neighborhood of the point (t_0, y_0) we can approximate the solution curve by its tangent y_0' and thereby move from (t_0, y_0) to the next point (t_1, y_1). Since this means that

$$y_0' = \frac{y_1 - y_0}{t_1 - t_0}$$

---

[1]The reader should be careful not to confuse y_i with $y(t_i)$. We use y_i for $i \neq 0$ to denote a *computed approximation* to the *exact value* $y(t_i)$. For $i = 0$, the initial condition gives $y_0 = y(t_0)$.

we obtain the formula

$$y_1 = y_0 + h_1 y_0' = y_0 + h_1 f(t_0, y_0), \qquad (8.22)$$

where $h_1 = t_1 - t_0$. This process may be repeated from the computed approximation y_1 to obtain

$$y_2 = y_1 + h_2 f(t_1, y_1) \qquad (8.23)$$

where $h_2 = t_2 - t_1$, and so on until the final destination is reached with

$$y_\eta = y_{\eta-1} + h_\eta f(t_{\eta-1}, y_{\eta-1}) \qquad (8.24)$$

with $h_\eta = t_\eta - t_{\eta-1}$. This method is called the *forward Euler method (FEM)* or *explicit Euler method.*

When the initial value and the approximations $y_0, y_1, \ldots, y_\eta$ are connected by straight lines, we obtain an *Euler polygon.* As the step sizes are altered, we generate different approximations y_i and obtain different Euler polygons. If the number of steps taken to reach y_η from y_0 is increased in such a way that the width of the largest panel tends to zero, then the Euler polygons tend to the unique solution of the differential equation (8.21). The following theorem is proven in [67].

Theorem 8.1 *Consider a domain*

$$D = \{(t, y) \mid t_0 \leq t \leq \tau, \ \|y - y_0\| \leq \beta\}.$$

If

- *$f(t, y)$ is continuous on D,*

- *$\|f(t, y)\|$ is bounded by $\mathcal{B}$ on D, and*

- *$f(t, y)$ satisfies a Lipschitz condition on D, i.e., if there exists a constant $L \neq 0$ such that*
  $$\|f(t, z) - f(t, y)\| \leq L\|z - y\|$$
  for all (t, y) and $(t, z) \in D$,

then, for $\tau - t_0 \leq \beta/\mathcal{B}$,

1. *The Euler polygons converge uniformly to a continuous function $\phi(t)$ as the width of the largest panel shrinks to zero.*

2. *The function $\phi(t)$ is continuously differentiable and is a solution of (8.21) on $t_0 \leq t \leq \tau$.*

3. *There exists no other solution of (8.21) on $t_0 \leq t \leq \tau$.*

The existence and uniqueness of the solution of (8.21), when $f(t, y)$ satisfies the three conditions of the theorem, follow immediately.

The fact that we generate an approximation y_i rather than the exact value $y(t_i)$ means that we are forced off the correct solution curve and onto a nearby one. It is interesting to observe the effect of this on the examples illustrated in Figures 8.2 and 8.3. In Figure 8.2 the solution curves *diverge* from each other with time, and hence a small error that throws us onto a different, nearby curve will increase unboundedly with time. On the other hand, an error made in the example illustrated by Figure 8.3 will decay with time. In this figure, the error curves *converge* on each other. This allows us to introduce the notion of *stability* in differential equations. The differential equation (8.21) is said to be *stable* if a small perturbation ϵ in the initial condition $y(t_0) = y_0$ produces a *bounded* change in the solution as $t \rightarrow \infty$. In other words, if $y(t, y_0)$ is the exact solution with initial condition $y(t_0) = y_0$, and $y(t, y_0 + \epsilon)$ is the exact solution using the initial condition $y(t_0) = y_0 + \epsilon$, then the differential equation is said to be stable if, for all t, we have

$$\|y(t, y_0) - y(t, y_0 + \epsilon)\| \leq M,$$

where $M > 0$ and is independent of t. Otherwise the ODE is said to be *unstable*[2].

In the case of $y' = \lambda y$, it is easy to verify that

$$|y(t, y_0) - y(t, y_0 + \epsilon)| = |\epsilon| e^{\lambda t},$$

which, for $\lambda > 0$, tends to infinity as $t \rightarrow \infty$ (hence unstable), and for $\lambda < 0$, tends to zero as $t \rightarrow \infty$ and is therefore stable. In fact, in this second case the ODE is said to be *asymptotically stable*, since solutions for all initial values tend to 0 as $t \rightarrow \infty$. Since the differential equation $y' = \lambda y$ covers a spectrum of stability conditions, it is used as a *test equation* for determining the stability of numerical solution methods.

For the special case of systems of linear differential equations with constant coefficients,

$$y' = Ay,$$

a necessary and sufficient condition for stability is that the eigenvalues λ of A satisfy

$$\text{Re}(\lambda) \leq 0 \qquad\qquad\qquad (8.25)$$

with strict inequality for multiple roots that engender Jordan blocks [67]. It follows that the Chapman–Kolmogorov differential equations arising from homogeneous Markov chains are stable, since they are linear with constant coefficients and the eigenvalues of Q satisfy equation (8.25).

---

[2]Note that this terminology is at odds with our definitions in Chapter 2. There we used the terms *well-conditioned* and *ill-conditioned* to describe characteristics of a problem (here an ODE equation) while reserving usage of *stability* and *instability* to designate properties of a numerical solution procedure. In the ODE literature the term *stability* is widely used to cover both instances, and we shall adhere to this convention.

8.4.3 Elementary Numerical Algorithms

8.4.3.1 The Method of Euler and Its Variants

As already mentioned, a numerical approach to solving the initial-value ordinary differential equation $y' = f(t, y);\ y(t_0) = y_0$ attempts to follow a unique solution curve from a given initial position. It does so in a step-by-step manner, approximating the exact solution $y(t_i)$ at each point t_i by a value y_i, for $i = 1, 2, \ldots, \eta$. In computing y_{i+1}, a method may incorporate the values of previously computed approximations y_j for $j = 0, 1, \ldots, i$, or even previous approximations to $y(t_{i+1})$. A method that uses only (t_i, y_i) to compute y_{i+1} is said to be an *explicit single-step method*. It is said to be a *multistep method* if it uses approximations at several previous steps to compute its new approximation. A method is said to be *implicit* if computing y_{i+1} requires a value for $y(t_{i+1})$; otherwise, it is said to be *explicit*.

From equations (8.22) through (8.24), the explicit Euler method may be written as

$$y_{i+1} = y_i + h_{i+1} f(t_i, y_i), \qquad i = 0, 1, \ldots, \eta - 1 \tag{8.26}$$

where $h_{i+1} = t_{i+1} - t_i$. An *implicit* or *backward Euler method* may also be defined by using the slope at the point (t_{i+1}, y_{i+1}). This gives

$$y_{i+1} = y_i + h_{i+1} f(t_{i+1}, y_{i+1}). \tag{8.27}$$

The *modified Euler method* incorporates the average of the slopes at both points under the assumption that this will provide a better average approximation of the slope over the entire panel $[t_i, t_{i+1}]$. The formula is given by

$$y_{i+1} = y_i + h_{i+1} \frac{f(t_i, y_i) + f(t_{i+1}, y_{i+1})}{2}. \tag{8.28}$$

This is also referred to as the *trapezoid rule*. A final formula in this category is the *implicit midpoint rule*, which involves the slope at the midpoint of the interval. The formula is given by

$$y_{i+1} = y_i + h_{i+1}\, f\left(t_i + \frac{h_i}{2}, \frac{y_i + y_{i+1}}{2}\right). \tag{8.29}$$

These are all single-step methods, because none of them uses information prior to the current position (t_i, y_i) to generate the next approximation y_{i+1}.

In general, implicit methods like those of equations (8.27) through (8.29) require the solution of a (system of) nonlinear equation(s). However, when f is linear in y, then the equations are linear and may be solved by the standard methods for linear systems such as Gaussian elimination or SOR. In many cases an approximation to $f(t_{i+1}, y_{i+1})$ is first obtained by an explicit method (called a *predictor*), and this value is then used in an implicit equation, (8.27) through

(8.29) (now called a *corrector*), to generate a better approximation. Such methods generally go under the name of *predictor-corrector* methods. The corrector may be applied several times, although it is generally recommended that if insufficient accuracy is achieved with one iteration, it is best to begin the predictor-corrector process all over again but with a smaller step size.

8.4.3.2 Stability

The stability of a solution method is often defined according to how it behaves when it is applied with a fixed step size h to the *Dahlquist test equation*,

$$y' = \lambda y, \qquad y(t_0) = y_0 = 1. \tag{8.30}$$

We have already seen how this differential equation may be used to generate a variety of stability conditions. More generally, when an ODE method is applied to a *system* of differential equations $y' = f(t, y)$, then it is the value of $\max_j |\text{Re}(\lambda_j)|$, where λ_j are the eigenvalues of the Jacobian $\partial f / \partial y$, that plays the role of λ in (8.30). For example, consider the system of equations $y' = Ay$, where A is of order n and has n distinct eigenvalues λ_i, $i = 1, \ldots, n$. There exists a nonsingular matrix S such that $SAS^{-1} = \Lambda \equiv \text{diag}\{\lambda_1, \lambda_2, \ldots, \lambda_n\}$. Thus,

$$Sy' = SAS^{-1} \, Sy,$$

and on substituting for $Sy = \hat{y}$ we find $\hat{y}' = \Lambda \hat{y}$, which separates to give $\hat{y}'_j = \lambda_j \hat{y}_j$ for $j = 1, 2, \ldots, n$, i.e., n equations identical to the test equation.

Applying the explicit Euler method with a fixed step size h to the Dahlquist test equation gives

$$y_{i+1} = y_i + h\lambda y_i = (1 + h\lambda)y_i = (1 + h\lambda)^i.$$

If the initial value is now perturbed to give $1 + \epsilon$, the explicit Euler method gives

$$\tilde{y}_{i+1} = (1 + h\lambda)^i (1 + \epsilon),$$

and hence

$$|y_{i+1} - \tilde{y}_{i+1}| = |1 + h\lambda|^i |\epsilon|,$$

which remains bounded, *independent* of i, if and only if

$$|1 + h\lambda| \leq 1. \tag{8.31}$$

The inequality (8.31) defines a circular region in the z-plane (complex plane), where $z = h\lambda$, of center -1 and radius 1. The function $R(z) = 1 + z$ is called the *stability function* of the explicit Euler method, and the region

$$S = \{z \in \mathcal{C}; \ |R(z)| \leq 1\}$$

is called the *stability domain* of the method.

Definition 8.1 (A-Stability) *A solution method is said to be absolutely stable or A-stable if its stability domain covers the entire negative half-plane, i.e., if*

$$S \supset \{z; \operatorname{Re}(z) \le 0\} = \mathcal{C}^{-}.$$

Note that this corresponds exactly with the region in which the differential equation itself, (8.30), is stable. This definition also implies that the explicit Euler method is not A-stable. It may be said to be *partially stable*, since there is a region of the negative half-plane, though not the entire half-plane, in which it is stable. The implicit Euler method, on the other hand, is A-stable. Applying this method to the test equation (8.30) yields

$$y_{i+1} = y_i + h\lambda y_{i+1}$$

and hence,

$$y_{i+1} = \left(\frac{1}{1-h\lambda}\right) y_i = \left(\frac{1}{1-h\lambda}\right)^{i} y_0.$$

This requires that $|(1-h\lambda)^{-1}| \le 1$, i.e., $|1-h\lambda| \ge 1$, $h\lambda \ne 1$. It follows that the stability function of the implicit Euler method is

$$R(z) = \frac{1}{1-z}$$

and the stability domain is the *exterior* of the circle with center 1 and radius 1. Since this more than covers the negative half-plane, the method is A-stable. The stability function for the trapezoid rule and the implicit midpoint rule is the same. It is given by

$$R(z) = \frac{1+z/2}{1-z/2}.$$

It follows that both are A-stable.

8.4.3.3 Application to Markov Chains

In Markov chain problems $f(t, y(t))$ is equal to $\pi(t)Q$, where $\pi(t)$ is the probability vector at time t and Q is the infinitesimal generator. This means that $f(t_j, y_j)$ must be set equal to $\pi_{(j)}Q$ in equations (8.26) through (8.29). Note that $\pi_{(j)}$ is the *state vector* of probabilities at time t_j. We use this notation, rather than π_j, so as not to confuse the jth component of the vector with the entire vector at time t_j. Thus, the explicit Euler method (Equation 8.26) is given by

$$\pi_{(i+1)} = \pi_{(i)} + h_{i+1}\pi_{(i)}Q,$$

i.e.,

$$\pi_{(i+1)} = \pi_{(i)}\left(I + h_{i+1}Q\right). \tag{8.32}$$

In this case, moving from one time step to the next is accomplished by a scalar-matrix product and a vector-matrix product. On the other hand, the modified Euler or trapezoid rule, equation (8.28), becomes

$$\pi_{(i+1)} = \pi_{(i)} + \frac{h_{i+1}}{2} \left(\pi_{(i)} Q + \pi_{(i+1)} Q \right),$$

i.e.,

$$\pi_{(i+1)} \left(I - \frac{h_{i+1}}{2} Q \right) = \pi_{(i)} \left(I + \frac{h_{i+1}}{2} Q \right), \tag{8.33}$$

which requires, in addition to the operations needed by the explicit Euler method, the solution of a system of equations at each step (plus a scalar-matrix product). These additional computations per step are offset to a certain extent by the better accuracy achieved with the trapezoid rule. If the step size is kept constant, the matrix

$$\left(I + \frac{h_{i+1}}{2} Q \right) \left(I - \frac{h_{i+1}}{2} Q \right)^{-1}$$

may be computed at the outset before beginning the stepping process, so that the computation per step required by the modified Euler method becomes identical to that of the explicit Euler. When the step size is not kept constant, each different value of h used requires that a system of linear equations be solved. Depending on the size and sparsity pattern of Q, the work required by the trapezoid rule to compute the solution to a specified precision may or may not be less than that required by the explicit Euler. The trade-off is between an implicit method requiring more computation per step but fewer steps and an explicit method requiring more steps but less work per step!

Consider as an example, the 3×3 infinitesimal generator given in Section 8.2.1. We have

$$Q = \begin{pmatrix} -2 & 1 & 1 \\ 3 & -8 & 5 \\ 1 & 2 & -3 \end{pmatrix}. \tag{8.34}$$

Let $\pi_{(0)} = (1, 0, 0)$ and the length of the interval of integration, $\tau = 1$. Table 8.1 shows the results obtained when equations (8.32) and (8.33) were implemented in MATLAB[3]. The first column gives the step size used throughout the range, the second is the 2-norm of the absolute error using the Euler method, and the third column gives the 2-norm of the absolute error when the trapezoid method is used. Observe that the accuracy achieved with the explicit Euler method has a direct relationship with the step size. Observe also the much more accurate results obtained with the trapezoid method. Since equal step sizes were used in these examples, an essentially identical amount of computation was required by both. The trapezoid rule needed only an additional (3×3) matrix inversion and a matrix-matrix product.

---

[3] A listing of all the MATLAB implementations of ODE methods used to generate results in this chapter is given at the end of Section 8.4.7.

Table 8.1: Euler and Trapezoid Method Results

Step Size	Explicit Euler $\|y(t_1) - y_1\|_2$	Trapezoid $\|y(t_1) - y_1\|_2$
$h = .1$	$0.133193e{-}01$	$0.788907e{-}03$
$h = .01$	$0.142577e{-}02$	$0.787774e{-}05$
$h = .001$	$0.143269e{-}03$	$0.787763e{-}07$
$h = .0001$	$0.143336e{-}04$	$0.787519e{-}09$
$h = .00001$	$0.143343e{-}05$	$0.719906e{-}11$

8.4.3.4 Taylor Series Methods

The forward Euler method corresponds to taking the first two terms of the Taylor series expansion of $y(t)$ around the current approximation point. Expanding $y(t)$ as a power series in h, we have

$$y(t + h) = y(t) + hy'(t) + \frac{h^2}{2!}y''(t) + \cdots. \tag{8.35}$$

If we know $y(t)$ exactly, we may compute as many terms in the series as we like by repeated differentiation of $y' = f(t, y)$, the only constraint being the differentiability of f. Different *Taylor series algorithms* are obtained by evaluating different numbers of terms in the series (8.35). For example, using the obvious notational correspondence with equation (8.35), the Taylor's algorithm of order 2 may be written

$$y_{i+1} = y_i + hT_2(t_i, y_i),$$

where

$$T_2(t_i, y_i) = f(t_i, y_i) + \frac{h}{2}f'(t_i, y_i).$$

The order-1 Taylor series algorithm corresponds to the FEM method, since we have

$$y_{i+1} = y_i + hT_1(t_i, y_i) = y_i + hf(t_i, y_i).$$

The chief utility of the Taylor series approach is that it provides a mechanism by which we may introduce the concept of *order of convergence*.

Definition 8.2 (Order of Convergence) *A numerical method is said to be convergent to order p if the sequence of approximations $\{y_0, y_1, y_2, \ldots, y_\eta\}$ generated by the method (with constant step size $h = \tau/\eta$) satisfies*

$$\|y_i - y(t_i)\| = O(h^p), \qquad i = 0, 1, \ldots, \eta.$$

Euler's method, described by equation (8.26), is convergent to order 1; the error behaves like Mh, where M is a constant that depends on the problem and h is the maximum step size. We have the following theorem.

Theorem 8.2 *Suppose y is a solution of (8.21) such that y'' is bounded on $[0, \tau]$, and f satisfies a Lipschitz condition in y and is continuous in t. Then the explicit Euler method is convergent to $O(h)$ accuracy on $[0, \tau]$ as $h \to 0$. In other words, there exists an $M > 0$, independent of i and h, such that $\|y_i - y(t_i)\| < Mh$ for all $i = 0, 1, \ldots, \tau/h$.*

This theorem is proved by bounding the error in the remainder term of the Taylor series expansion, viz., $h^2 y''(\xi)/2$, where $t_i \le \xi \le t_{i+1}$. Thus, it does not take into account any roundoff error that might occur; instead, it concerns only the truncation error made in truncating the infinite series (8.35). Indeed, as the truncation error tends to zero with h, the roundoff error has the opposite effect. The smaller the step size, the greater the number of steps that must be taken, and hence the greater the amount of computation that must be performed. With this greater amount of computation comes the inevitable increase in roundoff error. Thus, it is important that h not be chosen too small.

8.4.4   Stiff ODEs

An important characterization of initial-value ODEs has yet to be discussed — that of *stiffness*. Explicit methods have enormous difficulty solving stiff ODEs, indeed to such an extent that one definition of stiff equations is "problems for which explicit methods don't work" [68]! Many factors contribute to stiffness, including the eigenvalues of the Jacobian $\partial f/\partial y$ and the length of the interval of integration. The problems stem from the fact that the solutions of stiff systems of differential equations contain rapidly decaying transient terms.

As an example, consider the infinitesimal generator

$$Q = \begin{pmatrix} -1 & 1 \\ 100 & -100 \end{pmatrix} \tag{8.36}$$

with initial probability vector $\pi_{(0)} = (1, 0)$. Q has the decomposition $Q = S\Lambda S^{-1}$, where

$$S = \begin{pmatrix} 1 & -.01 \\ 1 & 1 \end{pmatrix} \quad \text{and} \quad \Lambda = \begin{pmatrix} 0 & 0 \\ 0 & -101 \end{pmatrix}.$$

Since $\pi_{(t)} = \pi_{(0)} e^{Qt} = \pi_{(0)} S e^{\Lambda t} S^{-1}$, we find that

$$\pi_{(t)} = \frac{1}{1.01} \left(e^0 + .01 e^{-101t}, \ .01 e^0 - .01 e^{-101t} \right).$$

The exponents e^{-101t} in this expression tend rapidly to zero, leaving the stationary probability distribution. In spite of this, however, when an explicit method is used, small step sizes must be used over the entire period of integration. It is not possible to increase the step size once the terms in e^{-101t} are effectively zero.

The classical definition asserts that a system of ODEs is stiff when certain eigenvalues of the Jacobian matrix (with elements $\partial f_i / \partial y_i$) have large negative real parts when compared to others. Using this definition, Markov chain problems (where the Jacobian is given by Q, the infinitesimal generator) are stiff when $\max_k |\text{Re}(\lambda_k)| \gg 0$. The Gerschgorin disk theorem is useful in bounding this quantity. This theorem states that the eigenvalues of any square matrix A of order n lie in the union of the n circular disks with centers $c_i = a_{ii}$ and radii $r_i = \sum_{j=1, j \neq i}^{n} |a_{ij}|$. From the special properties of infinitesimal generator matrices, this implies that

$$\max_k |\text{Re}(\lambda_k)| \leq 2 \max_j |q_{jj}|,$$

i.e., by twice the largest total exit rate from any one state in the Markov chain.

Reibman and Trivedi [127] point out that it is the presence of rapidly decaying solution components that affects the performance of numerical solution procedures, and since "rapid" is defined relative to the length of the solution interval, a computational definition of stiffness should include a dependence on the time scale over which the solution is sought. They recommend a definition in which the Chapman–Kolmogorov differential equations are said to be stiff on the interval $[0, \tau]$ if there exists a solution component that has a variation on that interval that is large compared to $1/\tau$. Thus, they recommend the use of $\tau \max_k |\text{Re}(\lambda_k)|$ as a measure of stiffness.

Now let us consider the effect of applying the explicit Euler, implicit Euler, and trapezoid methods to the 2×2 example, equation (8.36). From the stability functions of these methods, we need to take $h \leq \frac{2}{101}$ for the explicit Euler but are not restricted in our choice of h for the other two. Of course, these considerations are for stability reasons only, not for convergence purposes! Choosing $h = .1$ and $h = .06$, well outside the stability domain of the explicit Euler; $h = .025$ and $h = .021$, just outside; $h = .019$, just inside; and $h = .01$, well inside the Euler stability domain, we obtain the results shown in Table 8.2 for the error in the computed approximation at $t = 1$.

Observe the distinctly different behavior of the explicit Euler around the critical step size, $h = \frac{2}{101}$. Step sizes that produce values for which $|R(z)| > 1$ result in large errors in the computed solution. Indeed, not a single digit of accuracy is obtained in these cases, compared with the 13-decimal-place accuracy of the implicit Euler method. The reader may wish to verify that gross errors result from using the explicit Euler method with small step size $h = .01$ from $t = 0$ to $t = 1$ (at which point the stationary distribution has been achieved to machine accuracy) and then increasing the step size to $h = .1$ and continuing the solution until a later time. For example, continuing onto time $t = 10$ produces the result $\|y(t_{10}) - y_{10}\|_2 = 1.02597e + 68$!

Table 8.2: Stability of Numerical Methods on a Stiff Problem

Step Size	Explicit Euler $\|y(t_1) - y_1\|_2$	Implicit Euler $\|y(t_1) - y_1\|_2$	Trapezoid $\|y(t_1) - y_1\|_2$
$h = .1$	$0.545265e+08$	$0.559714e-12$	$0.253041e-03$
$h = .06$	$0.258584e+10$	$0.852669e-13$	$0.240575e-06$
$h = .025$	$0.299908e+06$	$0.937074e-13$	$0.998150e-13$
$h = .021$	$0.300371e+01$	$0.888216e-13$	$0.960393e-13$
$h = .019$	$0.173221e-03$	$0.951502e-13$	$0.899325e-13$
$h = .01$	$0.899325e-13$	$0.905987e-13$	$0.914869e-13$

Another feature of the results in this table deserves attention. Both the implicit Euler method and the trapezoid method are A-stable. Nevertheless, the implicit Euler performs better than the trapezoid rule for the larger values of h. This leads us to a further refinement of the definition of A-stability, called *L-stability*, which is particularly useful for stiff systems.

Definition 8.3 (L-Stability) *A solution method is said to be L-stable if it is A-stable and, in addition, its stability function is such that*

$$\lim_{z \to -\infty} R(z) = 0.$$

It may be readily verified that the implicit Euler method, with stability function $R(z) = 1/(1 - z)$, is L-stable, but the trapezoid method, with stability function $R(z) = (1 + z/2)/(1 - z/2)$, is not.

8.4.5 Single-Step Methods

All the methods that we have considered so far fall into the category of single-step methods. They use information at the current point (and at no points prior to this point) to determine the next approximation. Unlike multistep methods, single-step methods do not exhibit any numerical instability *for h sufficiently small*. Multistep methods may, in some cases, be unstable for all values of h!

From a numerical point of view a Taylor series approach is generally not practical. Taylor's algorithm of order 1 is equivalent to Euler's method and is not very effective, because a very small step size is needed to achieve reasonable accuracy. Taylor series algorithms of order greater than 1 involve the determination of higher derivatives of $y(t)$, i.e., the derivatives of $f(t, y)$, which may be very complicated. This leads to the development of methods for which the derivatives are not needed but which have the same order of convergence as Taylor series methods of order 2, 3, 4, etc. Runge–Kutta methods are such a class of methods. They are accurate one-step methods that are relatively inexpensive to use and require only additional functional values to improve the order of accuracy. They are the most effective of all single-step methods.

8.4.5.1 Runge–Kutta Methods

The purpose of a Runge–Kutta algorithm of order p is to obtain an accuracy comparable to a Taylor's series algorithm of order p, but without the need to determine and evaluate the derivatives $f', f'', \ldots, f^{(p-1)}$, requiring instead the evaluation of $f(t, y)$ at selected points. The derivation of an order p Runge–Kutta method is obtained from a comparison with the terms through h^p in the Taylor series method for the first step, i.e., the computation of y_1 from the initial condition (t_0, y_0). This is because the analysis assumes that the value around which the expansion is performed is known exactly. A Runge–Kutta method is said to be of order p if the Taylor series for the exact solution $y(t_0 + h)$ and the Taylor series for the computed solution y_1 coincide up to and including the term h^p.

For example, a Runge–Kutta method of order 2 (RK2) is designed to give the same order of convergence as a Taylor series method of order 2, without the need for differentiation. It is written as

$$y_{i+1} = y_i + h(ak_1 + bk_2) \tag{8.37}$$

where a and b are constants,

$$
\begin{aligned}
k_1 &= f(t_i, y_i) \\
k_2 &= f(t_i + \alpha h, y_i + h\beta k_1),
\end{aligned}
$$

and α and β are constants. The constants are chosen so that the Taylor series approach agrees with (8.37) for $i = 0$, through the term in h^2. Finding these constants requires computing the derivatives of (8.37), for $i = 0$, and then comparing coefficients with those of the first and second derivatives obtained from the exact Taylor series expansion. Equality of coefficients imposes the following restrictions on the constants a, b, α, β:

$$a + b = 1; \qquad \alpha b = \beta b = 1/2.$$

These are satisfied by the choice $a = b = \frac{1}{2}$ and $\alpha = \beta = 1$, so a resulting RK2 method is

$$y_{i+1} = y_i + \frac{h}{2} f(t_i, y_i) + \frac{h}{2} f(t_i + h, y_i + h f(t_i, y_i)).$$

Other choices for the constants are also possible.

The most widely used Runge–Kutta methods are of order 4. The standard RK4 method requires four function evaluations per step and is given by

$$y_{i+1} = y_i + \frac{h}{6}(k_1 + 2k_2 + 2k_3 + k_4) \tag{8.38}$$

where

$$
\begin{aligned}
k_1 &= f(t_i, y_i) \\
k_2 &= f(t_i + h/2, y_i + hk_1/2) \\
k_3 &= f(t_i + h/2, y_i + hk_2/2) \\
k_4 &= f(t_i + h, y_i + hk_3).
\end{aligned}
$$

Notice that each k_i corresponds to a point in the (t, y) plane at which the evaluation of f is required. Each is referred to as a *stage*. Thus, the method described by equation (8.38) is a fourth-order, four-stage, explicit Runge–Kutta method. Its stability function is given by

$$R(z) = 1 + z + \frac{z^2}{2!} + \frac{z^3}{3!} + \frac{z^4}{4!} \tag{8.39}$$

which implies that the method is partially stable rather than A-stable. For real λ, equation (8.39) implies that h should be chosen so that $-2.79 < h\lambda \leq 0$.

The most general form of an s-stage Runge–Kutta method may be written as

$$y_{i+1} = y_i + h \sum_{j=1}^{s} b_j k_j \tag{8.40}$$

where

$$k_j = f\left(t_i + c_j h,\ y_i + h \sum_{l=1}^{s} a_{jl} k_l\right), \qquad j = 1, 2, \ldots, s$$

and b_j and a_{jl} $(j, l = 1, 2, \ldots, s)$ are reals with $c_j = \sum_l a_{jl}$. This gives explicit Runge–Kutta methods when $a_{jl} = 0$ for $j \leq l$ and implicit methods otherwise. It is usual to write the coefficients a_{jl}, b_j, and c_j in tabular form as follows:

$$
\begin{array}{c|cccc}
c_1 & a_{11} & a_{12} & \cdots & a_{1s} \\
c_2 & a_{21} & a_{22} & \cdots & a_{2s} \\
\vdots & \vdots & \vdots & \ddots & \vdots \\
c_s & a_{s1} & a_{s2} & \cdots & a_{ss} \\
\hline
 & b_1 & b_2 & \cdots & b_s
\end{array}
$$

For example, the following is the tabular form of the standard explicit RK4 method given by equation (8.38).

$$
\begin{array}{c|cccc}
0 & & & & \\[2mm]
\frac{1}{2} & \frac{1}{2} & & & \\[2mm]
\frac{1}{2} & 0 & \frac{1}{2} & & \\[2mm]
1 & 0 & 0 & 1 & \\[2mm]
\hline \\[-2mm]
& \frac{1}{6} & \frac{2}{6} & \frac{2}{6} & \frac{1}{6}
\end{array}
\tag{8.41}
$$

It can be readily verified that substituting these values into equation (8.40) gives equation (8.38) directly.

Besides the standard explicit RK4 method given by equation (8.38), a number of variants are in current use. In particular, the *Runge–Kutta–Fehlberg (RKF)* method has been used for comparison purposes in Markov chain analysis [127]. However, experiments reported by Hairer, Norsett, and Wanner in [67] suggest that the fourth-order version of Dormand and Prince may be preferable. Explicit Runge–Kutta methods of order higher than 4 (in fact, up to order 10 [67]) have also been developed. These are recommended only when high precision is needed.

Among important implicit Runge–Kutta methods we would like to cite the *Radau IIA* methods [68]. These are based on the Radau quadrature formulae; s-stage Radau IIA methods are of order $2s - 1$. The coefficients of the order 3 method are given by

$$
\begin{array}{c|cc}
\frac{1}{3} & \frac{5}{12} & \frac{-1}{12} \\[2mm]
1 & \frac{3}{4} & \frac{1}{4} \\[2mm]
\hline \\[-2mm]
& \frac{3}{4} & \frac{1}{4}
\end{array}
$$

while those for the order 5 method are

$$
\begin{array}{c|ccc}
\frac{4-\sqrt{6}}{10} & \frac{88-7\sqrt{6}}{360} & \frac{296-169\sqrt{6}}{1800} & \frac{-2+3\sqrt{6}}{225} \\[3mm]
\frac{4+\sqrt{6}}{10} & \frac{296+169\sqrt{6}}{1800} & \frac{88+7\sqrt{6}}{360} & \frac{-2-3\sqrt{6}}{225} \\[3mm]
1 & \frac{16-\sqrt{6}}{36} & \frac{16+\sqrt{6}}{36} & \frac{1}{9} \\[3mm]
\hline \\[-2mm]
& \frac{16-\sqrt{6}}{36} & \frac{16+\sqrt{6}}{36} & \frac{1}{9}
\end{array}
$$

A listing of a Fortran implementation of the fifth-order method may be found in the appendix of [68].

It is shown in [68] that the stability function for RK methods of the general form (8.40) is given by

$$R(z) = 1 + zb^T(I - zA)^{-1}e, \tag{8.42}$$

where $b^T = (b_1, b_2, \ldots, b_s)$; $A = (a_{jl})^s_{j,l=1}$; and $e = (1, 1, \ldots, 1)^T$. The reader may wish to verify that the previous stability function for the fourth-order, four-stage Runge–Kutta method, equation (8.39), is obtained when the numerical values in (8.41) are substituted into (8.42). Note that in this case,

$$(I - zA)^{-1} = \begin{pmatrix} 1 & 0 & 0 & 0 \\ z/2 & 1 & 0 & 0 \\ z^2/4 & z/2 & 1 & 0 \\ z^3/4 & z^2/2 & z & 1 \end{pmatrix}.$$

It may be shown that the Radau IIA methods are not only A-stable but also L-stable.

8.4.5.2 Application to Markov Chains

When the standard explicit fourth-order Runge–Kutta method given by equation (8.38) is applied to the Chapman–Kolmogorov equations $\pi' = \pi Q$, the sequence of operations to be performed to move from $\pi_{(i)}$ to the next time step $\pi_{(i+1)}$ is as follows:

$$\pi_{(i+1)} = \pi_{(i)} + h(k_1 + 2k_2 + 2k_3 + k_4)/6 \tag{8.43}$$

where

$$
\begin{aligned}
k_1 &= \pi_{(i)}Q \\
k_2 &= (\pi_{(i)} + hk_1/2)Q \\
k_3 &= (\pi_{(i)} + hk_2/2)Q \\
k_4 &= (\pi_{(i)} + hk_3)Q
\end{aligned}
$$

As for implicit Runge–Kutta methods for Markov chains, the following has been recommended by Malhotra and Trivedi [89]:

$$\pi_{(i+1)}\sum_{j=0}^{s}\alpha_j(hQ)^j = \pi_{(i)}\sum_{j=0}^{s-1}\beta_j(hQ)^j.$$

Like Radau IIA, this implicit s-stage method is of order $2s - 1$. Note that higher-orders require higher powers of Q. Since taking higher powers of the (usually)

Table 8.3: Results of Runge–Kutta Methods

Step Size	Explicit RK4 $\|y(t_1) - y_1\|_2$	Implicit RK3 $\|y(t_1) - y_1\|_2$
$h = .1$	$0.112360e-04$	$0.399462e-04$
$h = .01$	$0.878797e-09$	$0.428946e-07$
$h = .001$	$0.842195e-13$	$0.432216e-10$
$h = .0001$	$0.188411e-14$	$0.513527e-12$
$h = .00001$	$0.179469e-14$	$0.187128e-11$

sparse matrix Q reduces its sparsity, Malhotra and Trivedi recommend taking $s = 2$, which gives the following third-order method:

$$\pi_{(i+1)} \left(I - 2hQ/3 + h^2 Q^2/6 \right) = \pi_{(i)} \left(I + hQ/3 \right). \tag{8.44}$$

Implementing equations (8.43) and (8.44) in MATLAB and applying them to the 3×3 infinitesimal generator given previously, equation (8.34), with the same initial condition, $\pi_{(0)} = (1, 0, 0)$, yields the results shown in Table 8.3.

8.4.5.3 Practicalities

Implementing an ODE solver based directly on an equation such as (8.38) may appear to be very straightforward. However, it hides two important details: the choice of the step size h and the estimation of the error that occurs. These go hand in hand, because the step size must be sufficiently small that the precision required of the solution is attained, but large enough to avoid the unneeded computation in producing a solution that is overly accurate for the application.

In single-step methods such as Runge–Kutta, the *local truncation error* (often abbreviated to *local error*) is the error that is made in one step of the method under the assumption that the approximation obtained from the previous step was correct. It does not include any roundoff error that might occur, but only the truncation error made in truncating an infinite series. It may be verified by means of Taylor expansions that the local error for a Runge–Kutta method of order p is given by

$$\|y(t_0 + h) - y_1\| = Ch^{p+1} + O(h^{p+2}) \tag{8.45}$$

for some constant C. The *global error* is the error in the computed solution after several steps, often at the terminal point τ, viz.: $\|y(\tau) - y_\tau\|$.

Two approaches have been used in estimating the local error. The first is *Richardson's extrapolation procedure*, and the second is the use of *embedded RK formulae.* The basic idea behind Richardson's extrapolation is to use two different approximations of a fixed RK method of order p to compute an estimate of the constant C in equation (8.45). Generally the method is applied over two steps, each of width h, and the result compared with that obtained when exactly that same method is applied over a single panel of width $2h$. It may be shown (see [67]), that the first gives a result y_{i+2}, for which

$$y(t_i + 2h) = y_{i+2} + 2Ch^{p+1} + O(h^{p+2}).$$

For the second (the application of the method over one large panel of width $2h$), we obtain an approximation $\bar{y}_{i+2}$ that satisfies

$$y(t_i + 2h) = \bar{y}_{i+2} + C(2h)^{p+1} + O(h^{p+2}).$$

These two equations allow us to solve for C and hence obtain not only an estimate for the error but also an approximation of order $p+1$ (rather than of order p) to $y(t_i + 2h)$. The following theorem to this effect is given in [67]:

Theorem 8.3 *Suppose that y_2 is the numerical result of two steps with step size h of a RK method of order p, and $\bar{y}_2$ is the result of one big step with step size $2h$. Then the error of y_2 can be extrapolated as*

$$y(t_0 + 2h) - y_2 = \frac{y_2 - \bar{y}_2}{2^p - 1} + O(h^{p+2}).$$

Furthermore,

$$\hat{y}_2 = y_2 + \frac{y_2 - \bar{y}_2}{2^p - 1}$$

is an approximation of order $p+1$ to $y(t_0 + 2h)$.

Embedded RK formulae for estimating the local error differ from the extrapolation approach in that the same formulation of the method is used to produce two approximations of different order, one to be used for estimating the result and the other for estimating the error. Thus, for a given s-stage RK method, coefficients are computed such that

$$
\begin{aligned}
y_{i+1} &= y_i + h(b_1 k_1 + \cdots + b_s k_s) \quad \text{is of order } p \\
\text{and } \hat{y}_{i+1} &= y_i + h(\hat{b}_1 k_1 + \cdots + \hat{b}_s k_s) \quad \text{is of order } q
\end{aligned}
$$

where $q = p+1$ or $q = p-1$. The RKF method is such an approach and is generally written as RKF4(5), signifying that the approximation to the result is of order 4 and that of the error is of order 5. The Dormand–Prince implementation adopts the opposite strategy, using the higher-order formula for the appoximation to the result and the lower-order for the error estimator. Hence, it is generally written as Dormand–Prince5(4). The first integer indicates the order of the approximation and the second the order of the error estimator. The coefficients of the Dormand–Prince5(4) methods are given in the following table.

0						
$\frac{1}{5}$	$\frac{1}{5}$					
$\frac{3}{10}$	$\frac{3}{40}$	$\frac{9}{40}$				
$\frac{4}{5}$	$\frac{44}{45}$	$-\frac{56}{15}$	$\frac{32}{9}$			
$\frac{8}{9}$	$\frac{19372}{6561}$	$-\frac{25360}{2187}$	$\frac{64448}{6561}$	$-\frac{212}{729}$		
1	$\frac{9017}{3168}$	$-\frac{355}{33}$	$\frac{46732}{5247}$	$\frac{49}{176}$	$-\frac{5103}{18656}$	
1	$\frac{35}{384}$	0	$\frac{500}{1113}$	$\frac{125}{192}$	$-\frac{2187}{6784}$	$\frac{11}{84}$
y_1	$\frac{35}{384}$	0	$\frac{500}{1113}$	$\frac{125}{192}$	$-\frac{2187}{6784}$	$\frac{11}{84}$	0
$\hat{y}_1$	$\frac{5179}{57600}$	0	$\frac{7571}{16695}$	$\frac{393}{640}$	$-\frac{92097}{339200}$	$\frac{187}{2100}$	$\frac{1}{40}$

The size of the error obtained with step size h may be used to estimate the step size needed to achieve a prescribed precision that is either larger or smaller than this error. If err is an estimate of the error that results with a step size h, i.e.,

$$err = \frac{\|y_{i+1} - \bar{y}_{i+1}\|}{2^p - 1}$$

and *tol* is the precision required, then the step size needed to achieve this tolerance is given by

$$h^* = h \left(\frac{tol}{err} \right)^{1/(p+1)}.$$

In application codes, h^* is generally multiplied by a safety factor and is not allowed to increase or decrease too quickly.

Estimations of the global error are more difficult to obtain. Often the user of a software package is requested to perform several experiments varying the tolerance request for local errors and to observe the resulting difference in the computed global solution. A crude estimate of the global error, obtained by summing the local errors, is sometimes used. The assumption is that bounding local errors somehow helps to bound the global error.

8.4.6   Multistep Methods

8.4.6.1   Milne, Adams, and BDF

We saw that the modified Euler method is given by

$$y_{i+1} \;=\; y_i + \frac{h}{2}(f(t_i, y_i) + f(t_{i+1}, y_{i+1})) \tag{8.46}$$

$$\;=\; y_i + \frac{h}{2}(f_i + f_{i+1}) \tag{8.47}$$

where we now write $f_i \equiv f(t_i, y_i)$ and $f_{i+1} \equiv f(t_{i+1}, y_{i+1})$. As previously mentioned, this is also referred to as the trapezoid rule, because of the relationship between the second term on the right-hand side and the trapezoid rule for numerical integration. Indeed, writing the differential equation (8.21) in integrated form, we have

$$y(t_{i+1}) = y(t_i) + \int_{t_i}^{t_{i+1}} f(t, y(t))dt. \tag{8.48}$$

Using the trapezoid rule to compute an approximation to the integral in (8.48) and substituting the approximations for $y(t_{i+1})$ and $y(t_i)$ yields (8.46) directly. It is now natural to ask whether it is possible to use more efficient quadrature rules, such as Simpson's rule, to obtain a better formula for numerically solving ODEs. The trapezoid rule essentially integrates the polynomial of degree 1 (straight line) that interpolates (t_i, y_i) and (t_{i+1}, y_{i+1}). Other methods, such as Simpson's, integrate higher-degree polynomials, which interpolate a greater number of points. This necessitates the use of more than the current approximation. For example, we have

$$y(t_{i+1}) = y(t_{i-1}) + \int_{t_{i-1}}^{t_{i+1}} f(t, y(t))dt$$

and using Simpson's rule, gives

$$y_{i+1} = y_{i-1} + \frac{h}{3}(f_{i-1} + 4f_i + f_{i+1}).$$

This is called *Milne's method*. Since it is an implicit formula, requiring y_{i+1} on the right-hand side, an initial approximation for y_{i+1} may be obtained from an explicit formula such as the basic Euler method. Milne, however, introduced his own predictor formula (an explicit multistep formula) given by

$$y_{i+1} = y_{i-3} + \frac{4h}{3}(2f_i - f_{i-1} + 2f_{i-2}).$$

However, we use these formulae for illustration purposes only, since they are not stable and therefore not recommended for general use.

Much more useful are the multistep methods (both explicit and implicit) of Adams. We assume that the steps are equally spaced of width h, i.e., $t_j = t_0 + jh$, and that approximations to y_j for $j = 0, 1, \ldots, i$ are already available. To obtain explicit Adams formulae, the integral in (8.48) is calculated by interpolating a polynomial through k previous points and extrapolating through to the $(i+1)^{th}$. It is usual to express the interpolating polynomial in terms of backward differences

$$\nabla^0 f_i = f_i; \qquad \nabla^{j+1} f_i = \nabla^j f_i - \nabla^j f_{i-1}.$$

The interpolating polynomial is given by

$$z(t) = z(t_i + sh) = \sum_{j=0}^{k-1} (-1)^j \begin{pmatrix} -s \\ j \end{pmatrix} \nabla^j f_i$$

and equation (8.48) becomes

$$y_{i+1} = y_i + \int_{t_i}^{t_{i+1}} z(t) dt.$$

This results in the *explicit Adams formulae*

$$y_{i+1} = y_i + h \sum_{j=0}^{k-1} \gamma_j \nabla^j f_i \quad \text{with} \quad \gamma_j = (-1)^j \int_0^1 \begin{pmatrix} -s \\ j \end{pmatrix} ds. \qquad (8.49)$$

The first two explicit Adams formulae ($k = 1, 2$) are given by

$$k = 1: \quad y_{i+1} = y_i + h f_i \qquad \text{(explicit Euler) and}$$
$$k = 2: \quad y_{i+1} = y_i + \frac{h}{2}(3f_i - f_{i-1}).$$

Implicit Adams formulae are obtained by making the interpolating polynomial pass through (t_{i+1}, y_{i+1}). In this case the formulae become

$$y_{i+1} = y_i + h \sum_{j=0}^{k} \rho_j \nabla^j f_{i+1} \quad \text{with} \quad \rho_j = (-1)^j \int_0^1 \begin{pmatrix} -s+1 \\ j \end{pmatrix} ds. \qquad (8.50)$$

Initial rules are

$$k = 0: \quad y_{i+1} = y_i + h f_{i+1} \qquad \text{(implicit Euler)}$$
$$k = 1: \quad y_{i+1} = y_i + \frac{h}{2}(f_{i+1} + f_i) \qquad \text{(modified Euler)}$$
$$k = 2: \quad y_{i+1} = y_i + \frac{h}{12}(5f_{i+1} + 8f_i - f_{i-1}). \qquad (8.51)$$

In general, these implicit formulae mean that a nonlinear equation has to be solved at each step. This may be accomplished by a fixed-point iteration. Let $y_{i+1}^{(0)}$ be an initial approximation to y_{i+1}, and let

$$y_{i+1}^{(l)} = y_i + h(\beta_k f(t_{i+1}, y_{i+1}^{(l-1)}) + \beta_{k-1} f_i + \cdots + \beta_0 f_{i-k+1}).$$

Then, if the step size h is sufficiently small, $y_{i+1}^{(l)}$ converges to y_{i+1} as $l \to \infty$. The reader should note that this fixed-point iteration turns the algorithm into an explicit method and can destroy the attractive stability properties of the implicit method. It is often preferable to use Newton's method to solve the implicit formula. We note also that in practice one often uses an explicit formula of a given order to obtain $y_{i+1}^{(0)}$ (the predictor), followed by one step of the implicit formula of the same order. Unfortunately, this also has the same detrimental effect on stability properties.

The so-called *BDF (backward differentiation formulae)* methods [50] are a class of multistep methods based on somewhat different principles. These are the most widely used methods for stiff systems. Only implicit versions are in current use, since low-order explicit versions correspond to the explicit Euler method ($k = 1$) and the midpoint rule ($k = 2$), and higher-order versions ($k \geq 3$) are not stable. Implicit versions are constructed by generating an interpolating polynomial $z(t)$ through the points (t_j, y_j) for $j = i - k + 1, \ldots, i + 1$. However, now y_{i+1} is determined so that $z(t)$ satisfies the differential equation at t_{i+1}, i.e., in such a way that $z'(t_{i+1}) = f(t_{i+1}, y_{i+1})$. The general formula for implicit BDF methods is

$$\sum_{j=1}^{k} \frac{1}{j} \nabla^j y_{i+1} = h f_{i+1}$$

and the first three rules are

$$
\begin{array}{lll}
k = 1: & y_{i+1} - y_i = h f_{i+1} & \text{(implicit Euler)} \\
k = 2: & 3y_{i+1}/2 - 2y_i + y_{i-1}/2 = h f_{i+1} & (8.52) \\
k = 3: & 11y_{i+1}/6 - 3y_i + 3y_{i-1}/2 - y_{i-2}/3 = h f_{i+1} &
\end{array}
$$

The BDF formulae are known to be stable for $k \leq 6$ and to be unstable for other values of k.

Explicit multistep methods require only a single function evaluation per step, compared, for example, with four for a fourth-order Runge–Kutta method. This makes the explicit Adams formulae attractive when the evaluation of the function is computationally expensive or when the interval of integration is long. For stiff systems, implicit methods must be used, and the BDF formulae are recommended. The major computational cost per step is then the evaluation and solution of the Jacobian. In many cases an approximation to the Jacobian is sufficient. For linear systems, such as Markov chain applications, the Jacobian is already available and the cost is proportional to that of solving a linear system of equations. Often software packages incorporate a direct solution procedure that allows for both a banded Jacobian and full two-dimensional storage. For large systems whose nonzero structure makes a direct solution method undesirable, iterative methods may have to be used.

In all the multistep formulae presented in this section, the step size is taken to have a constant value throughout the interval of integration. This keeps life simple. However, implementations are available that permit the step size to vary. They are called *variable-step-size* multistep methods. Also, when the local truncation error exceeds the specified tolerance criteria, instead of taking a smaller step size it is possible to apply a higher-order method to achieve the same effect. ODE methods that incorporate this feature are called *variable-order* multistep methods. The choice of the order of the method and the size of the step to be taken are related decisions. In some implementations, the order of the method is selected so that the largest step size possible can be taken.

8.4.6.2 Application to Markov Chains

This time we consider the application of an implicit Adams formula and an implicit BDF method, both for $k = 2$, equations (8.51) and (8.52), respectively, to the now familiar 3×3 example of equation (8.34). Applying equation (8.51) to the Chapman–Kolmogorov equations $\pi' = \pi Q$ gives

$$\pi_{(i+1)} = \pi_{(i)} + \frac{h}{12}\left(5\pi_{(i+1)}Q + 8\pi_{(i)}Q - \pi_{(i-1)}Q\right),$$

and therefore each step requires the solution of

$$\pi_{(i+1)}\left(I - \frac{5h}{12}Q\right) = \pi_{(i)}\left(I + \frac{2h}{3}Q\right) - \frac{h}{12}\pi_{(i-1)}Q. \tag{8.53}$$

For the BDF formula we find

$$3\pi_{(i+1)}/2 - 2\pi_{(i)} + \pi_{(i-1)}/2 = h\pi_{(i+1)}Q, \tag{8.54}$$

so that the system of equations to be solved at each step is

$$\pi_{(i+1)}\left(3I/2 - hQ\right) = 2\pi_{(i)} - \pi_{(i-1)}/2. \tag{8.55}$$

We also consider a method advocated by Reibman and Trivedi [127] for stiff Markov chains. This method, which is called TR-BDF2, is a combination of the trapezoid rule and the BDF formula with $k = 2$. Each step size h is divided into two, of size γh and $(1 - \gamma)h$, $0 < \gamma < 1$. The trapezoid rule is applied over the first part of the panel, starting from $\pi_{(i)}$ to obtain a value for $\pi_{(i+\gamma)}$, and then the BDF formula is applied over the remainder of the panel using the points $\pi_{(i)}$ and $\pi_{(i+\gamma)}$ to compute $\pi_{(i+1)}$. Thus, $\pi_{(i+\gamma)}$ is obtained from

$$\pi_{(i+\gamma)}\left(I - \frac{\gamma h}{2}Q\right) = \pi_{(i)} + \frac{\gamma h}{2}\pi_{(i)}Q,$$

and $\pi_{(i+1)}$ from

$$(2 - \gamma)\pi_{(i+1)} - \frac{1}{\gamma}\pi_{(i+\gamma)} + \frac{(1-\gamma)^2}{\gamma}\pi_{(i)} = (1 - \gamma)h\pi_{(i+1)}Q. \tag{8.56}$$

Table 8.4: Results of Multistep Methods

Step Size	Implicit Adams $\|y(t_1) - y_1\|_2$	Implicit BDF $\|y(t_1) - y_1\|_2$	TR-BDF2 $\|y(t_1) - y_1\|_2$
$h = .1$	$0.555672e{-}04$	$0.368695e{-}02$	$0.394442e{-}03$
$h = .01$	$0.515258e{-}07$	$0.319431e{-}04$	$0.383439e{-}05$
$h = .001$	$0.511108e{-}10$	$0.315530e{-}06$	$0.382401e{-}07$
$h = .0001$	$0.942115e{-}13$	$0.315167e{-}08$	$0.382414e{-}09$
$h = .00001$	$0.187264e{-}11$	$0.302695e{-}10$	$0.161891e{-}10$

A value of $\gamma = 2 - \sqrt{2}$ is recommended. Observe that substituting $\gamma = 0.5$ (equally spaced step sizes) and $h = 2h$ into equation (8.56) returns us directly to the original form of the BDF formula, equation (8.54).

When equations (8.53), (8.55), and (8.56) are implemented in MATLAB and applied to the 3×3 infinitesimal generator, we obtain the results shown in Table 8.4.

8.4.6.3   Stability, Consistency, and Convergence

Given real coefficients α_j, β_j, for $j = 0, 1, \ldots, k$, a general multistep method may be written as

$$\alpha_0 y_{i+1} + \alpha_1 y_i + \ldots + \alpha_k y_{i+1-k} = h[\beta_0 f_{i+1} + \beta_1 f_i + \ldots + \beta_k f_{i+1-k}] \quad (8.57)$$

where h denotes the step size (i.e., $t_i = t_0 + ih$), and $f_j = f(t_j, y_j)$. The method is explicit if $\beta_0 = 0$; otherwise it is implicit.

The local error is defined as $y(t_{i+1}) - y_{i+1}$, where y_{i+1} is the solution computed by equation (8.57) under the assumption that the values y_{i+1-j}, $j = 1, 2, \ldots, k$ are exact (i.e., $y_{i+1-j} = y(t_{i+1-j})$ for $j = 1, 2, \ldots, k$). Since $k = 1$ for single-step methods, this definition of the local error coincides with our earlier definition. When the local error is $O(h^{p+1})$, the multistep method (8.57) is said to be of order p.

Applying the general multistep method (Equation 8.57) to the test equation, $y' = \lambda y$, gives

$$\alpha_0 y_{i+1} + \alpha_1 y_i + \cdots + \alpha_k y_{i+1-k} = h\lambda[\beta_0 y_{i+1} + \beta_1 y_i + \cdots + \beta_k y_{i+1-k}]. \quad (8.58)$$

We assume solutions of the form $y_j = \zeta^j$ and substitute into equation (8.58). Then, dividing through by ζ^{i+1-k} and setting $z = h\lambda$ gives the *characteristic equation* of the multistep method (8.57),

$$(\alpha_0 - z\beta_0)\zeta^k + (\alpha_1 - z\beta_1)\zeta^{k-1} + \cdots + (\alpha_k - z\beta_k) = \rho(\zeta) - z\sigma(\zeta) = 0$$

where $\rho(\zeta)$ and $\sigma(\zeta)$ are polynomials defined as

$$\rho(\zeta) = \alpha_0\zeta^k + \alpha_1\zeta^{k-1} + \cdots + \alpha_{k-1}\zeta + \alpha_k, \qquad (8.59)$$
$$\sigma(\zeta) = \beta_0\zeta^k + \beta_1\zeta^{k-1} + \cdots + \beta_{k-1}\zeta + \beta_k. \qquad (8.60)$$

This allows us to define the stability domain of a multistep method. For a given $z \in \mathcal{C}$, let $\zeta_j(z)$ for $j = 1, 2, \ldots, k$ be the roots of $\rho(\zeta) - z\sigma(\zeta) = 0$. The stability domain of the multistep method (8.57) is then defined as the set

$$S = \{z \in \mathcal{C};\ |\zeta_j(z)| \leq 1;\ \text{with strict inequality for multiple roots}\}$$

Thus, when a multistep method is applied to $\pi' = \pi Q$ and the step size is such that $h\lambda \in S$ for all eigenvalues λ of Q, the stability of the method is guaranteed.

A multistep method is said to be *zero stable* when the roots of $\rho(\zeta) = 0$ lie on or within the unit circle, with multiple roots strictly within the unit circle. This is referred to as the *root condition* and corresponds to the case $z = h\lambda = 0$.

Finally, a multistep method is said to be *consistent* if it is exact for all first-degree polynomials (in the absence of roundoff errors). Thus, ODEs of the form $y' = c,\ y(t_0) = y_0$ for c constant, are solved exactly by consistent methods. Applying the general multistep method to the test equation with $\lambda = 0$, i.e., $y' = 0;\ y(t_0) = 1$, and assuming that it is exact in this case, implies that $\rho(1) = \sum \alpha_j = 0$. Furthermore, when the general multistep method is applied to $y' = c;\ y(t_0) = 1$, and the relationship $\rho(1) = 0$ employed, it is found that $\rho'(1) = \sigma(1)$. Thus, the condition for consistency in a multistep method may be written as

$$\rho(1) = 0; \qquad \rho'(1) = \sigma(1).$$

An important property is that a consistent and stable multistep method is convergent, and conversely, a multistep method that is convergent is both stable and consistent. Since it is straightforward to use the polynomials (8.59) and (8.60) to verify whether a multistep method is consistent, we are left with the problem of pronouncing on its stability. However, the determination of stability domains is beyond the scope of this book. We shall be content to make some observations and direct the reader interested in obtaining more information to the text [68], in which all the most commonplace multistep methods are analyzed.

- For explicit Adams methods, the case $k = 1$ corresponds to the explicit Euler method, and the stability domain is, as we have already seen, the unit circle with center at $(-1, 0)$. For $k = 1, 2, \ldots, 6$ the stability domains are of rapidly decreasing sizes, which eliminates such methods for stiff problems.

- The implicit Adams method with $k = 1$ corresponds to the trapezoid method and its region of stability covers the entire negative half-plane. It is A-stable. For $k = 2, 3, \ldots, 6$ the stability domains are larger than the corresponding stability regions of the explicit version, but nevertheless they also diminish in size with increasing k. The only A-stable implicit Adams method is the trapezoid method.

- The predictor-corrector approach to solving the implicit Adams formula, whereby the explicit formula with a certain value of k is used to provide an approximation to the corresponding implicit Adams formula, has the effect of reducing the stability region of the implicit Adams method. For example, for $k = 1$, the trapezoid rule becomes an explicit Runge–Kutta method of order 2, and the A-stability is lost.

- The same diminuation of stability regions with increasing k may also be observed with the BDF methods. For $k = 1$ and $k = 2$, the BDF methods correspond to the implicit Euler and the trapezoid rule respectively, and both are A-stable. The stability regions then shrink with increasing k. For $k \geq 7$, the BDF formulae are unstable even at the origin. For $k = 2, 3, \ldots, 6$ the stability region of the BDF formulae are considerably larger than their Adams counterparts. In fact, for $k = 3$ and 4, the BDF formulae are almost A-stable.

8.4.7   Software Sources and Comparisons

At the end of this section we present the MATLAB programs that were used to perform the numerical experiments given in the preceding sections. Our objective is to provide the reader who is not knowledgeable in ODE solution techniques with a quick means of experimenting on small Markov chain problems that may be of interest. Input to the programs are an infinitesimal generator Q and initial probability vector $y0$ (a row vector), an interval of integration t, and a step size h. The programs output the computed approximation y and the 2-norm of the error vector. The programs are by no means *industrial-strength*, nor are they meant to be. To solve real problems, the user will need to look farther afield.

The Numerical Algorithms Group (NAG) proposes three categories of method to its users: routines based on the fourth-order Runge–Kutta–Merson single-step method and variable-order, variable-step-size Adams (explicit) and BDF (implicit) multistep methods. The Runge–Kutta–Merson variants are recommended when the interval of integration is short, when the cost of evaluating the function f is not expensive, and when the tolerance requested is not high. When the interval of integration is long and accuracy requirements high, then the Adams variants are recommended. The BDF methods are recommended when the system of equations is stiff.

NAG is not the only source of reliable ODE software. Fortran source codes are listed in the appendices of the texts by Hairer et al. ([67, 68]). Furthermore, the code may be obtained directly by E-mail from the authors. Among the algorithms included are explicit Dormand–Prince Runge–Kutta variants. An order 5 method with error estimator of order 4 is recommended for relatively low tolerances (10^{-4} to 10^{-7}), while an order 8 method with order 7 error estimator is recommended when higher tolerances (10^{-7} to 10^{-13}) must be satisfied. The authors also include source code for an extrapolation method called *Odex* that is highly recommended when very high tolerances (10^{-13} to 10^{-30}) are needed. This is a variable-order, variable-step-size method that is based on the explicit midpoint rule. The second volume, which is dedicated to stiff and algebraic differential equations, contains source code listing for algorithms suitable for stiff ODEs. It includes a fifth-order implicit Radau IIA method that allows for both full and banded Jacobian. Various extrapolation procedures suitable for stiff systems are also given.

An ODE solution procedure with which we have experimented with some success is the *VODE* algorithm of Brown, Byrne, and Hindmarsh [15]. It is available from *netlib*. For nonstiff problems VODE uses an Adams method whose order varies between 1 and 12. For stiff problems the BDF method is used, and the order varies between 1 and 5. The coefficients of the methods (the α's and β's in equation 8.57) are computed as functions of the current and past step sizes. For the solution of the nonlinear equation inherent in implicit methods, a choice between a functional iteration and a modified Netwon iteration is offered. Possibilities for both banded and full Jacobians are provided.

MATLAB Program Listings

```
**********************************************************************
function [y,err] = euler(Q,y0,t,h);   %%%%%%%%%%%%%%%% Explicit Euler

[n,n] = size(Q);  I = eye(n);  eta = t/h;  y = y0;
yex = y0*expm(t*Q);              %%%% Exact soln using MATLAB fcn

R = I + h*Q;
for i=1:eta,
    y = y*R;
end
err = norm(y-yex,2);


**********************************************************************
function [y,err] = ieuler(Q,y0,t,h); %%%%%%%%%%%%%%%% Implicit Euler

[n,n] = size(Q);  I = eye(n);  eta = t/h;  y = y0;
yex = y0*expm(t*Q);              %%%% Exact soln using MATLAB fcn

R = inv(I-h*Q);
for i=1:eta,
    y = y*R;
end
err = norm(y-yex,2);


**********************************************************************
function [y,err] = trap(Q,y0,t,h);   %%%%%%% Trapezoid/Modified Euler
```

```
      [n,n] = size(Q);  I = eye(n);  eta = t/h;  y = y0;
      yex = y0*expm(t*Q);                    %%%%% Exact soln using MATLAB fcn


      R1 = I + h*Q/2;  R2 = inv(I-h*Q/2);  R = R1*R2;
      for i=1:eta,
          y = y*R;
      end
      err = norm(y-yex,2);


      ************************************************************************
      function [y,err] = rk4(Q,y0,t,h);   %  Explicit Runge-Kutta --- Order 4

      [n,n] = size(Q);  I = eye(n);  eta = t/h;   y = y0;
      yex = y0*expm(t*Q);                    % Exact soln using MATLAB fcn

      for i=1:eta,
          k1 = y*Q;
          k2 = (y + h*k1/2)*Q;
          k3 = (y + h*k2/2)*Q;
          k4 = (y + h*k3)*Q;
          y = y + h*(k1 + 2*k2 + 2*k3 + k4)/6;
      end
      err = norm(y-yex,2);


      ************************************************************************
      function [y,err] = irk(Q,y0,t,h);   %%%%%%%%%%%%  Implicit Runge-Kutta

      [n,n] = size(Q);  I = eye(n);  eta = t/h;  y = y0;
      yex = y0*expm(t*Q);                    %%%%%%% Exact soln using MATLAB fcn

      R1 = I + h*Q/3;  R2 = inv(I-2*h*Q/3+ h*h*Q*Q/6);  R = R1*R2;
      for i=1:eta,
          y = y*R;
      end
      err = norm(y-yex,2);


      ************************************************************************
      function [y,err] = adams(Q,y0,t,h);   %%%%%%%  Adams --- Implicit: k=2

      [n,n] = size(Q);  I = eye(n);  eta = t/h;
      yex = y0*expm(t*Q);                    %%%%% Exact soln using MATLAB fcn

      R1 = I + h*Q/2;  R2 = inv(I-h*Q/2);
      y1 = y0*R1*R2;                         %%%%%%% Trapezoid rule for Step 1

      S1 = (I+2*h*Q/3);  S2 =  h*Q/12;  S3 = inv(I-5*h*Q/12);
      for i=2:eta,                           %%%%%%%% Adams for steps 2 to end
          y = (y1*S1 - y0*S2)*S3;
          y0 = y1; y1 = y;
      end
      err = norm(y-yex,2);
```

```
**********************************************************************
function [y,err] = bdf(Q,y0,t,h);   %%%%%%%%%%% BDF --- Implicit: k=2

[n,n] = size(Q);  I = eye(n);  eta = t/h;
yex = y0*expm(t*Q);               %%%%%% Exact soln using MATLAB fcn

R1 = I + h*Q/2;  R2 = inv(I-h*Q/2);
y1 = y0*R1*R2;                    %%%%%%%% Trapezoid rule for Step 1

S1 = inv(3*I/2 - h*Q);
for i=2:eta,                      %%%%%%%%%%% BDF for steps 2 to end
    y = (2*y1 - y0/2)*S1;
    y0 = y1; y1 = y;
end
err = norm(y-yex,2);


**********************************************************************
function [y,err] = trbdf2(Q,y0,t,h);   %%%%%%%%%%%%%%%  TR-BDF2 Method

[n,n] = size(Q);  I = eye(n);  eta = t/h;
yex = y0*expm(t*Q);               %%% Exact soln using MATLAB fcn

gamma = 2-sqrt(2);  c = (1-gamma)^2/gamma;
R1 = I + gamma*h*Q/2;  R2 = inv(I-gamma*h*Q/2);  R = R1*R2;
S1 = inv((2-gamma)*I-(1-gamma)*h*Q);

for i=1:eta,
    ygamma = y0*R;
    y = (ygamma/gamma - c*y0)*S1;
    y0 = y;
end
err = norm(y-yex,2);
**********************************************************************
```

8.5 Krylov Subspace Methods

8.5.1 Introduction

Krylov subspace methods for computing stationary probability distributions of Markov chains were examined in detail in Chapter 4. We now return to them to conclude our discussions on the computation of transient distributions, for not only do such methods provide the basis for many of the most efficient approaches for solving large sparse systems of linear equations, but they have also been effectively used in other domains such as in the numerical solution of parabolic equations [49]; systems of stiff nonlinear differential equations [46]; and indeed in the evaluation of arbitrary functions of a matrix [37]. With respect to our current topic of interest, in [135] Krylov subspace methods are developed for the computation of the matrix exponential operator $e^A v$ for a given vector v. Of particular interest Philippe and Sidje [118] provide implementation details and report on Krylov subspace code that was developed for the express purpose of computing transient solutions of ergodic Markov chains.

As we mentioned in the introduction to this chapter, it is often appropriate to divide the interval of integration $[t_0, t]$ into subintervals $[t_i, t_{i+1}]$ for $i = 0, 1, \ldots, s$, where $t_{s+1} = t$, and to compute $\pi(t_i)e^{Q(t_{i+1}-t_i)}$ for $i = 0, 1, \ldots, s$, either because the analyst wishes to observe the evolution of certain measures of effectiveness with time, or because the size of $\|Qt\|$ is too large to permit the accurate computation of e^{Qt} directly. Thus, in considering Krylov subspace methods for transient solutions, we shall proceed in a step-by-step fashion much as in the manner of single-step ODE solvers like Runge–Kutta. The choice of a step size, which may be partially dictated by the user, is an important parameter of the algorithm. A second important parameter is the choice of the size of the Krylov subspace. Together, these two parameters determine the accuracy of the solution obtained.

8.5.2   The Basic Krylov Subspace Approach

We begin by examining the basic Krylov subspace approach to computing $\pi(t_{i+1}) = \pi(t_i)e^{Q(t_{i+1}-t_i)}$. For ease of notation, we shall set $A = Q^T$, $v^T = \pi(t_i)$, $w^T = \pi(t_{i+1})$, and $\tau_i = t_{i+1} - t_i$ and consider the computation of $w = e^A v$. Inserting the time constant τ_i into the computation to get $e^{A\tau_i}v$ poses no additional problem.

Our objective is the computation of an approximation of the form $p_{m-1}(A)v$ to the matrix exponential operation $e^A v$, where $p_{m-1}(A)$ is a polynomial of degree $m - 1$ in A. Notice that $p_{m-1}(A)$ is a linear combination of the vectors $v, Av, \ldots, A^{m-1}v$, i.e., it is an element of the Krylov subspace

$$K_m(A, v) \equiv \text{span}\left\{v, Av, \ldots, A^{m-1}v\right\}$$

and so our problem may be posed as that of finding the element of $K_m(A, v)$ that best approximates $w = e^A v$. If "best" is taken to mean in the least squares sense, then the best approximation $\hat{w}$ of w in the Krylov subspace satisfies

$$\|\hat{w} - w\|_2 = \min_{x \in K_m(A,v)} \|x - w\|_2 = \min_{y \in \Re^m} \|V_m y - e^A v\|_2,$$

where $V_m = [v_1, v_2, \ldots, v_m]$ is a set of basis vectors for $K_m(A, v)$. Since this is a *full-rank* least squares problem, we essentially solve a system of linear equations and obtain $\hat{w} = e^A v$. If we choose $v_1 = v/\beta$, where $\beta = \|v\|_2$, then $v = \beta V_m e_1$ and we have

$$
\begin{aligned}
e^A v \approx \hat{w} = V_m V_m^{-1} V_m^{-T} V_m^T e^A v &= V_m \left[(V_m^T V_m)^{-1} V_m^T e^A v\right] \\
&= V_m \left[(V_m^T V_m)^{-1} V_m^T e^A V_m\right] \beta e_1. \quad (8.61)
\end{aligned}
$$

Let us at this point turn our attention to the computation of the basis vectors V_m. One convenient means of doing so is the Arnoldi procedure already discussed in Chapter 4. Recall that the algorithm is as follows:

Arnoldi Process

1. Set $v_1 = v/\|v\|_2$.

2. For $j = 1, 2, \ldots, m$ do

   - Compute $z = Av_j$.
   - For $i = 1, 2, \ldots, j$ do
     - Compute $h_{ij} = v_i^T z$.
     - Compute $z = z - h_{ij}v_i$.
   - Compute $h_{j+1,j} = \|z\|_2$ and $v_{j+1} = z/h_{j+1,j}$.

Since this algorithm implements a modified Gram–Schmidt orthogonalization procedure, the computed basis V_m is orthonormal. The $m \times m$ upper Hessenberg matrix H_m formed from the coefficients h_{ij} satisfies the relationship

$$AV_m = V_m H_m + h_{m+1,m} v_{m+1} e_m^T, \tag{8.62}$$

which, from the orthogonality of V_m, yields

$$H_m = V_m^T A V_m.$$

In words, H_m represents the projection of the linear transformation A onto the subspace $K_m(A, v)$ with respect to the basis V_m.

Returning to our approximation $\hat{w}$ in equation (8.61), the orthonormality of the basis implies that

$$\hat{w} = \beta V_m (V_m^T e^A V_m) e_1.$$

However, this does not as yet appear to be very helpful, since it still requires e^A; the purpose of the Krylov subspace approach, namely to project the exponential of a large matrix approximately onto a small Krylov subspace, has yet to be achieved. This purpose is accomplished if we approximate $V_m^T e^A V_m$ by $e^{V_m^T A V_m} = e^{H_m}$, that is, if we replace the *restriction of the exponential operator* e^A onto the subspace $K_m(A, v)$ with respect to V_m with the *exponential of the restriction* of the operator A onto $K_m(A, v)$ with respect to V_m. This gives the approximation

$$\hat{w} \approx \tilde{w} = \beta V_m e^{H_m} e_1, \tag{8.63}$$

which still involves the evaluation of the exponential of a matrix, but this time of small dimension (m instead of n) and of a particular structure (upper Hessenberg). The methods of Section 8.3 recommended for small systems, such as the Schur decomposition and rational Padé approximants, may be used. We shall consider the latter in our analysis.

Since $V_m^T(A\tau)V_m = H_m\tau$, and the Krylov subspaces associated with A and $A\tau$ are identical, we have

$$e^{A\tau}v \approx \beta V_m e^{H_m\tau}e_1$$

for an arbitrary scalar τ. This justifies our earlier assertion that the time constant can be incorporated without difficulty. We shall therefore, for the sake of simplicity, continue to omit it. In fact, our only concerns with the parameter τ arise in the computation of $e^{H_m\tau}$, for which we have already discussed appropriate algorithms, and in the size of $\|A\tau\|$.

The alert reader at this point is perhaps becoming increasingly concerned by the apparent piling of approximation on approximation. Indeed, we make a first approximation in extracting $\hat{w}$ from a subspace of size m (instead of size n), a second in forming $\tilde{w}$ by replacing $V_m^T e^A V_m$ with $e^{V_m^T A V_m}$, and a third in recommending rational Padé *approximants* for the computation of e^{H_m}. However, Saad [135] shows that

$$\|e^A v - \beta V_m e^{H_m}e_1\|_2 \leq 2\beta\frac{\rho^m e^\rho}{m!},$$

where $\rho = \|A\|_2$ and $\beta = \|v\|_2$. Therefore, the approximations give fairly accurate results even for small-dimension subspaces, as long as $\|A\|_2$ is not large. For example, choosing a time step τ to give $\|A\tau\|_2 < .5$ and using a subspace of size 12 should give more than 10 decimal places accuracy.

To summarize, therefore, the basic Krylov subspace approach is to

- Use the Arnoldi process to construct the orthonormal basis V_m and the upper Hessenberg matrix H_m,

- Use rational Padé approximants to compute $e^{H_m\tau}$,

- Form an approximate solution as $\beta V_m e^{H_m\tau}e_1$.

The only interaction with the (usually large and sparse) matrix A ($= Q^T$) is the matrix vector multiply in the first line of step 2 of the Arnoldi process. Thus compact storage schemes may be conveniently implemented. However, to develop an industrial-strength production code, it is necessary to go beyond this basic implementation. A number of important questions still need to be answered, particularly those relating to error control, step size, and choice of the size of the subspace.

8.5.3 Corrected Schemes and Error Bounds

Notice that the Arnoldi process generates $m+1$ vectors, $v_1, v_2, \ldots, v_{m+1}$, but that only m of these are actually used. Following the analysis of Saad, we now show how this additional vector may be used to obtain an improved approximation at very little extra cost. Consider the function

$$\phi(z) = \frac{e^z - 1}{z}$$

and observe that

$$e^A v = v + (e^A - I)v = v + A\phi(A)v.$$

Since $\phi(A)v \approx V_m V_m^T \phi(A) V_m \beta e_1$, we approximate $\phi(A)v$ by

$$\phi(A)v \approx V_m \phi(H_m)\beta e_1.$$

Let us define

$$s_m = \phi(A)v_1 - V_m \phi(H_m)e_1.$$

Then, since $v = v_1\beta$, $s_m\beta$ is the error in this approximation of $\phi(A)v$. We have

$$e^A v_1 = v_1 + A\phi(A)v_1 = v_1 + A[V_m\phi(H_m)e_1 + s_m].$$

Substituting for AV_m from equation (8.62), we get

$$
\begin{aligned}
e^A v_1 &= v_1 + (V_m H_m + h_{m+1,m} v_{m+1} e_m^T)\phi(H_m)e_1 + As_m \\
&= V_m[e_1 + H_m\phi(H_m)e_1] + h_{m+1,m} \ e_m^T\phi(H_m)e_1 \ v_{m+1} + As_m
\end{aligned}
$$

from which it immediately follows that

$$e^A v = \beta e^A v_1 = \beta V_m e^{H_m} e_1 + \beta h_{m+1,m} \ e_m^T\phi(H_m)e_1 \ v_{m+1} + \beta As_m.$$

The first term on the right-hand side is the approximation obtained by the straightforward Krylov subspace method, equation (8.63). If instead, we use the approximation

$$e^A v \approx \beta[V_m e^{H_m} e_1 + h_{m+1,m} \ e_m^T\phi(H_m)e_1 \ v_{m+1}]$$

we get a *corrected scheme*, and the error in this scheme is given by βAs_m. Setting

$$\bar{H}_{m+1} \equiv \begin{pmatrix} H_m & 0 \\ h_{m+1,m}e_m^T & 0 \end{pmatrix},$$

the corrected scheme becomes equivalent to the approximation

$$e^A v \approx \beta V_{m+1} e^{\bar{H}_{m+1}} e_1$$

and has a computational cost close to that of the original. We have the following *a priori* error bound:

$$\|e^A v - \beta V_{m+1} e^{\bar{H}_{m+1}} e_1\|_2 \leq 2\beta \frac{\rho^{m+1} e^\rho}{(m+1)!}$$

Continuing in this fashion, *k-corrected schemes* may be developed. It is shown in [135] that the following expansion holds:

$$
\begin{aligned}
e^A v &= \beta V_m e^{H_m} e_1 + \beta h_{m+1,m} \sum_{j=1}^{k} e_m^T \phi_j(H_m) e_1 A^{j-1} v_{m+1} + \beta A^k s_m^k \\
&= \beta V_m e^{H_m} e_1 + \beta h_{m+1,m} \sum_{j=1}^{\infty} e_m^T \phi_j(H_m) e_1 A^{j-1} v_{m+1}
\end{aligned}
\tag{8.64}
$$

where $s_m^k \equiv \phi_k(A) v_1 - V_m \phi_k(H_m) e_1$ and the functions $\phi_k(z)$ are defined recursively as

$$\phi_0(z) \equiv e^z$$

$$\phi_k(z) \equiv \frac{\phi_{k-1}(z) - \phi_{k-1}(0)}{z}; \qquad k \geq 1.$$

Note that $\phi_1(z) = \phi(z)$. The error in the $k-$corrected scheme satisfies

$$\|\beta A^k s_m^k\|_2 \leq 2\beta \frac{\rho^{m+k} e^\rho}{(m+k)!}.$$

8.5.4   Error Estimates and Step Size Control

The Krylov subspace method is applied like any other single-step ODE solver. It is necessary at each time step t_i to choose values for the variables τ_i (the step size) and m (the size of the Krylov basis), compute the approximate solution at $t_i + \tau_i$, and estimate the local error that results in moving the solution from t_i to $t_i + \tau_i$. If the estimated local error lies within the specified tolerance *tol*, we proceed on to the next time step. Otherwise it is necessary to reject the computed solution and try again after reducing the step size, increasing the size of the basis, or both. When the local error estimate is smaller than the tolerance asked for, it is appropriate to consider increasing the step size, decreasing the size of the Krylov basis, or both. Control techniques for the parameters of single-step ODE solvers have been and continue to be an important area of research (see, for example, [62]). Here we follow the recommendations in [118].

Using the formulæ of the previous section, the local error may be approximated by the first omitted term in the expansion (8.64). When the step size is small, this is an adequate error measure. For the corrected scheme ($k = 1$), the estimated local error is given by

$$\epsilon_{i+1} = \beta|\tau_i h_{m+1,m} e_m^T \phi_2(\tau_i H_m) e_1| \; \|\tau_i A v_{m+1}\|_2$$

in which we have incorporated the time interval τ_i. Other cheaper estimates may be found [135], but this one appears to have a more regular behavior, which makes it more suitable for our purposes. If *tol* denotes the tolerance to be achieved, the step size is chosen as

$$\tau_i = \gamma \left(\frac{tol}{\epsilon_i}\right)^{1/m} \tau_{i-1}, \tag{8.65}$$

and the approximate solution is computed at time $t_{i+1} = t_i + \tau_i$. At this point the local error, ϵ_{i+1} at time t_{i+1}, is estimated, and the solution is rejected if $\epsilon_{i+1} > \delta \, tol$. The parameters γ and δ are safety factors; often values of $\gamma = 0.9$ and $\delta = 1.2$ are recommended. If a step size is rejected, the formula (8.65) is reapplied, this time replacing ϵ_i with ϵ_{i+1}, and the computation of the solution at $t_i + \tau_i$ is reinitiated. It may be necessary to repeat this procedure several times. At any step, the size of the Krylov subspace may also be readjusted (either upward or downward) according to the formula

$$m_i = m_{i+1} + \lceil \frac{\log(tol/\epsilon_i)}{\log \tau_{i-1}} \rceil.$$

Although there exist values for the step size and Krylov basis that minimize rounding error and execution cost with respect to the whole interval of integration, these values cannot be easily obtained, which explains the necessity for the heuristics just given.

An interesting feature of the Krylov subspace method is that the stationary probability vector is automatically detected by the algorithm. If a value of $h_{j+1,j}$ is computed for which $h_{j+1,j} \leq tol\|A\|$, an invariant subspace has been found, and the corresponding value of $\beta V_j e^{H_j} e_1$ is exact. This situation, sometimes referred to as a *happy breakdown*, lets us know that the computed solution satisfies $\pi Q = 0$ and that continuing to run the algorithm beyond this point is unnecessary.

A further interesting feature of the method is that if the stationary solution is known beforehand, it may be incorporated into the basis. In numerical tests conducted by Philippe and Sidje, this always led to a gain in accuracy in the computed approximation. Indeed, it is a topic of current discussion whether the first step in computing transient solutions should not be the computation of the stationary distribution.

8.6   Selected Experiments

To conclude this chapter, we apply some of the methods described earlier to the queueing network model of a multiprogrammed, time-shared computer system. This model was first introduced in Chapter 2 and used again in Chapter 6. The reader will recall that the parameters may be chosen to yield an NCD or non-NCD infinitesimal generator. Experiments relating to transient solutions are described here for both cases. For additional information concerning the model, the reader should refer to Section 6.3.8.1 of Chapter 6.

Table 8.5 shows the computation time in seconds on a Cray-2 computer when a variety of numerical methods is applied to both the NCD and non-NCD case. The numerical methods examined are the uniformization method ($Unif$, Section 8.2); the Uniformized Power method ($U.P.$, Section 8.3.2.1); Krylov subspace methods with subspaces of size 20 and 30 (Section 8.5); and the three NAG algorithms, Runge–Kutta ($R.K.$, Section 8.4.5), Adams (Section 8.4.6.1), and BDF (Section 8.4.6.1). The uniformized power method and BDF were applied only to the smallest example, since memory requirements became excessive in the other cases.

In this table, n denotes the order of the infinitesimal generator, and n_z the number of nonzero elements that it possesses. Experiments were conducted to determine the transient solution at different times from $t = 1$ to $t = 1,000$ depending on the order of Q. The initial state is taken at time $t = 0$. In presenting this table, we have not intended to recommend a particular solution approach but simply to show how some selected methods behave on one particular model.

Table 8.5: Experiments with Selected Methods

Results for NCD Example								
n	Time	Unif.	U.P.	Krylov		NAG		
(n_z)	t			$m = 20$	$m = 30$	R.K.	Adams	BDF
286	1	0.02	3.36	0.06	—	0.03	0.05	3.38
(1,606)	10	—	17.54	0.06	0.09	0.11	0.10	4.05
	100	0.44	42.94	0.68	0.97	0.57	0.94	9.34
	1000	9.73	95.52	3.13	2.93	11.09	15.27	12.60
1,771	1	0.18	—	0.31	0.49	0.20	0.34	—
(11,011)	10	0.78	—	0.31	0.49	0.74	0.70	—
	100	7.36	—	1.78	1.80	7.92	10.01	—
	1000	75.71	—	21.98	20.92	171.47	240.43	—
5,456	1	0.74	—	0.88	1.92	1.88	2.37	—
(35,216)	10	8.16	—	1.23	2.25	3.86	4.21	—
	100	69.45	—	12.75	9.81	69.35	122.65	—
12,341	1	7.29	—	9.99	12.71	6.58	11.60	—
(81,221)	10	66.64	—	10.41	11.69	23.75	17.83	—
	100	836.04	—	73.30	70.13	271.71	—	—
23,426	1	16.32	—	13.07	21.37	11.43	24.70	—
(156,026)	10	172.91	—	68.02	25.04	61.51	51.96	—

Results for Non-NCD Example								
n	Time	Unif.	U.P.	Krylov		NAG		
(n_z)	t			$m = 20$	$m = 30$	R.K.	Adams	BDF
286	1	0.01	5.53	0.40	0.09	0.99	0.15	4.89
(1,606)	10	0.04	23.11	1.74	0.62	1.20	2.38	18.27
	100	1.27	129.32	1.82	0.37	3.76	5.11	60.78
	1000	4.31	255.12	5.11	1.12	16.13	20.86	51.19
1,771	1	0.09	—	0.29	0.48	0.90	0.75	—
(11,011)	10	0.29	—	0.59	0.74	6.23	4.12	—
	100	2.78	—	5.07	5.77	22.47	15.94	—
	1000	32.27	—	13.81	13.99	56.16	110.74	—
5,456	1	0.29	—	0.93	3.13	6.10	4.33	—
(35,216)	10	1.85	—	3.75	4.23	18.01	10.82	—
	100	9.99	—	9.55	12.18	59.19	63.99	—
12,341	1	0.68	—	4.16	6.57	12.61	10.56	—
(81,221)	10	2.79	—	7.44	9.13	44.28	25.65	—
	100	22.02	—	22.91	23.55	146.37	127.64	—
23,426	1	1.21	—	13.44	24.83	38.88	31.68	—
(156,026)	10	11.80	—	27.27	31.86	131.00	76.38	—

Chapter 9

Stochastic Automata Networks

9.1 Introduction

In complex systems different components may lend themselves more naturally to one type of modelling technique than to another. For example, queueing network models are well suited for analyzing resource contention and allocation (where processes are forced to queue for limited resources); generalized Petri nets provide an excellent vehicle for modelling process synchronization; an explicit Markov state space representation is sometimes the modelling stratagem of choice, such as for example, in modelling system reliability. *Stochastic automata networks (SANs)* provide a convenient methodology that allows us to combine these different modelling techniques. Additionally, SANs by themselves have proven to be a useful vehicle for system modelling [120]. They appear to be particularly useful in modelling parallel systems such as communicating processes, concurrent processors, or shared memory. The advantage that the SAN approach has over generalized stochastic Petri nets, and indeed over any Markovian analysis that requires the generation of a transition matrix, is that its representation remains compact even as the number of states in the underlying Markov chain begins to explode.

A stochastic automata network consists of a number of individual stochastic automata that operate more or less independently of each other. An automaton is represented by a number of states and probabilistic rules that govern the manner in which it moves from one state to the next. The state of an automaton at any time t is just the state it occupies at time t, and the state of the SAN at time t is given by the state of each of its constituent automata.

463

9.2 Noninteracting Stochastic Automata

9.2.1 Independent, Two-Dimensional Markov Chains

Consider the case of a system that is modelled by two completely independent stochastic automata, each of which is represented by a discrete-time Markov chain. Let us assume that the first automaton, denoted $A^{(1)}$, has n_1 states and stochastic transition probability matrix given by $P^{(1)} \in \Re^{n_1 \times n_1}$. Similarly, let $A^{(2)}$ denote the second automaton; n_2, the number of states in its representation; and $P^{(2)} \in \Re^{n_2 \times n_2}$, its stochastic transition probability matrix. The state of the overall (two-dimensional) system may be represented by the pair (i, j), where $i \in \{1, 2, \ldots, n_1\}$ and $j \in \{1, 2, \ldots, n_2\}$. Indeed, the stochastic transition probability matrix of the two-dimensional system is given by the *tensor product* of the matrices $P^{(1)}$ and $P^{(2)}$.

If, instead of being represented by two *discrete-time* Markov chains, the stochastic automata are characterized by *continuous-time* Markov chains with infinitesimal generators, $Q^{(1)}$ and $Q^{(2)}$ respectively, the infinitesimal generator of the two-dimensional system is given by the *tensor sum* of $Q^{(1)}$ and $Q^{(2)}$.

9.2.2 Basic Properties of Tensor Algebra

It is useful at this point to recall the definitions of tensor product and tensor addition. Let us take the following matrices as an example:

$$A = \begin{pmatrix} a_{11} & a_{12} \\ a_{21} & a_{22} \end{pmatrix} \quad \text{and} \quad B = \begin{pmatrix} b_{11} & b_{12} & b_{13} & b_{14} \\ b_{21} & b_{22} & b_{23} & b_{24} \\ b_{31} & b_{32} & b_{33} & b_{34} \end{pmatrix}.$$

The *tensor product* $C = A \otimes B$ is given by

$$C = \begin{pmatrix} a_{11}B & a_{12}B \\ a_{21}B & a_{22}B \end{pmatrix}$$

$$= \left(\begin{array}{cccc|cccc} a_{11}b_{11} & a_{11}b_{12} & a_{11}b_{13} & a_{11}b_{14} & a_{12}b_{11} & a_{12}b_{12} & a_{12}b_{13} & a_{12}b_{14} \\ a_{11}b_{21} & a_{11}b_{22} & a_{11}b_{23} & a_{11}b_{24} & a_{12}b_{21} & a_{12}b_{22} & a_{12}b_{23} & a_{12}b_{24} \\ a_{11}b_{31} & a_{11}b_{32} & a_{11}b_{33} & a_{11}b_{34} & a_{12}b_{31} & a_{12}b_{32} & a_{12}b_{33} & a_{12}b_{34} \\ \hline a_{21}b_{11} & a_{21}b_{12} & a_{21}b_{13} & a_{21}b_{14} & a_{22}b_{11} & a_{22}b_{12} & a_{22}b_{13} & a_{22}b_{14} \\ a_{21}b_{21} & a_{21}b_{22} & a_{21}b_{23} & a_{21}b_{24} & a_{22}b_{21} & a_{22}b_{22} & a_{22}b_{23} & a_{22}b_{24} \\ a_{21}b_{31} & a_{21}b_{32} & a_{21}b_{33} & a_{21}b_{34} & a_{22}b_{31} & a_{22}b_{32} & a_{22}b_{33} & a_{22}b_{34} \end{array} \right). \quad (9.1)$$

In general, to define the tensor product of two matrices, A of dimensions $(\rho_1 \times \gamma_1)$ and B of dimensions $(\rho_2 \times \gamma_2)$, it is convenient to observe that the tensor product matrix has dimensions $(\rho_1 \rho_2 \times \gamma_1 \gamma_2)$ and may be considered as consisting of $\rho_1 \gamma_1$ blocks, each having dimensions $(\rho_2 \times \gamma_2)$, i.e., the dimensions of B. To specify a particular element, it suffices to specify the block in which the element occurs and the position within that block of the element under consideration. Thus, in the example the element c_{47} $(= a_{22}b_{13})$ is in the $(2,2)$ block and at position $(1,3)$ of that block. The tensor product $C = A \otimes B$ is defined by assigning the element of C that is in the (i_2, j_2) position of block (i_1, j_1) the value $a_{i_1 j_1} b_{i_2 j_2}$. We shall write this as

$$c_{\{(i_1,j_1);(i_2,j_2)\}} = a_{i_1 j_1} b_{i_2 j_2}.$$

The *tensor sum* of two *square* matrices A and B is defined in terms of tensor products as

$$A \oplus B = A \otimes I_{n_2} + I_{n_1} \otimes B \tag{9.2}$$

where n_1 is the order of A; n_2 the order of B; I_{n_i} the identity matrix of order n_i; and "+" represents the usual operation of matrix addition. Since both sides of this operation (matrix addition) must have identical dimensions, it follows that tensor addition is defined for square matrices only. For example, with

$$A = \begin{pmatrix} a_{11} & a_{12} \\ a_{21} & a_{22} \end{pmatrix} \quad \text{and} \quad B = \begin{pmatrix} b_{11} & b_{12} & b_{13} \\ b_{21} & b_{22} & b_{23} \\ b_{31} & b_{32} & b_{33} \end{pmatrix}, \tag{9.3}$$

the tensor sum $C = A \oplus B$ is given by

$$C = \left(\begin{array}{ccc|ccc} a_{11} + b_{11} & b_{12} & b_{13} & a_{12} & 0 & 0 \\ b_{21} & a_{11} + b_{22} & b_{23} & 0 & a_{12} & 0 \\ b_{31} & b_{32} & a_{11} + b_{33} & 0 & 0 & a_{12} \\ \hline a_{21} & 0 & 0 & a_{22} + b_{11} & b_{12} & b_{13} \\ 0 & a_{21} & 0 & b_{21} & a_{22} + b_{22} & b_{23} \\ 0 & 0 & a_{21} & b_{31} & b_{32} & a_{22} + b_{33} \end{array} \right).$$

Some important properties of tensor products and additions are

- Associativity:
  $A \otimes (B \otimes C) = (A \otimes B) \otimes C$ and $A \oplus (B \oplus C) = (A \oplus B) \oplus C$.

- Distributivity over (ordinary matrix) addition:
  $(A + B) \otimes (C + D) = A \otimes C + B \otimes C + A \otimes D + B \otimes D$.

- Compatibility with (ordinary matrix) multiplication:
  $(AB) \otimes (CD) = (A \otimes C)(B \otimes D)$.

- Compatibility with (ordinary matrix) inversion:
  $(A \otimes B)^{-1} = A^{-1} \otimes B^{-1}$.

The associativity property implies that the operations

$$\bigotimes_{i=1}^{N} A^{(i)} \quad \text{and} \quad \bigoplus_{i=1}^{N} A^{(i)}$$

are well defined. In particular, observe that the tensor sum of N independent stochastic automata may be written as the (usual) matrix sum of N terms, each term consisting of an N-fold tensor product. We have

$$\bigoplus_{i=1}^{N} A^{(i)} = \sum_{i=1}^{N} I_{n_1} \otimes \cdots \otimes I_{n_{i-1}} \otimes A^{(i)} \otimes I_{n_{i+1}} \otimes \cdots \otimes I_{n_N}, \qquad (9.4)$$

where n_k is the number of states in the kth automaton and I_{n_k} is the identity matrix of order n_k. The operators $\otimes$ and $\oplus$ are not commutative. More information concerning the properties of tensor algebra may be found in [35].

9.2.3 Probability Distributions

Let us return to the system modelled as two independent stochastic automata. Let $\pi_i^{(1)}(t)$ be the probability that the first automaton is in state i at time t, and let $\pi_j^{(2)}(t)$ the probability that the second is in state j at time t. Then the probability that at time t the first automaton is in state i *and* the second is in state j is simply the product $\pi_i^{(1)}(t) \times \pi_j^{(2)}(t)$. Furthermore, the probability vector of the overall (two-dimensional) system is given by the tensor product of the two individual probability vectors, $\pi^{(1)} \in \Re^{n_1}$ and $\pi^{(2)} \in \Re^{n_2}$, viz.: $\pi^{(1)} \otimes \pi^{(2)}$.

In the case of a system modelled by N independent stochastic automata, the probability that the system is in state $(i_1, i_2, \ldots, i_N)$ at time t, where i_k is the state of the kth automaton at time t, is given by $\prod_{k=1}^{N} \pi_{i_k}^{(k)}(t)$, where $\pi_{i_k}^{(k)}(t)$ is the probability that the kth automaton is in state i_k at time t. Furthermore, the probability distribution of the N-dimensional system $\pi(t)$ is given by the tensor product of the probability vectors of the individual automata at time t, i.e.,

$$\pi(t) = \bigotimes_{k=1}^{N} \pi^{(k)}(t). \qquad (9.5)$$

To solve N-dimensional systems that are formed from independent stochastic automata is therefore very simple. It suffices to solve for the probability distributions of the individual stochastic automata and to form the tensor product of these distributions. Although such systems may exist, the more usual case occurs when the transitions of one automaton may depend on the state of a second. It is to this topic that we now turn.

9.3   Interacting Stochastic Automata

Among the ways in which stochastic automata may interact, we consider two.

1. The rate at which a transition may occur in one automaton may be a *function* of the state of other automata. We shall say that such transitions are *functional* transitions (as opposed to *constant*-rate transitions).

2. A transition in one automaton may *force* a transition to occur in one or more other automata. We shall refer to this type of transition as a *synchronizing* transition or *synchronizing event.*

9.3.1   A Simple Example

Consider as an example a simple queueing network consisting of two service centers in tandem and an arrival process that is Poisson at rate λ. Each service center consists of an infinite queue and a single server. The service time distribution of the first server is assumed to be exponential at fixed rate μ, while the service time distribution at the second is taken to be exponential with a rate ν that varies with the number and distribution of customers in the network. Since a state of the network is completely described by the pair (η_1, η_2), where η_1 denotes the number of customers at station 1 and η_2 the number at station 2, the service rate at station 2 is more properly written as $\nu(\eta_1, \eta_2)$.

We may define two stochastic automata $A^{(1)}$ and $A^{(2)}$ corresponding to the two different service centers. The state space of each is given by the set of nonnegative integers $\{0, 1, 2, \ldots\}$, since any nonnegative number of customers may be in either station. It is evident that the first automaton $A^{(1)}$ is completely independent of the second. On the other hand, transitions in $A^{(2)}$ depend on the first automaton in two ways. First, the rate at which customers are served in the second station depends on the number of customers in the network, and hence on the number at the first station in particular. Thus $A^{(2)}$ contains functional transition rates, $\nu(\eta_1, \eta_2)$. Second, when a departure occurs from the first station, a customer enters the second and therefore instantaneously forces a transition to occur within the second automaton. The state of the second automaton is instantaneously changed from η_2 to $\eta_2 + 1$. This entails transitions of the second type, namely, synchronizing transitions.

9.3.2   Functional Transition Rates and Synchronizing Events

Let us examine how these two different types of interaction may be specified in a stochastic automata network. Consider first the case of constant and functional transition rates; normally an automaton will contain both. The elements in the infinitesimal generator of any single stochastic automaton are either

- Constants, i.e., nonnegative real numbers, or

- Functions from the global state space to the nonnegative reals.

Transition rates that depend only on the state of the automaton itself and not on the state of any other automaton are to all intents and purposes constant transition rates. This is the case if the rate of transition of the second exponential server in the two-station queueing network example is *load-dependent*, i.e., depending only on the number of customers present in station 2. The *load-dependent* value is instantiated at each state and taken to be that constant value in the rest of the analysis. Constant transition rates obviously pose no problem.

If a given state of a stochastic automaton occasions functional transitions, the rate of these transitions must be evaluated according to their defining formulae and the current global state. In certain instances it may be advantageous to evaluate these functional rates only once and to store the values in an array, from which they may be retrieved as and when they are needed. In other cases it may be better to leave them in their original form and to reevaluate the formula each time the rate is needed.

Whether the transition rates are constant or functional, it is important to note that only the state of the *local* automaton is affected. Therefore, all the information concerning constant and functional transition rates within an automaton can be handled within that automaton (assuming only that that automaton has a knowledge of the global system state). *Functional transitions affect the global system only in that they change the state of a single automaton.*

Now consider synchronizing events. The two-station queueing network example, given previously, clearly shows that the first of the two stochastic automata is independent of the second. However, this does not mean that the complete set of information needed to specify the synchronizing event can be confined to $A^{(2)}$. It is not sufficient for $A^{(2)}$ to realize that it is susceptible to instantaneous transitions forced upon it by another automaton. Additionally $A^{(1)}$, although completely independent, needs to participate in a mechanism to disseminate the fact that it executes synchronizing transitions. One way to implement this is to generate a list of all possible synchronizing events that can occur in a SAN. This list needs to provide a unique name for each synchronizing event, the manner in which it occurs, and its effect on other stochastic automata. In contrast to functional transitions, *synchronizing events affect the global system by altering the state of possibly many automata.*

Given this information on functional transition rates and synchronizing events, it is possible to *formally* define a network of stochastic automata. This has been done by Atif and Plateau [6], and we refer the reader to their work for further information on this subject. We shall continue by considering the effects of both synchronizing events and functional transition rates on our ability to efficiently solve networks of stochastic automata. We begin with synchronizing events.

9.3.3 The Effect of Synchronizing Events

We begin with a small example of two[1] interacting stochastic automata, $A^{(1)}$ and $A^{(2)}$, whose infinitesimal generator matrices are given by

$$Q^{(1)} = \begin{pmatrix} -\lambda_1 & \lambda_1 \\ \lambda_2 & -\lambda_2 \end{pmatrix} \quad \text{and} \quad Q^{(2)} = \begin{pmatrix} -\mu_1 & \mu_1 & 0 \\ 0 & -\mu_2 & \mu_2 \\ \mu_3 & 0 & -\mu_3 \end{pmatrix}$$

respectively. At the moment, neither contains synchronizing events or functional transition rates. The infinitesimal generator of the global, two-dimensional system is therefore given as

$$Q^{(1)} \oplus Q^{(2)} =$$

$$\left(\begin{array}{ccc|ccc} -(\lambda_1 + \mu_1) & \mu_1 & 0 & \lambda_1 & 0 & 0 \\ 0 & -(\lambda_1 + \mu_2) & \mu_2 & 0 & \lambda_1 & 0 \\ \mu_3 & 0 & -(\lambda_1 + \mu_3) & 0 & 0 & \lambda_1 \\ \hline \lambda_2 & 0 & 0 & -(\lambda_2 + \mu_1) & \mu_1 & 0 \\ 0 & \lambda_2 & 0 & 0 & -(\lambda_2 + \mu_2) & \mu_2 \\ 0 & 0 & \lambda_2 & \mu_3 & 0 & -(\lambda_2 + \mu_3) \end{array} \right).$$

$$(9.6)$$

Its probability distribution vector (at time t) is obtained by forming the tensor product of the individual probability distribution vectors of all the stochastic automata at time t.

Let us now observe the effect of introducing synchronizing events. Suppose that each time automaton $A^{(1)}$ generates a transition from state 2 to state 1 (at rate λ_2), it forces the second automaton into state 1. It may be readily verified that the global generator matrix is given by

$$\left(\begin{array}{ccc|ccc} -(\lambda_1 + \mu_1) & \mu_1 & 0 & \lambda_1 & 0 & 0 \\ 0 & -(\lambda_1 + \mu_2) & \mu_2 & 0 & \lambda_1 & 0 \\ \mu_3 & 0 & -(\lambda_1 + \mu_3) & 0 & 0 & \lambda_1 \\ \hline \lambda_2 & 0 & 0 & -(\lambda_2 + \mu_1) & \mu_1 & 0 \\ \lambda_2 & 0 & 0 & 0 & -(\lambda_2 + \mu_2) & \mu_2 \\ \lambda_2 & 0 & 0 & \mu_3 & 0 & -(\lambda_2 + \mu_3) \end{array} \right).$$

If, in addition, the second automaton $A^{(2)}$ initiates a synchronizing event each time it moves from state 3 to state 1 (at rate μ_3), such as by forcing the first automaton into state 1, we obtain the following global generator.

---

[1] The extension to more than two is immediate.

$$\begin{pmatrix}
-(\lambda_1 + \mu_1) & \mu_1 & 0 & \lambda_1 & 0 & 0 \\
0 & -(\lambda_1 + \mu_2) & \mu_2 & 0 & \lambda_1 & 0 \\
\mu_3 & 0 & -(\lambda_1 + \mu_3) & 0 & 0 & \lambda_1 \\
\lambda_2 & 0 & 0 & -(\lambda_2 + \mu_1) & \mu_1 & 0 \\
\lambda_2 & 0 & 0 & 0 & -(\lambda_2 + \mu_2) & \mu_2 \\
\lambda_2 + \mu_3 & 0 & 0 & 0 & 0 & -(\lambda_2 + \mu_3)
\end{pmatrix}.$$

Note that synchronizing events need to be well defined, in the sense that they should not lead to unstable behavior. Thus, for example, it would not be appropriate to define synchronizing events that instantaneously move the system from one global state to a second global state and then instantaneously back to the first again.

Our immediate reaction in observing these altered matrices may be to assume that a major disadvantage of incorporating synchronizing transitions is to remove the possibility of representing the global transition rate matrix as (a sum of) tensor products. However, Plateau [120] has shown that if local transitions are separated from synchronizing transitions, this is not necessarily so; the global transition rate matrix can still be written as (a sum of) tensor products. To observe this we proceed as follows.

The transitions at rates λ_1, μ_1, and μ_2 are not synchronizing events, but rather *local* transitions. The part of the global generator that consists uniquely of local transitions may be obtained by forming the tensor sum of infinitesimal generators $Q_l^{(1)}$ and $Q_l^{(2)}$ that represent only local transitions, viz.:

$$Q_l^{(1)} = \begin{pmatrix} -\lambda_1 & \lambda_1 \\ 0 & 0 \end{pmatrix} \quad \text{and} \quad Q_l^{(2)} = \begin{pmatrix} -\mu_1 & \mu_1 & 0 \\ 0 & -\mu_2 & \mu_2 \\ 0 & 0 & 0 \end{pmatrix},$$

with tensor sum

$$Q_l = Q_l^{(1)} \oplus Q_l^{(2)} =$$

$$\begin{pmatrix}
-(\lambda_1 + \mu_1) & \mu_1 & 0 & \lambda_1 & 0 & 0 \\
0 & -(\lambda_1 + \mu_2) & \mu_2 & 0 & \lambda_1 & 0 \\
0 & 0 & -\lambda_1 & 0 & 0 & \lambda_1 \\
0 & 0 & 0 & -\mu_1 & \mu_1 & 0 \\
0 & 0 & 0 & 0 & -\mu_2 & \mu_2 \\
0 & 0 & 0 & 0 & 0 & 0
\end{pmatrix}.$$

The rates λ_2 and μ_3 are associated with two synchronizing events, which we call e_1 and e_2, respectively. The part of the global generator that is due to the first synchronizing event is given by

$$
Q_{e_1} = \left(\begin{array}{ccc|ccc}
0 & 0 & 0 & 0 & 0 & 0 \\
0 & 0 & 0 & 0 & 0 & 0 \\
0 & 0 & 0 & 0 & 0 & 0 \\
\hline
\lambda_2 & 0 & 0 & -\lambda_2 & 0 & 0 \\
\lambda_2 & 0 & 0 & 0 & -\lambda_2 & 0 \\
\lambda_2 & 0 & 0 & 0 & 0 & -\lambda_2
\end{array} \right)
$$

which is the (ordinary) matrix sum of two tensor products, viz.:

$$
Q_{e_1} = \left(\begin{array}{cc} 0 & 0 \\ \lambda_2 & 0 \end{array} \right) \otimes \left(\begin{array}{ccc} 1 & 0 & 0 \\ 1 & 0 & 0 \\ 1 & 0 & 0 \end{array} \right) + \left(\begin{array}{cc} 0 & 0 \\ 0 & -\lambda_2 \end{array} \right) \otimes \left(\begin{array}{ccc} 1 & 0 & 0 \\ 0 & 1 & 0 \\ 0 & 0 & 1 \end{array} \right).
$$

Similarly, the part of the global generator due to synchronizing event e_2 is

$$
Q_{e_2} = \left(\begin{array}{ccc|ccc}
0 & 0 & 0 & 0 & 0 & 0 \\
0 & 0 & 0 & 0 & 0 & 0 \\
\mu_3 & 0 & -\mu_3 & 0 & 0 & 0 \\
\hline
0 & 0 & 0 & 0 & 0 & 0 \\
0 & 0 & 0 & 0 & 0 & 0 \\
\mu_3 & 0 & 0 & 0 & 0 & -\mu_3
\end{array} \right)
$$

which may be obtained from a sum of tensor products as

$$
Q_{e_2} = \left(\begin{array}{cc} 1 & 0 \\ 1 & 0 \end{array} \right) \otimes \left(\begin{array}{ccc} 0 & 0 & 0 \\ 0 & 0 & 0 \\ \mu_3 & 0 & 0 \end{array} \right) + \left(\begin{array}{cc} 1 & 0 \\ 0 & 1 \end{array} \right) \otimes \left(\begin{array}{ccc} 0 & 0 & 0 \\ 0 & 0 & 0 \\ 0 & 0 & -\mu_3 \end{array} \right).
$$

Observe that the global infinitesimal generator is now given by

$$
Q = Q_l + Q_{e_1} + Q_{e_2}.
$$

Although we considered only a simple example, the above approach has been shown by Plateau and her coworkers to be applicable in general. Stochastic automata networks that contain synchronizing transitions may always be treated by separating out the local transitions, handling these in the usual fashion by means of a tensor sum, and then incorporating the sum of two additional tensor products per synchronizing event. Furthermore, since tensor sums are defined in terms of the (usual) matrix sum of tensor products (see equation (9.4)), the infinitesimal generator of a system containing N stochastic automata with E synchronizing events (and no functional transition rates) may be written as

$$
\sum_{j=1}^{2E+N} \bigotimes_{i=1}^{N} Q_j^{(i)}. \tag{9.7}
$$

This quantity is sometimes referred to as the *descriptor* of the stochastic automata network. Notice that many of the individual $Q_j^{(i)}$ will be extremely sparse. Furthermore, a large number of them will contain simply a single integer (1) in each row.

The computational burden imposed by synchronizing events is now apparent and is twofold. First, the number of terms in the descriptor is increased — two for each synchronizing event. This suggests that the SAN approach is not well suited to models in which there are many synchronizing events. On the other hand, it may still be useful for systems that can be modelled with several stochastic automata that operate mostly independently and need only infrequently synchronize their operations, such as those found in many models of highly parallel machines.

A second and even greater burden is that the simple form of the solution, equation (9.5), no longer holds. Although we have been successful in writing the descriptor in a compact form as a sum of tensor products, the solution is not simply the sum of the tensor products of the vector solutions of the individual $Q_j^{(i)}$. Other methods for computing solutions must be found. The usefulness of the SAN approach will be determined uniquely by our ability to solve this problem. This is taken up in Section 9.6. In the meantime, we turn our attention to functional transition rates, for these may appear not only in local transitions but also in synchronizing events.

9.3.4   The Effect of Functional Transition Rates

We return to the two original automata given in equation (9.6) and consider what happens when one of the transition rates of the second automaton becomes a functional transition rate. Suppose, for example, that the rate of transition from state 2 to state 3 in the second stochastic automaton is $\hat{\mu}_2$, when the first is in state 1, and $\tilde{\mu}_2$, when the first is in state 2. The global infinitesimal generator is now

$$\left(\begin{array}{ccc|ccc} -(\lambda_1 + \mu_1) & \mu_1 & 0 & \lambda_1 & 0 & 0 \\ 0 & -(\lambda_1 + \hat{\mu}_2) & \hat{\mu}_2 & 0 & \lambda_1 & 0 \\ \mu_3 & 0 & -(\lambda_1 + \mu_3) & 0 & 0 & \lambda_1 \\ \hline \lambda_2 & 0 & 0 & -(\lambda_2 + \mu_1) & \mu_1 & 0 \\ 0 & \lambda_2 & 0 & 0 & -(\lambda_2 + \tilde{\mu}_2) & \tilde{\mu}_2 \\ 0 & 0 & \lambda_2 & \mu_3 & 0 & -(\lambda_2 + \mu_3) \end{array} \right).$$

$$(9.8)$$

If, in addition, the rate at which the first stochastic automaton produces transitions from state 1 to state 2 is $\bar{\lambda}_1$, $\hat{\lambda}_1$, or $\tilde{\lambda}_1$ depending on whether the second automaton is in state 1, 2, or 3, the two-dimensional infinitesimal generator becomes

$$\left(\begin{array}{ccc|ccc} -(\bar{\lambda}_1 + \mu_1) & \mu_1 & 0 & \bar{\lambda}_1 & 0 & 0 \\ 0 & -(\hat{\lambda}_1 + \hat{\mu}_2) & \hat{\mu}_2 & 0 & \hat{\lambda}_1 & 0 \\ \mu_3 & 0 & -(\tilde{\lambda}_1 + \mu_3) & 0 & 0 & \tilde{\lambda}_1 \\ \hline \lambda_2 & 0 & 0 & -(\lambda_2 + \mu_1) & \mu_1 & 0 \\ 0 & \lambda_2 & 0 & 0 & -(\lambda_2 + \tilde{\mu}_2) & \tilde{\mu}_2 \\ 0 & 0 & \lambda_2 & \mu_3 & 0 & -(\lambda_2 + \mu_3) \end{array}\right).$$

It is apparent that the introduction of functional transition rates has no effect on the *structure* of the global transition rate matrix. Also, it is equally obvious that the actual values of the nonzero elements prevent us from writing the global matrix directly as a (sum of) tensor product(s). We may conclude, as we did for synchronizing transitions, that the introduction of functional transition rates prevents us from writing the solution in the simple form of equation (9.5). However, it is still possible to profit from the fact that the nonzero structure is unchanged. This is essentially what Plateau has done in her extension of the classical tensor algebraic concepts [122]. The descriptor is still written as in equation (9.7), but now the elements of $Q_j^{(i)}$ may be functions. This is sometimes referred to as a *generalized* tensor product. It is necessary to track elements that are functions and to substitute the appropriate numerical value each time the functional rate is needed.

An alternative approach to that of using an extended tensor algebra to handle functional elements is to use the classical approach and replace each tensor product that incorporates matrices having functional entries with a *sum* of tensor products of matrices that incorporate only numerical entries. This is the approach that is used in the software package PEPS [121]. In this case, equation (9.7) with functional entries may be written as

$$\sum_{j=1}^{2E+N} \bigotimes_{i=1}^{N} Q_j^{(i)} = \sum_{j=1}^{T} \bigotimes_{i=1}^{N} \bar{Q}_j^{(i)}$$

where $\bar{Q}$ contains only numerical values and the size of T depends on $2E + N$ and on $\prod_{i \in F} n_i$, where F is the set of automata whose state variables appear as arguments in functional transition rates. For example, equation (9.8), in which the parameter μ_2 is functional, may be written as a sum of tensor products as

$$\left(\begin{array}{cc} -\lambda_1 & \lambda_1 \\ \lambda_2 & -\lambda_2 \end{array}\right) \otimes I_3 + I_2 \otimes \left(\begin{array}{ccc} -\mu_1 & \mu_1 & 0 \\ 0 & 0 & 0 \\ \mu_3 & 0 & -\mu_3 \end{array}\right) +$$

$$\left(\begin{array}{cc} 1 & 0 \\ 0 & 0 \end{array}\right) \otimes \left(\begin{array}{ccc} 0 & 0 & 0 \\ 0 & -\hat{\mu}_2 & \hat{\mu}_2 \\ 0 & 0 & 0 \end{array}\right) + \left(\begin{array}{cc} 0 & 0 \\ 0 & 1 \end{array}\right) \otimes \left(\begin{array}{ccc} 0 & 0 & 0 \\ 0 & -\tilde{\mu}_2 & \tilde{\mu}_2 \\ 0 & 0 & 0 \end{array}\right).$$

Although T may be large, it may be shown to be bounded by $T \leq (2E + N) \times \prod_{i \in F} n_i$. In what follows, we shall omit the bar over $\bar{Q}$.

9.4 Computing Probability Distributions

It follows from the foregoing discussion that in stochastic automata networks that contain functional transitions or synchronizing events, it is no longer possible to construct the global probability distribution vector directly from the distributions of the individual stochastic automata. We therefore ask ourselves what can be salvaged. Perhaps a preliminary question to pose is, "What are the alternatives?" In certain circumstances it may be appropriate to replace functional rates with constant rates that reflect an average behavior and to use the efficient tensor product mechanism to obtain an approximate solution. Certain experimental studies have shown this to give reasonably accurate results when the number of functional transitions is not large and those that do exist are not vastly different [5]. If this is not acceptable, then it may be possible to generate the global infinitesimal generator and to apply one of the standard numerical solution methods. However, if the objective is to analyze several modest-sized stochastic automata for which the transition matrices are dense, a quick analysis will reveal that the amount of memory needed to store the expanded matrix may exceed the capacity of the computer destined to perform the computation. For example, 10 stochastic automata, each with 10 states and a dense transition matrix, give rise to a global transition matrix with 10^{10} elements!

If we make the assumption that it is either not possible or too inefficient to generate the global matrix, we must then turn to methods of computing the global probability distribution from the representations of the individual automata *without* generating the global matrix. In this respect, we may regard the representations of the individual stochastic automata as a compact storage mechanism for the global system. Therefore, given that the global infinitesimal generator is available as the descriptor of a SAN, we ask what types of numerical methods may be used to obtain stationary probability distributions.

Let Q be the descriptor of a stochastic automata network, i.e.,

$$Q = \sum_{j=1}^{T} \bigotimes_{i=1}^{N} Q_j^{(i)}.$$

We wish to obtain the vector π such that $\pi Q = 0$ and $\pi e = 1$. We may immediately eliminate methods that are based on LU decompositions, since the "descriptor" approach to storing the infinitesimal generator is very unsuitable for direct solution methods. Additionally, the size of the matrix will likely eliminate the possibility of using direct methods. This leaves iterative methods, of which there is a wide variety. However, later we shall see that the constraints imposed by this method of storing the matrix limits our choice of iterative method. For the moment we turn our attention to observe how an operation basic to all matrix iterative algorithms — the product of a vector with a matrix — may be carried out when the matrix is given in the form of a SAN descriptor. Since

$$xQ = x \sum_{j=1}^{T} \bigotimes_{i=1}^{N} Q_j^{(i)} = \sum_{j=1}^{T} x \bigotimes_{i=1}^{N} Q_j^{(i)}, \tag{9.9}$$

the basic operation is $x \bigotimes_{i=1}^{N} Q_j^{(i)}$.

9.5 Multiplication with a Vector

Given N stochastic automata $A^{(1)}, A^{(2)}, \ldots, A^{(N)}$, with $n_1, n_2, \ldots, n_N$ states, respectively, and infinitesimal generators $Q^{(1)}, Q^{(2)}, \ldots, Q^{(N)}$, our concern is the computation of

$$x \bigotimes_{i=1}^{N} Q^{(i)}$$

where x is a vector of length $\prod_{i=1}^{N} n_i$. We consider the case when the $Q^{(i)}$ are full matrices; the results need to be modified when the individual infinitesimal generators are sparse and stored in compact form. We shall begin with the case when the $Q^{(i)}$ contain only constant (nonfunctional) transition rates. This includes models in which

1. The stochastic automata are independent (in equation (9.9), $T = N$ and the SAN contains neither synchronizing nor functional transitions);

2. The SAN contains synchronizing events but no functional transitions (in equation (9.9), $T = 2E + N$);

3. The SAN contains functional transitions (and possibly synchronizing events), and the functional elements are handled by expansion ($T > 2E + N$).

Later we shall consider the effect of introducing functional transition rates directly into the tensor product, thereby making use of generalized tensor product concepts.

9.5.1   The Nonfunctional Case

Theorem 9.1 *The product*

$$x \bigotimes_{i=1}^{N} Q^{(i)},$$

where x is a real vector of length $\prod_{i=1}^{N} n_i$ and the $Q^{(i)}$ are the infinitesimal generators of N stochastic automata as described above, may be computed in ρ_N multiplications, where

$$\rho_N = n_N \times \left(\rho_{N-1} + \prod_{i=1}^{N} n_i \right) \;=\; \prod_{i=1}^{N} n_i \times \sum_{i=1}^{N} n_i. \tag{9.10}$$

This complexity result is valid only when the $Q^{(i)}$ contain nonfunctional rates; that is, when generalized tensor products are *not* used. It is valid when functional transitions are handled by expanding the number of terms in the summation of the SAN descriptor from $2E + N$ to T, for then the individual matrices contain only numerical values. To compare ρ_N with the number needed when advantage is not taken of the special structure resulting from the tensor product, observe that to naïvely expand $\bigotimes_{i=1}^{N} Q^{(i)}$ requires $\left(\prod_{i=1}^{N} n_i \right)^2$ multiplications and that this same number is required each time we form the product $x \ \bigotimes_{i=1}^{N} Q^{(i)}$.

Some examples will help us see this more clearly and at the same time show us where extra multiplications are needed when generalized tensor products are used to handle functional transitions. Consider two stochastic automata, the first A with $n_1 = 2$ states and the second B with $n_2 = 3$ states, as in equation (9.3). The product $y = x(A \otimes B)$ may be obtained as

$$y_{1_1} = a_{11}(x_{1_1}b_{11} + x_{1_2}b_{21} + x_{1_3}b_{31}) + a_{21}(x_{2_1}b_{11} + x_{2_2}b_{21} + x_{2_3}b_{31})$$
$$y_{1_2} = a_{11}(x_{1_1}b_{12} + x_{1_2}b_{22} + x_{1_3}b_{32}) + a_{21}(x_{2_1}b_{12} + x_{2_2}b_{22} + x_{2_3}b_{32})$$
$$y_{1_3} = a_{11}(x_{1_1}b_{13} + x_{1_2}b_{23} + x_{1_3}b_{33}) + a_{21}(x_{2_1}b_{13} + x_{2_2}b_{23} + x_{2_3}b_{33})$$

$$\rule{10cm}{0.4pt} \tag{9.11}$$

$$y_{2_1} = a_{12}(x_{1_1}b_{11} + x_{1_2}b_{21} + x_{1_3}b_{31}) + a_{22}(x_{2_1}b_{11} + x_{2_2}b_{21} + x_{2_3}b_{31})$$
$$y_{2_2} = a_{12}(x_{1_1}b_{12} + x_{1_2}b_{22} + x_{1_3}b_{32}) + a_{22}(x_{2_1}b_{12} + x_{2_2}b_{22} + x_{2_3}b_{32})$$
$$y_{2_3} = a_{12}(x_{1_1}b_{13} + x_{1_2}b_{23} + x_{1_3}b_{33}) + a_{22}(x_{2_1}b_{13} + x_{2_2}b_{23} + x_{2_3}b_{33})$$

Note that each inner product (the terms within the parentheses) above the horizontal line is also to be found below the line, so each needs to be computed only once. The evaluation of each inner product requires n_2 $(= 3)$ multiplications, and since there is a total of $n_1 n_2$ $(= 2 \times 3)$ distinct inner products (those above the line), the total number of multiplications needed to evaluate all the inner products is $n_1 n_2^2$. To complete the computation of y, each of the $n_1 n_2$ distinct inner products must be multiplied by n_1 different elements of the matrix A. The total number of multiplications needed to compute the product $y = x(A \otimes B)$ is thus $n_1 n_2^2 + n_1^2 n_2 = n_1 n_2 (n_1 + n_2)$.

Now consider the effect of increasing the number of stochastic automata by including a third, C, with $n_3 = 2$ states. We wish to compute $y = x(A \otimes B \otimes C)$, where x and y are both of length $n_1 n_2 n_3$. The first six elements of y are

$$
\begin{aligned}
y_{1_{1_1}} &= a_{11}[b_{11}(x_{1_{1_1}}c_{11} + x_{1_{1_2}}c_{21}) + b_{21}(x_{1_{2_1}}c_{11} + x_{1_{2_2}}c_{21}) + b_{31}(x_{1_{2_1}}c_{11} + x_{1_{2_2}}c_{21})] + \\
&\quad a_{21}[b_{11}(x_{2_{1_1}}c_{11} + x_{2_{1_2}}c_{21}) + b_{21}(x_{2_{2_1}}c_{11} + x_{2_{2_2}}c_{21}) + b_{31}(x_{2_{2_1}}c_{11} + x_{2_{2_2}}c_{21})] \\
y_{1_{1_2}} &= a_{11}[b_{11}(x_{1_{1_1}}c_{12} + x_{1_{1_2}}c_{22}) + b_{21}(x_{1_{2_1}}c_{12} + x_{1_{2_2}}c_{22}) + b_{31}(x_{1_{2_1}}c_{12} + x_{1_{2_2}}c_{22})] + \\
&\quad a_{21}[b_{11}(x_{2_{1_1}}c_{12} + x_{2_{1_2}}c_{22}) + b_{21}(x_{2_{2_1}}c_{12} + x_{2_{2_2}}c_{22}) + b_{31}(x_{2_{2_1}}c_{12} + x_{2_{2_2}}c_{22})]
\end{aligned}
$$

$$
\begin{aligned}
y_{1_{2_1}} &= a_{11}[b_{12}(x_{1_{1_1}}c_{11} + x_{1_{1_2}}c_{21}) + b_{22}(x_{1_{2_1}}c_{11} + x_{1_{2_2}}c_{21}) + b_{32}(x_{1_{2_1}}c_{11} + x_{1_{2_2}}c_{21})] + \\
&\quad a_{21}[b_{12}(x_{2_{1_1}}c_{11} + x_{2_{1_2}}c_{21}) + b_{22}(x_{2_{2_1}}c_{11} + x_{2_{2_2}}c_{21}) + b_{32}(x_{2_{2_1}}c_{11} + x_{2_{2_2}}c_{21})] \\
y_{1_{2_2}} &= a_{11}[b_{12}(x_{1_{1_1}}c_{12} + x_{1_{1_2}}c_{22}) + b_{22}(x_{1_{2_1}}c_{12} + x_{1_{2_2}}c_{22}) + b_{32}(x_{1_{2_1}}c_{12} + x_{1_{2_2}}c_{22})] + \\
&\quad a_{21}[b_{12}(x_{2_{1_1}}c_{12} + x_{2_{1_2}}c_{22}) + b_{22}(x_{2_{2_1}}c_{12} + x_{2_{2_2}}c_{22}) + b_{32}(x_{2_{2_1}}c_{12} + x_{2_{2_2}}c_{22})]
\end{aligned}
$$

$$
\begin{aligned}
y_{1_{3_1}} &= a_{11}[b_{13}(x_{1_{1_1}}c_{11} + x_{1_{1_2}}c_{21}) + b_{23}(x_{1_{2_1}}c_{11} + x_{1_{2_2}}c_{21}) + b_{33}(x_{1_{2_1}}c_{11} + x_{1_{2_2}}c_{21})] + \\
&\quad a_{21}[b_{13}(x_{2_{1_1}}c_{11} + x_{2_{1_2}}c_{21}) + b_{23}(x_{2_{2_1}}c_{11} + x_{2_{2_2}}c_{21}) + b_{33}(x_{2_{2_1}}c_{11} + x_{2_{2_2}}c_{21})] \\
y_{1_{3_2}} &= a_{11}[b_{13}(x_{1_{1_1}}c_{12} + x_{1_{1_2}}c_{22}) + b_{23}(x_{1_{2_1}}c_{12} + x_{1_{2_2}}c_{22}) + b_{33}(x_{1_{2_1}}c_{12} + x_{1_{2_2}}c_{22})] + \\
&\quad a_{21}[b_{13}(x_{2_{1_1}}c_{12} + x_{2_{1_2}}c_{22}) + b_{23}(x_{2_{2_1}}c_{12} + x_{2_{2_2}}c_{22}) + b_{33}(x_{2_{2_1}}c_{12} + x_{2_{2_2}}c_{22})]
\end{aligned}
$$

Elements $y_{2_{1_1}}$ through $y_{2_{3_2}}$ are similarly defined. In fact, in the above equations it suffices to replace all a_{11} with a_{12} and a_{21} with a_{22}.

To determine the number of multiplications involved in computing $y = x(A \otimes B \otimes C)$, notice that the effect of incorporating the additional automaton is twofold — the number of equations is increased *and* the number of multiplications within each equation is increased. Let us consider these separately.

- Firstly, the number of equations is increased by a factor of n_3 (indeed the length of the vector x (and y) increases from $n_1 n_2$ to $n_1 n_2 n_3$). If this were the only effect of adding a third automaton, the total number of multiplications would be $n_3 \rho_2$, where $\rho_2 = n_1 n_2 (n_1 + n_2)$ is the number of multiplications needed with two automata.

- The second effect of incorporating automaton C is to add a further $n_1 n_2 n_3$ multiplications into each equation. To see this, observe that in the case of two automata, equation (9.11), each component b_{ij} of B is multiplied by the scalar x_{j_i}. In the case of three automata, each of these scalars

x_{j_i} becomes an inner product (in parentheses), and each needs $n_3 \, (= 2)$ multiplications to be evaluated. Since there are $n_1 n_2$ terms of B in each equation, it follows that each equation incorporates an additional $(n_1 n_2)n_3$ multiplications. However, considerable savings may be effected since the same inner products occur many times. The inner products to be found in computing $y_{1_{1_1}}$ and $y_{1_{1_2}}$ are identical to those in $y_{i_{j_1}}$ and $y_{i_{j_2}}$ for all i and j, so that the $n_1 n_2 n_3$ additional multiplications per equation need only be computed for n_3 equations (and not $n_1 n_2 n_3$ equations).

Together, these two effects imply that the number of multiplications needed with the three automata is given by $n_3 \rho_2 + n_3(n_1 n_2 n_3) = n_1 n_2 n_3(n_1 + n_2 + n_3)$.

Proof We now prove Theorem 9.1 by induction on the number of stochastic automata. With 2 automata, we have

$$x \left(Q^{(1)} \otimes Q^{(2)} \right) = x \begin{pmatrix} q_{11}^{(1)} Q^{(2)} & q_{12}^{(1)} Q^{(2)} & \cdots & q_{1n_1}^{(1)} Q^{(2)} \\ q_{21}^{(1)} Q^{(2)} & q_{22}^{(1)} Q^{(2)} & \cdots & q_{2n_1}^{(1)} Q^{(2)} \\ \vdots & \vdots & \ddots & \vdots \\ q_{n_1 1}^{(1)} Q^{(2)} & q_{n_1 2}^{(1)} Q^{(2)} & \cdots & q_{n_1 n_1}^{(1)} Q^{(2)} \end{pmatrix} = (y_1, y_2, \ldots, y_{n_1}),$$

where

$$y_i = q_{1i}^{(1)} x_1 Q^{(2)} + q_{2i}^{(1)} x_2 Q^{(2)} + \cdots + q_{n_1 i}^{(1)} x_{n_1} Q^{(2)}, \qquad i = 1, 2, \ldots n_1,$$

and x_i is a row vector of length n_2. Each $x_j Q^{(2)}$, for $j = 1, 2, \ldots, n_1$, is a row vector of length n_2 and can be computed in n_2^2 multiplications. All n_1 of them can therefore be found in $n_1 n_2^2$ multiplications. Each of the n_1 row vectors y_i (of length n_2) can subsequently be computed in an additional $n_1 n_2$ multiplications, and all n_1 of them in $n_1^2 n_2$ multiplications. The total number of multiplications in computing $x \left(Q^{(1)} \otimes Q^{(2)} \right)$ is therefore $n_1 n_2^2 + n_1^2 n_2 = n_1 n_2(n_1 + n_2)$, and thus, the theorem is true for 2 automata.

Assume now that the theorem is true for $N - 1$ automata. We wish to show that it is also true for N. Since $y = x Q^{(1)} \otimes Q^{(2)} \otimes \cdots \otimes Q^{(N)} = x Q^{(1)} \otimes \left(Q^{(2)} \otimes \cdots \otimes Q^{(N)} \right)$, we find that

$$y = x \begin{pmatrix} q_{11}^{(1)} B & q_{12}^{(1)} B & \cdots & q_{1n_1}^{(1)} B \\ q_{21}^{(1)} B & q_{22}^{(1)} B & \cdots & q_{2n_1}^{(1)} B \\ \vdots & \vdots & \ddots & \vdots \\ q_{n_1 1}^{(1)} B & q_{n_1 2}^{(1)} B & \cdots & q_{n_1 n_1}^{(1)} B \end{pmatrix} = (y_1, y_2, \ldots, y_{n_1}),$$

where

$$B = Q^{(2)} \otimes \cdots \otimes Q^{(N)},$$

$$y_i = q_{1i}^{(1)} x_1 B + q_{2i}^{(1)} x_2 B + \cdots + q_{n_1 i}^{(1)} x_{n_1} B,$$

and y_i and x_i are row vectors of length $\prod_{j=2}^{N} n_j$. From the inductive hypothesis, each row vector $x_i B$ (of length $\prod_{j=2}^{N} n_j$) can be computed in $\rho_{N-1} = \prod_{j=2}^{N} n_j \times \sum_{j=2}^{N} n_j$ multiplications. Hence, we need $n_1 \rho_{N-1}$ multiplications to compute all $x_i B$, for $i = 1, 2, \ldots, n_1$. When this has been completed, a further $n_1 \prod_{j=2}^{N} n_j$ multiplications are needed to form each y_i, and since there are n_1 of them, all the y_i can be found in $n_1^2 \prod_{j=2}^{N} n_j$ multiplications. This, of course, is over and above the number needed to compute the $x_i B$. Hence, the total number of multiplications needed to form $y = x Q^{(1)} \otimes Q^{(2)} \otimes \cdots \otimes Q^{(N)}$ is given by

$$n_1 \rho_{N-1} + n_1^2 \prod_{j=2}^{N} n_j = n_1 \prod_{j=2}^{N} n_j \times \sum_{j=2}^{N} n_j + n_1^2 \prod_{j=2}^{N} n_j = \prod_{j=1}^{N} n_j \times \sum_{j=1}^{N} n_j.$$

□

9.5.2 Multiplication in the Presence of Functional Elements

We are now ready to examine the effect of handling functional transitions by means of an extended tensor algebra rather than by expansion. The savings made previously in the computation of $x \bigotimes_{i=1}^{N} Q^{(i)}$ are due to the fact that once a product is formed, it may be used in several places without having to redo the multiplication. With functional rates, the elements in the matrices may change according to their context, so the same savings are not always possible. For example, if in forming $x(A \otimes B)$ the element b_{ij} is a functional rate, it will have n_1 different values depending on the state of automaton A (recall that the state of automaton B is automatically accounted for in the fact that the b_{ij} are distinct).

We shall return to the case of two automata A and B as given by equation (9.3). Note first that if B possesses functional transition rates, the number of multiplications remains identical. For example, if $b_{ij} = \hat{b}_{ij}$ when the first automaton is in state 1 and $b_{ij} = \tilde{b}_{ij}$ when the first automaton is in state 2, the first three equations become

$$y_{1_1} = a_{11}(x_{1_1}\hat{b}_{11} + x_{1_2}\hat{b}_{21} + x_{1_3}\hat{b}_{31}) + a_{21}(x_{2_1}\tilde{b}_{11} + x_{2_2}\tilde{b}_{21} + x_{2_3}\tilde{b}_{31})$$
$$y_{1_2} = a_{11}(x_{1_1}\hat{b}_{12} + x_{1_2}\hat{b}_{22} + x_{1_3}\hat{b}_{32}) + a_{21}(x_{2_1}\tilde{b}_{12} + x_{2_2}\tilde{b}_{22} + x_{2_3}\tilde{b}_{32})$$
$$y_{1_3} = a_{11}(x_{1_1}\hat{b}_{13} + x_{1_2}\hat{b}_{23} + x_{1_3}\hat{b}_{33}) + a_{21}(x_{2_1}\tilde{b}_{13} + x_{2_2}\tilde{b}_{23} + x_{2_3}\tilde{b}_{33})$$

The final three terms are similar; in particular, the inner products are identical. Thus, there is no increase in the number of multiplications needed when B has rates that depend on the state of A. However, if A contains functional transition rates, this means of computing the product $x(A \otimes B)$ is no longer valid. The equations used were generated by factoring out common terms, a_{ij}. However, these terms are no longer common (or constant) but functional. We need to adopt a different strategy.

9.5.3   The Use of Symmetric Permutations

For the example that produced equation (9.1), it is apparent that $A \otimes B$ is, up to a symmetric permutation, equal to $B \otimes A$. Given the permutation matrix

$$M = \begin{pmatrix} 1 & 0 & 0 & 0 & 0 & 0 \\ 0 & 0 & 0 & 1 & 0 & 0 \\ 0 & 1 & 0 & 0 & 0 & 0 \\ 0 & 0 & 0 & 0 & 1 & 0 \\ 0 & 0 & 1 & 0 & 0 & 0 \\ 0 & 0 & 0 & 0 & 0 & 1 \end{pmatrix},$$

it is easy to verify that

$$\begin{align} B \otimes A &= M \ (A \otimes B)M^T \tag{9.12} \\ A \otimes B &= M^T(B \otimes A)M \tag{9.13} \end{align}$$

It may be shown that this property holds in general, i.e., that for all matrices $A \in \Re^{n_1 \times n_1}$ and $B \in \Re^{n_2 \times n_2}$ there exists a permutation matrix M of order $n_1 n_2$ such that $A \otimes B = M^T(B \otimes A)M$. Indeed, the following theorem is proven in [122].

Theorem 9.2 *Let* $X = (X^{(1)}, X^{(2)}, \ldots, X^{(N)})$ *be an* N-*dimensional Markov chain with infinitesimal generator* $\bigotimes_{i=1}^{N} Q^{(i)}$ *and stationary probability vector* π. *Let* σ *be a permutation of the integers* $[1, 2, \ldots, N]$. *Then:*

1. *The infinitesimal generator of the permuted Markov chain,* $X_\sigma = (X^{(\sigma_1)}, X^{(\sigma_2)}, \ldots, X^{(\sigma_N)})$, *is given by* $\bigotimes_{i=1}^{N} Q^{(\sigma_i)}$.

2. *There exists a permutation matrix* P_σ *of order* $\prod_{i=1}^{N} n_i$, *such that*

   - *The stationary distribution of* X_σ *is* $\pi_\sigma = \pi P_\sigma$, *and*
   - $\bigotimes_{i=1}^{N} Q^{(i)} = P_\sigma \left(\bigotimes_{i=1}^{N} Q^{(\sigma_i)} \right) P_\sigma^T$.

In terms of our algorithm for computing $y = x(A \otimes B)$ when only A contains functional transition rates, it is therefore possible to perform the permutation specified in equation (9.12) and compute the product with the same number of multiplications as before. However, Theorem 9.2 allows us to go farther than this. In forming the tensor product of N stochastic automata, we may arrange the automata in any order we wish, to try to minimize the number of multiplications that must be performed. For example, in the case of multiple stochastic automata in which each contains rates that are functions of the state of a single specific (e.g., master) automaton, it may be advantageous to permute the automata so that the master automaton is the first.

Another theorem relating to the use of a certain class of permutation matrices (those that perform right circular shifts) has also been provided by Plateau and Fourneau [122].

Theorem 9.3 *Let ς_i be the right circular shift of the integers $[1, 2, \ldots, N]$ that maps i onto N. Let $\eta_i = \prod_{k=1, k \neq i}^{N} n_k$. Then, if $Q^{(i)}$ for $i = 1, 2, \ldots, N$ are infinitesimal generators that contain only nonfunctional transition rates,*

$$\bigotimes_{i=1}^{N} Q^{(i)} = \prod_{i=1}^{N} P_{\varsigma_i} \left(I_{\eta_i} \otimes Q^{(i)} \right) P_{\varsigma_i}^T.$$

Note that each of the matrices $I_{\eta_i} \otimes Q^{(i)}$ is a block diagonal matrix, all the diagonal blocks being of size $n_i \times n_i$ (and in fact all equal to $Q^{(i)}$). Therefore, to form the product $x(I_{\eta_i} \otimes Q^{(i)})$ requires $\eta_i n_i^2 = n_i \prod_{k=1}^{N} n_k$ multiplications, which in turn means that $x \bigotimes_{i=1}^{N} Q^{(i)}$ may be computed in $\sum_{i=1}^{N} \left(n_i \prod_{k=1}^{N} n_k \right)$, the same as that obtained previously in equation (9.10). Unfortunately, the proof of Theorem 9.3 depends on the compatibility of the tensor product and regular matrix product, which is valid only when the elements of the stochastic automata contain nonfunctional transition rates. Theorem 9.2, on the other hand, is valid in the presence of both functional and nonfunctional transition rates.

9.5.4 When All Else Fails

The problem we now address is that of finding the best means of performing the computation when *both* A and B contain functional transition rates. Consider the worst possible case, when all of the elements of A and B are state-dependent transition rates. In our two-automata example, we have

$$y_{1_1} = \hat{a}_{11}\hat{b}_{11}x_{1_1} + \tilde{a}_{11}\hat{b}_{21}x_{1_2} + \bar{a}_{11}\hat{b}_{31}x_{1_3} + \hat{a}_{21}\tilde{b}_{11}x_{2_1} + \tilde{a}_{21}\tilde{b}_{21}x_{2_2} + \bar{a}_{21}\tilde{b}_{31}x_{2_3}$$
$$y_{1_2} = \hat{a}_{11}\hat{b}_{12}x_{1_1} + \tilde{a}_{11}\hat{b}_{22}x_{1_2} + \bar{a}_{11}\hat{b}_{32}x_{1_3} + \hat{a}_{21}\tilde{b}_{12}x_{2_1} + \tilde{a}_{21}\tilde{b}_{22}x_{2_2} + \bar{a}_{21}\tilde{b}_{32}x_{2_3}$$
$$y_{1_3} = \hat{a}_{11}\hat{b}_{13}x_{1_1} + \tilde{a}_{11}\hat{b}_{23}x_{1_2} + \bar{a}_{11}\hat{b}_{33}x_{1_3} + \hat{a}_{21}\tilde{b}_{13}x_{2_1} + \tilde{a}_{21}\tilde{b}_{23}x_{2_2} + \bar{a}_{21}\tilde{b}_{33}x_{2_3}$$

$$y_{2_1} = \hat{a}_{12}\hat{b}_{11}x_{1_1} + \tilde{a}_{12}\hat{b}_{21}x_{1_2} + \bar{a}_{12}\hat{b}_{31}x_{1_3} + \hat{a}_{22}\tilde{b}_{11}x_{2_1} + \tilde{a}_{22}\tilde{b}_{21}x_{2_2} + \bar{a}_{22}\tilde{b}_{31}x_{2_3}$$
$$y_{2_2} = \hat{a}_{12}\hat{b}_{12}x_{1_1} + \tilde{a}_{12}\hat{b}_{22}x_{1_2} + \bar{a}_{12}\hat{b}_{32}x_{1_3} + \hat{a}_{22}\tilde{b}_{12}x_{2_1} + \tilde{a}_{22}\tilde{b}_{22}x_{2_2} + \bar{a}_{22}\tilde{b}_{32}x_{2_3}$$
$$y_{2_3} = \hat{a}_{12}\hat{b}_{13}x_{1_1} + \tilde{a}_{12}\hat{b}_{23}x_{1_2} + \bar{a}_{12}\hat{b}_{33}x_{1_3} + \hat{a}_{22}\tilde{b}_{13}x_{2_1} + \tilde{a}_{22}\tilde{b}_{23}x_{2_2} + \bar{a}_{22}\tilde{b}_{33}x_{2_3}$$

and some small savings may be effected by storing intermediate results. Observe, for example, that the product $\hat{a}_{11}x_{1_1}$ is used in each of the first three equations. The same is true of $\tilde{a}_{11}x_{1_2}$, $\bar{a}_{11}x_{1_3}$, $\hat{a}_{21}x_{2_1}$, etc. If these intermediate values are stored, they may be used in later calculations. In general, the number of multiplications needed to form all the products that involve the elements of A and x is $n_1^2 n_2$. A further $n_1^2 n_2^2$ is needed to complete the calculation. The total is therefore $n_1^2 n_2^2 + n_1^2 n_2$. However, when we compare this with $n_1^2 n_2^2$, the number of multiplications needed if the global generator is expanded, we see that the only advantage of the generalized tensor product approach lies in its sparse storage mechanism. In an iterative method this storage advantage will be paid for by an additional $n_1^2 n_2$ multiplications per iteration. When more than two automata are involved, the greater savings in space are paid for by an even greater increase in multiplications per iteration.

The above discussion leads us to conclude that the generalized tensor product approach will seldom be satisfactory in the presence of many synchronizing events and state-dependent transition rates. On the other hand, writing generalized tensor products in terms of ordinary tensor products may considerably increase the number of terms in the summation of the SAN descriptor. Under these conditions it is perhaps better to use two completely different means of storing the underlying infinitesimal generator. Let Q, the infinitesimal generator of the SAN, be written as

$$Q = Q_{nf} + Q_f$$

where Q_{nf} contains all nonfunctional transitions (both local and synchronizing) and is stored in the usual highly efficient form of a classical tensor product, and Q_f contains only functional rates and is stored in the usual compact storage scheme for sparse matrices. Matrix products xQ may then be carried out in two independent operations xQ_{nf} and xQ_f and the results simply added to compute the final result. When the system being modelled consists of many almost-independent components, the number of nonzero functional elements will be small, and the size of the arrays needed to store Q_f will not be large. In this case the SAN modelling approach can be very powerful.

As a final comment, we note that considerable research still remains to be carried out into efficient ways of computing the product $x \bigotimes_{i=1}^{N} Q^{(i)}$ when the $Q^{(i)}$ are sparse, as well as into ways of determining an optimal (or nearly optimal) ordering of the automata as a function of both sparsity patterns and the presence of functional transition rates.

9.6 Iterative Solution Methods

Consider a stochastic automata network consisting of N automata. Let Q be its descriptor, i.e.,

$$Q = \sum_{j=1}^{T} \bigotimes_{i=1}^{N} Q_j^{(i)}.$$

Our goal is to find a vector π such that $\pi Q = 0$ and $\pi e = 1$. Furthermore, we wish to use iterative methods that do not require the explicit construction of the matrix Q but rather can work with the descriptor in its compact form.

9.6.1 Unpreconditioned Methods

In Section 9.5 we saw how to compute the product of an arbitrary vector with the compact descriptor of a SAN. This means that numerical iterative Markov chain methods whose only interaction with the coefficient matrix is to compute a vector-matrix product may be used to compute the stationary probability vector of a SAN. Thus, possible candidates are the power method, simultaneous iteration, the method of Arnoldi, and GMRES. Jacobi, Gauss–Seidel, and SOR cannot easily be used, because they do not work with the transition matrix directly. Neither can the iterative aggregation/disaggregation methods be used, since these methods work with submatrices of the transition matrix which are not easily extracted from the SAN descriptor.

The four methods suggested have all been examined in previous chapters. The power method and simultaneous iteration both require a discretized Markov chain. This we may obtain from Q by writing

$$P = I + \Delta t Q \tag{9.14}$$

where $\Delta t \leq 1/\max_i |q_{ii}|$ and may be computed without too much difficulty. Notice that P may be written as a sum of tensor products, since

$$I + \Delta t Q = \bigotimes_{i=1}^{N} I_{n_i} + \sum_{j=1}^{T} \Delta t \bigotimes_{i=1}^{N} Q_j^{(i)}.$$

Thus, for example, the power method becomes

$$\pi^{(l+1)} = \pi^{(l)}(I + \Delta t Q) = \pi^{(l)} + \Delta t \pi^{(l)} \left(\sum_{j=1}^{T} \bigotimes_{i=1}^{N} Q_j^{(i)} \right), \qquad l = 0, 1, 2, \ldots.$$

Similarly, it suffices to replace P for the coefficient matrix in the description of simultaneous iteration and Q for the coefficient matrix in the descriptions of the method of Arnoldi and GMRES.

9.6.2   Preconditioning

The objective of preconditioning is to modify the eigenvalue distribution of the iteration matrix so that convergence onto the solution vector may be attained more quickly. In a Markov chain context, with n states and stochastic transition probability matrix denoted by P, this can imply finding a matrix M^{-1} so that $I - (I - P)M^{-1}$ possesses one unit eigenvalue and $n - 1$ others that are close to zero (see Chapter 3, Section 3.4). As an example, a preconditioned power method can be written as

$$\pi^{(l+1)} = \pi^{(l)}(I - (I - P)M^{-1}) = \pi^{(l)} - \pi^{(l)}(I - P)M^{-1}. \qquad (9.15)$$

In a certain sense, M^{-1} may be regarded as an approximate inverse of $(I - P)$. When the Markov chain is ergodic, $(I - P)$ has a one-dimensional null space, so the product $(I - P)M^{-1}$ will yield a matrix that is close to

$$\begin{pmatrix} 1 & & & & \\ & 1 & & & \\ & & \ddots & & \\ & & & 1 & \\ & & & & 0 \end{pmatrix}$$

rather than one that is close to the identity matrix. Thus, $I - (I - P)M^{-1}$ will be close to

$$\begin{pmatrix} 0 & & & & \\ & 0 & & & \\ & & \ddots & & \\ & & & 0 & \\ & & & & 1 \end{pmatrix},$$

and now, since $|\lambda_2| \approx 0$, convergence will be rapid.

The problem is one of finding a suitable matrix M^{-1}. Incomplete LU factorizations have been successfully used in Markov chain contexts and have been discussed in previous chapters. These methods essentially follow a Gaussian elimination procedure but drop off elements at various points according to some specific criteria. The result is a lower triangular matrix $\tilde{L}$, an upper triangular matrix $\tilde{U}$, and a matrix $\tilde{E}$ (preferably small) such that $(I - P) = \tilde{L}\tilde{U} + \tilde{E}$. The preconditioner is taken to be $\tilde{U}^{-1}\tilde{L}^{-1}$. Unfortunately, in the context of stochastic automata networks, where the infinitesimal generator is in the form of a sum of tensor products, this approach is not possible. We need to adopt a different strategy.

From equation (9.14), we have $I - P = -\Delta t Q$. Furthermore, it is shown in [97] that

$$(I - P)^{\#} = \sum_{k=0}^{\infty} (P^k - L) \tag{9.16}$$

where the superscript $\#$ denotes the generalized inverse, and L is an $n \times n$ matrix whose rows are all equal to the stationary probability vector π. Hence, an approximation to the (generalized) inverse of $I - P$ may be obtained by summing the right-hand side of equation (9.16) over K terms and by substituting into each row of L the most recent approximation available for π, i.e., we take

$$M^{-1} = \sum_{k=0}^{K} (P^k - \tilde{L}),$$

where $\tilde{L}_{jk} = \pi_k^{(l)}$ for all j. In this case, the preconditioned power method takes the form

$$
\begin{aligned}
\pi^{(l+1)} &= \pi^{(l)} + \Delta t \pi^{(l)} Q \left(\sum_{k=0}^{K} P^k - \tilde{L} \right) \\
&= \pi^{(l)} + \Delta t \pi^{(l)} Q \left(\sum_{k=0}^{K} P^k \right) - \Delta t \pi^{(l)} Q \tilde{L}. \tag{9.17}
\end{aligned}
$$

However, in the last term $\pi^{(l)} Q \tilde{L}$ is zero, since for each element k we have

$$
\begin{aligned}
\left(\pi^{(l)} Q \tilde{L} \right)_k &= \sum_{i=1}^{n} \sum_{j=1}^{n} \pi_i^{(l)} Q_{ij} \tilde{L}_{jk} \\
&= \tilde{L}_{jk} \sum_{i=1}^{n} \pi_i^{(l)} \sum_{j=1}^{n} Q_{ij} \\
&= 0.
\end{aligned}
$$

We are able to take $\tilde{L}_{jk}$ outside the summation over j because $\tilde{L}_{jk} = \pi_k^{(l)}$ for all j; also, since Q is an infinitesimal generator, the sum of elements in any row is zero, and this then forces the last term of equation (9.17) to zero. Thus, with this means of preconditioning, the preconditioned power method is given by

$$\pi^{(l+1)} = \pi^{(l)} + \Delta t \pi^{(l)} Q \left(\sum_{k=0}^{K} P^k \right). \tag{9.18}$$

This same preconditioning is obtained if we approximate the inverse of $I - P$ directly by a polynomial series. It is known that for any matrix A for which

$\|A\| \leq 1$, the inverse of $I - A$ may be written in a Neumann series ([60]) as

$$(I - A)^{-1} = \sum_{k=0}^{\infty} A^k.$$

Since P is a stochastic matrix, $\|P\| \leq 1$, so an approximate inverse of $-\Delta t Q = I - P$ is given by

$$M^{-1} = \sum_{k=0}^{K} P^k.$$

When this is substituted into equation (9.15), it yields equation (9.18) directly. Since Q and P may be expressed as a sum of tensor products, all numerical operations may be carried out without having to expand the descriptor of the SAN. When written out in full, the preconditioned power method is given by

$$\pi^{(l+1)} = \pi^{(l)} + \Delta t \pi^{(l)} \left(\sum_{j=1}^{T} \bigotimes_{i=1}^{N} Q_j^{(i)} \right) \left[\sum_{k=0}^{K} \left(\bigotimes_{i=1}^{N} I_{n_i} + \sum_{j=1}^{T} \Delta t \bigotimes_{i=1}^{N} Q_j^{(i)} \right)^k \right]. \quad (9.19)$$

Similar expressions may be written for preconditioned Arnoldi and GMRES.

The advantage of this means of preconditioning is that, by increasing the number of terms in the Neumann expansion, we can obtain an accurate preconditioner and hence expect to converge in relatively few iterations. Unfortunately, the downside is that, as is apparent from equation (9.19), it is computationally very expensive to calculate an accurate preconditioner. This is illustrated by the example that follows.

9.7   A Queueing Network Example

In this section our purpose is twofold. First, we would like to use a queueing model to demonstrate the effect of incorporating both synchronizing events and functional transitions and to show how the structure of the SAN descriptor (i.e., as a sum of tensor products) may be maintained. Indeed, the particularity of queueing networks is that each transition of a customer from one station to another is a synchronizing event. The second purpose is to examine the effectiveness of the preconditioner suggested in the previous section.

As an example, let us consider an open queueing network consisting of two exponential, finite-capacity, single-server stations arranged in tandem. The (external) arrival process to station 1 is Poisson with rate λ, and the service rates at the exponential servers are μ and ν, respectively. On exiting from station 1, customers proceed to station 2 with probability p or exit from the queueing system with probability $1 - p$. Customers that leave station 2 exit from the system.

Station 1 can hold at most C_1 customers, while the maximum at station 2 is C_2. Customers who arrive at a full station are lost rather than blocked. This gives a two-dimensional Markov chain whose states may be written as (η_1, η_2), where η_1 denotes the number of customers in the first station and η_2 the number in the second. For the special case of $C_1 = 1$ and $C_2 = 2$, it is easy to verify that the total number of states is 6 and that the infinitesimal generator is given by

$$
Q = \begin{pmatrix}
 & (0,0) & (0,1) & (0,2) & (1,0) & (1,1) & (1,2) \\
 & * & 0 & 0 & \lambda & 0 & 0 \\
 & \nu & * & 0 & 0 & \lambda & 0 \\
 & 0 & \nu & * & 0 & 0 & \lambda \\
\hline
 & \mu(1-p) & \mu p & 0 & * & 0 & 0 \\
 & 0 & \mu(1-p) & \mu p & \nu & * & 0 \\
 & 0 & 0 & \mu(1-p) & 0 & \nu & *
\end{pmatrix}
$$

where in each row the diagonal element is denoted by $*$ and is equal to the negated sum of the off-diagonal elements in that row. For the moment this generator contains no functional transition rates.

Two stochastic automata may be used to model this system: one for each of the two service centers. The transitions λ and $\mu(1 - p)$ are local to the first automaton, while ν is local to the second. There is only one synchronized event, which occurs when a customer leaves the first station and enters the second. This transition, at rate μp, has the effect of moving the first automaton from state η_1 to $\eta_1 - 1$ and at the same time forcing the second automaton from state η_2 to $\eta_2 + 1$. Note that this transition occurs only when the second station is not full, for otherwise the transiting customer is lost.

The generators containing only local transitions for the two automata are therefore given by

$$
Q^{(1)} = \begin{pmatrix} -\lambda & \lambda \\ \mu(1-p) & -\mu(1-p) \end{pmatrix} \quad \text{and} \quad Q^{(2)} = \begin{pmatrix} 0 & 0 & 0 \\ \mu & -\mu & 0 \\ 0 & \mu & -\mu \end{pmatrix},
$$

and the (6×6) matrix formed from $Q^{(1)} \otimes I_3 + I_2 \otimes Q^{(2)}$ is

$$
\begin{pmatrix}
* & 0 & 0 & \lambda & 0 & 0 \\
\nu & * & 0 & 0 & \lambda & 0 \\
0 & \nu & * & 0 & 0 & \lambda \\
\hline
\mu(1-p) & 0 & 0 & * & 0 & 0 \\
0 & \mu(1-p) & 0 & \nu & * & 0 \\
0 & 0 & \mu(1-p) & 0 & \nu & *
\end{pmatrix},
$$

i.e., the infinitesimal generator Q without the synchronizing transitions μp. The synchronizing transitions are included by adding to this matrix the matrix

$$
\begin{pmatrix} 0 & 0 \\ 1 & 0 \end{pmatrix} \otimes \begin{pmatrix} 0 & \mu p & 0 \\ 0 & 0 & \mu p \\ 0 & 0 & 0 \end{pmatrix} + \begin{pmatrix} 0 & 0 \\ 0 & 1 \end{pmatrix} \otimes \begin{pmatrix} -\mu p & 0 & 0 \\ 0 & -\mu p & 0 \\ 0 & 0 & 0 \end{pmatrix}. \qquad (9.20)
$$

These matrices may be extended in a straightforward way to take care of the case of larger values of C_1 and C_2.

Let us turn now to the case in which we wish to include functional transition rates. Suppose, for example, that the service rate at station 1 is a function of the number of customers at the second station. Let μ_0, μ_1, and μ_2 denote the service rates at station 1 when the second station contains 0, 1, and 2 customers, respectively. This now means that not only are some of the local transitions in the first automaton functional, but the synchronizing transition is also a functional transition. The generators for the automata that contain only local, nonfunctional transition rates are now given by

$$
Q^{(1')} = \begin{pmatrix} -\lambda & \lambda \\ 0 & 0 \end{pmatrix} \quad \text{and} \quad Q^{(2')} = \begin{pmatrix} 0 & 0 & 0 \\ \mu & -\mu & 0 \\ 0 & \mu & -\mu \end{pmatrix}
$$

and as before, the part of the infinitesimal generator corresponding to such transitions is given by $Q^{(1')} \otimes I_3 + I_2 \otimes Q^{(2')}$. To this we need to add the local functional transitions and the synchronizing functional transitions. For each possible value of the functional rate in the local transitions, $\mu_i(1 - p)$, we must add a tensor product. Since there are three possible values, we need to add the following terms:

$$
\begin{pmatrix} 0 & 0 \\ 1 & -1 \end{pmatrix} \otimes \begin{pmatrix} \mu_0(1-p) & 0 & 0 \\ 0 & 0 & 0 \\ 0 & 0 & 0 \end{pmatrix}
$$

$$
\begin{pmatrix} 0 & 0 \\ 1 & -1 \end{pmatrix} \otimes \begin{pmatrix} 0 & 0 & 0 \\ 0 & \mu_1(1-p) & 0 \\ 0 & 0 & 0 \end{pmatrix}
$$

and

$$
\begin{pmatrix} 0 & 0 \\ 1 & -1 \end{pmatrix} \otimes \begin{pmatrix} 0 & 0 & 0 \\ 0 & 0 & 0 \\ 0 & 0 & \mu_2(1-p) \end{pmatrix}.
$$

Finally, to account for the functional synchronizing transition, the values of μ in equation (9.20) are functional, and each different possible value must be treated separately. Equation (9.20) is replaced by

$$
\begin{pmatrix} 0 & 0 \\ 1 & 0 \end{pmatrix} \otimes \begin{pmatrix} 0 & \mu_0 p & 0 \\ 0 & 0 & \mu_1 p \\ 0 & 0 & 0 \end{pmatrix} + \begin{pmatrix} 0 & 0 \\ 0 & 1 \end{pmatrix} \otimes \begin{pmatrix} -\mu_0 p & 0 & 0 \\ 0 & -\mu_1 p & 0 \\ 0 & 0 & 0 \end{pmatrix}.
$$

This, therefore, allows us to maintain the special structure of the descriptor in the presence of both functional and synchronizing events. However, the cost of this approach is also evident. For this particular case, the number of terms in the summation of the SAN descriptor rises from $N + 2E = 4$, in the nonfunctional case, to $T = 7$ when only one of the service rates becomes functional. Note that this exceeds the number of states in the model!

Let us now turn to our second objective: that of experimenting with preconditioning. The following numerical values were used:

$$\lambda = 13, \qquad \mu = 15, \qquad \nu = 11, \qquad p = 0.7, \qquad C_1 = 9, \qquad C_2 = 7.$$

These result in a model with a total of 80 states and $T = N + 2E = 4$, as the transition rates are not functional. The experiments were conducted on a multiuser Sun 4/690 with arithmetic coprocessor. The methods tested were the power method, the method of Arnoldi (for eigenvalues), and GMRES. In these last two methods the iterative variant with a subspace fixed at $m = 10$ was used. In all cases, the iterative procedure was halted at the first iteration for which the computed residual was less than 10^{-10}; i.e., given that the approximation at iteration k is $\pi^{(k)}$, the process is stopped once $\|\pi^{(k)}Q\| \leq 10^{-10}$. Although the size of the problem is such that methods that require an expanded infinitesimal generator can be used, and probably quite efficiently, we do not compare our results with these methods, because we are simply interested in comparing and contrasting methods that can work with the SAN descriptor in its compact form.

Table 9.1: Preconditioning Experiments with the Queueing Model; Iteration and Time Requirements

	Power		Arnoldi		GMRES	
K	Iterations	Time	Iterations	Time	Iterations	Time
0	748	16.62	130	3.73	120	3.51
1	375	16.26	80	4.25	75	4.10
2	250	37.23	45	8.17	45	8.16
3	188	127.44	41	33.95	41	33.80
4	151	534.90	31	124.96	32	133.16

Tests were conducted to observe the effectiveness of the preconditioner. Table 9.1 shows the number of iterations and the time (in seconds) needed for convergence when all three methods are applied to the test problem. The parameter K denotes the number of terms taken in the expansion of the preconditioner; see equation (9.19). The value $K = 0$ reflects the unpreconditioned methods.

As is to be expected, the number of iterations decreases the more accurately the preconditioner represents a (generalized) inverse of Q. However, the cost of determining this better preconditioner is evident in the total time needed by the algorithms. Indeed, these results clearly indicate that no real benefit is derived from this preconditioner. Much more research needs to be conducted into finding preconditioning techniques that can be applied to SAN descriptors *without* the need to expand the global generator. On the other hand, the results clearly demonstrate the success of the Arnoldi and GMRES methods over the power method, in terms of both number of iterations and total time.

Chapter 10

Software

10.1 The Categories of Available Software

10.1.1 Introduction

Our goals in this chapter are twofold. First, we wish to provide a global overview of the different types of software that the reader might find valuable. These range from isolated routines that solve systems of linear equations to complete packages that incorporate many different approaches for the solution of Markov models. In this vein we shall examine libraries of general numerical algorithms; software that is geared to the solution of queueing networks, stochastic Petri nets and stochastic automata networks; and packages designed for specific applications, such as the analysis of telecommunication systems. It is obviously impossible to give a detailed description of each of these possibilities or even to cite all the software that is available. We shall be content to mention briefly some of the better known.

Our second objective is to examine some packages in sufficient detail to allow the reader to become more aware of the considerations that enter into their development. We shall consider the software package MARCA, which contains a large number of matrix manipulation and numerical solution procedures, and XMARCA, a graphical interface for MARCA that has been specifically designed for the analysis of queueing networks. The convenience of the graphical interface comes at the expense of the great freedom that is available in MARCA to generate and solve any type of Markov chain and not just those that are derived from queueing networks. We have chosen these two packages, since they are the ones with which we are most familiar.

10.1.2   Libraries of Numerical Algorithms

Many sources provide numerical software that spans the entire mathematical spectrum. Some sources such as IMSL (International Mathematical and Statistical Library [75]) and NAG (Numerical Algorithms Group [108]) are maintained and provide a service at charge. Others are free for the taking, with the caveat "Anything free comes with no guarantee." Perhaps the best source for free and usually very reliable numerical software is *netlib*. Software is obtained by making a request to *netlib@research.att.com*, a gateway at AT&T Bell Labs in Murray Hill, NJ. To receive an index of the different libraries available from *netlib*, it suffices to send a message to the above address containing only the words *send index*. In short order, an index of all the libraries, as well as information on how to use *netlib* will be sent. Some of the libraries available in *netlib* that will be of particular interest to readers of this text include *Eispack, Linpack, Itpack*, and *Slap*.

Eispack is a suite of Fortran programs designed to compute the eigenvalues and eigenvectors of nine classes of matrices. It may be used to determine the eigensystems of complex general, complex Hermitian, real general, real symmetric, real symmetric band, real symmetric tridiagonal, special real tridiagonal, generalized real, and generalized real symmetric matrices. *Eispack* is an exceptionally reliable software library.

The *Linpack* software library is geared to the solution of linear equations and linear least squares problems. It may be applied to linear systems in which the coefficient matrices are general, banded, symmetric indefinite, symmetric positive-definite, triangular, and tridiagonal square. In addition, the package computes the QR and singular value decompositions of rectangular matrices and applies them to least squares problems. Like *Eispack*, *Linpack* is a very reliable software library.

As its name suggests, *Itpack* is a collection of iterative subroutines for solving linear systems of equations. It includes the methods of Jacobi, SOR, and SSOR, combined when possible with conjugate gradient or Chebyshev (semi-iterative) acceleration procedures. Major features include the automatic selection of the acceleration parameters and the use of accurate stopping criteria. Its authors state that although the package can be called with any linear system containing positive diagonal elements, the package is most successful in solving systems with symmetric positive-definite or mildly nonsymmetric coefficient matrices. *Itpack* is written in Fortran.

Slap (Sparse Linear Algebra Package) is a package for the solution of symmetric and nonsymmetric linear systems by iterative methods. Included in this package are core routines that perform iterative refinement, iteration on the normal equations, preconditioned conjugate gradient, preconditioned biconjugate gradient, preconditioned biconjugate gradient squared, Orthomin, and generalized minimum residual iterations.

Besides these well-known libraries of numerical software, *netlib* is also a source for all the software that has appeared in the ACM–TOMS (Transactions on Mathematical Software); the algorithms in the text *Computer Methods for Mathematical Computations* by Forsythe, Malcolm, and Moler, etc. Some special packages that may be of interest include

- *lanczos* — a few eigenvalues/eigenvectors of a large sparse symmetric matrix,

- *odepack* — a package for solving ordinary differential equations,

- *sparse* — the solution of large sparse systems of linear equations using LU factorization,

- *y12m* — another package for sparse linear systems.

Finally, *netlib* may be used to access the archives of the numerical analysis mailing group. This regular mailing is used as a forum for all who have an interest in scientific computing. As it reaches a wide audience, it is often used by newcomers to seek advice from "old hands."

10.1.3 Queueing Networks

Queueing networks have long been used to model a variety of systems. Indeed, the research of Erlang early in the 20th century was dedicated to the analysis of telephone systems and served to initiate an interest in queueing networks. In the 1960s Scherr [140] applied a simple queueing model to analyze a time-shared computer system and found it to give very satisfactory results, and a period of considerable research in queueing networks ensued.

Currently, several packages are available that are geared to the solution of queueing networks. They include XMARCA [82], QNAP2 [171], MGMtool [70], Panacea [124], and Q-lib [166], to name but a few. Most packages for solving queueing networks do not include a Markov chain solver. Rather, they consist of implementations that evaluate analytical expressions for approximate or exact solutions. In some cases they also include simulations. In this section, we consider only those that include a numerical Markov chain solver. We briefly consider QNAP2 and MGMtool, since they range from the very large to the rather limited. In Section 10.3, we shall consider the software package XMARCA in some detail.

10.1.3.1 QNAP2

QNAP2, (Queueing Network Analysis Package [171]) was developed by INRIA and Bull, France, under the guidance of Dominique Potier and Michel Veran and with the participation of many other researchers. It was built to provide a portable modelling environment with specific facilities for designing and solving queueing network models. An underlying criterion was that users be able to specify systems in high-level terms as a queueing network model but deal as little as possible with the intricacies of the techniques used to solve the models. This is achieved by providing a user interface based on a high-level object-oriented language with specific features for building and handling large models. QNAP2 is composed of

- A collection of solution methods including discrete event simulation, exact product form algorithms, approximation procedures, and a Markovian solver.

- A common user interface for model description, analysis control, and result presentation.

To support the modelling framework, QNAP2 offers a set of predefined primitives and mechanisms for defining routing rules, queueing disciplines, and server allocation policies. Although this set is limited to the most standard modelling mechanisms, it can be extended by user-defined mechanisms using the QNAP2 language. The three types of solvers are

1. Analytical solvers (convolution algorithms, mean value analysis, and approximate methods for product form networks);

2. Discrete event simulation with run length control facilities;

3. Markovian solver.

The solvers are accessed from the same language interface in order to achieve independence between the description of the model and its analysis. The technical intracies of the model are therefore transparent to the user. The activation of a solver is made in a specific section of a QNAP2 program that consists of a block of algorithmic statements in which the activation is specified as a procedure call.

The QNAP2 language is a high-level, object-oriented, interactive language. It includes two levels of specification:

1. A command language, consisting of a limited set of parametrized commands, which correspond to the main functions of QNAP2:

   - Declaration of the objects and variables manipulated in a model
   - Specification of the service stations
   - Analysis control and interactive dialog specification

2. An algorithmic language, derived from Pascal and Simula, which allows

   - Extension and definition of object types

- Specification of services involving complex mechanisms
- Activation of the solvers
- Definition of specific interfaces tailored to the user's needs

Of particular interest to us is the Markovian solver. This solver, which is based on an early version of MARCA, transforms a model described in the language of QNAP2 into a first-order Markovian process and then computes the steady-state solution of the process. The state descriptor takes the full configuration of each queue of the network into account: the class of the customer occupying the ith location of the queue and the current stage of service for this customer. The following operations are performed by the solver:

- Verification that assumptions are met and definition of state descriptor

- Generation of all the states of the model and their transition probabilities

- Computation of steady-state probabilities and derivation of performance measures

The states of the Markov chain are generated from an initial state. All possible transitions from this state are considered, analyzed and recorded. When this step is completed, the current state is marked, and one of the newly obtained states becomes the current state. The generation process terminates when all the states in the list have been marked and no new state is obtained. The numerical solution method used is the method of Arnoldi. No preconditioning is performed in the current version of QNAP2.

10.1.3.2   MGMtool

Unlike QNAP2, which is a very large package with many different approaches for solving queueing networks MGMtool [70] is geared specifically to the solution of $Ph/Ph/c$ queues. It allows a user to specify interarrival and service time distributions and converts these to a Markov chain whose stationary probability distribution possesses a matrix-geometric structure. The specifications are made by means of C language procedure calls. The package has facilities for most of the standard phase-type distributions, such as Erlang, hypoexponential, hyperexponential, and Coxian, as well as exponential. Additionally, a feature is included to allow more complex phase-type distributions to be defined.

The method used to compute Neuts's R matrix is the successive substitution approach described in Chapter 4. The rapidly converging approach of Latouche and Ramaswami (with its greater computational complexity and larger memory requirements per iteration) has not been implemented, since this approach was unknown when MGMtool was developed.

10.1.4   Stochastic Petri Nets

Along with queueing networks, stochastic Petri nets are often used to model computer and computer communication systems. In many respects these approaches complement each other. Petri nets are particularly appropriate for modelling synchronization phenomena, simultaneous resource possession and fork and join behavior, all of which are more difficult to incorporate into a queueing network model. On the other hand, Petri nets are less adept at handling queues, the "raison d'être" of queueing models. We do not intend to provide a comprehensive review of Petri nets, or even an extended overview as we did for queueing networks in Chapter 1. Rather we shall very briefly outline their major features and point the reader interested in obtaining more information to some of the excellent papers in the area, viz.: ([2], [24], [27], [73], [103], [104], [105], [172]).

10.1.4.1   Definitions

A *Petri net* is a directed bipartite graph with two different types of nodes, called *places* and *transitions*. Places are connected to transitions by means of *input arcs*, and transitions are connected to places by means of *output arcs*. A place is said to be an *input place* of a transition when there is an (input) arc connecting it to the transition. Similarly, a place is called an *output place* when there is an (output) arc that leads from the transition to the place. Each place may contain an integer number (possibly zero) of *tokens*, and the distribution of tokens among the places constitutes a *marking*. In our previous terminology, a marking is simply a *state of the Petri net*. A Petri net is defined by the quadruple $PN \equiv (P, T, A, M_0)$ where P is the set of places, T is the set of transitions, A is the set of input and output arcs, and M_0 is an initial marking.

A transition is said to be *enabled* if each of its input places contains at least one token. An *inhibitor arc* is an extension to the basic definition and adds Turing completeness to the Petri net. If an inhibitor arc connects a place to a transition, then the transition is enabled only if the place is empty. When a transition is enabled, it can *fire* and move the Petri net into a different marking. This new marking is obtained by removing one token from each of the input places of the transition that fired and adding one token to each of its output places. A marking is said to be *reachable* if, from the initial marking, there exists a sequence of transitions that can (eventually) fire and leave the Petri net in that marking. The set of all reachable markings is called the *reachability set* of the Petri net.

A *generalized stochastic Petri net (GSPN)* is a Petri net in which *firing times* are associated with the transitions. If these firing times are all exponentially distributed (called *timed transitions*), the Petri net is said to be a *stochastic Petri net*. When both *immediate transitions* and exponentially distributed firing times are permitted, it is said to be a GSPN. More recently, deterministic firing times have been introduced. The resulting Petri nets are called DSPNs [3] when both exponential and deterministic firing times are allowed, and DTPNs (deterministically timed Petri nets [73]) when the timings are all deterministic.

If an immediate transition is enabled in a marking, it will fire instantaneously. The marking is said to be *vanishing*, and the probability that the GSPN is in a vanishing marking is therefore zero. Since it is not permissible for several immediate transitions to fire at the same time, a *firing probability* must be associated each time more than one immediate firing is enabled in a marking. A nonvanishing marking is said to be *tangible*.

It is apparent that a GSPN implicitly defines a stochastic process that will become a continuous-time Markov chain once the vanishing markings are removed. One of the disadvantages of current implementations is that these vanishing markings are actually all generated and then subsequently removed. This is in contrast to the software package MARCA [157], which permits *instantaneous transitions* to occur but does not generate the intermediate vanishing states. When the CTMC has been obtained, numerical solution methods are usually applied to determine the steady-state probability of each marking. This is turn is used to compute the expected number (or distribution) of tokens in a given place or the throughput of a transition. We shall now consider three of the software packages that are currently available.

10.1.4.2   GreatSPN

GreatSPN (Graphical Editor and Analyzer for Timed and Stochastic Petri Nets) [25] was developed at the University of Torino, Italy, by Giovanni Chiola. It has gone through several revisions and is now at version 1.5. Is is perhaps the most widely distributed software package for stochastic Petri nets. A major advantage of this package is its graphical user interface, derived from Molloy's SPAN graphic tool [102], which allows a user to validate a model interactively. The graphical interface runs under SunView.

Version 1.4 contains solution procedures for both stationary and transient solutions of GSPNs and for stationary solutions of DSPNs. In the latest version, the solution procedure for DSPNs has been removed and a software module for interactive timed simulation added. The approach used to compute the stationary distribution of a GSPN is to apply the method of Gauss–Seidel to the embedded Markov chain, i.e., the Markov chain obtained by observing the CTMC at state departure instants. The stationary distribution of the CTMC is then recovered using the mean sojourn time in the markings as the rewards in a Markov reward model. Transient solutions at a given time are obtained from a truncated Taylor series expansion of the CTMC matrix exponential.

10.1.4.3   SPNP

SPNP (Stochastic Petri Net Package) [26] is a software package developed at Duke University by Kishor Trivedi and two of his students, Gianfranco Ciardo and

Jogesh Muppala. The package is written entirely in the C programming language, and data structures are managed dynamically. SPNP contains features that are particularly attractive to users whose areas of interest lie in reliability modelling. Thus, it can process very general reward specifications, and it incorporates an automatic sensitivity analysis of the model to its parameters. However, it does not provide a graphical user interface (one is in the works). Instead, users of SPNP have to specify their GSPN in terms of a C-based specification language.

SPNP computes both stationary and transient solutions of GSPNs. The successive overrelaxation method (SOR) is used to compute the stationary distribution, and randomization is used to find transient solutions. It does not handle DSPNs.

10.1.4.4 DSPNexpress

DSPNexpress [87] was developed at the University of Berlin, Germany, by Lindemann and some of his colleagues and arose from its authors' perceived lack of viable tools for deterministic SPNs. DSPNexpress has a graphical interface running under X11. The software components for solving the linear systems of equations are written in Fortran77; the remaining modules are all written in C.

Stationary distributions are computed by means of the SOR method provided in *Itpack2c*. If it is observed that convergence is not being achieved, DSPNexpress automatically switches to a sparse Gaussian elimination procedure. Transient solutions are obtained by means of the randomization method.

10.1.5 Other Software

A number of software packages have been developed to solve Markov chain problems with a specific application in mind. For example, the SAVE (System Availability Estimator) package [56], SHARPE (Symbolic Hierarchical Automated Reliability/Performance Evaluator [139], and SURE-2 Reliability Analysis Program [16] have all been developed for the analysis of reliability models; PEPS [121], although more widely applicable, was initially developed for the analysis of parallel systems, and MACOM [145] for the analysis of telecommunication systems. We shall briefly consider these last two.

10.1.5.1 PEPS

The software package PEPS [121] (Package for the Evaluation of Parallel Systems) is the work of Brigitte Plateau and her coresearchers, especially J.M. Fourneau and K. Atif. It is written in Pascal. As its name indicates, it was designed as an aid in the performance evaluation of parallel systems, but in fact it is applicable to a much larger class of model, since it is based on the concept of networks of stochastic automata. Stochastic automata networks (SANs) are frequently used to model systems of interacting modules, such as communicating processes, concurrent processors, shared memory, etc. The dynamic behavior of a given module is represented by a single automaton. This automaton has a certain number of states, and an action performed by or to the module is represented by a transition of the automaton from one state to another. Information concerning the interaction among the different modules, such as synchronization among processes or resource contention, must also be specified. The advantage of working with these SANs is that the global descriptor of the underlying Markov chain may be written in an extremely compact tensor form.

PEPS allows a user to describe the individual stochastic automata and their interactions and synchronizations, then to generate the compact tensor product form of the infinitesimal generator of the underlying Markov chain, and finally to compute its stationary probability vector. Thus, PEPS is composed of three modules: an editor, a compiler, and a solver. The goal of the editor is to build a compact descriptor of the generator of a multidimensional Markov chain; the compiler translates this descriptor into a form that is executable by the solver; and the solver computes the stationary distribution.

The editor is a text editor based on a hierarchy of menus and is logically divided into an object editor, a model editor, and a verification module. The object editor allows the manipulation of objects such as their creation, deletion, modification, and visualization. The model editor is used to specify the model type (discrete or continuous time), the performance measures desired, etc. Finally, the verification module checks that all names used have been defined and that functions and relations are not recursive.

The compiler takes the symbolic form of the matrix descriptor provided by the editor and translates it into a form usable by the solver. Thus, it replaces generalized tensor products by ordinary tensor products; tags special matrices such as the identity matrix to help reduce the number of numerical operations; etc. It also generates an initial vector for the numerical iterative procedure.

Currently, the solver executes the power method. However, recent experiments with more powerful iterative methods, such as the method of Arnoldi and GM-RES, have proven to be beneficial and will be incorporated into future versions [159]. A problem that remains is the choice of suitable preconditioners for these more advanced methods. One other note on the future evolution of PEPS is that it is currently being given a graphical interface.

10.1.5.2   MACOM

MACOM [145] is a software package designed specificially for the analysis of telecommunication models. It was developed in Beilner's research group at the University of Dortmund, Germany, with the support of the Deutsche Bundespost. Among its principal authors, we cite Buchholz, Krieger, Muller-Clostermann, Russman, Sczittnick, Stahl, Verwohlt, and Zaske. MACOM is queueing network-based and includes provision for features such as

- Limited capacity service centers;

- Simultaneous arrivals;

- State-dependent routing;

- Simultaneous resource possession;

- Phase-type distributions.

A graphical interface built upon the SunView window system is provided to facilitate the specification of telecommunication models. A model is built by selecting icons representing telecommunication features and moving them to an appropriate position on the screen. MACOM distinguishes between *machine* elements and *load* elements. Machines describe the resources of a physical system (both hardware and software), whereas the loads are created by tasks that move through the machine and require the allocation of resources to accomplish a specific objective. The machine part of the model is built from *service centers*. The parameters of a service center are its name, its scheduling discipline, number of processors, rate of service, and capacity. Load units are routed according to chains, and each unit belongs to exactly one chain. They enter the network through *sources* and exit through *sinks*, or *loss exits*. Completeness and (syntactical) correctness of the specification of the model are checked automatically.

MACOM computes both stationary and transient probabilities. The transition matrix for the Markov chain is stored in a compact form. When the matrix is small or when it possesses a small bandwidth, a direct solution method is used. In most cases, however, MACOM uses successive overrelaxation. Where appropriate, an aggregation/disaggregation step is included. The technique used by the package to compute transient solutions is based on the determination of distributional characteristics of *time to absorption* in an absorbing Markov chain.

10.2 MARCA — MARkov Chain Analyzer

10.2.1 Basic Concepts and Terminology

MARCA, MARkov Chain Analyzer, is a software package designed to facilitate the generation of large Markov chains and to compute both transient and stationary probability distributions. It is written entirely in Fortran.

10.2.1.1 State Space Representation and Transitions

The state of any Markov chain may be represented as an integer-valued row vector, and this is the means of representation adopted by MARCA. In MARCA, each component of this *state descriptor vector* is referred to as a *bucket*. Each bucket contains a number of *balls* that is equal to the numeric value of the corresponding vector component. At any moment, the state of the model is completely specified by the number of balls in each of the buckets. The evolution of the Markov chain is represented by the movement of balls among the buckets. In MARCA, it is not necessary for the total number of balls in the description of one state to be equal to that in another; balls may be created or destroyed according to the model characteristics.

The movement of a ball from one bucket to another is called a *transition* from a *source bucket* to a *destination bucket*, and the *rate of transition*, which may be a function of the state itself, is defined as the rate at which the source bucket loses balls to the destination bucket. In MARCA it is possible for the source bucket and the destination bucket to be the same bucket. A transition of particular interest is an *instantaneous transition*, which means, as its name implies, that the actual movement takes no time at all.

10.2.1.2 State Space Generation

As the states of the Markov chain are generated by MARCA, they are stored contiguously in a one-dimensional array called *LIST*. Simultaneously, the matrix of transition rates is constructed and stored in compact form.

The following is the procedure used to generate the state space. An initial feasible state, supplied by the user, is taken to be the *current state* and is examined to determine which states it can reach in a single-step transition. For each possible pair of source and destination buckets in this state, a subroutine (called *RATE* and written by the user) is used to determine the rate at which transitions to its destination state, *newstate,* occur. If the transition rate is zero or negative, the destination state is ignored. On the other hand, if the transition rate is strictly positive, a second user-written subroutine, called *INSTANT*, is invoked to see whether an instantaneous transition is made from *newstate*. If so, the result of the instantaneous transition becomes *newstate,* so that it is a single step from *currentstate.* The list of states is now searched to determine whether it already contains *newstate.* If *newstate* is not in the list of states, then it is added to the end of *LIST*, and a pointer, *LIMIT*, which is set equal to the number of states in the list, is incremented by 1. The rate of transition obtained by *RATE*, and its location in the transition matrix, are stored in one-dimensional arrays.

When all destination states that emanate from *currentstate* have been so treated, the diagonal element of the transition rate matrix is computed as the negated sum of all the rates from *currentstate* and is also stored along with its position. The next state in *LIST*, which is flagged by a variable *MARKER*, becomes *currentstate*, and the process is repeated. When the value of *MARKER* exceeds that of *LIMIT*, all the states of an ergodic Markov chain will have been generated.

10.2.1.3   Decomposability

After the states of the system and the matrix of transition rates have been generated, MARCA initiates a procedure to determine the degree of near-decomposability of the matrix. A parameter γ, called the *decomposability factor*, is varied from 10^{-10} to 10^{-1}. For each different value, MARCA zeros out of (a copy of) the stochastic transition probability matrix all elements that are less than this value. The resulting matrix is treated as a directed graph, and a search is conducted for its strongly connected components. The magnitude of the nonzero elements plays no role in this search. As the value of γ increases, the strongly connected component search is applied to each of the strongly connected components found at the previous level. Strongly connected components must necessarily be nested as γ increases. MARCA produces a table that, for each value of γ, displays the number of disjoint groups and their average size and standard deviation.

10.2.1.4   Numerical Solutions

After the states of the system and the matrix of transition rates have been generated, control is passed to one of several numerical solution procedures for the determination of the stationary probability vector. MARCA offers a variety of different types of algorithm for determining this vector, including direct methods, fixed-point iterative methods, projection methods, and iterative aggregation/disaggregation. A particular method must be chosen by the user.

After the stationary probability vector has been computed, MARCA determines the marginal probability distribution of balls in each of the buckets and the mean and standard deviation of these distributions. However, these may not be sufficient. A user may wish to examine or manipulate the stationary probabilities belonging to a very specific subset of states, as in Markov reward models. A facility included in MARCA allows such interaction by providing the user with access to the list of states, the stationary probability vector, the hashing functions used to find the position of a given state in the list of states, and both double-precision and integer work arrays.

To compute transient solutions, the user is requested to input the time at which the probability distribution is required, an initial state, and an accuracy parameter. A particular solution method must then be chosen. Methods available include randomization, matrix powering for small-dimensional problems, and ODE solvers. The transient distribution is computed at the time specified, and from it the marginal probabilities. The user may ask to be given the probability of finding the system in specific states at that time or for MARCA to compute the transient solution at a different point of time.

10.2.2   Model Definition in MARCA

In order to be able to supply MARCA with the information needed to generate and analyze a Markov chain model, a user must first have a clear idea of the model expressed in terms of *balls and buckets*. More precisely, a user must define a state descriptor vector and understand the ways in which the different components of this vector interact. Once this is accomplished, it is relatively easy to generate the data file and Fortran subroutines required by the software package.

The state descriptor vector and the interactions that are possible among the buckets of the descriptor are provided to MARCA by means of an input data file called *in*. This file provides such information as the number of buckets in the descriptor, a maximum value for each bucket, an initial distribution of balls across the buckets, and a list of transitions that can occur among the buckets. It may also be used to provide common data needed by the subroutines *RATE* and *INSTANT*.

When constructing the state space, MARCA considers each state in turn as a source state and determines all the destination states that can be reached in a single step from that source state. It does so by examining the individual buckets of the source state and determining whether the rules that define the model permit a transition to occur. These rules are the constraints specified in the input data file and in *RATE*.

When examining a source state, for each possible pair of buckets (l_1, l_2), MARCA determines whether it is possible to transfer a ball from l_1 to l_2. Bucket l_1 is a candidate *source bucket* and bucket l_2, a candidate *destination bucket*. MARCA first checks to see whether l_1 and l_2 satisfy the constraints in the input data file. These are satisfied if

- The contents of l_1 are strictly greater than zero.

- The contents of l_2 are strictly less than its maximum value.

- A transfer is permissible from l_1 to l_2.

The subroutine *RATE* is called for all pairs (l_1, l_2) that satisfy these conditions. It is supplied with these values as well as with the complete source state, and it is required to return the rate at which this transition occurs (it may be zero) and the destination state. Notice that the user is free to define the destination state completely and is not restricted to making changes only to the source and destination buckets.

The subroutine *INSTANT* is called each time *RATE* returns a destination state and a strictly positive transition rate. The purpose of *INSTANT* is to examine this destination state and determine whether an instantaneous transition occurs from it to any other state. Instantaneous transitions often occur as the result of certain scheduling policies imposed on the system and which the user wishes to consider as taking no time at all to implement. For example, a user may use *INSTANT* to model the case in which possession of a CPU is instantaneously taken from a low-priority process and given to a higher-priority one. In this case it is known from the state of the system (i.e., the low-priority process computing and the high-priority process waiting) that an instantaneous transition will occur. Furthermore, the state to which the instantaneous transition is destined may be easily determined. It suffices to put the high-priority task into service and return the low-priority task to the waiting state.

Consider, however, the case in which a CPU has just finished performing a task for a high-priority job and finds that not just one, but several jobs having the same priority are waiting for it. It is known in advance that an instantaneous transition will occur, for the CPU will be allocated to one of the waiting processes. However, it may not be known with certainty which particular process this will be. In this case, it is necessary for the subroutine *INSTANT* to return a list of all possible destination states and associate with each the probability that it will be the result of the instantaneous transition.

10.2.3   Numerical Methods

10.2.3.1   Stationary Distributions

The choice of a particular solution method for the computation of stationary distributions and of the numerical values of its parameters is left to the user. They may be entered following prompts by MARCA or read automatically from the input file. The methods which are currently available are

1. SOR (successive overrelaxation)
2. SSOR (symmetric SOR)
3. POWER (the standard power method)
4. ARNOLDI (the Arnoldi projection method)
5. GE (compact Gaussian elimination)
6. ILU-GMRES (GMRES with incomplete factorization)
7. ILU-ARNOLDI (Arnoldi with incomplete factorization)
8. SORGMR (GMRES with SOR acceleration)
9. SSORGMR (GMRES with SSOR acceleration)
10. FXPTIT (fixed-point iterations with preconditioning)
11. CGS (conjugate gradient squared)
12. NCDSOLVER (iterative agregation/disaggregation)

10.2.3.2   Transient Distributions

In MARCA all interactions concerning transient solutions are performed interactively. The user is requested to input a time t at which the probability distribution is to be found, an initial starting state for the chain (MARCA verifies that the state actually exists), and the precision that is required of the solution. The available solution methods include uniformization and, for small-dimensioned problems, the matrix-powering approach. Additionally, for users who have access to the NAG subroutine library, links have been created for Adams and Runge/Kutta methods.

When a particular method has been chosen, the transient solution is computed, and the user is asked to enter any states for which the probability at time t is sought. The user may also request transient solutions at different times and may choose a different solution method on each occasion.

10.3 XMARCA: A Graphical Interface for Queueing Networks

10.3.1 Introduction

XMARCA is an X Window System[1] interface to the MARCA software package discussed in the previous section. In its current version it provides a graphical interface for building queueing models that are subsequently solved by MARCA. It is envisaged that later versions will include interfaces that are specially designed for specific applications such as communication modelling, reliability modelling, etc.

XMARCA was written using the Motif Widget Set for the X Window System [182]. To compile XMARCA's source code, it is necessary to have X11 (revision 3 or higher) running on the target machine and to have the libraries and header files for both Motif and X11 [182]. XMARCA also requires a Fortran and a C compiler. At North Carolina State University, XMARCA runs on a Sun SPARCstation 2 with 16 MB of RAM.

It is obvious from the discussions in the previous section, that a user of MARCA has to prepare two Fortran subroutines, *RATE* and *INSTANT*, and to provide additional information such as the number of balls, buckets, and possible transitions. The role of XMARCA is to relieve the user from this task when the system being modelled can be represented as a queueing model. XMARCA allows a user to build queueing network models by using a mouse to click icons of queueing components, such as stations, queues, servers, and connectors, and to combine these in an arbitrary manner. Thus, a queueing model can be created and incorporated into MARCA without requiring the user to write any programs at all. The following is a partial list of queueing network features that can be handled by XMARCA.

---

[1] X Window System is a trademark of Massachusetts Institute of Technology.

- Exponential stations;
- Infinite (exponential) server stations;
- Hyperexponential stations;
- Erlang stations;
- Coxian stations;
- Finite and infinite queues;
- Blocking mechanisms:
  - Blocking before service
  - Blocking after service
  - Repetitive service;
- Priority-based service;
- Priority-based routing;
- Probabilistic routing.

XMARCA also provides a library feature whereby entire models may be saved and loaded as needed.

A user interacts with XMARCA through two windows that are automatically generated by the program. The first is called the XMARCA *Main Window* (Figure 10.1) and is used for building the queueing network model and initiating the conversion process, including all linking and loading with other solution modules. The second window is called the *Solve Window*, and it is here that the user instructs XMARCA to generate transition matrices, manipulate these matrices, compute stationary and transient distributions, and display results.

10.3.2 Building a Queueing Network Model

A user builds a queueing network model in XMARCA by using a mouse to place icons of queueing system components, such as queues and servers, onto a graphical display. The components are selected from a palette of queueing system components by clicking on the icon. This palette is called the Visual Menu, or Vmenu. After a component type is selected, the user can use the mouse to add instances of that component to the Drawing Area, which is a two-dimensional virtual sheet of paper on which the queueing network is designed. Each type of component has a specific attribute dialog box associated with it and is used for setting attributes such as the maximum length of finite queues, service rate functions, priorities, and phase characteristics.

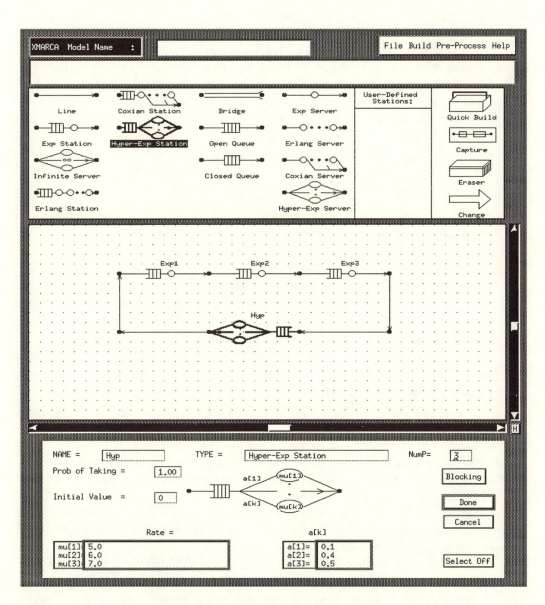

Figure 10.1: The XMARCA Main Window showing the Vmenu, a queueing model in the Drawing Area, and a hyperexponential dialog box.

10.3.2.1   The Vmenu

A queueing model can be created or modified only when the Vmenu is visible. It is vertically divided into three sections:

- Icons in the left portion of the Vmenu are queueing components that can be added to the drawing. Notice that there are four columns of icons. The first two columns consist of icons representing both a queue and a server (except for the "line" icon). These are called *predefined stations*. The last two contain queues and servers that can be combined to build more sophisicated queueing stations. Such stations are called *custom-built stations*. The centerpiece of a custom-built station is a "bridge" item that connnects queues to servers. A queue may be an input to only one bridge, and a server may be an output of only one bridge.

- The center section is used to hold the names of *user-defined stations (UDS's)*.

- Items in the right section are drawing functions. They include features for changing and erasing components that are already in the drawing area. A "QuickBuild" feature allows users to create customized stations automatically by specifying the numbers and types of queues and servers in a station. The user is also permitted to capture groups of components from the drawing and identify that group as a new type of component referenced by a single name. The groups of components, called user-defined stations, are stored for use at a later time.

10.3.2.2   Scheduling: Priority and Probabilistic

In a station that contains multiple queues or multiple servers, important scheduling decisions must be made. For example, when a server in a multiqueue station becomes free, a decision must be made regarding the choice of the queue from which the next customer is selected. Again, in a multiserver station, a newly arriving customer may find several servers free and must decide which one to choose. These scheduling decisions may be either deterministic or probabilistic. XMARCA allows for both possibilities. When the scheduling is deterministic, the user must assign a priority ordering to all the possibilities. This is simply an integer, and the selection made is the one that corresponds to the lowest integer among all *available* choices. When the scheduling is probabilistic, real numbers are initially assigned to all possibilities. Later, when a scheduling decision must be made from among a subset of *all* possibilities, the real numbers assigned to all *available* choices are normalized to sum to 1, and each available choice is assigned its appropriate probability. Scheduling decisions are made by the servers, the bridges and the queues.

- The dialog box of a server incorporates the scheduling mechanism for choosing a customer from among several competing, nonempty, queues. This implies that different servers may have different priorities or probabilities in choosing which queue to serve.

- On the other hand, the scheduling concerning the choice of a server from among several idle servers is made, not by the queue containing the customer that enters into service, but by the bridge item. This implicitly means that all the queues connected to a given bridge choose a server according to the same fixed priority or probabilistic scheduling.

- The decision concerning the choice of server was taken out of the hands of the queue (with the resulting possible inconvenience of all queues having the same policy), because queues need to include a scheduling policy under different circumstances. It sometimes happens that when a slot becomes available in a finite queue, that queue may need to select from among several possible sources that are blocked and waiting to enter. This is a scheduling decision that is left to the queue.

10.3.3   Converting the Graphical Representation

Before XMARCA can begin to generate the Markov chain corresponding to the graphical representation of a queueing network, a number of preprocessing steps must be undertaken. These involve creating rate and instant subroutines and linking them with the rest of the software package. The *rate subroutine* contains rules for determining the rates of transition among states, and the rate functions, defined in the dialog boxes of the model components, become part of this subroutine. The *instant subroutine* contains rules for handling instantaneous transitions such as the reloading and selection of servers. Unlike MARCA, these subroutines are generated automatically and in the C programming language rather than in Fortran. If the C compiler detects an error in one of the rate functions, it will display an error message to that effect and bring up the dialog box for the item whose rate function is incorrect.

After the rate and instant subroutines have been generated and compiled, they are linked to the rest of the program. If the linking/loading process is error-free, then a "Solve Window" is automatically brought up and partially overlies the XMARCA main window.

10.3.4   Matrix Generation and Numerical Solutions

The XMARCA Solve Window appears automatically once the linking process is completed. It is this window that controls the generation of the Markov chain and its transition matrix, the choice of a particular solution method, and the viewing of results. It is in this window that XMARCA

- Provides information concerning states and transitions;
- Draws the structure of the transition matrix;
- Determines nearly completely decomposable components;
- Determines the periodicity of the embedded chain;
- Permutes the matrix according to periodicity, if desired;
- Writes the compact form of matrix to file, if desired.

During the generation process, XMARCA continuously updates and displays the number of states generated to that point, and when generation is completed, the matrix's nonzero structure is displayed graphically. Figure 10.2 provides an example of the structure of a transition rate matrix as initially generated by XMARCA. The black areas show the position of the nonzero elements in the matrix.

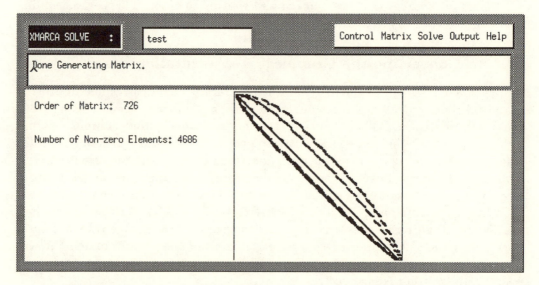

Figure 10.2: Graphical representation of a transition matrix.

After a matrix is permuted to normal periodic form, XMARCA may be used to observe the effect of this permutation. In a slightly modified version of the example in Figure 10.2, we find that the structure of the permuted matrix is as shown in Figure 10.3.

An XMARCA model is solved by calling subroutines from the original MARCA software package. The user enters solution parameters interactively and visually monitors the progress of the solution process. Methods for both stationary and transient distributions are available. These include

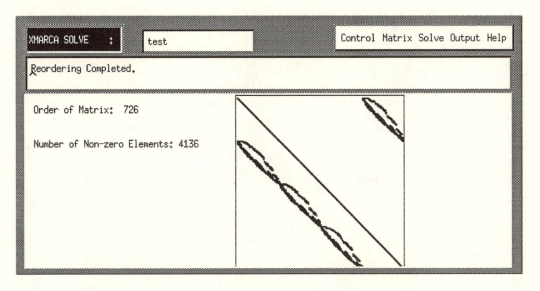

Figure 10.3: Transition matrix permuted to periodic form.

- Stationary solution methods:
  - Sparse Gaussian elimination
  - Single-vector iterations (SOR, SSOR, power, etc.)
  - Iterative aggregation/disaggregation
  - Projection methods (GMRES, Arnoldi, etc.)
  - Preconditioned variants
- Transient solution methods:
  - Randomization
  - ODE solvers (NAG's Runge-Kutta and Adams methods)
  - Matrix powering

When a particular method had been chosen, a dialog box for entering parameters needed by the method appears in the middle of the output window. Most of the methods require some solution parameters such as tolerance or maximum number of iterations. Default values are given in all cases. For direct methods, as the solution is being computed, the message area in the "Solve" window displays the row number currently being reduced. This gives the user an idea of the length of time that will be needed to find the solution. For iterative methods, the current iteration number and the current residual is displayed in the message area. When the model has been solved, a "Solution Completed" message is displayed, and information about the solution process appears in the output area.

10.3.5   Displaying Results

XMARCA has three ways of displaying results: a text version of marginal probability distributions; graphical plots of marginal probability distributions (Figure 10.4); and, to provide a mechanism for allowing Markov reward models to be handled, a relational database system for determining probabilities of the system being in a state or subset of states that satisfy constraints specified by the user.

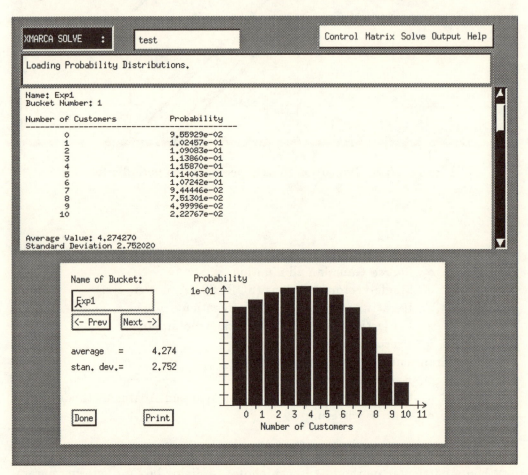

Figure 10.4: Marginals (text and graph) showing stationary distributions.

Bibliography

[1] A.V. Aho, J.E. Hopcroft, and J.D. Ullman. *The Design and Analysis of Computer Algorithms*. Addison-Wesley, Reading, MA, 1974.

[2] M. Ajmone-Marsan, G. Balbo, and G. Conte. A class of generalized stochastic Petri nets for the performance evaluation of multi-processor systems. *ACM Transactions on Computing Systems*, 2(2):93–122, 1984.

[3] M. Ajmone-Marsan and G. Chiola. On Petri nets with deterministic and exponentially distributed firing times. In G. Rozenberg, ed. *Advances in Petri Nets — Lecture Notes in Computer Science*, vol. 266, pp. 132–145. Springer-Verlag, Berlin, Heidelberg, New York, 1987.

[4] W.E. Arnoldi. The principle of minimized iteration in the solution of the matrix eigenvalue problem. *Quarterly for Applied Mathematics*, 9:17–29, 1951.

[5] K. Atif. *Modélisation du Parallèlisme et de la Synchronisation*. Ph.D. thesis, L'Institut National Polytechnique de Grenoble, France, September 1992.

[6] K. Atif and B. Plateau. Stochastic automata networks for modelling parallel systems. *IEEE Transactions on Software Engineering*, 17(10):1093–1108, 1991.

[7] O. Axelsson and V.A. Barker. *Finite Element Solution of Boundary Value Problems*. Academic Press, Orlando, FL, 1984.

[8] G.A. Baker. *Essentials of Padé Approximants*. Academic Press, New York, 1975.

[9] F.L. Bauer. Das Verfahren der Treppeniteration und Verwandte Verfahren zur lösung algebraischer Eigenwertprobleme. *Zeitschrift für Angewandte Mathematik und Physik*, 8:214–235, 1957.

[10] A. Berman and R.J. Plemmons. *Nonnegative Matrices in the Mathematical Sciences*. Academic Press, New York, 1979.

[11] A. Bobbio and K.S. Trivedi. An aggregation technique for the transient analysis of stiff Markov chains. *IEEE Transactions on Computers*, C-35(9): 803–814, 1986.

[12] F. Bonhoure. *Solution numérique de chaines de Markov particulieres pour l'étude de systemes à évenements discrets.* Ph.D. thesis, Université Pierre et Marie Curie, Paris, 1990.

[13] F. Bonhoure, Y. Dallery, and W.J. Stewart. Algorithms for periodic Markov chains. In *IMA Volumes in Mathematics and Its Applications*, vol. 48: *Linear Algebra, Markov Chains and Queueing Models*, pp. 71–88. Springer-Verlag, Berlin, Heidelberg, New York, 1993.

[14] F. Bonhoure, Y. Dallery, and W.J. Stewart. On the use of periodicity properties for the efficient numerical solution of certain Markov chains. *Journal of Numerical Linear Algebra with Applications*, 1994. To appear.

[15] P.N. Brown, G.D. Byrne, and A.C. Hindmarsh. VODE: A variable coefficient ODE solver. *SIAM Journal of Scientific and Statistical Computing*, 10:1038–1051, 1989.

[16] R.W. Butler. *The SURE-2 Reliability Analysis Program.* Technical report, NASA Langley Research Center, January 1985.

[17] S.L. Campbell and C.D. Meyer. *Generalized Inverses of Linear Transformations.* Pitman Publishing Ltd., London, 1979.

[18] W.-L. Cao and W.J. Stewart. Iterative aggregation/disaggregation techniques for nearly-uncoupled Markov chains. *Journal of the ACM*, 32(3):702–719, 1985.

[19] J.L. Carroll, A. Van der Liefvoort, and L. Lipsky. Solutions of $M/G/1//N$-type loops with extensions to $M/G/1$ and $GI/M/1$ queues. *Operations Research*, 30:490–514, 1982.

[20] K.M. Chandy. The analysis and solution of general queueing networks. In *Proceedings of the Sixth Princeton Conference on Information Sciences and Systems*, pp. 224–228. Princeton, NJ, 1972.

[21] K.M. Chandy, U. Herzog, and L. Woo. Approximate analysis of queueing networks. *IBM Journal of Research and Development*, 19(1):43–49, 1975.

[22] F. Chatelin and W.L. Miranker. Acceleration by aggregation successive approximation methods. *Linear Algebra and Its Applications*, 43:17–47, 1982.

[23] Y.T. Chen. *Iterative Methods for Least Squares Problems.* Ph.D. thesis, University of Waterloo, Ontario, Canada, 1975.

[24] G. Chiola. A graphical Petri net tool for performance analysis. In *Proceedings of the AFCET International Workshop on Modelling Techniques and Performance Evaluation*, Paris, March 1987.

[25] G. Chiola. GreatSPN 1.5, Software Architecture. In *Proceedings of the Fifth International Conference on Modelling Techniques and Tools for Performance Analysis*, pp. 117–132. Torino, Italy, 1991.

[26] G. Ciardo, J. Muppala, and K.S. Trivedi. SPNP — Stochastic Petri Net Package. In *Proceedings of the Third International Workshop on Petri Nets and Performance Models*, pp. 142–151. Kyoto, Japan, 1989.

[27] G. Ciardo and K.S. Trivedi. Solution of large GSPN models. In William J. Stewart, ed. *Numerical Solution of Markov Chains*, pp. 565–595. Marcel Dekker Inc., New York, 1991.

[28] P.J. Courtois. *Decomposability: Queueing and Computer System Applications*. Academic Press, New York, 1977.

[29] P.J. Courtois and P. Semal. Block decomposition and iteration in stochastic matrices. *Philips Journal of Research*, 39:178–194, 1984.

[30] P.J. Courtois and P. Semal. Bounds for the positive eigenvectors of nonnegative matrices and their approximation by decomposition. *Journal of the ACM*, 31:804–825, 1984.

[31] P.J. Courtois and P. Semal. Block iterative algorithms for stochastic matrices. *Linear Algebra and Its Applications*, 70:59–76, 1986.

[32] P.J. Courtois and P. Semal. Computable bounds for conditional steady-state probabilities in large Markov chains and queueing models. *IEEE Journal on Selected Areas in Communications*, SAC-4(6), 1986.

[33] D.R. Cox and H.D. Miller. *The Theory of Stochastic Processes*. John Wiley, New York, 1965.

[34] J.N. Daigle and D.M. Lucantoni. Queueing systems having phase-dependent arrival and service rates. In William J. Stewart, ed. *Numerical Solution of Markov Chains*, pp. 161–202. Marcel Dekker Inc., New York, 1991.

[35] M. Davio. Kronecker products and shuffle algebra. *IEEE Transactions on Computers*, C-30(2), 1981.

[36] T. Dayar and W.J. Stewart. Using the GTH procedure in iterative aggregation/dis-aggregation methods. In *Colorado Mountain Conference on Iterative Methods*, Breckenridge, Colorado, April 1994.

[37] H. Van der Vorst. An iterative solution method for solving $f(A) = b$ using Krylov subspace information obtained for the symmetric positive definite matrix, A. *Journal of Computing and Applied Mathematics*, 18:249–263, 1987.

[38] M. Eiermann, W. Niethammer, and A. Ruttan. Optimal successive overrelaxation iterative methods for p-cyclic matrices. *Numerische Mathematica*, 57(6/7):593–606, 1990.

[39] H.C. Elman. *Iterative Methods for Large Sparse Nonsymmetric Systems of Linear Equations*. Ph.D. thesis, Yale University, New Haven, CT, 1982.

[40] A.I. Elwalid, D. Mitra, and T.E. Stern. A theory of statistical multiplexing of Markovian sources: spectral expansions and algorithms. In William J. Stewart, ed. *Numerical Solution of Markov Chains*, pp. 223–238. Marcel Dekker Inc., New York, 1991.

[41] R.V. Evans. Geometric distribution in some two-dimensional queueing systems. *Operations Research*, 15:830–846, 1967.

[42] B.N. Feinberg and S.S. Chiu. A method to calculate steady state distributions of large Markov chains. *Operations Research*, 35(2):282–290, 1987.

[43] R. Fletcher. Conjugate gradient methods for indefinite systems. In *Lecture Notes in Mathematics 506*, pp. 73–89. Springer-Verlag, Berlin, Heidelberg, New York, 1976.

[44] J.G.F. Francis. The QR transformation: A unitary analogue to the LR transformation. Parts 1 and 2. *Computing Journal*, 4:265–271 and 332–345, 1961–1962.

[45] R.W. Freund, M.H. Gutknecht, and N.M. Nachtigal. *An implementation of the look-ahead Lanczos algorithm for non-Hermitian matrices*. Technical report, ETH-Zentrum, Swiss Federal Institute of Technology, Zurich, 1991.

[46] R.A. Friesner, L.S. Tuckerman, B.C. Dornblaser, and T.V. Russo. A method for exponential propagation of large systems of stiff nonlinear differential equations. *Journal of Scientific Computing*, 4:327–354, 1989.

[47] R.E. Funderlic and C.D. Meyer. Sensitivity of the stationary distribution vector for an ergodic Markov chain. *Linear Algebra and Its Applications*, 76:1–17, 1986.

[48] R.E. Funderlic, M. Neumann, and R.J. Plemmons. LU decompositions of generalized diagonally dominant matrices. *Numerische Mathematica*, 40:57–69, 1982.

[49] E. Gallopoulos and Y. Saad. Efficient solution of parabolic equations by Krylov approximation methods. *SIAM Journal of Scientific and Statistical Computing*, 13(5):1236–1264, 1992.

[50] C.W. Gear. *Numerical Initial Value Problems in Ordinary Differential Equations*. Prentice Hall, Englewood Cliffs, NJ, 1971.

[51] E. Gelenbe. On approximate computer system models. *Journal of the ACM*, 22(2):261–269, 1975.

[52] S.B. Gershwin and I.C. Schick. Modelling and analysis of three-stage transfer lines with unreliable machines and finite buffers. *Operations Research*, 31(2):354–380, 1983.

[53] G.H. Golub and J. de Pillis. Toward an effective two-parameter SOR method, 1986. Preprint.

[54] G.H. Golub and C.F. Van Loan. *Matrix Computations*, The Johns Hopkins University Press, Baltimore, 1989.

[55] G.H. Golub and C.D. Meyer. Using the QR factorization and group inversion to compute, differentiate and estimate the sensitivity of stationary probabilities for Markov chains. *SIAM Journal of Algebraic and Discrete Mathematics*, 7(2):273–281, 1986.

[56] A. Goyal, W.C. Carter, E. de Souze e Silva, S.S. Lavenberg, and K.S. Trivedi. The system availability estimator. In *Proceedings of the 16th International Symposium on Fault-Tolerant Computing*, pp. 84–89. Vienna, Austria, 1986.

[57] W. Grassmann. Transient solutions in Markovian queueing systems. *Computers and Operations Research*, 4:47–66, 1977.

[58] W. Grassmann. Transient solutions in Markovian queues. *European Journal of Operations Research*, 1:396–402, 1977.

[59] W.K. Grassmann, M.I. Taskar, and D.P. Heyman. Regenerative analysis and steady state distributions for Markov chains. *Operations Research*, 33(5):1107–1116, 1985.

[60] A. Greenbaum, P.F. Dubois, and G.H. Rodrique. Approximating the inverse of a matrix for use in iterative algorithms on vector processors. *Computing*, 22:257–268, 1979.

[61] D. Gross and D.R. Miller. The randomization technique as a modelling tool and solution procedure for transient Markov processes. *Operations Research*, 32:343–361, 1984.

[62] K. Gustafsson. Control theoretic techniques for stepsize selection in explicit Runge-Kutta methods. *ACM Transactions on Mathematical Software*, 17(4):533–554, 1991.

[63] M. Gutknecht, W. Niethammer, and R.S. Varga. K-step iterative methods for solving nonlinear systems of equations. *Numerische Mathematica*, 48:699–712, 1986.

[64] A. Hadjidimos. On the optimization of the classical iterative schemes for the solution of complex singular linear systems. *SIAM Journal of Algebraic and Discrete Mathematics*, 6(4):555–566, 1985.

[65] A. Hadjidimos and R.J. Plemmons. Analysis of p-cyclic iterations for Markov chains. In *IMA Volumes in Mathematics and Its Applications*, vol. 48: *Linear Algebra, Markov Chains and Queueing Models*, pp. 111–124. Springer-Verlag, Berlin, Heidelberg, New York, 1993.

[66] A.L. Hageman and D.M. Young. *Applied Iterative Methods*. Academic Press, New York, 1981.

[67] E. Hairer, S.P. Norsett, and G. Wanner. *Solving Ordinary Differential Equations I: Nonstiff Problems*. Series in Computational Mathematics. Springer-Verlag, Berlin, Heidelberg, New York, 1987.

[68] E. Hairer and G. Wanner. *Solving Ordinary Differential Equations II: Stiff and Differential-Algebraic Problems*. Series in Computational Mathematics. Springer-Verlag, Berlin, Heidelberg, New York, 1991.

[69] W.J. Harrod and R.J. Plemmons. Comparison of some direct methods for computing stationary distributions of Markov chains. *SIAM Journal of Scientific and Statistical Computing*, 5(2):453–469, 1984.

[70] B.R. Haverkort, A.P.A. van Moorsel, and A. Dijkstra. MGMtool: A performance modelling tool based on matrix geometric techniques. In R. Pooley and J. Hillston, eds. *Computer Performance Evaluation '92, Modelling Techniques and Tools*, pp. 397–401. Antony Rowe Ltd., Chippenham, England, 1992.

[71] M. Haviv. Aggregation/disaggregation methods for computing the stationary distribution of a Markov chain. *SIAM Journal on Numerical Analysis*, 24:952–966, 1987.

[72] M.R. Hestenes and E. Stiefel. Methods of conjugate gradients for solving linear systems. *Journal of Research of the National Bureau of Standards*, 49:409–435, 1952.

[73] M.A. Holliday and M.K. Vernon. *The GTPN Analyzer: Numerical Methods and User Interface.* Technical Report 639, Computer Science Department, University of Wisconsin, Madison, April 1986.

[74] J.E. Hopcroft and R.J. Tarjan. Efficient algorithms for graph manipulation. *Communications of the ACM*, 16(6):372–378, 1973.

[75] *IMSL — International Mathematical and Statistical Library.* IMSL Inc., NBC Building, 7500 Bellaire Boulevard, Houston, TX 77036.

[76] I. Ipsen and C.D. Meyer. *Uniform stability of Markov chains.* Technical Report 19201, Department of Mathematics, North Carolina State University, Raleigh, 1992.

[77] M.A. Jafari and J.G. Shanthikumar. *Finite State Spatially Non-Homogeneous Quasi-Birth-Death Processes.* Technical Report 85-009, Department of Industrial Engineering and Operations Research, Syracuse University, Syracuse, New York 13210, 1985.

[78] A. Jennings and W.J. Stewart. Simultaneous iteration for partial eigensolution of real matrices. *Journal of the Institute of Mathematics and Its Applications*, 15:351–361, 1975.

[79] A. Jensen. Markoff chains as an aid in the study of Markoff processes. *Skandinavisk Aktuarietidskrift*, 36:87–91, 1953.

[80] L. Kaufman. Matrix methods for queueing problems. *SIAM Journal of Scientific and Statistical Computing*, 4:453–552, 1984.

[81] D.S. Kim and R.L. Smith. *An exact aggregation algorithm for a special class of Markov chains.* Technical Report 89-2, Department of Industrial and Operations Engineering, University of Michigan, Ann Arbor, 1989.

[82] R.L. Klevans and W.J. Stewart. XMARCA: An X-windows interface for queueing network modelling. In R. Pooley and J. Hillston, eds. *Computer Performance Evaluation '92, Modelling Techniques and Tools*, pp. 385–389. Antony Rowe Ltd., Chippenham, England, 1992.

[83] H. Kobayashi. Application of the diffusion approximation to queueing networks. I: Equilibrium queue distributions. *Journal of the ACM*, 21(2):316–328, 1974.

[84] K. Kontovasilis, R.J. Plemmons, and W.J. Stewart. Block SOR for the computation of the steady state distribution of finite Markov chains with p-cyclic infinitesimal generators. *Linear Algebra and Its Applications*, 154:145–223, 1991.

[85] R. Koury, D.F. McAllister, and W.J. Stewart. Methods for computing stationary distributions of nearly-completely-decomposable Markov chains. *SIAM Journal of Algebraic and Discrete Mathematics*, 5(2):164–186, 1984.

[86] G. Latouche and Y. Ramaswami. A logarithmic reduction algorithm for quasi-birth-and-death processes. *Journal of Applied Probability*, 30:650–674, 1993.

[87] C. Lindemann. DSPNexpress: A software package for the efficient solution of deterministic and stochastic Petri nets. In R. Pooley and J. Hillston, eds. *Computer Performance Evaluation '92, Modelling Techniques and Tools*, pp. 13–30. Antony Rowe Ltd., Chippenham, England, 1992.

[88] C.F. Van Loan. The sensitivity of the matrix exponential. *SIAM Journal on Numerical Analysis*, 14(6):97–101, 1973.

[89] M. Malhotra and K.S. Trivedi. *Higher Order Methods for Transient Analysis of Stiff Markov Chains*. Technical report, Department of Electrical and Computer Engineering, Duke University, Durham, North Carolina, 1991.

[90] R.A. Marie. An approximate analytic method for general queueing networks. *IEEE Transactions on Software Engineering*, SE-5(5):530–538, 1979.

[91] R.A. Marie. Transient numerical solutions of stiff Markov chains. In *20th International Symposium on Automative Technology*, Florence, Italy, May 1989.

[92] R.A. Marie and J.M. Pellaumail. Steady state probabilities for a queue with a general service time distribution and state dependent arrivals. *IEEE Transactions on Software Engineering*, SE-9:109–113, 1979.

[93] R.A. Marie, A.L. Reibman, and K.S Trivedi. Transient solutions of acyclic Markov chains. *Performance Evaluation*, 7(3):175–194, 1987.

[94] D.F. McAllister, G.W. Stewart, and W.J. Stewart. On a Rayleigh-Ritz refinement technique for nearly-uncoupled stochastic matrices. *Linear Algebra and Its Applications*, 60:1–25, 1984.

[95] J.A. Meijerink and H.A. van der Vorst. An iterative solution method for linear systems of which the coefficient matrix is a symmetric M-matrix. *Mathematics of Computation*, 31(137):240–272, 1977.

[96] B. Melamed and M. Yadin. The randomization procedure in the computation of cumulative time distributions over discrete state Markov processes. *Operations Research*, 32:929–943, 1984.

[97] C.D. Meyer. The role of the group generalized inverse in the theory of finite Markov chains. *SIAM Review*, 17(3), 1975.

[98] C.D. Meyer. The condition of a finite Markov chain and perturbation bounds for the limiting probabilities. *SIAM Journal of Algebraic and Discrete Mathematics*, 1(3):273–283, 1980.

[99] C.D. Meyer. Stochastic complementation, uncoupling Markov chains and the theory of nearly reducible systems. *SIAM Review*, 31(2):240–272, 1989.

[100] D. Mitra and P. Tsoucas. Relaxations for the numerical solutions of some stochastic problems. *Communications in Statistics: Stochastic Models*, 4(3):387–419, 1988.

[101] C. Moler and C. Van Loan. Nineteen dubious ways to compute the exponential of a matrix. *SIAM Review*, 20(4):801–836, 1978.

[102] M.K. Molloy. A CAD tool for stochastic Petri nets. In *Proceedings of the ACM-IEEE-CS Fall Joint Computer Conference*, pp. 1082–1091, Dallas, TX, November 1986.

[103] M.K. Molloy. Structurally bounded stochastic Petri nets. In *International Workshop on Petri Nets and Performance Models*, pp. 156–163, Madison, WI, August 1987.

[104] M.K. Molloy. Performance analysis using stochastic Petri nets. *IEEE Transactions on Computers*, C-31(9):913–917, 1992.

[105] T. Murata. Modelling and analysis of concurrent systems. In C.R. Vick and C.V. Ramamoorthy, eds. *Handbook of Software Engineering*, pp. 39–63. Van Nostrand Reinhold, New York, 1984.

[106] E.J. Muth and S. Yeralan. Effect of buffer size on productivity of work stations that are subject to breakdowns. In *The 20th IEEE Conference on Decision and Control*, pp. 643–648, December 1981.

[107] N.M. Nachtigal, S.C. Reddy, and L.N. Trefethen. How fast are nonsymmetric matrix iterations? In *Copper Mountain Conference on Iterative Methods*, April 1990.

[108] *NAG — Numerical Algorithms Group*. NAG (USA) Inc., 1250 Grace Court, Downer's Grove, IL 60515.

[109] M. Neumann and R.J. Plemmons. Convergent nonnegative matrices and iterative methods for consistent linear systems. *Numerische Mathematica*, 31:265–279, 1978.

[110] M.F. Neuts. *Matrix Geometric Solutions in Stochastic Models — An Algorithmic Approach.* Johns Hopkins University Press, Baltimore, 1981.

[111] M.F. Neuts and D.M. Lucantoni. A Markovian queue with N servers subject to breakdown and repair. *Management Science*, 25:849–861, 1979.

[112] W. Niethammer. Relaxation bei Matrizen mit der Eigenschaft "A". *Zeitschrift für Angewandte Mathematik und Mechanik*, 44:49–52, 1964.

[113] W. Niethammer, J. de Pillis, and R.S. Varga. Convergence of block iterative methods applied to sparse least squares problems. *Linear Algebra and Its Applications*, 58:327–341, 1984.

[114] C.C. Paige. *The Computation of Eigenvalues and Eigenvectors of Very Large Sparse Matrices.* Ph.D. thesis, University of London, 1971.

[115] C.C. Paige and M.A. Saunders. Solution of sparse indefinite systems of linear equations. *SIAM Journal on Numerical Analysis*, 12:617–624, 1975.

[116] B.N. Parlett. A recurrence among the elements of functions of triangular matrices. *Linear Algebra and Its Applications*, 14:117–121, 1976.

[117] B.N. Parlett, D.R. Taylor, and Z.S. Liu. A look-ahead Lanczos algorithm for non-symmetric matrices. *Mathematics of Computation*, 44:105–124, 1985.

[118] B. Philippe and B. Sidje. *Transient Solutions of Markov Processes by Krylov Subspaces.* Technical report, IRISA, Campus de Beaulieu, 35042 Rennes, France, May 1993.

[119] B. Philippe, Y.Saad, and W.J. Stewart. Numerical methods in Markov chain modelling. *Operations Research*, 40(6):1156–1179, 1992.

[120] B. Plateau. On the stochastic structure of parallelism and synchronization models for distributed algorithms. In *Proceedings of the ACM Sigmetrics Conference on Measurement and Modelling of Computer Systems*, Austin, TX, August 1985.

[121] B. Plateau. PEPS: A package for solving complex Markov models of parallel systems. In R. Puigjaner and D. Potier, eds. *Modelling Techniques and Tools for Computer Performance Evaluation*, pp. 291–306, Plenum Press, New York, 1990.

[122] B. Plateau and J.M. Fourneau. A methodology for solving Markov models of parallel systems. *Journal of Parallel and Distributed Computing*, 12, 1991.

[123] R.J. Plemmons. Adjustments by least squares in geodesy using block iterative methods for sparse matrices. In *Proceedings of the Annual U.S. Army Conference on Numerical Analysis and Computation*, pp. 151–186, 1979.

[124] K.G. Ramakrishnan and D. Mitra. Panacea: An integrated set of tools for performance analysis. In R. Puigjaner and D. Potier, eds. *Modelling Techniques and Tools for Computer Performance Evaluation*, pp. 25–40, 1988.

[125] V. Ramaswami. Nonlinear matrix equations in applied probability — solution techniques and open problems. *SIAM Review*, 30(2):256–263, 1988.

[126] V. Ramaswami and G. Latouche. A general class of Markov processes with explicit matrix-geometric solutions. *OR Spektrum*, 8:209–218, 1986.

[127] A.L. Reibman and K.S. Trivedi. Numerical transient analysis of Markov models. *Computers and Operations Research*, 15(1):19–36, 1988.

[128] A.L. Reibman, K.S. Trivedi, and R. Smith. Markov and Markov reward model transient analysis: an overview of numerical approaches. *European Journal of Operations Research*, 40(2):257–267, 1989.

[129] V.I. Romanovsky. *Discrete Markov Chains*. Wolters-Noordhoff, Groningen, The Netherlands, 1970.

[130] D.J. Rose. Convergent regular splittings for singular M-matrices. *SIAM Journal of Algebraic and Discrete Mathematics*, 5(1):133–144, 1984.

[131] Jr. S. Stidham. *Stable recursive procedures for numerical computations in Markov models*. Technical Report 85-10, Department of Industrial Engineering, North Carolina State University, Raleigh, N.C., 1985.

[132] Y. Saad. Variations on Arnoldi's method for computing eigenelements of large unsymmetric matrices. *Linear Algebra and Its Applications*, 34:269–295, 1980.

[133] Y. Saad. Krylov subspace methods for solving unsymmetric linear systems. *Mathematics of Computation*, 37:105–126, 1981.

[134] Y. Saad. Chebyshev acceleration techniques for solving non-symmetric eigenvalue problems. *Mathematics of Computation*, 42:567–588, 1984.

[135] Y. Saad. Analysis of some Krylov subspace approximations to the matrix exponential operator. *SIAM Journal on Numerical Analysis*, 29(1):209–228, 1992.

[136] Y. Saad and M.H. Schultz. Conjugate gradient–like algorithms for solving nonsymmetric linear systems. *Mathematics of Computation*, 44:417–424, 1985.

[137] Y. Saad and M.H. Schultz. GMRES: A generalized minimal residual algorithm for solving non-symmetric linear systems. *SIAM Journal of Scientific and Statistical Computing*, 7:856–869, 1986.

[138] E.B. Saff. On the degree of the best rational approximation to the exponential function. *Journal of Approximation Theory*, 9:97–101, 1973.

[139] R. Sahner and K.S. Trivedi. Reliability modelling using SHARPE. *IEEE Transactions on Reliability*, R-36(2):186–193, 1987.

[140] A.L. Scherr. *An Analysis of Time-Shared Computer Systems*. M.I.T. Press, Cambridge, MA, 1967.

[141] H. Schneider. Theorems on M-splittings of a singular M-matrix which depends on graph structure. *Linear Algebra and Its Applications*, 58:407–424, 1984.

[142] P.J. Schweitzer. Aggregation methods for large Markov chains. In *International Workshop on Applied Mathematics and Performance Reliability Models of Computer Communication Systems*, pp. 225–234, University of Pisa, Italy, 1983.

[143] P.J. Schweitzer. Aggregation methods for large Markov chains. In G. Iazeolla, P.J. Courtois, and A. Hordijk, eds. *Mathematical Computer Performance and Reliability*. Elsevier Science Publishers, Amsterdam, 1984.

[144] P.J. Schweitzer. A survey of aggregation-disaggregation in large Markov chains. In William J. Stewart, ed. *Numerical Solution of Markov Chains*, pp. 63–88. Marcel Dekker Inc., New York, 1991.

[145] M. Sczittnick and B. Muller-Clostermann. MACOM: A tool for the Markovian analysis of communication systems. In G. Pujolle and R. Puigjaner, eds. *Proceedings of the Fourth International Conference on Data Communication Systems and Their Performance*, pp. 485–498. Elsevier Science Publishers, 1991.

[146] T.J. Sheskin. A Markov chain partitioning algorithm for computing steady state probabilities. *Operations Research*, 33(1):228–235, 1985.

[147] H.A. Simon and A. Ando. Aggregation of variables in dynamic systems. *Econometrica*, 29:111–138, 1961.

[148] P.M. Snyder and W.J. Stewart. Explicit and iterative numerical approaches to solving queueing models. *Operations Research*, 33(1):183–202, 1985.

[149] P. Sonneveld. CGS, a fast Lanczos-type solver for nonsymmetric linear systems. *SIAM Journal of Scientific and Statistical Computing*, 10(1):36–52, 1989.

[150] G.W. Stewart. *Introduction to Matrix Computations*. Academic Press, New York, 1973.

[151] G.W. Stewart. Simultaneous iteration for computing invariant subspaces of non-Hermitian matrices. *Numerische Mathematica*, 25:123–136, 1976.

[152] G.W. Stewart. Computable error bounds for aggregated Markov chains. *Journal of the ACM*, 30:271–285, 1983.

[153] G.W. Stewart. On the sensitivity of nearly uncoupled Markov chains. In William J. Stewart, ed. *Numerical Solution of Markov Chains*, pp. 105–119. Marcel Dekker Inc., New York, 1991.

[154] G.W. Stewart, W.J. Stewart, and D.F. McAllister. A two stage iteration for solving nearly uncoupled Markov chains. In *IMA Volumes in Mathematics and Its Applications*, vol. 60:1 *Recent Advances in Iterative Methods*, pp. 201–216. Springer-Verlag, Berlin, Heidelberg, New York, 1993.

[155] W.J. Stewart. A comparison of numerical techniques in Markov modelling. *Communications of the ACM*, 21(2):144–152, 1978.

[156] W.J. Stewart. On computing the stationary probability vector of a network of two Coxian servers. In *Applied Probability – Computer Science: The Interface*, vol. 1, pp. 275–296. Birkhauser, Boston, 1982.

[157] W.J. Stewart. MARCA: Markov chain analyzer. A software package for Markov modelling. In William J. Stewart, ed. *Numerical Solution of Markov Chains*, pp. 37–62. Marcel Dekker Inc., New York, 1991.

[158] W.J. Stewart. On the use of numerical methods for ATM models. In H. Perros, G. Pujolle, and Y. Takahashi, eds. *Modelling and Performance Evaluation of ATM Technology*, pp. 375–396. North-Holland, Amsterdam, 1993.

[159] W.J. Stewart, K. Atif, and B. Plateau. Efficient numerical methods for stochastic automata networks. To appear in: *European Journal of Operations Research*, 1994.

[160] W.J. Stewart and A. Jennings. A simultaneous iteration algorithm for real matrices. *ACM Transactions on Mathematical Software*, 7(2):184–198, 1981.

[161] W.J. Stewart and W. Wu. Numerical experiments with iteration and aggregation for Markov chains. *ORSA Journal on Computing*, 4(3):336–350, 1992.

[162] J. Stoer and R. Bulirsch. *Introduction to Numerical Analysis.* Springer-Verlag, New York, 1980.

[163] U. Sumita and M. Rieders. Numerical comparison of the replacement process approach with the aggregation-disaggregation algorithm for row-continuous Markov chains. In William J. Stewart, ed. *Numerical Solution of Markov Chains*, pp. 287–302. Marcel Dekker Inc., New York, 1991.

[164] Y. Takahashi. *A Lumping Method for Numerical Calculation of Stationary Distributions of Markov Chains.* Technical Report B-18, Department of Information Sciences, Tokyo Institute of Technology, Tokyo, Japan, June 1975.

[165] R.J. Tarjan. Depth first search and linear graph algorithms. *SIAM Journal of Computing*, 1(2):146–160, 1972.

[166] H.C. Tijms. Q-lib: A software package for queueing models. In William J. Stewart, ed. *Numerical Solution of Markov Chains*, pp. 691–692, Marcel Dekker Inc., New York, 1991.

[167] K. Trivedi, A.L. Reibman, and R. Smith. Transient analysis of Markov and Markov reward models. In G. Iazeolla, P.J. Courtois, and O.J. Boxma, eds. *Computer Performance and Reliability*, pp. 535–545. North-Holland, Amsterdam, 1988.

[168] H. Vantilborgh. *The Error of Aggregation in Decomposable Systems.* Technical Report R453, Philips Research Laboratory, Brussels, Belgium, 1981.

[169] R.S. Varga. *P*-cyclic matrices: A generalization of the Young-Frankel successive overrelaxation scheme. *Pacific Journal of Mathematics*, 9:617–628, 1959.

[170] R.S. Varga. *Matrix Iterative Analysis.* Prentice Hall, Englewood Cliffs, NJ, 1962.

[171] M. Veran and D. Potier. QNAP2: A portable environment for queueing systems modelling. In D. Potier, ed. *Modelling and Tools for Performance Analysis*, pp. 25–63. Elsevier Science Publishers, 1985.

[172] M. Vernon, J. Zahorjan, and E.D. Lazowska. A comparison of performance Petri nets and queueing network models. In *Proceedings of the AFCET International Workshop on Modelling Techniques and Performance Evaluation*, Paris, March 1987.

[173] V.L. Wallace. *The Solution of Quasi-Birth-and-Death Processes Arising from Multiple Access Computer Systems.* Ph.D. thesis, University of Michigan, Systems Engineering Laboratory, 1969.

[174] V.L. Wallace and R.S. Rosenberg. *RQA-1, The Recursive Queue Analyzer.* Technical report, University of Michigan, Ann Arbor, 1966.

[175] R.C. Ward. Numerical computation of the matrix exponential with accuracy estimate. *SIAM Journal on Numerical Analysis,* 14(4):600–610, 1977.

[176] P. Wild and W. Niethammer. Over and underrelaxation for linear systems with weakly cyclic Jacobi matrices of index p. *Linear Algebra and Its Applications,* 91:29–52, 1987.

[177] J.H. Wilkinson. *Rounding Errors in Algebraic Processes.* Her Majesty's Stationery Office, Notes on Applied Science, 32, London, 1963.

[178] J.H. Wilkinson. *The Algebraic Eigenvalue Problem.* Clarendon Press, Oxford, England, 1965.

[179] R.W. Wolff. *Stochastic Modeling and the Theory of Queues.* Series in Industrial and Systems Engineering. Prentice Hall, Englewood Cliffs, NJ, 1988.

[180] B. Wong, W. Giffin, and R.L. Disney. Two finite $M/M/1$ queues in tandem: A matrix solution for the steady state. *OPSEARCH,* 14(1):1–18, 1977.

[181] J. Ye and S.Q. Li. Analysis of multi-media traffic queues with finite buffer and overload control — Part I: Algorithm. In *Proceedings of INFOCOM '91,* pp. 1464–1474, 1991.

[182] D.A. Young. *The X Window System – Programming and Applications with Xt – OSF/Motif Edition.* Prentice Hall, Englewood Cliffs, NJ, 1990.

[183] D.M. Young. Iterative methods for solving partial differential equations of elliptic type. *Transactions of the American Mathematical Society,* 76:92–111, 1954.

[184] D.M. Young. *Iterative Solution of Large Linear Systems.* Academic Press, New York, 1971.

Index

WILLIAM J. STEWART IS PROFESSOR OF COMPUTER SCIENCE AT NORTH CAROLINA STATE UNIVERSITY.